COSMIC PERSPECTIVES IN SPACE PHYSICS

ASTROPHYSICS AND SPACE SCIENCE LIBRARY

VOLUME 242

COSMIC PERSPECTIVES IN SPACE PHYSICS

by

SUKUMAR BISWAS

Formerly, Senior Professor & Head, Cosmic Ray-Space Physics Group, Tata Institute of Fundamental Research, Bombay, India

KLUWER ACADEMIC PUBLISHERS

DORDRECHT / BOSTON / LONDON

A C.I.P. Catalogue record for this book is available from the Library of Congress.

ISBN 0-7923-5813-9

Published by Kluwer Academic Publishers,
P.O. Box 17, 3300 AA Dordrecht, The Netherlands.

Sold and distributed in North, Central and South America
by Kluwer Academic Publishers,
101 Philip Drive, Norwell, MA 02061, U.S.A.

In all other countries, sold and distributed
by Kluwer Academic Publishers,
P.O. Box 322, 3300 AH Dordrecht, The Netherlands.

Printed on acid-free paper

Cover figure:
An artist's drawing of a gamma-ray 'BLAZAR' – the class of Active Galactic Nuclei which emit enormous amounts of energy in gamma-rays and other radiations in the form of jets. These AGN, which are believed to be powered by super-massive black holes at the centre of the galaxies, are the most energetic events in the universe.

This book is dedicated to the memory of my late parents,
Satyendra Prasad and Suhasini Biswas

University of Durham
Department of Physics,
and
The Royal Institution

Sir Arnold Wolfendale, *FRS*

FOREWORD

It is with considerable pleasure that I write this brief Foreword for Sukumar Biswas's book on Space Physics. I can think of no one more able to tackle this daunting task, by virtue of his wide experience both as a teacher and as a researcher.

The title "Cosmic Perspectives ...." is well-chosen in that the cosmic, and all-embracing, nature of the physics involved is beautifully brought out.

Of course, the text will be required reading for all involved in Space Physics, at whatever level, but I feel that it will have a great appeal to many undergraduates and graduates in Physics who are specialising in other areas, too.

Here we have many, many examples of "physics in action" on all sorts of scales. As a way of illustrating the application of the techniques of physics, often in extreme situations, space physics is second-to-none.

I commend this book very warmly.

A.W.

February 1999
Durham, U.K.

Preface

In the early years of the twentieth century, Victor Hess of Germany flew instruments in balloons and so discovered in 1912 that an extra-terrestial radiation of unknown origin is incident on the earth with an almost constant intensity at all times. These penetrating non-solar radiations which were called Cosmic Rays by Millikan, USA, opened the new frontier of space physics and many leading scientists were attracted to it. At the end of World War II a number of space vehicles, e.g. stratospheric balloons, rockets and satellites were developed. In 1950 and onwards, these vehicles enabled spectacular advances in space physics and space astrophysics. New horizons were opened in the explorations of cosmic rays, the earth's magnetosphere, the Sun and the heliosphere, the moon and the planets. Using space-borne instruments, exciting discoveries were made of stars, and galaxies in the infra-red, ultra violet, x-ray and gamma-ray wavelengths.

In this text book these fascinating new findings are presented in depth and on a level suitable for senior undergraduate and graduate students, research scientists and scientists of other disciplines. Although there are several excellent books and monographs on different aspects, most of these deal with specific areas. In this text book the findings of space physics and astrophysics are presented in an integrated manner with proper introductions to the fundamental aspects, and these are supplemented by relevant ground based observations.

The relevant topics are organised in various chapters of the book as follows: In the introductory Chapter 1, a brief historical account of space research is given, followed by short reviews of space vehicles used and some aspects of the early years of space research. In Chapter 2, on "The Earth's Magnetosphere" readers are introduced to the dramatic discoveries of the Radiation Belts and the Magnetosphere of the earth, with the advent of earth satellites. Chapter 3 on "The Sun and the Heliosphere" begins with the phenomena of the quiet Sun followed by processes in the active Sun. The latter topics reveal new spectacular facets of the Sun in solar flares, solar radio bursts and solar X-ray, gamma-ray and neutron emission. In Chapters 4 and 5, attention is turned to the physical features and properties of large bodies and aggregates in the matter of the solar system; - in Chapter 4 on "The Moon and Planets and Asteroids", and in Chapter 5 on "Comets and Meteorites". New discoveries in these areas obtained from spacecraft are emphasised, e.g. the paleo- magnetism of the moon, mysteries of the planetary atmospheres, immense magnetospheres of Jupiter and Saturn, spacecraft rendezvous with Comet Halley and the complexities of the origin of the solar system. In Chapter 6 "Solar Energetic Particles" are presented, bringing out their dual role - firstly as a sample of accelerated matter from the Sun and secondly as a sensitive probe of the interplanetary fields in the heliosphere.

We leave the heliosphere with Chapter 6 and enter the inter-stellar medium in Chapter 7 on "Galactic Cosmic Rays". We present basic features and main findings of cosmic rays which pervade interplanetary space, the interstellar medium, the galaxy and possibly beyond. We include the highlights of ultra high energy cosmic rays, some of which probably originate from violent events in the vast depths of the cosmos such as exploding galactic nuclei and in collisions of galaxies. In Chapter 8 on "Space Astrophysics" we look through the new windows of the universe from space and present exciting new results of the new astronomies of Extreme UV, Infra-red, X-rays and gamma-rays from space. These are followed by main outlines of the new vistas of cosmological problems as revealed by the results on cosmic X-ray background, cosmic gamma-ray background and cosmic microwave background. In Chapter 9, which is the last topical chapter, the readers are introduced to the physical and dynamical properties of the Interstellar Medium. We conclude this chapter with the astrophysical aspects of the fascinating subject of the search for the origin of life. In the concluding Chapter 10, on "Looking to the Future" some general trends and future directions are briefly presented.

The level of presentation in the book is aimed at intermediate level so that it is suitable for students who are familiar with introductory physics including general physics, classical mechanics and electrodynamics, special theory of relativity, and some Introductory Astronomy. Thus senior under-graduate students interested in space physics and space astrophysics will be able to readily follow most of the topics covered in the book. Graduate students will find more advanced topics discussed in their underlying features so that by understanding the basic ideas they would be able to pursue more complex topics for which a large number of references to books and original papers are given. Each chapter is followed by a number of problems; the answering of these will enable the students to have a good grasp of the topics. The teachers are expected to find the book as a suitable resource book for their lectures. Materials presented in the book can be used for one or two semester courses on physics and astrophysics in space.

In the two companion books the author's colleague Professor S.N. Ghosh has presented graduate level text books on "The Upper Atmosphere", and "The Ionized Atmosphere". Thus the present book and these two form a trilogy on the physics and astrophysics in space, the upper atmosphere and the ionosphere.

February, 1999 SUKUMAR BISWAS
Bombay

ACKNOWLEDGEMENTS

The author expresses his grateful thanks to Sir Arnold Wolfendale, FRS, for writing a nice Foreword of the book. The author wishes to express his sincere thanks to Drs. Carl E. Fichtel and J.F. Ormes (NASA-GSFC), Drs. M.A. Shea and D.F. Smart (AFGL), and Dr. B. Klecker (MPI-Garching), for sending the copies of their recent papers, including some prior to publication. The author is very thankful to his colleagues, Profs. K.C. Anand, S.V. Damle, K.S. Krishnaswamy, R.K. Manchanda and S. Ramadurai of T.I.F.R., Prof. J.N. Goswami of PRL, and Prof. Manashi Roy of IIG for their help and valuable suggestions which were very useful in the various phases in the preparation of the manuscript. The author is very thankful to Prof. M.G.K. Menon, FRS, and Prof. B.V. Sreekantan, for helpful suggestions and to Prof. D. Lal, FRS, for encouraging comments. The author expresses his thanks to his colleague, Prof. S.N. Ghosh of Calcutta for helpful discussions.

The author is thankful to Sir Dorabji Tata Trust for providing some valuable support for secretarial work, to Mr. Saleem Akhtar for his very fine and patient typing work and to Mr. R.K. Talekar and his colleagues for the careful tracing of the diagrams. Lastly, I thank my wife Reba for her patient support.

February 1999 S.B.
Bombay

ACKNOWLEDGEMENTS

[illegible]

Table of Contents

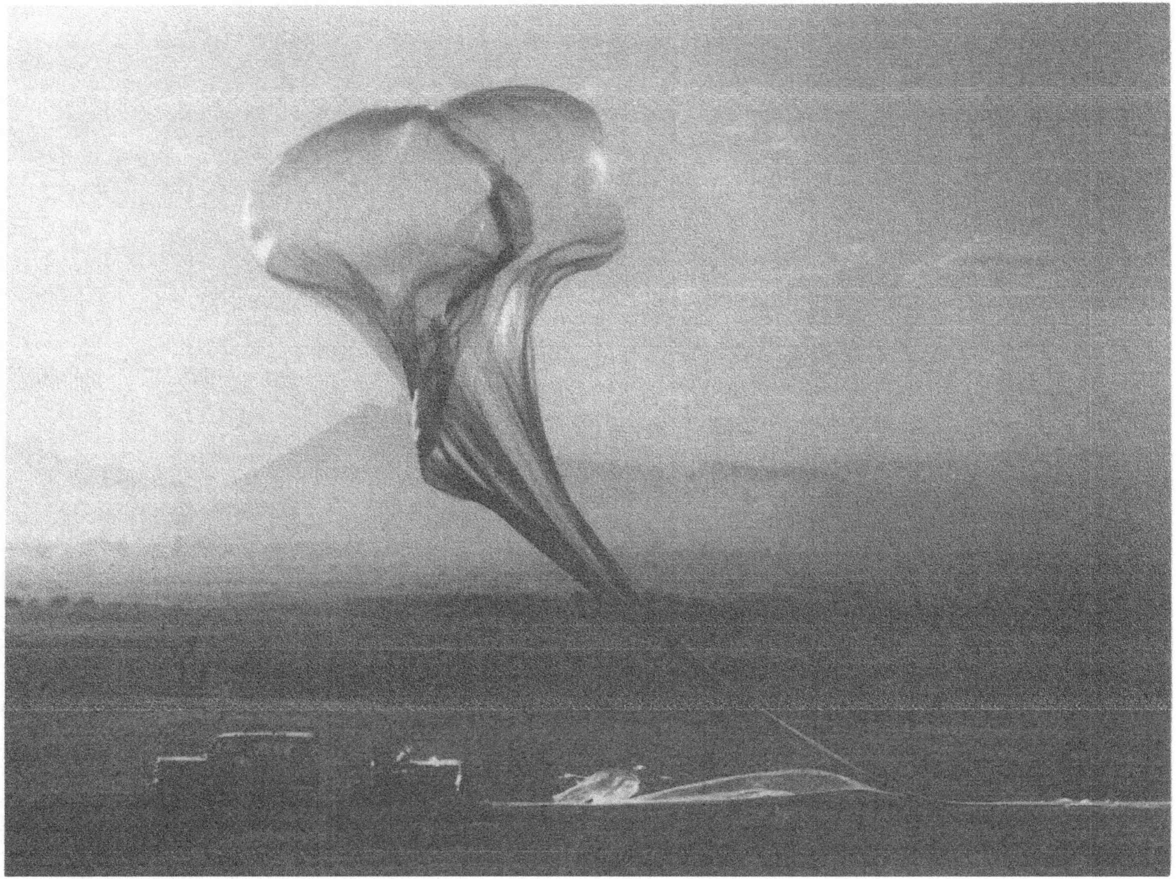

A *stratospheric balloon being launched in the early dawn carrying cosmic ray experiments' payload* (Courtesy : National Balloon Facility, Hyderabad, India).

The Indian cosmic-ray instrument, `ANURADHA' which was flown in space in the Space-Shuttle-Spacelab-3 mission of NASA.

The Space-Shuttle Challenger, in the orbit of the 350 km with Spacelab-3 instruments in the cargo-bay is seen in the gravity-gradient orientation in space. The Anuradha instrument of India, and the Atmos' of NASA are seen on the external support structure.

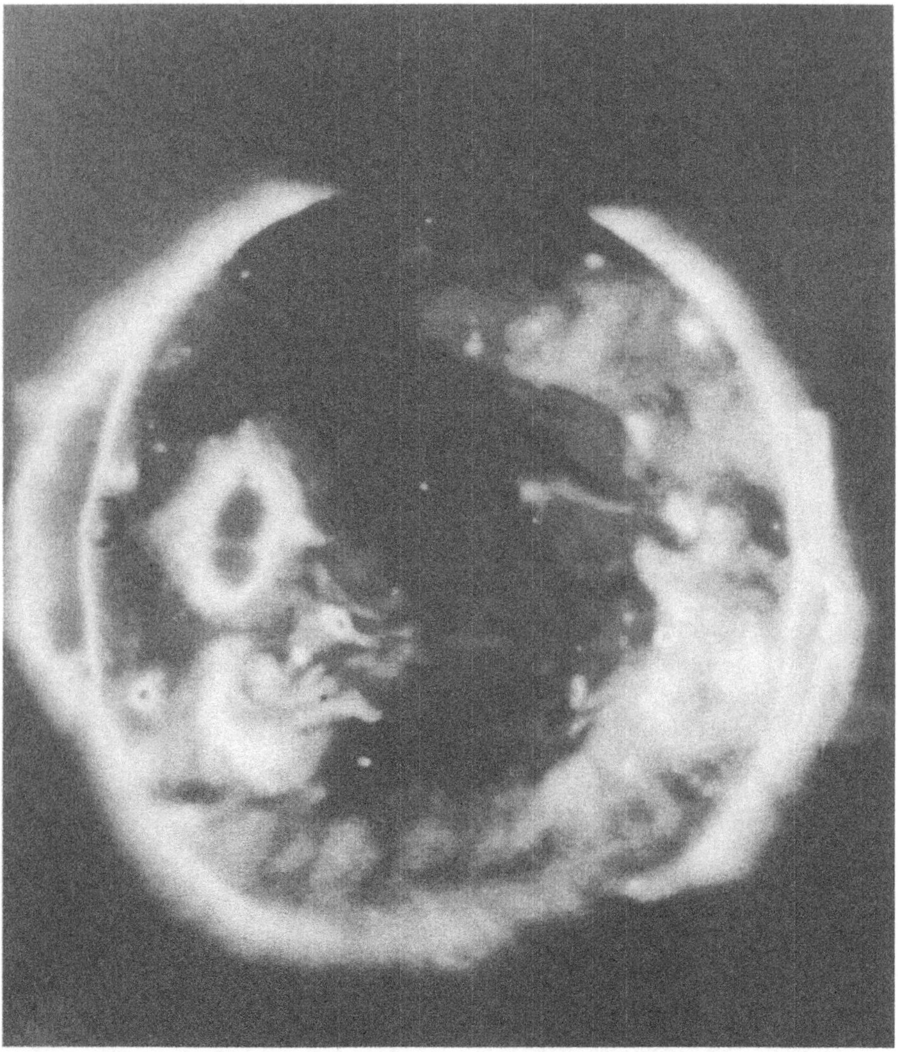

A soft X-ray photograph of the Sun taken by NASA's Apollo Telescope mount in the Skylab mission showing the large coronal hole as a huge dark patch extending from the solar north polar region (top) to the equatorial and southern latitudes. The coronal hole is the seat of high speed solar wind. The bright knots are very high temperature regions of the corona.

CHAPTER 1
INTRODUCTION TO SPACE RESEARCH

"The mysterious cloud of appearance is hard to pass beyond, but those who come in truth go beyond the world of shadows ..."

- *Bhagawat Gita*

1.1 Early History of Space Physics

Our knowledge of the universe comes from the signals we receive at the earth in many forms. For centuries the signals have been detected in the form of visible light, and optical telescopes looking out into space - to the Sun, the moon, the planets and myriads of stars and galaxies have opened up new vistas of the universe. To these another kind of signal was added in the 1930's with the discovery of radio waves coming from astronomical objects which ushered the new branch of radio astronomy. However, it was realized that besides the optical and radio bands received at the surface of the earth, the vast expanse of the electromagnetic spectrum remained hidden from us because of the protective blanket of the earth's atmosphere which absorbs most of these electromagnetic waves. In the past decades scientists were able to send instruments in space vehicles to progressively higher altitudes in the atmosphere and finally with satellites completely outside the earth's atmosphere. These unveiled dramatically new vistas of the near-earth space, the interplanetary medium and of the distant interstellar space, which reveiled to be populated by new kinds of plasmas, particles and magnetic fields, as well as energetic nuclei, ions and cosmic rays. Also scientists were able to look at the Sun, the planets, the stars and the galaxies at several different wavelengths of the electromagnetic spectrum such as ultra-violet soft x-rays, hard x-rays and gamma rays and thus obtain a strikingly new picture of them. Thus new astronomies were born, i.e.. ultra-violet, x-ray and gamma ray astronomies which rapidly expanded in many dimensions and became new disciplines of astrophysics and space science. We shall briefly trace the growth and developments of these exciting new achievements in space physics and space astrophysics in this chapter and thus bring to focus new perspectives in these cosmic phenomena.

One usually reckons the period at the end of World War II as the beginning period of space research, when giant war-time rockets were available for sending instruments to high altitudes. However we can trace the beginning of space research to the early part of this century with the discovery of cosmic rays in 1912. Scientists in early this century noticed the curious phenomenon that even when a charged electrometer was completely shielded from the radioactivity of the earth there was always a steady leakage of charge due to ionization of the air by some unknown source. The nature of this unknown source of ionizing radiation remained a mystery. To explore this problem, in 1912 the Austrian scientist, Victor Hess, flew his instruments in the gondola of balloons which he himself piloted. In course of six balloon flights he noticed that above 650 m, the intensity of the ionizing radiation increased rapidly to the highest altitude of $\sim$ 6 km attained by the balloons and the intensity remained the same both in day and in night time. Hess[(1)] concluded that a highly penetrating ionizing radiation of extra-terrestrial origin, but not from the Sun was incident on the earth. This new discovery was followed up, after the end of the first world war, as several scientists, including Robert Millikan[(2)] of the USA, made balloon flight experiments. The name cosmic rays was given by Millikan to this penetrating radiation which was thought to be high energy gamma rays at that time. We may, hence, say that the discovery of cosmic rays in 1912 in balloon flights was the beginning of space physics which developed into many streams in the following decades.

After the end of World War I during the 1920's and early 1930's the Geiger-Müller counter was invented and it opened new horizons. Using these detectors it was discovered that cosmic ray intensities at the surface of the earth varied significantly due to the deflection by the magnetic field of the earth. Thus it was established that cosmic rays are charged particles and not gamma rays. Further studies with these detectors showed that cosmic ray intensity from the western direction is larger than that from the east. Therefore from the direction of the deflection of cosmic ray particles in the magnetic field of the earth it was concluded that the large majority of cosmic rays are positively charged particles such as high energy protons and helium nuclei. These discoveries paved the way for the further explorations of this penetrating radiations from outer space.

In the years intervening the two world wars scientists discussed different ideas for space research using rocket propulsion. A major landmark was established by Dr. Robert H. Goddard of USA who in 1926 had constructed and tested successfully the first rocket using liquid fuel. This historic contribution, however, was not sustained or supported in the USA during the 1930's on account of economic depression. During World War II Germany developed huge V-2 missiles using liquid fuel propulsion technology which in the subsequent years led to military applications.

Although sounding rockets reached the altitudes of 100-200 km and opened new areas of knowledge of the upper atmosphere, these were inadequate for some space experiments such as for cosmic ray studies, because of the short duration of about several minutes above 50 km altitude for these rocket flights and of the low intensity of the primary cosmic ray particles. In the 1930's and 1940's scientists employed several dozens of rubber balloons filled with hydrogen gas to carry light-weight scientific instrument to high altitudes of 30-50,000 feet altitudes for a few hours of exposure in the thin upper atmosphere. During 1948-50 the scientists at the Universities of Minnesota and Rochester, USA with the assistance of the Department of Navy flew for the first time several large polythene balloons filled with helium gas to lift several packages of scientific instruments to the height of about 30 km where these floated at nearly constant level for many hours. These balloons code named as "Skyhook" left behind almost 90% of the atmosphere below it and had only 10% above, so that nearly 'pure' or primary cosmic ray particles with only a small contamination of 'secondary' particles produced by the primaries in interactions with the atoms of the air, could reach the detectors. In these experiments in 1948-50 the USA scientists Freier et al.[(3)] and Bradt and Peters[(4)] made a striking discovery that the primary cosmic rays arriving from the deep space consist of, in addition to hydrogen and helium, energetic atomic nuclei of practically all elements, e.g. Li, Be, B, C, N, O., etc. up to Fe, of the periodic table. This discovery opened a new window of space astrophysics that had been pursued for subsequent decades, with improved detectors and new space vehicles.

After the end of World War II some seventy long range V-2 missiles were captured from Germany and brought to the USA; similarly a number of these missiles were taken to the USSR. These weapons of war were used by the scientists and engineers of the USA and the USSR to fly scientific instruments for space research to as high as 180 km. Large efforts were made by several countries of the world in the post war period to design and develop high performance rockets to carry scientific payloads for space exploration and these ultimately led to the launching of the satellites in the orbit of the earth.

1.2 The New Space Age and Global Plans

1.2.1 The IGY and International Collaborations in Space Research

As high altitude rockets and stratospheric balloons were being used for upper atmosphere and space research in the late 1940's after World War II and in the early 1950's, the inadequacy of these space vehicles for collecting comprehensive data was realized by scientists. It is shown in Figure 1.2.1 that the absorption of the atmosphere shields and obscures a large portion of electromagnetic radiation from space. The stratospheric balloons, although they provided observations of several hours, were confined to altitudes of 30 to 35 km only, while one needed *in situ* observations at altitudes of 150 km or more. Although sounding rockets reached ~100-150 km altitudes, the duration of exposure above ~ 50 km altitude was limited to three to six minutes only. So small groups of scientists in several countries, notably in the USA and USSR began thinking about earth satellites which will carry instruments in the high altitudes above 150 km for a prolonged period. For example, J.A. Van Allen of the USA who made many sounding rocket flights presented a paper at the 1948 Conference of International Union of Geophysics and Geodesy (IUGG) in Oslo in which he summarized the American rocket program by saying "serious attention is being given to the development of a satellite missile which will continuously orbit around the earth at a distance of, say, 1000 km". In November 1953, A.N. Nesmeyanov, President of the Academy of Sciences of USSR said in the World Peace Council meeting in Vienna, "science has reached a state when it is feasible to create an artificial satellite of the earth". In the Congress of the International Astronautical Federation in Zurich in 1953, S.F. Signer of USA suggested the "MOUSE", the acronym of Minimum Orbit Unmanned Satellite of the Earth. This was to be an instrumented sphere to collect scientific data.

The International Council of Scientific Unions (ICSU) coordinated the diverse areas of global cooperation in scientific research. In 1931 ICSU

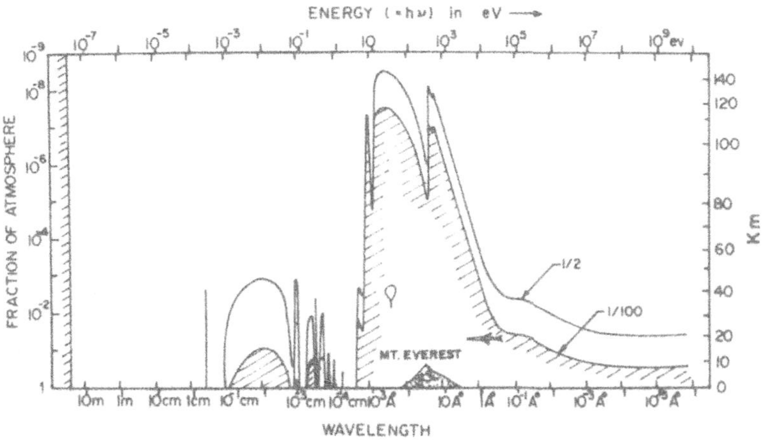

Figure 1.2.1: Absorption of electromagnetic radiation in the earth's atmosphere. The lines marked 1/2 and 1/100 indicate the altitudes at which incoming radiation is attenuated by the above factors

was formed as a federation of thirteen scientific unions. At the end of World War II new scientific organizations were set up and many of these were brought under the ICSU. In Washington in 1951 the ICSU executive committee representing all unions endorsed the plan of a new polar year proposed by the mixed commission of the Union Radio Science International (URSI), the International Astronomical Union (IAU), and the IUGG, and created a special committee to work on the Third Polar Year. In the ICSU General Assembly at Amsterdam in October 1952, Sidney Chapman proposed to change the name to International Geophysical Year (IGY) which was approved and a special committee for IGY, called Comite Special del'Anne Geophysique Internationale (CSAGI) was set up. S. Chapman of the UK was elected the Chairman of the Committee, and with L. Berkner of the USA as Vice-President and N. Nicolet of Belgium as Secretary-General formed the initial Bureau of CSAGI, having its headquarters in Brussels. It was decided that the International Geophysical Year would run for eighteen months, from July 1, 1957 to December 1958. This period was chosen because of its proximity to the period of maximum solar activity. By 1954 twenty-two nations formed IGY National Committees and made research plans. The US National Academy of Sciences made a massive US program with several hundred scientists in government agencies, universities and private research institutions and this included rocket explorations of the upper atmosphere using 36 Aerobee and 37 Rockoon firings. In the Plenary meeting of the Special Committee of IGY in Rome in September 1954, the Soviet scientists joined the IGY committee and Soviet scientists who made important contributions in these field, showed high regard for the implementation of the IGY projects. These were planned through a special committee of the Academy of Science of the USSR in 1957. By early 1957, about twenty-eight institutes, observatories, societies etc. of the USSR were preparing for the IGY studies, many other countries joined in the IGY and about sixty-four countries were participating in the global plans at the beginning of IGY in July, 1957 in which an estimated 4000 stations were taking part. The Rome Conference of IGY set in motion the exploration of the two uncharted regions - namely Antarctica and Outer Space - using the new technologies of the mid-twentieth century .

1.2.2 Development of Launch Vehicles for Satellites

The major goal of the exploration of the new frontier - outer space - required new technologies, the high altitude sounding rockets and new plans for multistage rockets capable of placing instrumented satellites orbiting the earth. The technology of powerful launch vehicles in the early 1950's were used for military applications of long range missiles and so the information were of a classified nature. As mentioned earlier, the first successful liquid fuel rocket flight by R.H. Goodard of the USA in 1926 did not find support in the USA of the 1930's. However these new concepts found fertile ground in Europe. In 1927 the Verein für Raum - Schiffahrt (Society for Space Navigation) was formed in Germany where active programs of research in rocket propulsion were conducted by the German scientists and engineers led by Hermann Oberth, a pioneer in the rocket theory. In 1931 the German Rocket Society members achieved the first European flight of a liquid-fueled rocket to several hundred feet altitude. By the end of that year about one hundred rocket flights were successfully made by them. Although the depression of the 1930's brought an end to the Verein activities, the German Army was attracted to this technology of rocketry. In 1933 Captain (later Major General) Walter R. Dornberger was placed in charge of the Army's rocket project, and he selected an assistant, a former Director of Vernier - a twenty-one year old student named Werner von Braun. In the subsequent years larger and more powerful rockets were developed by the German Army and these were called A-1, A-2 and A-3. For long range rocket flights the project was moved to Peenemunde at the Baltic sea coast. After two failures, the giant A-4 rocket was successfully test launched in October 1942, a few years after the onset of World War II. This monstrous rocket, 47 feet long and weighing 12 tons on the launch pad, including nine tons of propellant that burnt in the first minute of flight, could carry one ton of high explosive explosives. It had a range of about 190 km. These rocket weapons, called Vergeltungs Waffe II by the Nazi Germany was first fired at London in 1944 and in the six months of the remaining period of the war, 3600 V-2 rockets were launched to London and other western Europe cities by Nazi Germany. After the end of the war, the captured V-2 rockets were taken to the USA and these together with about 120 rocket experts of Peenemunde led by Werner von Braun formed the core group of the USA Army's rocket center at Huntsville, Alabama. In 1955 von Braun became an US citizen and the Director of the US Army Ballistic Missile Program.

The Soviet Union carried out important research on rocket propulsion in the period between the two world wars of which very little was known to the western world. The Russian scientist, K.E. Tsiolkovsky made such pioneering contributions on rocket propulsion at the turn of this century that he is called 'the father of the science of astronautics'. Two years after the formation of the Verein in Germany, the Russians formed the society called GIRD which is the acronym in Russian of the Study of Reactive Motion. This group achieved a good amount of sophisticated knowledge on rocket research and space flight. During the war the Russians had developed advanced technologies for a large and extensive ballistic missile program much of which was unknown to the west during that time.

On July 29, 1955, the US administration made public the joint announcement by the National Academy of Sciences and the National Science Foundation that the USA was intended to launch a satellite during the International Geophysical Year. This was also communicated to the IGY Chairman, Sir Sydney Chapman and the objective would be to gather significant scientific data in the fields of geodesy, atmospheric physics, ionosphere physics, solar physics and cosmic radiation. This was welcomed by the IGY organization and the Bureau of CSAGI which then included in addition of Chapman, Berker and Nicolet, the Russian member Beloussov and Jean Coulomb of France.

On April 16, 1955, three months prior to the US announcement, the Soviet Union published in a Moscow newspaper the Soviet plan to launch a satellite vehicle. It described an establishment of a permanent commission on Interplanetary Communication under the Academy of Sciences of the USSR, with Leonid I. Sedov as Chairman. This commission included among its 27 members, the best known scientists of the USSR such as Ambartsumyan, a famous astronomer, Kapitsa, who once headed the atomic research laboratory at the Cambridge University, and Bogolyubov, a mathematician turned atomic scientist. The inclusion of the scientists of such eminence signified the high importance given to space research by the USSR. This commission was entrusted to organize with high priority the construction of a scientific earth satellite and the tasks of the satellite were to be observations of ultraviolet and x-ray emissions from the Sun and stars, the studies of the ionosphere on radio signals from the satellite, the biological studies of effects of weightlessness and others. This news report from Moscow was almost unnoticed in the western world. A year later at the CSAGI Conference of 1956 in Barcelona, Sedov, Chairman of the Soviet IGY Committee announced USSR's intention to launch a satellite for measurements of the atmospheric composition, cosmic rays, micro-meteorites, the geomagnetic field and of solar radiation. The CSAGI Conference in Barcelona further adopted the resolution that satellite launching countries should provide to any country requesting it the full information for its optical and radio tracking. Furthermore, it recommended that all IGY satellites should use the same radio frequency of 108 megacycles as announced by the USA so that the same ground based equipment can be used by tracking stations of different countries to follow the orbits of the satellites. The Soviet scientists however preferred to use frequencies of 20 and 40 megacycles for the tracking of their satellites and they gave long explanations at the conferences on the aspects of radio propagation that they considered as advantages on scientific considerations. At the IGY Conference in Belgium in 1957 an important landmark was established with the resolution on free scientific data exchange not only between the IGY countries, but has to be available to scientists of all countries. It was agreed to establish three World Data Centers, one in the USA, one in the USSR and the third in Western Europe-Japan. The British Hydrographer Vice-Admiral, Sir Archibald Day was appointed the IGY coordinator and he was entrusted to see the establishment of the three World Data Centers (WDC). The location of three WDC's would enable quick data transmission in a region and also ensured against accidental data loss in any Center. Thus IGY was all set to commence its activities which was described by Chapman as "the greatest example of world wide scientific cooperation in the history of our race".

1.2.3 The Sputnik and the Explorer

On October 4, 1957, Radio Moscow made an announcement which electrified the scientific world and had a profound effect on the people of all countries. It said that the Russians had placed a satellite in orbit around the earth at an altitude of 900 km. This satellite called "Sputnik" was a metallic sphere of 58 cm diameter weighting 400 kg and orbited the earth once in 93 minutes, while the inclination of the orbit was 65° to the equator. The highest point of the orbit or the apogee was located at about 980 km and the lowest point or the perigee was at about 240 km altitude from the surface of the earth. Four whip-like antennas of lengths of about 1.4 to 2.9 m were fixed on the polished metallic surface of the sphere of the satellite and transmitted the data in two frequencies, 20 and 40 Mc/s. The telemetry system used both the change of frequency method, called the frequency modulation mode and the length of the pulses and the pauses between them, called the pulse code modulation mode. These variations of the signals were encoded to provide the temperature inside the sphere and other data. The polished surface of the aluminium alloy was coated with reflecting and absorbing substances in such a way that in spite of a very high temperature on the outside surface at the sunlit side of the earth and of a very low temperature at the night side of the orbit, the temperature inside the sphere was maintained within the operating range of the scientific instruments. The airtight sphere of the Sputnik was filled with dry nitrogen gas for thermal circulation. The telemetered signals of the newly launched satellite - the "beep" could be picked up from many locations in Europe and the USA besides in the USSR. In the dawn and the dusk, when the observer on the surface was in darkness, the satellite, lit up by sunlight, could be seen by the unaided eye as a 'star' moving slowly among the fixed stars. The first successful launching of a satellite of the earth was an epoch making event in human history and it was hailed by people all over the world. The IGY Chairman, Sydney Chapman congratulated the Russian scientists for their "magnificent achievement" and said this settled USSR as the "first winner in the cooperative race to enrich geophysical knowledge by means of satellites".

Sputnik-I was not an instrumented satellite of the IGY type. It was a test vehicle designed to establish the capabilities of the launching system and the tracking and telemetry procedures of the earth satellite. One month later on November 3, 1957 USSR achieved further distinctions by launching a huge satellite called Sputnik-II, having a total weight of about 4 tons including the payload of over 600 kg in an orbit around the earth similar to the previous one. This contained in the payload a living passenger - a dog called Laika, with its life support system, together with instruments to measure cosmic rays, solar ultraviolet and x-ray emissions. This massive satellite which was active in orbit for one week firmly placed the USSR as leader in the space race.

The US scientists although surprised at the phenomenal advance of the Soviet space technology congratulated the USSR for their splendid success and they set out to establish an USA satellite as quickly as possible. Till then the US plans progressed on three lines. The US technical committee for earth satellite for the IGY was inclined to give the first preference to Navy's 'Vanguard' project. This was based on the research rockets of the 'Naval Research Laboratory' (NRL) of Washington D.C., the Viking which was a large rocket, about half the weight of the V-2 and had a peak altitude of about 250 km, and the Aerobee-Hi, many of which were to be flown for the IGY. It was planned to use a three-stage configuration with Viking and Aerobee-II as the first and second stage respectively, for the US satellite project of the IGY. The other proposal of the US Army was put forward by Von Braun and his team of the Redstone Arsenal of Huntsville, Alabama. In this plan a Redstone rocket - a direct successor of the German V-2 rocket - but much larger and having more range and payload - was to be used as the first stage booster - and the Sergeant rocket as the second stage followed by a smaller upper stage. It was calculated that in this arrangement the Redstone assembly could place a 28 kg satellite in an earth orbit at about 350 km altitude. The propulsion and guidance of the upper stages to place a satellite in orbit presented several problems for which some tentative solutions were worked out in both the projects of the Navy and Army of the USA. In the Vanguard project the third stage rocket was to be imparted spin so that it is injected in the correct orientation into the orbit. As different components of the satellite launch vehicles of the Vanguard project of the Navy and of the Redstone system of the Army were being tested, the tussle between the Navy and the Army continued for the first choice for the US satellite launch vehicle for the IGY. The special group finally tilted in favor of the Navy's Vanguard project in 1955 but this project was beset with many difficulties during 1955-57. In October 1957, when the USSR Sputnik was in orbit, the Vanguard project was caught unaware as several of its systems were running behind schedule. There was great pressure on the Vanguard project from all quarters. To alleviate the public disappointment in the USA for being in the second position in the space race, President Eisenhower stated on October 9, 1957 that he considered the US satellite program well designed and properly scheduled to achieve its scientific purpose and the first test vehicle was planned to be launched in December 1957. This boosted the morale of the Vanguard team and the preparations were rushed for the first launch. On December 6, 1957 the much awaited and much publicized US launching of the first Vanguard ended in complete failure as the first stage rocket exploded on the launch pad on ignition, not allowing the test flight of second and third stage rockets. This event brought dismay and gloom to all around. The next test flight of Vanguard-II made on February 5, 1958 again failed as the Viking lost control and broke apart 60 sec after launch. With these set-backs, the Army was summoned to go ahead to prepare for the satellite launching. Von Braun and his team was waiting for this opportunity as they had already successfully test launched the first and second stages in long range missions for nose-cone testings. For the satellite launching they planned the Redstone as the first stage with a guidance system, and this was topped by a cylinder containing a cluster of fifteen Sergeant rockets for the upper stages. Of these eleven Sergeant rockets were arranged in the outer ring and would be used for the second stage. Three rockets would be as the third stage and a single rocket protruding in the center had on its top an instrumented section so that it served both as the forth stage and the satellite. Before launch the cluster of Sergeant rockets was to be spinned to 400 to 750 r.p.m. by electric motors so as to provide spin stabilization of their flight. The ignition of the second, third and the fourth stages would be done by radio command by the ground controller at the correct times. Such a complex system required split second accuracy in the rocket firings and there were several other problems. The first launching of the Redstone assembly was made on January 31, 1958 and the huge Redstone rocket perfectly lifted the vehicle in the right path. After firing of the upper stages from the ground command the vehicle disappeared in the night sky. After 100 minutes the tracking station in California received the signal from the spacecraft and it was announced that an USA satellite has been successfully launched in the orbit. This was named 'Explorer-I', amid great rejoicing.

Explorer-I was a needle shaped satellite with a length of about 2 m and a diameter of 15 cm and had a total orbit weight of 14 kg of which the instrumented payload took 8.2 kg. It had four whip antennae each of 1.26 m in length placed in the middle portion of the satellite and it had a high power transmitter of 60 milliwatts power transmitting at 108.03 MHz and a low power one of 10 milliwatts at 108 MHz. The electronic components were in the form of three disks with miniaturized circuits which were potted in plastic foams to protect them against vibrations, shock etc. A tiny tape recorder which advanced in step once a second so that its 90 cm long tape could operate for about 2 hrs to record data, was built; but it was left out in Explorer-I ,but was included in the second Explorer. The scientific instruments consisted of a Geiger counter to record cosmic ray intensity at satellite altitudes and some recorders for micro-meteorites. The cosmic ray counts were scaled down by a factor of 16 and were transmitted as a modulated audio frequency signal on the high frequency carrier. Explorer-I had an elliptical orbit around the earth with its high point, apogee at 2500 km and its low point, perigee at 360 km, the orbital inclination to the equator was 33.3^o, and it made one revolution around the earth once in 116 sec. As the satellite spent about one third of its time at the night side of the earth at very low temperature and at the day side in the scorching Sun, it was specially painted in stripes. This provided good thermal control of the instruments inside where the temperature was maintained between 0° to 35^oC.

The US Navy made the second attempt with Vanguard on February 5, 1958 which again failed. Its third attempt on May 17, 1958 was successful as it placed a small sphere of 15 cm diameter weighting 1.5 kg was in a high altitude orbit. This however did not have science instruments.

Although the second Explorer flight on March 5, 1958 by US Army failed due to failure of the fourth stage rocket, the next mission Explorer-III was a highly successful one. It orbited the instruments same as in Explorer-I; in addition it had for the first time a small tape recorder which recorded cosmic ray intensity for the entire orbit of 116 min and then played back the data, on radio command, to the ground receiving station in the USA. Both the Explorers, I and III with simple instruments explored the new area of cosmic rays in space and thus made historic discoveries of the radiation belts around the earth by Van Allen and his team of the Iowa University, USA, as will be discussed later.

On December 6, 1958 the US Army made a bold attempt to launch a spacecraft to fly-by the moon with the massive Jupiter rocket of the intercontinental ballistic missile, instead of the Redstone rocket, the entire vehicle weighting about 60 tons. The last stage reached to 105,000 km altitude but failed its mission and fell back to the earth. This mission, Pioneer III, nevertheless, gave valuable new data of radiation in space with improved instruments of the type of Explorer I and III.

In May 1958, the USSR launched Sputnik-III, with a massive payload of 1350 kg including a variety of scientific instruments for measuring atmospheric properties, geomagnetism, solar radiations, cosmic rays, electrical field of the earth etc. Thus when IGY formally ended in December 1958, there were 28 attempts for satellite launching, 10 by the USSR and 18 by the USA; of these the successful missions were 3 by the USSR and 4 by the USA. The new frontiers of international collaboration in space physics which were opened up in the IGY were carried out further in the following years as the International Geophysical Co-operation (IGC). Many countries established their national Space Research Organizations to undertake space explorations. In the USA a new organization was formed in 1958, called the National Aeronautics and Space Administration (NASA) by the government and it was entrusted with the large scale exploration of space. In order to promote international cooperation in space research and for scientific data exchange, ICSU formed a new international organization, Committee on Space Research (COSPAR) which had its initial meeting in London in November 1958 and the first general meeting in Nice in 1960.

Thus we come to the end of the early period and the dramatic and historical account of the beginning of space research. In the 1960's and 1970's space research flourished in many directions and in many countries using space vehicles like stratospheric balloons, sounding rockets, earth satellites and space probes. By the 1970's earth satellites for space research were launched successfully by six countries, USSR, USA, France, China, Japan and India. Several spacecraft were launched that escaped the earth's gravity and made lunar and planetary explorations and reached the outer regions of the solar system, bringing to us a fascinating new view of the solar system and the universe around us. An account of the early years of space exploration is given by D.P. Le Galley.

1.3 Space Vehicles and Space Physics

In this section we shall briefly discuss the general features of different types of vehicles of space research and their applications in space physics. This is intended to provide the readers an introductory and perspective view of the space research vehicles which have been used to obtain major new information on space physics and space science. These are (a) sub-orbital space vehicles, which include, balloons, sounding rockets and (b) orbital space vehicles like satellites, space stations and (c) planetary probes and deep space probes.

1.3.1 Balloons for space research

Early Studies

In the early years balloons made of synthetic rubber or neoprene, commonly used for weather studies were employed to lift instruments for measuring cosmic ray intensities in the atmosphere. Normally a number of these rubber balloons filled with hydrogen gas were used in tandem, sometimes as many as 30 to 50. With a load line the instrument package was attached with a parachute to the balloon. In this manner about 10-30 kg instrument payload could be exposed to 20-25 km altitude for a few hours. Optical theodolites from two or three stations could observe the balloons and their altitude and location were determined by triangulation. As several balloons burst, the balloon train slowly descended. When the instrument package was cut-off a by timer device it descended to the ground by parachute and was recovered. Using this method pioneering studies on cosmic rays were made in several countries, including India in the 1940's by Millikan, Neher and Pickering[(2)].

In the late 1940's and early 1950's cosmic ray studies were carried out with the simple methods of using clusters of rubber balloons by several countries including India, and by using small plastic balloons by the University of Bristol, England. These led to major discoveries in elementary particle physics such as the discovery of several types of K-mesons and hyperons by the

Bristol group led by Prof. C.F. Powell[5]. In India during 1951-53 scientific teams of the Tata Institute of Fundamental Research (TIFR), Bombay made discoveries of new features of heavy unstable particles in cosmic rays such as the accurate mass measurements of K-mesons, identifying their unique mass by Biswas et al.[6] and of their decay properties and associate production by Lal et al.[7]. These illustrate some of the major advancements made in cosmic ray physics with early balloon flights. The rubber balloons had the limitations of lifting only relatively light payloads and the rather short time at a ceiling altitude which was also uncertain. To overcome these difficulties large polyethylene balloons came into use in the 1950's.

Plastic Balloons for Space Research

As noted earlier large plastic balloons made from thin polythene films developed by the office of Naval Research of the USA were used for cosmic ray experiments in 1948-50 so that instruments could reach high altitudes of 25-30 km and float there at constant altitude for several hours and then the instruments were released and recovered. Using this technique scientists of the University of Minnesota and the University of Rochester, USA, made the major new discovery of the presence of heavy nuclei of helium to iron in the primary cosmic rays[3,4]. However, large plastic balloon techniques developed by the US Navy were classified for a long time. So these were developed independently by several groups e.g. the University of Bristol, England and the Italian groups in 1950's using available information and balloon flights from Sicily were made. The balloon group of the Tata Institute of Fundamental Research, Bombay started the development of large plastic balloon flight technique, more or less independently in 1956 using available information. In the following decades this grew up into the TIFR National Balloon Facility located at Hyderabad, India which undertakes the design and fabrication of large plastic balloons, and the launching, tracking and recovering of the payload for scientific communities of India and other countries. Several groups of the USA, Europe, Japan and Australia participated in the balloon flight experiments from the Hyderabad Balloon Center, which is the only permanent balloon facility located near the equator ($\lambda = 19^o$). Since the 1970's many improvements were made in the techniques of plastic balloons for space research by scientists of the USA Balloon Flight Center at Palestine, Texas and of many other countries including England, USSR, Japan, France, India, Australia etc. The scientific ballooning became an actively pursued scientific and technical discipline during the past decades a review of which is given by Damle[8]. In this section we shall briefly discuss, the basic features and methods of the normal and the long duration flights of scientific balloons for space research.

Basic Features of Plastic Balloon Flights

The objectives of the plastic balloon technology are as follows: (i) to lift scientific payloads of $\sim$ 50 to 1000 kg in general to the high altitudes of 30-40 km where only a small amount of 0.5% or less of the residual atmosphere lies above the balloon; (ii) at this ceiling altitude the balloon with the scientific instruments is to float at a constant level for several hours, from $\sim$ 6 to 24 hours or more as required by the scientific experiments so that radiation and particles coming from outer space can reach the instruments with minimum attenuation by the small mass of air above the balloon. The final ceiling pressure altitude P_f is determined such that at this low pressure P_f the initial volume V_i of the gas completely fills up the balloon volume V_f. Any further rise of the balloon will make the gas leak out of the escape tubes of the balloon so that it maintains a constant pressure altitude; and (iii) finally to release the instrument package from the balloon by radio command so that the payload descends by a parachute to the earth to be recovered by the flight team and the scientific data recorded in the instruments can be retrieved. The above basic objectives are achieved by the balloon techniques.

Thus technology of the plastic balloon flights for high altitude space physics experiments involve the following aspects. (a) Balloon design, balloon material and fabrication, (b) balloon launching, tracking, telemetry and flight control and (c) recovery of payload.

(a) Balloon design

Plastic balloons for research purpose can be either designed and fabricated by research laboratories, as noted earlier or can be purchased from commercial companies like Winzen Inc. in the USA. The balloon design has evolved through many stages, and the type commonly favored is the inverted 'tear-drop' design or natural shape design. In this design at the ceiling altitude the fully inflated balloon assumes the shape of an inverted 'tear-drop' or natural shape. There are many variations in the shape of balloon designs.

The balloon material is normally made of special types of polyethylene films manufactured by several companies having film thickness of 12 μm to 35 μm which can withstand very low temperatures specially at the tropopause region of the atmosphere. In the equatorial region the tropopause region is located at 16-18 km (and at $\sim$ 25 km at mid latitudes) where temperatures are as low as -90°C in contrast to the mid-latitude tropopause at 25 km with a temperature of about -60° C. In the 1960's the Indian group developed one type of a balloon fabric, using carbon black to tint the polythene film, with about 12% absorption of solar radiation; this heating by solar radiation enabled the balloon material to withstand its passage through the very cold region of the tropopause, in which normal white balloon fabric became brittle and the balloons used to burst. The tinted balloons could be used only during day-time flights, so it was necessary to have a superior type of white or transparent balloon material. In the late 1960's Winzen International Inc. of the USA developed a special type of transparent material for the balloon film with a very low brittle point. This film, called "Strato-film" has been used for balloon making for many years since then. In 1992 the Indian Balloon Facility group developed a new kind of transparent balloon fabric, called 'Anterix', which was found to be as good as Strato-film[(8)]. Balloons fabricated from the India made fabric 'Anterix' were successfully tested for equatorial balloon flights with heavy payloads. This new balloon film with a thickness of 20 μm is now available with a cost factor a few times less than the Strato-film of the USA.

Balloon Fabrication - Special grade balloon films with a thickness of 12-37 μm, available in rolls with standard widths of 1.32 meters are laid flat on a long table with a length of 100 m or more and are cut into sections. The seams are heat-sealed with a sealing unit and are cooled with an air flow. The sealing unit travels with the desired speed along the table, so that the balloon takes the designed shape with sealed segments like orange peel sections. Load bearing plastic tapes are then applied along the seams, and after adding several items like end fittings, escape tubes etc., the balloon is made ready for launch and flight operations.

(b) Balloon launching, tracking and flight operations

The suitable time for launching a scientific balloon is when the speed of the upper wind at 30-35 km altitude, where the balloon floats, is low, say

5-20 km/hr. This usually occurs for low and mid-latitudes during the period when the sun is close to the equinoxes i.e. during March-April and September-October. For example in the low latitude region in India such as Hyderbad ($\lambda = 17.5^o$, $\phi = 78.5^o$ E) at 30 km height the winter time easterly wind changes over to the westerly wind of the summer pattern sometime close to the equinox; during this change period the upper wind speeds often drops to low values of 5-15 km/hr. This period is identified by test flights of rubber balloons and is suitable for balloon flights. This is so because at 10 km/hr wind speed the balloon can provide exposure for 15 hrs at the ceiling altitude and still be at a distance of $\sim$ 150-200 km from the launch station permitting convenient recovery of the scientific payload on cut down.

The period, one or two hours before sun-rise, when ground-wind speeds are very small, is generally chosen for the launching of scientific balloons. A flat open field of at least $\sim$ 500 m in diameter is normally used. A truck slowly unloads the balloon - about 100 meters long - on a canvas cloth strip laid on the ground. Then the line with flight operation items are attached. These are as follows, starting from the lower end of the balloon: (1) Payload release loop, with cutters. This is a device which cuts off the load line and release the parachute and payload, on receiving a command signal by radio from the ground through the telecommand receiver packages; (2) Parachute; (3) Radiosonde which provides pressure altitude signals; (4) Telecommand-Telemetry package. This contains the telecommand receiver channels for ballast release and for cut down at flight termination. It also includes a standard air-borne PCM telemetry encoder for the use by scientific experiments; (5) Radar reflectors: Used for Radar tracking of the balloon; (6) Red warning light; (7) Payload attachment plate and the main experiment gondola; (8) Radio beacon and (10) Ballast. A schematic diagram of the typical load line for balloon launching at Hyderabad, India is shown in Figure 1.3.1

The balloon is filled with hydrogen (or helium) gas to the desired volume, which is usually 1% to 5% of the maximum balloon volume, so as to provide the required free lift and this is measured by an weighing mechanism or by measuring the volume of the gas by a gas-meter. The bubble of gas in the balloon is held by a smooth roller and the load line is kept anchored to a heavy truck on which the scientific payload is mounted. Two small polythene balloons each with 25 km lift are provided on the load line control instrument packages to assist a smooth launch. The balloon and the load line

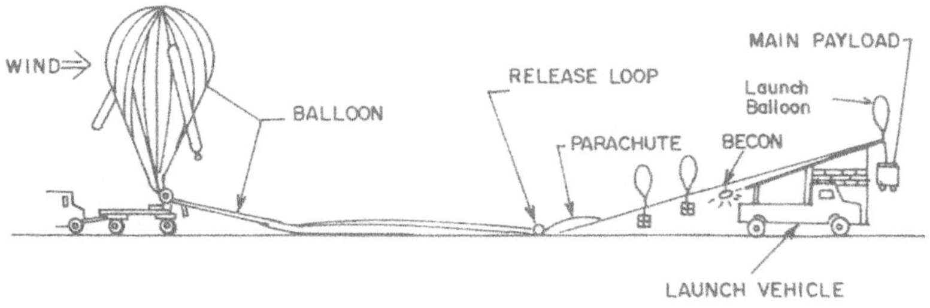

Figure 1.3.1: A sketch of a typical load line (not to scale) of a balloon launching by the TIFR-Indian National Balloon Facility at Hyderabad. (Courtesy: S.V. Damle)

is laid in the direction of ground wind (Figure 1.3.1). When all systems are ready, command is given for the launching. The clamp holding the balloon bubble is released and the bubble rises up rapidly and moves towards the launch truck. The latter is slowly maneuvered so that the balloon and the load line is stretched vertically overhead. At that point in time the scientific payload is released and it is smoothly lifted up vertically without pendulum motion. The balloon is tracked by radar and telemetered data from the radio-sonde provides the information about the height of the balloon. If the rate of rise is found to be slower than the expected limits, the flight control center issues radio commands which release the required amount of ballasts over a period of time and the rate of rise is rectified. Normally it takes 2.5-3 hrs for the balloon to reach the ceiling altitude of 30-35 km. The fully inflated balloon floats at a constant altitude for the desired period. Scientific data is telemetered to the ground and also recorded on an onboard recorder. At the end of the flight a telecommand signal given by the flight controller releases the instrument packages which descend by parachute to the ground.

(c) Recovery of the payloads

A recovery vehicle travels to the region where the balloon payload descended and awaits for the information from the people of the locality on the exact location of the instrument landing. On receiving the position, usually in a few days time, the recovery of the instrument is made and a suitable reward is paid to the finder of the instrument. In some cases, particularly for the long duration flights, the balloon is followed by a light aircraft and the impact location of the payload is determined from the aircraft when the instrument lands in an uninhabited area.

Long duration Balloon Flights

For special studies e.g. for very low intensity high energy cosmic rays and for rare events it is desirable to have long duration balloon flights for several days or weeks. These have been achieved by balloon technologists by special procedures and devices as follows.

As noted earlier, during a certain period of the year e.g. near the spring equinox, the wind speed drops to nearly zero at the altitude of 30-40 km for a brief period of one week or so. By conducting test flight with smaller balloons or sounding rockets, the balloonist identifies the period correctly and the scientific instruments are launched with a large balloon promptly. At near zero speed of the upper wind, the balloon remains at the floating altitude for many hours or days before cut-down. Thus balloonists have achieved flights of 24 hrs to about 90 hrs.

In countries with small land mass and surrounded by sea, e.g. in Japan, the recovery of balloon payload from sea with an ocean going ship is too expensive for a regular operation. So a novel technique has been developed. When the balloon reaches a level, say, 15 km altitude, a valve at the top is opened by radio command and the balloon is made to stop ascending and it floats at this altitude. Then the balloon travels fast in the strong westerly wind. As it reaches the reception limit of the ground station, the balloon is made to rise to, say, 30 km by dropping sufficient ballast, and then it floats at the ceiling altitude, traveling back westward in the easterly wind towards the launch site and is recovered near the shore.

Special long duration flight campaigns are made by balloon groups as needed by launching balloons either from an island or from the ship and recovering the payload from land or the sea with a recovery ship. Such expeditions were made in the Mediterranean sea by groups of France, Italy and Spain and in the Atlantic by the USA groups. Russian groups conducted trans-Siberian flights with launching from the eastern part of the country, Kamchatka and terminating in the western region of Russia. They achieved a long duration flight of about 150 hours at a float altitude of 30 km. These flights require large weights of ballast to maintain float altitude and long range radio communication equipments.

Attempts are being made to perfect a special type of balloon called super pressure balloon which is made of a high strength material. The objective is to make long duration balloon flights which will travel round the world in 10 to 20 days. The southern hemisphere launching either from Australia or Brazil at a latitude of 22° is preferred as the balloon trajectory passes over only three countries including South Africa. A few successful flights from Antarctica have been made with normal and new types of balloons.

1.3.2 Sounding rockets for space research

Before the advent of satellites and its early years sounding rockets were extensively used for space research in the 1960's and 1970's. With more countries launching satellites in the 1980's and 1990's, the sounding rockets have receded to the background now, and are rarely used at present. So a brief account of this method is presented for the historical perspective.

Sounding rockets carry the fuel and the oxidizer, either separately as in liquid fuel rockets, or in combined form, as in solid fuel rockets, and burn them quickly so as to produce gas at high temperature and pressure. The exit of the gas at high speed through the nozzle of the rocket motor imparts opposite reaction and momentum to the rocket which is thus propelled upward with high speed. After the complete burning of the rocket fuel of the first stage in about 10-30 sec the rocket reaches a height of $\sim$ 20 km and it coasts with nearly the same velocity for a short time which is slightly reduced by air drag; then the first stage of the rocket drops out and the second stage of the rocket ignites. The burning of the second stage imparts large momentum and high velocity to the rocket in the thin rarefied upper atmosphere with low air resistance. As a result the rocket ascends a to peak altitude of about 150 km. At the end of the burning of the second stage rocket motor, the second stage is usually separated from the payload by a timer activated mechanism and it drops out, while the payload travels in a ballistic trajectory reaching peak altitude and then enter the atmosphere. For some experiments where recovery of the rocket payload is necessary, special parachutes are released to slow down the descent of the payload which is located with radio beacons and is retrieved. For others, the data recorded by the instruments in the rocket payload are telemetered to the ground during its flight at above 60 km or so, and then the rocket payload burns out on re-entry. Typical two stage rockets like Nike-Cajun, Nike-Apache etc. of NASA, USA were flown extensively for many years for space physics studies. For medium weight payloads bigger sounding rockets were used, e.g. Aerobee of the USA, Skylark of the UK etc. For still heavier payloads and higher peak altitudes, bigger rocket systems were used such as Aerobee-Hi of the USA, Centaur of France etc. The perfection of sounding rocket technology and the fabrication of bigger and more powerful booster rockets led to the launching of earth satellites and the beginning of the full scale space age.

1.3.3 Satellites for space research

Satellites and space vehicles are the major tools in carrying the scientific instruments above the earth's atmospheric - into the magnetosphere and in interplanetary space - for prolonged periods of time for conduction space research and explorations. Satellites and spacecraft are launched in space by means of multi stage rockets, the science and technology of which represent one of the major human achievements. The details of these technologies involve various branches of aerospace engineering as well as material science, control engineering and several other branches such as celestial mechanics etc. so only a brief account of the basic physical features is included here.

Basic Features:

Rocket engines carry with them both the fuel and the oxidizer, which are allowed to burn very quickly in a combustion chamber thereby producing a very hot gaseous mixture at a high pressure that escapes through a nozzle forming a high velocity, hot jet of gas. The opposite reaction of this gas jet imparts an upward impulse to the rocket during the period of burning of the fuel that lifts the rocket and propels it through the atmosphere to high altitudes. At this stage different types of control gas jets are applied to counteract the deflecting forces on the rocket such as winds etc. and steer it to its designated path. As the first stage of the rocket travels through the dense part of the atmosphere that offer maximum resistance and it has to lift its own weight as well as the weights of the multi stage rockets above it, the first stage requires the maximum amount of fuel and oxidizer to provide the required thrust and hence it is the bulkiest.

The rocket engines are basically of two types, the liquid fuel and the solid fuel ones. In the liquid fuel rocket motors the fuel and the oxidizer are stored in separate chambers and these are pumped in a controlled manner into the combustion chamber where they burn and produce high velocity gas. The thrust produced by a rocket motor depends on the type of propellant, and the specific impulse of a propellant is defined as the thrust produced by unit weight of propellant mixture. Commonly used as fuel and oxidizer are kerosene-liquid and liquid-oxygen respectively. The cryogenic rocket engine of liquid hydrogen as fuel and liquid oxygen as oxidizer has almost the highest specific impulse. This propellant combination is used in the sophisticated rocket engines such as the main engine of the Space Shuttle. In the solid fuel rockets, on the other hand, both the fuel and the oxidizer are combined in a granular form. These normally stable solid propellants burn rapidly when ignited. The solid propellants such as HTPB (Hydroxy Terminated Poly Butadine) are usually cast in form of a cylinder which fits into the rocket casing with its igniter. When it is fired the solid propellant starts burning from the center to the outer periphery till the entire propellant is burnt out producing a hot gas at high pressure in a short time, say $\sim$ 30-50 sec, which produces the desired thrust. Solid fuel rockets are normally employed in pairs or in a cluster of 4 to 6 in the first stage of the satellite launch vehicle (SLV), so as to produce the maximum thrust to lift the entire assembly of SLV weighting typically 200-300 tons or more. In the designs of some of the SLV with very large lift-off weights, large solid booster rockets together with liquid fuel engines like used with the Space Shuttle, with a gross lift-off of 2 million kg.

Liquid fueled motors have the advantage of the flexibility - it can be switched on and off as desired, while the solid fuel rocket engines operate on a one time ignition principle. Liquid fuel engines are complex machines involving many valves etc. whereas solid fuel engines have no moving parts and hence the reliability factor of these is higher than that of liquid fuel rocket motors.

Principles of Earth Orbiting Satellites

We now briefly discuss the basic parameters which determine the conditions for the earth orbiting satellites. Let us consider a satellite of mass, m, which is to be placed in a circular orbit at an altitude of h (say 500 km) around the earth. What should the final velocity imparted to it by the last stage rocket be at say 500 km altitude? This is derived from the basic principle that in a circular orbit around the earth, the centrifugal force on the satellite of mass, m (kg) moving with velocity V, km/sec must balance the gravitational force of the earth. Thus we can write

$$\frac{mV^2}{(R+h)} = \frac{m \cdot G \cdot M}{(R+h)^2} \tag{1.3.1}$$

or

$$V = \left(\frac{G \cdot M}{(R+h)}\right)^{\frac{1}{2}} \tag{1.3.2}$$

where R is the radius of the earth (= 6378 km), M the mass of the earth (= $6 \cdot 10^{24}$ kg) and G the gravitational constant (= $6.67 \cdot 10^{-11}$ $N \cdot m^2/(kg)^2$). By substituting the values in the above equation, we get V = 7.6 km/sec or equal to 27,410 km/hr. If the final velocity is smaller than the above, the spacecraft will re-enter the earth's atmosphere with a large elliptical trajectory. If the velocity is higher than the above the spacecraft will escape the gravitational force of the earth and will eventually be a sun-orbiting spacecraft. From more detailed calculations it is seen that the final velocity of the Satellite V = 8.3 km/sec would lead to its elliptical orbit, and V = 11 km/sec corresponds to the escape velocity from the earth's gravitational pull. The time period, T, of the above earth orbiting satellite can be derived from equation (1.3.2) and is given by

$$T^2 = \frac{4\pi^2}{GM}(R+h)^3 \tag{1.3.3}$$

This is of the same form as the Keplers law of motion of planets in the solar system which states that the square of the time period of a planet is proportional to the cube of its distance from the Sun.

Launching of a Satellite into Orbit

Multi stage rockets - in three or four stages are assembled and integrated in an appropriate configuration to launch earth orbiting satellites. A typical sequence of operation for a ~ 800 kg satellite launched into an orbit at ~ 500 km altitude will be briefly described.

1. At countdown time 0 sec, the two strap-on boosters and the first stage rocket motor are ignited and these three solid fuel rockets providing a total lift-off thrust of 200,000 kg, lift the whole assembly in its proper path against external disturbances, such as wind etc. The strap-on boosters burn typically for $\sim$ 50 sec after which they are separated and plunge into the sea to be retrieved and reused. The first stage separates after the burn out at $\sim$ 5 km height where-after the second stage ignites.

2. The second stage of solid fuel burns for about 40 sec and its path is directed by the inertial guidance system. This guidance system, which is of crucial significance, controls and maintains the second and subsequent stages in their proper path until the satellite is injected into the orbit. The inertial guidance system consists of an inertial platform employing three gyroscopes and three acceleration-meters which measure the parameters of the vehicle along three inertial axes. The on-board computer system computes the necessary data and controls the flight path.

3. After the burn-out of the second stage it is separated at $\sim$ 110 km and the third stage is ignited. The third stage rocket burns for $\sim$ 50 sec during which the spacecraft reaches the orbital altitude of 450 km and the inertial guidance system adjusts the trajectory such as it is parallel to the surface of the earth within an accuracy of a few degrees. As the third stage burn-out is over, it is separated and the forth stage rocket plus the satellite is given spin so that its direction is spin stabilized. The inertial guidance system ends with the separation of the third stage.

4. The final fourth stage rocket which may be of liquid or solid fuel is activated such that the spacecraft attains the correct velocity of 7.5 km/sec and is moving in the proper direction i.e. parallel to the earth's surface. At the end of the fourth stage burn-out the fourth stage motor plus the satellite is injected into a satellite orbit. Then the fourth stage rocket is separated from the satellite and the protective cover of the satellite is removed. The injected satellite is then activated with its

power-supplies, and its trajectory is followed by receiving signals from its transmitters.

The above description is given for some typical launching of a low mass satellite. The rocket power and other parameters will vary depending on the payload weight and the altitude.

Satellites in a Geostationary Orbit

For global communications, television transmission and weather forecasting, satellites at geostationary orbits play an unique role. One can calculate from equation 1.3.3 that when a satellite is placed at an altitude of 36,000 km above the surface its period of rotation around the earth is the same as the rotational period of the earth around its axis. As a result such a satellite would appear to be stationary with respect to the earth's surface. Launching of spacecraft into a geostationary orbit is carried out in several steps as follows. The spacecraft is first placed in a near earth circular orbit at a small inclination to the earth's equator say at an altitude of 400 km. The spacecraft in orbit contains a rocket motor which is activated by ground command so that it is ejected from the circular orbit and the added impulse makes the orbit elliptical. After several such orbits, the spacecraft, is oriented in the calculated direction and is given the required thrust at the right time at the apogee so that the orbit becomes close to a circular one. After several tests and trials, the spacecraft is brought at the correct altitude and location so that it is geostationary at a given longitude. Geostationary satellites are used in space physics, as a continuously monitoring platform in space.

Lifetime and Active Life of a Satellite

It may be noted that the lifetime of a satellite in near earth space say at 400 km altitude is determined by the air drag which depends on the ambient density of residual air at the orbital altitude. This air drag slowly reduces the altitude of the orbit and when the satellite reaches the denser portion of the atmosphere, at < 100 km altitude, its orbit rapidly sinks and the satellite enters into the atmosphere where the heat of friction with the air burns out the spacecraft. Satellites at an orbital altitude of 300 to 600 km generally have a lifetime in orbit ranging from ~ 10 to 1000 years or more. A more important factor is the active lifetime of a satellite which is determined by the lifetime of the electrical power system and the amount of gas available to keep the spacecraft in the proper orientation in space. When either of them is exhausted, the active and useful lifetime of a satellite is over. Typically this is 5-10 years for a normal satellite.

Reusable Space Vehicle System

In the normal procedure as a spacecraft is launched into orbit, all parts of the launch vehicle including the guidance and control systems employing high technology are destroyed. Because of these the costs of space vehicles and the launching of spacecraft into orbit had become very high and time consuming. In order to make space flights less costly and less time consuming, NASA of the USA successfully developed the reusable space transportation system called the Space Shuttle. Four Space Shuttles constructed by NASA have successfully completed a large number of space missions in the 1980's and 1990's. So far these carried out the orbiting missions of about ten highly complex Spacelab missions, each comprising of 15-30 space science experiments for one week or more and then returning these to Earth. A Space Shuttle launches one or more large mass satellites into space in one flight. Space Shuttles have been used for the rendezvous with disabled large spacecraft in space, bringing them in the Space Shuttle, repairing them and placing them in the orbit again so as to give them a new life. The European Space Agency is planning to build a reusable space vehicle system that should be operational in the early 21st century. The Space Shuttle of NASA has been designed so that it combines and performs the functions of a rocket propelled vehicle, a satellite in the orbit of the earth and a space-plane which reenters the atmosphere and lands on an air field like a normal aircraft.

The salient features of a Space Shuttle launching and reentry are as follows. The Space Shuttle consists of (i) the orbiter (34 m long) with a crew of seven or more, a cargo bay which houses the Spacelab or other large spacecraft and the main engines, (ii) the external tank containing fuel of liquid hydrogen and liquid oxygen as oxidizer for the main engine and (iii) two solid rocket boosters strapped to the external tank.

1. As the two solid rocket boosters and the main engine are ignited, they slowly lifts from the ground a gross weight of 2 million kg i.e. 2000 tons. Each of the two solid boosters generate a thrust of about 1.2 million kg

and the main engines of 0.2 million kg and thus a total thrust of 2.5 million kg is generated.

2. At the height of 45 km, the burnt-out solid boosters are separated and land in sea from which they are recovered for reuse. Shortly afterwards, when the fuel of the external tank is fully used up, the external tank is separated and jettisoned to the sea. This is the only component (only an housing) which is not recovered.

3. The orbiter with its main engines is inserted in an orbit at $\sim$ 185 km where its velocity is $28.3 \cdot 10^3$ km/hr. At a height of 300 to 350 km the orbiter opens it cargo bay doors and conducts scientific missions e.g. the Spacelab missions, etc. for a week or so. A crew of about seven astronauts participates in each mission. Three crew members control and operate the Space Shuttle and the other four conduct scientific experiments in the Spacelab module and also on the exterior as needed. A typical mission lasts for 7-10 days. At the end of the mission the cargo bay doors are closed and the shuttle is prepared for reentry.

4. As the Shuttle reenters the atmosphere with a velocity of $\sim$ 28,000 km/hr at about 125 km altitude, the air friction raises the surface temperature of the shuttle to a few thousand degrees Celsius. The Shuttle is covered with slabs or bricks of a specially devised protective insulating material so that the inside temperature is maintained at a normal value. The Shuttle utilizes aerodynamic braking to slow down and to land on a long air strip. On cooling down, the protective thermal tiles regain their normal properties and can be used for another flight. The orbiter has a landing weight of about $100 \cdot 10^3$ kg. After unloading the experiments the Space Shuttle is prepared for the next flight with a new set of scientific and technology missions.

There are diverse types of space flights to meet different types of space science missions such as interplanetary probes, planetary missions, solar missions etc. and these will be briefly discussed in the respective chapters.

Space Stations

Another kind of space research facility developed by the USA and the USSR involves a large space vehicle in orbit fitted with various space research instruments and laboratory facilities. This semi-permanent space station is located at $\sim$ 500 km altitude and is visited by teams of crew members in manned space vehicles which dock with the space station. The earlier team of crew members returns to the earth while a new team of crew takes over the charge of maintaining the space station and conducting scientific experiments. The historic SKYLAB, the first space station of NASA, USA, was placed in an orbit of 450 km in 1973 and was visited by three teams of astronauts in three spacecraft called the Skylab-I, II and III missions and a large number of scientific experiments were successfully conducted in space during 1973-74. Several years later in 1979, the abandoned Skylab plummeted back into the atmosphere and burnt out in huge fire balls in desolate parts of Central Australia. The 'SOYUZ' Space Station of the USSR was in operation in the 1970-80's. An Apollo mission of the USA docked with SOYUZ in the 1970's to conduct a joint mission. Then a larger USSR space station called 'MIR' has been in operation for space science experiments in the 1990's. The Space Shuttle Atlantis of NASA had a rendezvous with the 'MIR' Space Station of Russia and they conducted joint science experiments. European crews participated in scientific missions in the Space Shuttle of NASA and in the MIR space station in programs of international collaboration.

In the 1990's plans are in progress by NASA of the USA for an international collaboration for establishing a large "SPACE STATION" in the early years of 2000 for conducting new kinds of space experiments jointly with several other countries. The building of this "Space Station" has started in 1998.

1.4 Problems on Chapter 1

1. Estimate the minimum altitude at which space astronomy detectors are to be flown to record cosmic X-rays of (i) 1-2 keV and (ii) 20-50 keV, with 50% attenuation in the earth's atmosphere. Explain what type of space vehicles - rockets, balloons, etc. are suitable for the measurements of (i) and (ii), and why?

2. Explain the principles and operations of zero pressure balloon flights for space physics experiments. What are the critical requirements of the plastic films used for these balloons?

3. Describe the main methods used for long duration balloon flights.

4. Calculate the orbital period of an earth satellite placed in a circular orbit at an altitude of 1000 km. [Hints: use the parameters given with equation 1.3.1]

5. Calculate the altitude above the earth's equator of a geo-stationary earth satellite. [Use the constants given with equation 1.3.2]

CHAPTER 2
THE MAGNETOSPHERE OF THE EARTH

2.1 Early Concepts

The phenomenon of the magnetism of the earth was known in ancient times, but its nature remained obscure for many centuries. When Queen Elizabeth-I ruled England in the 16th Century, her personal physician, William Gilbert, spent his spare time for studying magnetism. He shaped a load-stone in the form of the earth, which he called 'terrella' and found that the magnetic forces on its surface were distributed somewhat similar to that of the earth. This led Gilbert to propose that the earth resembled a giant magnet and that a small and powerful bar magnet placed near the center of the earth in roughly north-south direction would roughly reproduce the behavior of the compass needle on the earth's surface. Explorers and navigators found that the compass needle did not point to the north pole, but to another direction at some distance from it. Edmund Halley - of the famous Comet of Halley - made long sea voyages in the North and South Atlantic during which he made systematic studies of the displacements of the compass needles with respect to the north-south axis of the earth and this led to the discovery of magnetic poles. Halley also noted the connection between the disturbances of earth's magnetic force and the appearance of the aurora. The Swedish physicist A. Celsius recorded in 1741, the observations that a bright aurora was accompanied by a large displacements of compass needle indicating the disturbed nature of the earth's magnetic field. In later years and in the first half of the twentieth century, improved magnetometers with higher sensitivities were developed and the properties of the earth's magnetic field on the surface of the earth were measured in various locations. These measurements included the total intensity - its magnitude and direction, the horizontal component of the magnetic field, the declination (the angle between the direction of the horizontal component and the north-south direction on the surface of the earth), the inclination (the angle between the direction of the total intensity and the local vertical) and their variations with time - periodic and non-periodic modes and both in short as well as long time scale. These observations led to the fundamental properties of the terrestrial magnetism. (See e.g. Chapman and Bartel's book on Geomagnetism).

2.2 Experimental Methods in Space Studies and Observations

The advent of space age in the 1950-60's opened up the new dimensions in the studies of the magnetic field of the earth in the near earth space. For more than a century scientists wondered on the basic question, - how far does the magnetic field of the earth extend in space? The answer could come only from detailed measurements for about a decade in the space age. We first discuss the basic ideas of experimental methods of measurements of the geomagnetic field in space in two aspects - the magnetometers and the orbital features of the spacecraft employed for these studies. Prior to the satellite borne experiments, rocket borne magnetometers were flown in the 1950's and 60's to about 120 km. Some of these flights were made near the equator and observed an abrupt change in the magnetic field at an altitude of $\sim$ 100 km leading to the discovery of an equatorial electrojet.

The early satellites, Sputnik-III, Lunik-I and II of the USSR and Pioneer-I and V and Explorer-VI of the USA flown in 1958-59 carried magnetometers in space and gave information on the geomagnetic field far away from the earth. However systematic data on the magnetic field in space was first collected with Explorer-X, XII and XIV and IMP-1 and other satellites in 1961-63. The instruments and the experimental methods used are briefly described.

Tri-axial Fluxgate Magnetometers: Three fluxgate magnetometers mounted in the spacecraft in three mutually perpendicular directions were employed in these investigations. Fluxgate magnetometers were standard electronic instruments used extensively in a aerial magnetic surveys, balloon borne experiments and satellites. It had a sensing element of permalloy or similar material which becomes magnetically saturated in a very low magnetic field. A coil surrounding the core excites it to near saturation at a frequency of about 400 Hz. In the absence of an external applied magnetic field the induced flux in the core is symmetrical in two directions. But if an external steady magnetic field is applied along its axis, the saturation occurs more quickly during the half of the cycle, the resulting flux becomes asymmetric. Thus the output pulses telemetered to the ground station give the magnitude of the external field along the axial direction using ground calibrations. The measured magnitudes of the magnetic field in the three orthogonal directions are combined to determine the total magnetic field and its direction. For reliable measurements of the magnetic fields in space several precautions are essential. The housing of the magnetometers must be of nonmagnetic material and the electrical and electronic instruments in the spacecraft must not interfere with the operation of the magnetometers. In order to avoid the local fields produced by the spacecraft the magnetometers are placed at the end of a sufficiently long broom.

Proton Precession Magnetometer: Proton precession magnetometers are based on the principle of nuclear magnetic resonance (NMR) and are widely used in air-borne and other types of magnetic surveys. To observe NMR signals, a sample such as water containing protons is placed between the poles of an electromagnet which carried in addition to the main coil a small auxiliary sweep coil. A coil which surrounds the sample is connected to a radio frequency oscillator and a second is placed around the sample such that it is perpendicular to both the oscillator and sweep coils. The oscillator frequency is fixed and the magnetic field is varied continuously with the sweep coil. Protons in the sample, which are tiny nuclear magnets precess around the magnetic field H with angular frequency ω and when this matches with the radio frequency, transitions take place from $+\frac{1}{2}$ to $-\frac{1}{2}$ spin states, which induce a signal in the receiver coil. The resonance condition is given by $\omega = \gamma$H where γ is the gyromagnetic ratio. For protons in magnetic field of 10,000 Gauss the resonance frequency is 42.6 MHz. Since the resonance lines are very sharp, NMR can be used to measure magnetic field with great precision. This instrument is adopted to the measurement of the total vector magnetic field of the earth. Proton precession magnetometers provide absolute values of the magnetic field, whereas the fluxgate magnetometers are relative types.

Based on the NMR principle rubidium vapour magnetometers which employ optical pumping are also sometimes used for magnetic field measurements in space.

Coordinate Systems for Magnetic Field Measurements: As the magnetic field in the heliosphere is controlled by the Sun and the solar wind plasma, it is of advantage to use the solar ecliptical coordinate system as shown in Figure 2.2.1. The center of this coordinate system is the earth and the X-axis is directed from the earth to the Sun at all times, as shown by X_{se}. The X_{se} and Y_{se} are in the plane of the ecliptic and the Z_{xe} direction is normal to the ecliptic plane. Thus these form a right handed coordinate

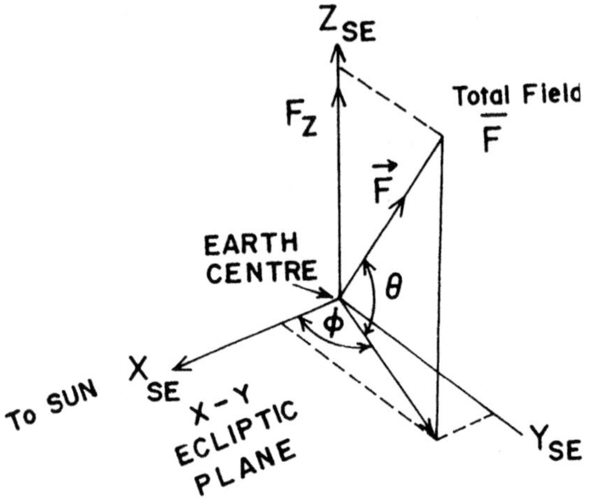

Figure 2.2.1: Solar-ecliptic coordinate system

system. The direction of the magnetic field vector $\vec{F}$ is given by the angle ϕ, measured from X_{se} to Y_{se} in the plane of the ecliptic and the angle θ measured from the ecliptic plane component to $\vec{F}$.

As the termination of the geomagnetic field at the day side of the earth occurs at $\sim$ 8-10 R_e (R_e radius of the earth, 6400 km), it is necessary to have satellites with highly elliptical orbits with apogee of $>$ 10 R_e so that it can provide information from a wide region of space. Also the satellite should be launched with it apogee close to the local noon side. The satellite launching and its orbital properties are discussed to illustrate the special requirements for measurements of the geomagnetic field.

Observations of the Magnetosphere

Explorers-XII and XIV launched in 1961 and 1962 in highly elliptical orbits had apogees of about 14 and 17 R_e respectively and these provided, from many crossing of the boundary regions by the satellite orbits conclusive evidence of the bow shock front and the confinement of the earth's magnetosphere. Following these successes, further improvements were

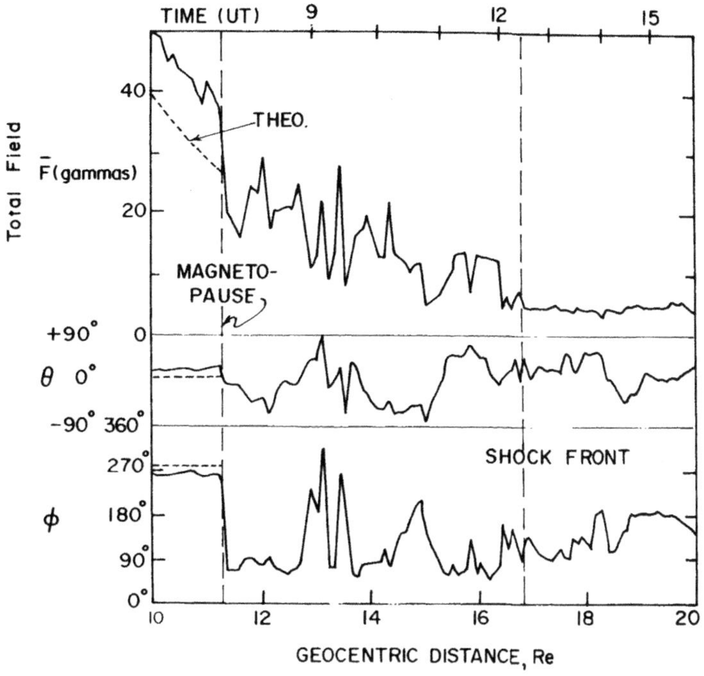

Figure 2.2.2: A sample of magnetic field data from IMP-1 Orbit 1

made in the instrumentations and launching of Explorer-XVIII, which was called Interplanetary Monitoring Platform-1, (IMP-1). It was launched on November 27, 1963, at an orbital inclination of 33° to the equator in an highly elliptical orbit of apogee 197,000 km (= 31 r_e) and perigee 192 km, the period of the orbit being about 94 hours. It should be noted that due to a highly elliptical orbit, although about 25% of the orbital path of the satellite is in the interplanetary medium, the satellite spends about 2/3th of the orbital time in interplanetary space. This is a consequence of Kepler's law of planetary motion that equal areas are swept out in equal times.

In Figure 2.2.2 we show the typical magnetic field data obtained in orbit-1 by the IMP-1 satellite[(1)]. The top figure shows that $\vec{F}$, the total magnetic field undergoes a sudden large discontinuity at 11.3 R_e which is determined as the magnetopause (MP) boundary. At this point the directions of the field as given by θ and ϕ values undergo sudden changes and the irregularities in the field value of $\vec{F}$ and its direction continues till 16.8 R_E which is identified as the shock wave (SW) boundary. Beyond this an almost constant magnetic

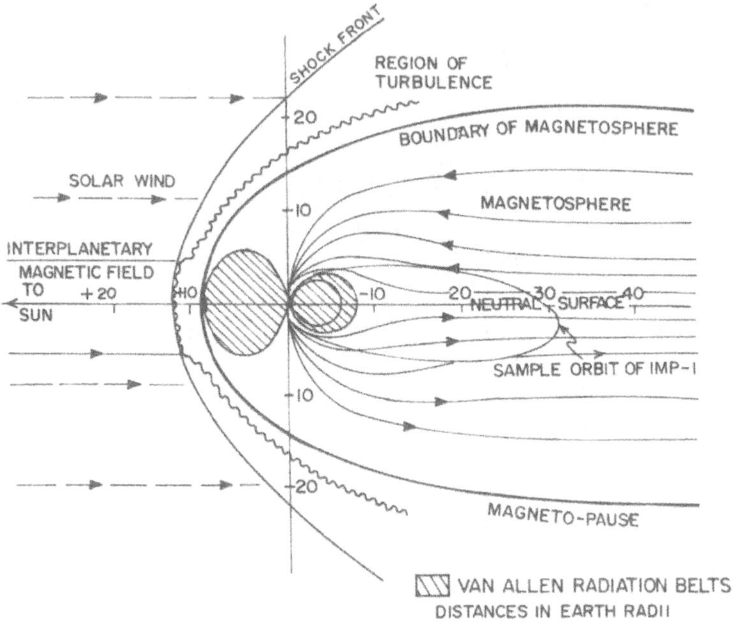

Figure 2.2.3: Results from the IMP-1 magnetic field experiment

field of 5γ ($\gamma = 10^{-5}$ Gauss) signifies that the spacecraft is sampling the interplanetary field. Each orbit of the spacecraft, IMP-1, thus provided two values of boundary crossings(2).

The cumulative data of the boundary crossings by many orbits of IMP-1 and by some orbits of Explorer-X are shown in Figure 2.2.3. These results yield the shape of the boundaries of the shock front and the magnetopause and their locations.

2.3 The Structure of the Magnetosphere

The magnetosphere of the earth is defined as the region of near earth space where the magnetic field of the earth plays the dominant role in the physical processes occurring there. The regions of the magnetosphere and their shapes and dimensions are shown in Figure 2.3.1. This sketch (2) shows the cross-section in the meridian plane of N-S direction and the earth-sun line. The magnetic equator is inclined by $\sim 30^o$ to the equatorial plane of the earth.

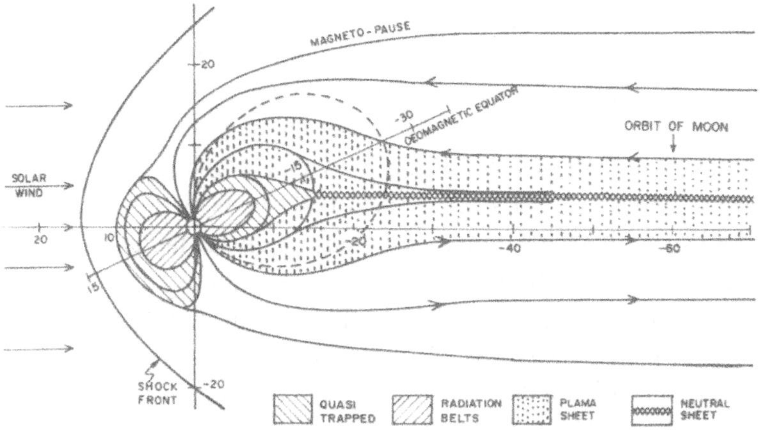

Figure 2.3.1: A sketch of the magnetosphere of the earth, when the magnetic equator is inclined by 30° to the earth-sun line

As a result of the interactions of high speed plasma from the Sun, known as solar wind (Chapter 3.3) with the distant region of the earth's magnetic field, there are three distinct domains of the magnetic field which are separated by two boundary regions. These are as follows:

(I) **The region of interplanetary magnetic field:** Here the steady, low value of the interplanetary field of about 5γ occurs, undisturbed by the presence of the earth and its magnetic field.

(II) **The transition region or magnetosheath:** This is the region with a turbulent magnetic field due to the interactions of solar wind plasma and the interplanetary field with the geomagnetic field.

(III) **The magnetosphere:** The region or the cavity in which the earth's magnetic field is confined.

The two boundary regions separating these three regions are:

(a) **The shock front:** This separates the interplanetary magnetic field with the transition region.

(b) **The magnetopause:** This forms the boundary between the turbulent region and the magnetosphere.

The physical properties of these regions are briefly discussed here. The solar wind plasma carries with it the interplanetary magnetic field which travels with an average speed of 300 km/s, at the orbit of the earth. This speed is called 'supersonic' as it greatly exceeds the Alfven velocity of the magnetohydrodynamic waves in the interplanetary medium (Chapter 3.4.4). The interaction of the solar wind plasma with the geomagnetic field generate a 'supersonic' shock-front in the form of a bow shock. Inside the bow shock we find the transition region where a turbulent magnetic field and the solar wind are present. In the earth-sun direction, the magnetopause occurs at about 10 R_e (R_e = earth radius = 6400 km). This location of the boundary of the magnetopause is determined by the balance between the pressure exerted by the solar wind on the outside and the magnetic pressure inside. The pressure, p, of the solar wind in the magnetosheath is given by

$$p \simeq n \cdot m \cdot V^2 \cdot \cos^2 \theta \tag{2.3.1}$$

where n is the number density, m the mass and V the velocity of the solar wind protons and θ is the angle between the solar wind direction and the normal to the shock. The magnetic field strength B just inside the magnetopause on the forward side is typically about $40 \cdot 10^{-5}$ Gauss which is the measured value from IMP-1 (Figure 2.2.2). This value is about twice that due to the earth's dipole field, as the enhancement occurs because of ring currents flowing in the magnetosphere. The position of the magnetopause is thus approximately estimated by equating the magnetic pressure, $B^2/8\pi$ (Chapter 3.3.4) with solar wind pressure

$$p = B^2/8\pi \tag{2.3.2}$$

Using the mean value of solar wind density as $n \approx 5$ cm^{-3} which is usually in the range of $n \simeq$ 3-10 cm^{-3}, the mass $m = 1.6 \cdot 10^{-24}$g and the mean velocity of the solar wind, $V \simeq 300$ km·s^{-1}, we obtain $p \simeq 7 \cdot 10^{-9}$ dyne. For $B \approx 40 \cdot 10^{-5}$ Gauss, the magnetic field pressure is estimated as $B^2/8\pi \approx 7 \cdot 10^{-9}$ dyne. Thus approximately the equality of eqn. 2.3.1 and 2.3.2 is established.

Knowing the magnetopause boundary, the expected position of the shockwave wave boundary was calculated by using gas dynamics models. The theoretical and observed positions are in good agreement as shown in Figure 2.2.3.

The striking feature of the shock front is that no particle to particle collision is involved, as the mean free path of solar wind particle at the orbit of the earth is of the order of 10^{13}cm whereas the thickness of the shockfront does exceed several hundred km. The gyroradius of a solar wind proton in the interplanetary magnetic field is nearly the same as the thickness of the shock front. Therefore this shock front is a clear example of a collision-free shockwave.

The magnetosphere and the geomagnetic tail: The pressure of the solar wind causes the compression of the earth's magnetic field lines at the Sun side and the expansion at the night side; hence the dipole field lines of the earth become somewhat distorted. This is shown in Figure 2.3.1 for the case where the magnetic north-south axis is inclined to the earth-sun line by about 30°. Inside the magnetosphere, the magnetic field lines are of two kinds: the closed field lines as shown by the shaded region and the open field lines in the rest. At the dawn and dusk meridians the magnetopause it located roughly at 15 R_e. Inside the closed field lines lie the Van Allen radiation belts - which contain trapped electrons and protons. The discovery of the trapped radiation belts is one of the major landmarks in the space age and is discussed in Chapter 2.5. The regions of quasi-trapped particles are indicated. The closed field lines in the northern-most and southern-most regions comprise the auroral zones. The magnetic field originating near the poles of the earth are swept back into the night-side of the earth and some of these field lines extend to a very long distance at the midnight side of the earth. forming the geomagnetic tail. How far does the tail of the earth's magnetosphere extend? This question has been the subject of interesting studies both theoretically and experimentally but no definite answer is available. It is believed that the geomagnetic tail extends to a distance beyond the orbit of the moon (which is 60 R_e) to about 100 R_e or more (R_e = 6400 km). The observations of the spacecraft Pioneer-6, indicated that the wake or influence of the earth's geomagnetic tail was detectable to a distance of about 800 R_e ($\cong 5{\cdot}10^5$ km). The magnetospheric plasma is presumed to be present in the night-side region of the magnetosphere as shown in the figure.

The geomagnetic tail has several striking features. The tail contains a substantial portion of the field lines originated near the poles. The magnetic field lines in the tail do not curve round and connect the two polar regions; instead these lines are almost parallel to the sun-earth line at the midnight side of the earth. In the northern half of the tail the field lines are directed to the earth and in southern half, away from the earth. The parts of the magnetosphere tail are separated by a thin neutral sheet or plane with very low magnetic field which contains plasma with enough pressure to keep the oppositely directed magnetic field lines apart. The presence of such a long appendage to the geomagnetic field in the form of a tail was theoretically supported by the early works of Parker(3), Axford(4) et al., Dessler(5) and several others. What happens to the magnetic field lines of the tail? It is found that while some of the field lines of the tail do reconnect across the neutral sheet, called the magnetic field reconnection process, most of the field lines extend to interplanetary medium. The evidence of this is provided by the observations in spacecraft of low energy ($\sim$ 40 keV) electrons emitted by solar flares which move towards the earth from the anti-solar direction, signifying the linkage between the interplanetary magnetic field with some of the field lines of the geomagnetic tail.

The shapes of the magnetosphere and the tail shown in Figure 2.3.1 represent the case when the earth-sun line is in the plane of the earth's equator. In general, this is not so and the shapes of the magnetic field lines in the magnetosphere become more distorted. As the earth rotates around its axis, the magnetic field lines and the geomagnetic tail wobble. The occurrence of the neutral sheet in the tail implies the presence of a quasi-steady electric field which is believed to be the caused by the acceleration of particles in the magnetosphere.

2.4 Motion of Charged Particles in the Magnetosphere

2.4.1 Störmer Theory, Main Cone and Shadow Cone

Historical Introduction: The study of the motion of charged particles in the dipole field of the earth began at the end of the 19th century in order to understand the polar aurora. In 1890 Birkland of Sweden noted the similarities between the glow of cathode rays in a gas discharge tube and the aurora, and correctly guessed that like cathode rays, electrons impinging on the rarefied upper atmosphere produced the glow of the aurora. To study this phenomenon he constructed a small metal sphere and magnetized in such a way as to be similar to that of the earth. This sphere, called "terrela" was placed in near vacuum and he allowed beams of electrons to impinge on it in a darkened laboratory. He observed spectacular results of curtains of lights and luminous rings similar to auroral displays. The type of display varied with the angle incidence of the electron beam. From this experiment Birkland proposed the theory that incoming electrons and protons from the Sun after entering the earth's magnetic field some way, impinged on the rarefied upper atmosphere and produced the aurora. Carl Störmer of Sweden was immediately attracted to Birkland's results published in 1901 and he began calculating mathematically the trajectories of electrons in the earth's magnetic field in order to understand the mysteries of the auroral phenomena.

Carl Störmer, assisted by two research students made painstaking calculations of the complex trajectories of electrons in the geomagnetic field and solved difficult equations numerically for almost forty years in search of the answers to the problems of the aurora. He published in 1955 his celebrated book 'The Polar Aurora'(6) covering his works of over forty years. He believed that there is a ring current which affected the position of the aurora. However, the true picture of the origin of the aurora only emerged with the advent of the space age and earth satellites.

The importance of Störmers work was realized as soon as the variation of cosmic ray intensity with latitude was discovered and the charged particle nature was established. The observations of the increase in cosmic ray intensity from the west as compared to that from the east - called east-west effect, proved that positively charged protons are the main constituent of cosmic rays. Thus Störmer theory of the motion of charged particles in the geomagnetic field became a major tool in understanding the properties of the primary cosmic rays incident at the top of the atmosphere.

Störmer Theory: In the first approximation the geomagnetic field can be represented by placing a dipole of magnetic moment M at the center of the earth. The coordinate system used is shown in Figure 2.4.1; spherical and cylindrical systems are used. Let the dipole be placed at the center of the earth and thus the coordinate system, having a dipole moment, (M =

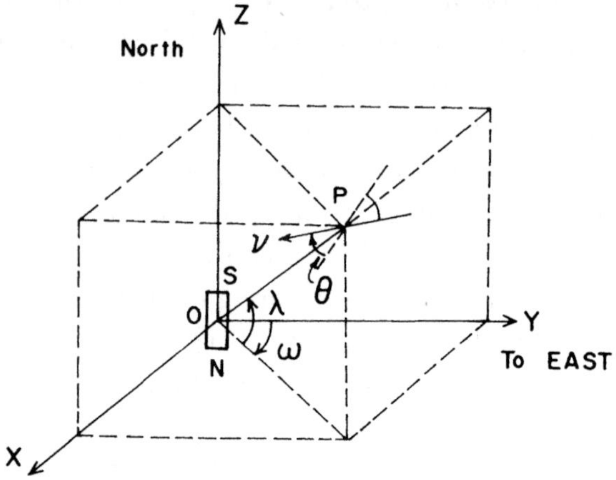

Figure 2.4.1: The coordinate system used in Störmer theory

$8.1 \cdot 10^{25}$ Gauss·cm) along the Z-axis with its south-pole in the +Z-direction. Let P be the position of a particle when it crosses the meridian plane. Let

v($\vec{r}$,t) = instantaneous velocity of the particle at time t, as it crosses the meridian plane at point P having coordinates, r, λ, ω.

r = the radial distance, OP.

λ = the latitude of P measured from the equatorial plane.

ω = the longitude of P measured from east to west.

R = the projection of $\vec{r}$ on the equatorial plane.

θ = the angle between the velocity vector $\vec{v}$, and its projection on the meridian plane, θ being positive when the particle crosses the meridian plane from east to west.

The Lagrangian of a relativistic charged particle in an electromagnetic field is given by

$$L = -m_o c^2 \sqrt{1-\beta^2} + (e/c)^2 \cdot \vec{A} \cdot \vec{v} - e\phi \tag{2.4.1}$$

where m_o is the rest mass of the particle, $\vec{v}(\vec{r}$,t) its velocity, $\beta = $ v/c, c being the velocity of light, e the electronic charge, $\vec{A}(\vec{r}, t)$ the vector potential of the magnetic field and $\phi(\vec{r}, t)$ = the scalar potential. For the derivation of equation 2.4.1, see e.g. Classical Mechanics by Goldstein. The equation of motion of the charged particle is given by the Lagrangian equation

$$\frac{d}{dt}\left(\frac{\partial L}{\partial \dot{q}_i}\right) - \frac{\partial L}{\partial q_i} = 0 \tag{2.4.2}$$

In the dipole field, $\phi = 0$, and the vector potential $\vec{A}$ is given by

$$\vec{A} = \vec{M} \times \vec{r}/r^3 = -M \cdot \vec{1}_z \times \left(R \cdot \vec{1}_R + Z \cdot \vec{1}_z\right)/\left(R^2 + Z^2\right)^{3/2}$$

$$= \left[0, 0, MR/\left(R^2 + Z^2\right)^{3/2}\right] \tag{2.4.3}$$

$$\vec{v} = d\vec{r}/dt = (\dot{R}, \dot{Z}, R\dot{\omega}) \tag{2.4.4}$$

Therefore

$$\vec{A} \cdot \vec{v} = MR^2\dot{\omega}/\left(R^2 + Z^2\right)^{3/2} \tag{2.4.5}$$

Substituting the equation (2.4.5) in 2.4.1, the Lagrangian is,

$$L = m_o c^2 \sqrt{1 - v^2/c^2} + (e/c) \cdot MR^2\dot{\omega}/\left(R^2 + Z^2\right)^{3/2} \tag{2.4.6}$$

As ω does not appear explicitly in L because of axial symmetry, therefore corresponding to the ω coordinate we get

$$\frac{d}{dt}\left(\frac{\partial L}{\partial \dot{\omega}}\right) = \frac{\partial L}{\partial \omega} = 0 \tag{2.4.7}$$

differentiating L in 2.4.6 with respect to $\dot{\omega}$, we get 2.4.7 as

$$\frac{d}{dt}\left[mR^2\dot{\omega} + (e/c)MR^2/\left(R^2 + Z^2\right)^{3/2}\right] = 0 \tag{2.4.8}$$

where

$$m = m_o/\sqrt{1 - \beta^2}$$

Störmer Unit: The analysis is simplified by combining the magnetic rigidity (= $\frac{pc}{e}$ where p is the momentum and e the charge) of the particle with the magnetic moment of the earth, M, so as to form a length C, in units of which all distances are measured. This is

$$C^2 = \frac{M}{pc/e} = \frac{eM}{pc} = \left[\frac{charge \times charge \times length}{energy}\right]$$

$$= \left[\frac{(charge)^2 \times (length)}{(charge)^2/(length)}\right] = [length]^2 \qquad (2.4.9)$$

The physical meaning of C is given as follows. When a particle of rigidity mvc/e is moving in a circular orbit in the equatorial plane of the earth, its radius of curvature r is equal to C. This is derived as

$$\frac{mv}{r} = (e/c)\vec{v} \times \vec{H} = \frac{e}{c} \cdot M/R^3$$

Therefore

$$r^2 = \frac{eM}{mvc} = C^2, \; or \;\; r = C \qquad (2.4.10)$$

Now we introduce a variable r_o, in units of C,

$$r_o = a/C \;\; \equiv \;\; Störmer \;\; Unit \qquad (2.4.11)$$

Which is a function of the rigidity of the particle only. With a the radius of the earth. Thus the rigidity of a particle pc/e is related to r_o as

$$r_o^2 = a^2/C^2 = (pc/e)\frac{a^2}{M}$$

Therefore

$$pc/e = Mr_o^2/a^2 \qquad (2.4.12a)$$

For numerical evaluation, M = $8.1 \cdot 10^{25}$ Gauss·cm and a = $6.37 \cdot 10^8$ cm, and the rigidity expressed in eV, we have

$$pc/e = (pc/e)_{ineV} = \frac{1.6 \times 10^{-12} ergs}{4.8 \times 10^{-10}} = \frac{(pc/e)_{eV}}{300}$$

Therefore

$$(pc/e)_{eV} = 300\frac{Mr_o^2}{a^2} = (59.6 \times 10^9 r_o^2)_{eV}$$

Therefore

$$Rigidity = 56.9 r_o^2 \; GV \qquad (2.4.12b)$$

Further simplification of the analysis is made by making the equations dimensionless by replacing time t by the path length s = v·t, measured in units of C, as t = s/v. Then the velocity $\dot{R}^2 + \dot{Z}^2 + (R\dot{\omega})^2 = v^2$ is reduced to

$$\left(\frac{ds}{dt}\right)^2 \left[\left(\frac{dR}{ds}\right)^2 + \left(\frac{dz}{ds}\right)^2 + R^2\left(\frac{d\omega}{ds}\right)^2\right] = v^2$$

or

$$\dot{R}^2 + \dot{Z}^2 + (R\dot{\omega})^2 = 1 \tag{2.4.12c}$$

where the dot denotes d/ds.

Then the axial momentum integral of equation 2.4.8 is rewritten as

$$\left[mr^2\dot{\omega} + \frac{e}{c}\frac{MR^2}{(R^2+Z^2)^{2/3}}\right] = const. \quad , \quad where \;\; \dot{\omega} = d\omega/dt \tag{2.4.13}$$

Replacing

$$\frac{d\omega}{dt} = \frac{d\omega}{ds}\cdot\frac{ds}{dt} = \frac{d\omega}{ds}\cdot v$$

we have

$$mR^2\cdot\dot{\omega}v + \frac{e}{c}\frac{MR^2}{(R^2+Z^2)^{2/3}} = const. \quad , \quad where \;\; \dot{\omega} = \frac{d\omega}{ds}$$

or

$$R^2\cdot\dot{\omega} + \frac{eM}{mvc}\cdot\frac{R^2}{(R^2+Z^2)^{2/3}} = const. \tag{2.4.14}$$

or

$$C^2\left(\frac{R}{C}\right)^2\dot{\omega} + C^2\cdot\frac{C^2\cdot R^2/C^2}{C^3\,(R^2/C+Z^2/)^{3/2}} = const.$$

or

$$\left(\frac{R}{C}\right)^2\dot{\omega} + \frac{1}{C}\frac{(R/C)^2}{\left(\frac{R^2}{C^2}+\frac{Z^2}{C^2}\right)^{3/2}} = const.$$

or

$$\left(\frac{R}{C}\right)^2 \cdot (C\dot{\omega}) + \frac{(R/C)^2}{[(R/C)^2 + (Z/C)^2]^{3/2}} = const.$$

Expressing R, Z and $\dot{\omega}$ in units of C, we have

$$R^2 \cdot \dot{\omega} + \frac{R^2}{(R^2 + Z^2)^{3/2}} = const. = \ -2\gamma \tag{2.4.15}$$

as

$$\dot{\omega} = \frac{d\omega}{ds} = \frac{d\omega}{ds'} \cdot \frac{ds'}{ds} = \frac{1}{C} \cdot \frac{d\omega}{ds'} \quad , \ where \ \ s' = s/C.$$

Because time is replaced by arc length, the components of the velocity have the following meanings:

$$d\vec{r} = ds \cdot \vec{1}_T = [dR, dZ, Rd\omega]$$

where $\vec{1}_T$ = unit vector along the tangent to the path. Then

$$\vec{1}_T = [\dot{R}, \dot{Z}, R\dot{\omega}] \tag{2.4.16}$$

and $\dot{R}\vec{1}_T$ and $\dot{Z}\vec{1}_Z$ are components of $\vec{1}_T$ in the meridian plane. As θ is defined as the angle between the velocity vector and its projection on the meridian plane, we have

$$R\dot{\omega}/1 = \sin\theta$$

or

$$\theta = \sin^{-1}(R\dot{\omega}) \tag{2.4.17}$$

Thus equation 2.4.15 is written as

$$2\gamma + \frac{R^2}{(R^2 + Z^2)^{3/2}} = R\dot{\omega} = \sin\theta$$

Thus, in cylindrical coordinates,

$$2\gamma/R + R/\left(R^2 + Z^2\right)^{3/2} = -\sin\theta \tag{2.4.18a}$$

and, in spherical coordinates,

$$2\gamma/r\cos\lambda + \cos\lambda/r^2 = -\sin\theta \tag{2.4.18b}$$

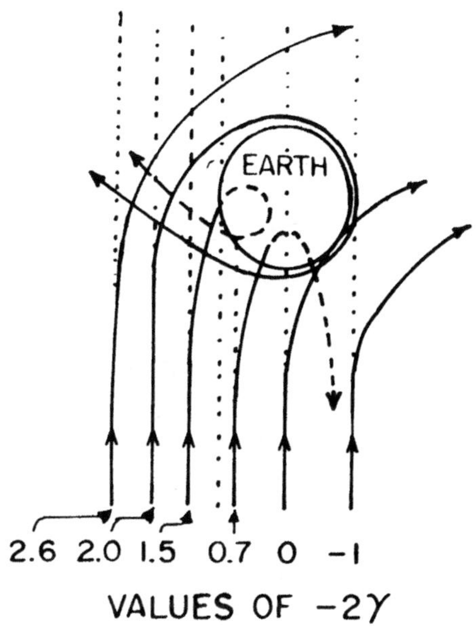

Figure 2.4.2: A sketch of proton trajectories in the equatorial plane showing the role of the impact parameter -2γ.

These are the general Störmer equations of motion of charged particles in the geomagnetic field. We evaluate the physical meaning of these, firstly in the case of motions in the equatorial plane.

Motion in the Equatorial Plane: To obtain a physical idea how certain regions become forbidden to particles from infinity, let us first consider their motions in the equatorial plane. First, we consider protons of rigidity 59.6 GV, for which, the radius of the earth r_o becomes equal to 1, as given in equation 2.4.12. As noted earlier, in Störmer theory the rigidity of the particle is replaced by a variable r_o. In Figure 2.4.2 we show the proton trajectories with the radius of the earth as $r_o = 1$ in the equatorial plane with the N magnetic pole up from the plane of the paper. The values of -2γ = -1, 0, +0.7 and +2.6 represent the impact parameters of particles coming from infinity. The effect of the magnetic field of the earth is to bend the particle trajectories from the west to east. For -2γ = -1, the particle does not reach the earth while for -2γ = 0 the trajectory touches the earth indicating that it can come from west to east i.e. θ = -90°. For -2γ = 0 to 2.0, the trajectories

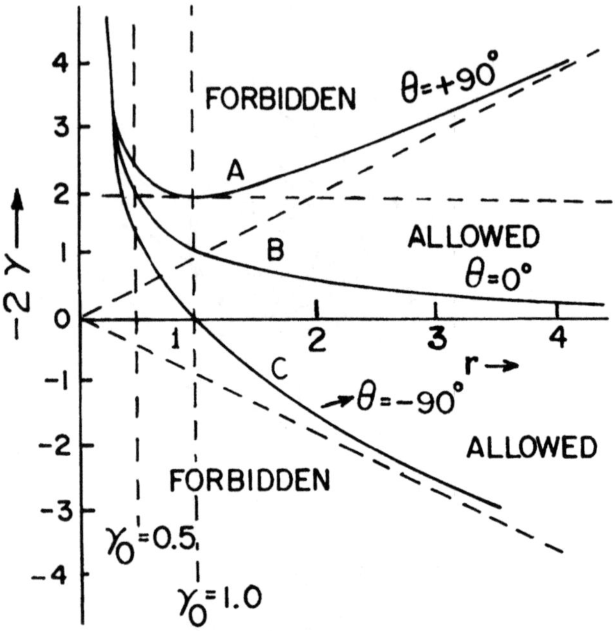

Figure 2.4.3: The plot of r vs -2γ, for $\theta = 90^o$, and -90°

reach the earth. In case of -2γ = +2.0 the particle travels in a circular path and comes from the east. Thus the particle of rigidity 59.6 GV can reach the equatorial plane from all direction. Now we consider the allowed and forbidden regions.

For $\lambda = 0$, the equation 2.4.18 becomes

$$2\gamma/r + 1/r = -\sin\theta$$

or

$$-2\gamma = r\sin\theta + 1/r \tag{2.4.19}$$

In Figure 2.4.3 we show the plot of -2γ vs r for $\theta = 90^o$, 0° and -90°, as given by the relations,

$$\theta = +90^o, -2\gamma = r + 1/r$$

$$\theta = 0^o, -2\gamma = 1/r$$

$$\theta = -90^o, -2\gamma = -r + 1/r \qquad (2.4.20)$$

The vertical line through $r_o = 1$ defines the surface of the earth for protons of 59.6 GV rigidity whereas that through $r_o = 0.5$ defines the surface of the earth for rigidity 14.9 GV. The motion of the particle with a given value of -2γ is a horizontal line with large value of r which decreases until it intersects the curve $\theta = +90^o$ or $\theta = 90^o$, or the vertical line through r_o, which is the closest approach. Then the point moves outwards again for large values of r. For $-2\gamma = 2.6$ (say), the horizontal line of $-2\gamma = 2.6$ intersects the curve $\theta = +90^o$ at r = 2.15 at closest approach. This distance is too far from the surface of the earth (r = 1 for 59.6 GV) so that it does not reach the earth. As seen in Figure 2.4.3 the trajectory with $-2\gamma = 2$ will reach the earth from $\theta = 90^o$ i.e. for the grazing incidence from the east and for $-2\gamma = 1$ from $\theta = 0^o$ i.e. from the vertical direction, and $-2\gamma = 0$ from $\theta = -90^o$ i.e. from the grazing incidence from the west. Thus a 59.6 GV proton can reach the earth from all directions of the sky.

For $r_o = 0.5$ i.e. for protons of 14.9 GV rigidity, the trajectory with $-2\gamma = 2$ cannot strike the earth, as the closest approach point is r = 1, which is too far out. For -2γ slightly less that 2, the trajectory can reach the earth at $\theta = 0$ i.e. from vertically downwards. When $-2\gamma = 1.5$, it reaches the earth at $\theta = -90^o$ i.e. from the west. Hence for this particle ($r_o = 0.5$), the eastern half of the sky is forbidden and it is allowed to reach the earth from all directions in the western half of the sky.

The limiting angle i.e. largest algebraic angle of incidence for the particle reaching the earth is determined as $\theta_c = 0$ for $-2\gamma = 2$ as seen in the figure. Analytically we obtain θ_c from equation 2.4.19, as given by $-2\gamma = r_c \cdot \sin\theta_c + 1/r_c$. By substituting $-2\gamma = 2$ and $r_c = 0.5$, we get $\theta_c = 0$. It is also seen in the figure that for a r_o less than 0.414 a particle from infinity cannot reach the surface at all at $\lambda = 0$.

In Figure 2.4.4 we show the allowed and forbidden regions for $-2\gamma = 1.0$ in the cylindrical coordinate system. The surface of the earth is now a quarter circle of $r_o = 0.5$, 1.0 and 1.5 as determined by the rigidity of the particle. The curves of $\theta = +90^o$ are drawn from equation 2.4.18a.

Störmer had given a set of figures in R, Z coordinates showing the allowed

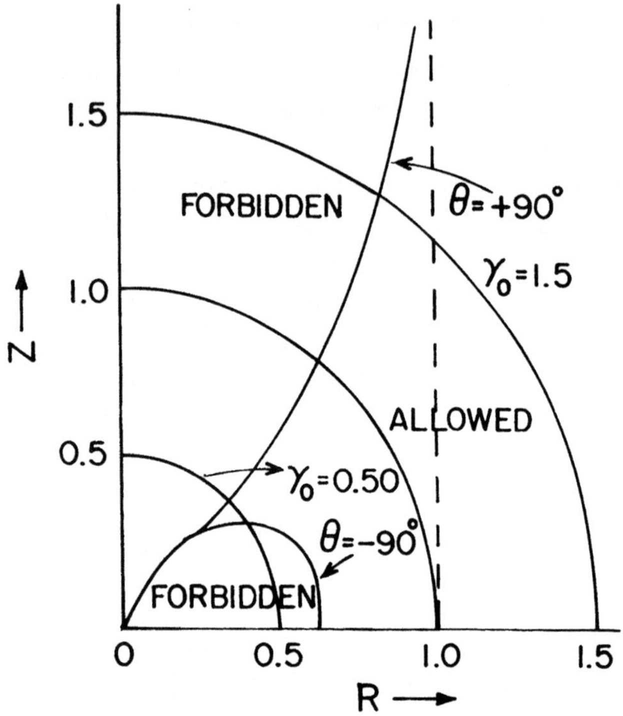

Figure 2.4.4: Allowed and forbidden regions for -2γ = 1.0 in cylindrical coordinates for r_o = 0.5, 1.0 and 1.5

and forbidden regions for various values of -2γ = 2.032, 1.94, 1.0, 0.1, 0.06 and 0.04. It was noted by him that for -2γ = 2.0, the allowed regions are separated in two sections, the inner region being bounded by forbidden regions. This inner region constitute the trapped particle zone, where particles cannot reach from infinity, but if somehow particles could be injected in this region, the trapped particles will be possible. This is shown in Chapter 2.5 in Figure 2.5.1.

Allowed and Forbidden Cones in Störmer Theory: In the general case where $\lambda = 0$, the equation of motion as derived earlier (in equation 2.4.18b) is

$$2\gamma/(r\cos\lambda) + (\cos\lambda)/r^2 = \sin\theta$$

or

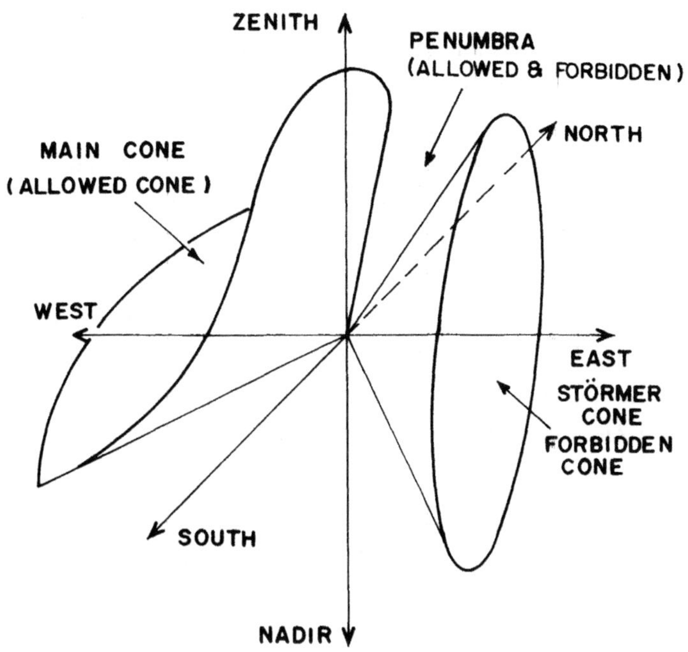

Figure 2.4.5: A sketch showing the Störmer cone, the main cone and the penumbra. (Courtesey: M.A. Shea and D.F. Smart)

$$r \sin\theta \cos\lambda + \frac{\cos^2\lambda}{r} = 2\gamma \tag{2.4.21}$$

From this we can calculate the angle θ_c of the limiting cone of allowed directions. For example, for $\lambda = 30^o$ and $r_o = 0.4$, and using $-2\gamma = 2$ to get the maximum value of the angle θ, we get $\theta_c = +21.7^o$. One can plot the curves for r vs -2γ similar to those in Figure 2.4.3 showing the allowed and forbidden regions for $\lambda = 30^o$.

Let us consider the numerical example given above for $\theta_c = +21.7^o$. This is the maximum angle for the allowed direction. Therefore, the forbidden directions are given by the angle 90°-θ_c. In a given location on the surface of the earth, the Störmer cone and the allowed cone are shown in the sketch in Figure 2.4.5. The X-axis is taken towards the east and the Y-axis to the north. The XY plane is the horizontal plane on the earth's surface, Z-axis being in the zenith direction. With θ_c calculated as above, we draw a half-cone with its apex at 0 and its axis along the east direction, the semi-vertical angle of the cone being 90°-θ_c, as shown in Figure 2.4.5. This half-cone with its axis in the east direction as shown, is called the Störmer cone within which the directions are forbidden for particles of a given rigidity (i.e. of r_o value) at the given latitude. Outside the Störmer cone, lie the allowed directions.

Cut-off Rigidity: The cut-off rigidity, R_c, is defined as the minimum rigidity, (= p·c/z) of a charged particle coming from infinity, which can reach the observer at a given location on the surface of the earth at a given latitude λ and from a given direction ξ, ϕ of the sky, where ξ and ϕ are the zenith angle and azimuth angle of the incident direction of the particle respectively. The concept of cut-off rigidity or threshold rigidity is widely applied for cosmic rays and solar energetic particles and this is given by Störmer theory. In equation 2.4.21, the rigidity of a given particle is minimal when $-2\gamma = 2$. Therefore, we have

$$\left(\frac{1}{r_c}\right)^2 \cos\lambda - \left(\frac{1}{r_c}\right)\frac{2}{\cos\lambda} + \sin\theta = 0$$

Therefore

$$\frac{1}{r_c} = \frac{1+\sqrt{1-\sin\theta\cos^3\lambda}}{\cos^2\lambda}$$

or

$$r_c = \frac{\cos^2\lambda}{1+\sqrt{1-\sin\theta\cos^3\lambda}}$$

Now the Rigidity, $R = (M/a^2)\cdot r_o^2 = 59.6\ r_o^2$ GV. Hence the cut-off rigidity,

$$R_c = 59.6\frac{\cos^4\lambda}{\left[1+\sqrt{1-\sin\theta\cos^3\lambda}\right]^2}GV \tag{2.4.22}$$

By replacing the angle θ, by the zenith angle ξ and azimuth angle ϕ as given by the relation $\cos(90-\theta) = \cos(90-\xi)\cos\phi$ i.e. $\sin\theta = \sin\xi\cos\phi$, we can write

$$R_c = 59.6\frac{\cos^4\lambda}{\left[1+\sqrt{1-\sin\xi\cos\phi\cos^3\lambda}\right]^2}GV \tag{2.4.23}$$

For vertical incidence, $\xi = 0$,

$$(R_c)_{vertical} = \frac{59.6}{4}\cdot\cos^4\lambda = 14.9\cos^4\lambda GV \tag{2.4.24}$$

The Main Cone, Penumbra and the Shadow Cone: The detailed calculations of Vallarta and co-workers of Mexico and others showed that for particles of a given rigidity at a particular geomagnetic latitude, the sky may be divided into four regions as follows:

1. **The Störmer Cone** - This is the forbidden region in which no particle of the given rigidity can enter.

2. **The Main Cone** - This defines the cone in which all particles of the given rigidity are allowed to enter.

3. **The Penumbra** - This region, which lie between the Störmer cone and the Main cone, consists of alternate bands of allowed and forbidden regions. At low latitudes, near the equator the penumbra is covered almost by forbidden regions i.e. it is almost completely "dark". Hence the cut-off rigidity is determined primarily by the main cone. At high latitudes the penumbra is almost 'transparent' so that the Störmer cone mainly determines the cut-off rigidity. In the intermediate latitudes, $\lambda \simeq 20^o$-35° the penumbra is most complex with many alternate bands of allowed and forbidden regions over a fairly large rigidity range. Here the effective cut-off rigidity is determined by taking into account weighted average of the transmissions by the penumbra bands.

4. **Shadow Cone** - The regions lying near the horizon towards the north and south directions in which the trajectories are obstructed by the presence of the solid earth are called shadow cones.

In Figure 2.4.5, we show a sketch illustrating the relative locations of the above regions. Here the observer is located at 0 and the N-S and E-W directions shown lie in the horizontal plane. The Störmer cone is open towards the east, and the main cone to the west, with a penumbra region in between. At the surface of the earth, the portions of the cones lying below the horizon becomes forbidden by the solid earth. As the observer (or the detector) travels in space away from the earth, the obscuration by the solid earth becomes reduced and the cones below the horizon progressively open up.

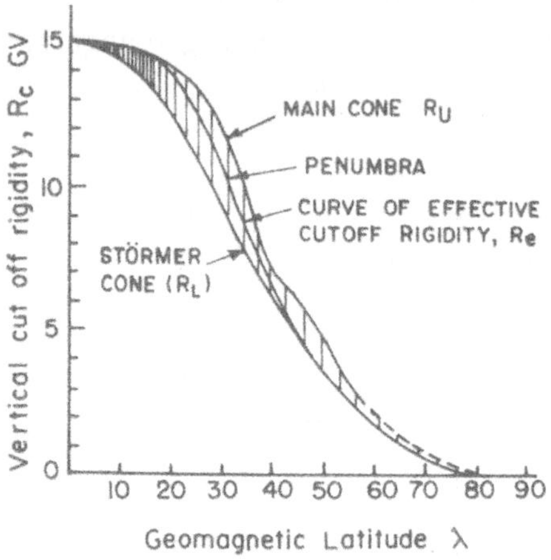

Figure 2.4.6: Vertical cut-off rigidity, R_c, as a function of geomagnetic latitude

The Vertical Cut-off Rigidity: This is given by equation 2.4.24 in the Störmer theory as function of geomagnetic latitude, λ. As discussed above the fraction of particles allowed by the penumbra varies with latitude. In the geomagnetic of latitude of 0-20°, 20°-40°, 40°-60° and >60°, the transparency of the penumbra is approximately 0, 30, 60 and 100% respectively. We show in Figure 2.4.6 the plot of the vertical cut-off Rigidity, R_c, for the Störmer cone, the Main cone and the effective rigidity, as a function of the geomagnetic latitude λ.

2.4.2 Cosmic Ray Trajectories in the Geomagnetic Field

Non-Dipole Effects and Cosmic Ray Cut-off Rigidities

In the foregoing calculations, the earth's magnetic field was assumed to be of dipole nature which is correct in first approximation. Detailed studies carried out by various investigators experimentally and theoretically showed the effect of higher order terms. Groups of scientists of different parts of the world undertook long aeroplane flights and sea voyages whereon continuous measurements of the vertical intensity of cosmic rays were conducted particularly near the equator at various longitudes of the earth. The geographical latitudes and longitudes at different parts of the world were determined to find where the vertical cosmic ray intensity was at a minimum.

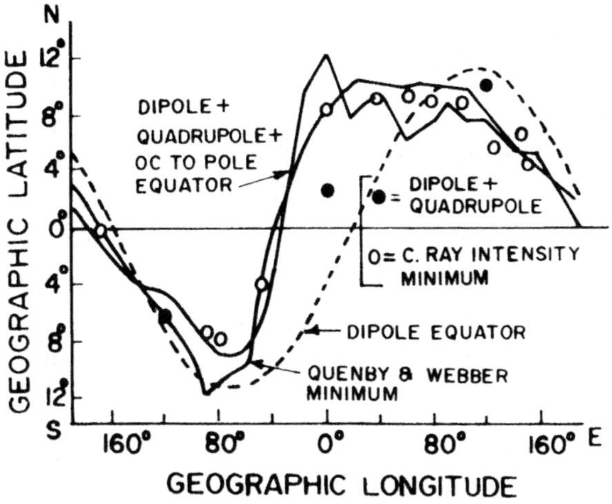

Figure 2.4.7: Comparisons of the cosmic ray equator from (o) air-craft measurements[7], with (a) calculations of Queenby and Webber[8], and (b) of Kellogg and Schwartz[9] using dipole, quadrupole and octopole terms

Comparing these data with the location of the geomagnetic equator it was noted that the 'cosmic ray equator' did not agree with the geomagnetic equator over a wide longitude region and the difference being as much as $\sim 10^o$ in the zero longitude region. These are shown in Figure 2.4.7. This data provided direct and quantitative evidence of non-dipole effects in the cosmic ray cut-off rigidities[7,8,9]. Another set of measurements of the vertical intensity of cosmic ray helium nuclei as a function of energy in balloon flights from Minneapolis in the USA and in South England showed that although both locations have the same geomagnetic latitude, the vertical cut-off energy over South England was $\sim$ 150 MeV/n higher than that measured over Minneapolis. It was also found that at latitudes, $\lambda > 55^o$, the cut-off rigidities varied with time, which indicated that in addition to non-dipole terms of internal origin there is a contribution of the external current system on the cut-off rigidity. A westward flowing ring current at several earth radii will decrease the magnetic moment of the earth and hence reduce the cut-off rigidities at latitudes $\geq 60^o$.

Trajectory Tracing Techniques: With the advent of high speed digital computers, the computations of the trajectories of cosmic ray particles in the magnetic field of the earth are carried out with high degree of accuracy by machine computations and the relevant properties of cosmic ray particles are determined. The objectives of this technique are : (a) For a given location on the earth (or in near earth space) and for a given direction (e.g. local vertical direction), what are the lower cut-off R_L (i.e. Störmer cut-off) and upper cut-off, R_U (Main cone cut-off) and what are the values for the penumbra bands? Knowing these one can calculate the effective cut-off rigidities; (b) For cosmic ray particles of a given rigidity arriving at a given location from the given direction, what is its original direction in space i.e. "the asymptotic direction"? (c) For the worldwide network of neutron monitors (Chapter 6.2) continuously recording cosmic ray intensities, what are their correct effective threshold rigidities and their asymptotic arrival direction in space?

The multipole expansion model of the geomagnetic field used here was introduced by Chapman and Bartles[10] in 1940. The geomagnetic field due to sources internal to the earth is represented by a potential U in spherical polar coordinates, (r, θ, ϕ) using a spherical multipole expansion of the form

$$U(r,\theta,\phi) = \sum_{n=0}^{n} \sum_{m=0}^{n} (g_n^m \cos m\phi + h_n^m \sin m\phi)\, P_n^m(\cos\theta) \left(\frac{a}{r}\right)^{n+1} \qquad (2.4.25)$$

where a = mean radius of the earth, r = radial distance from the center of the earth, g_n^m and h_n^m are Gauss coefficients, p_n^m is the associated Legendre function of degree n and order m of the Schmit quasi-normalized type. For m = 0, p_n^m reduces to the Legendre Polynomial, P_n.

The equation of motion of a particle can be written as

$$\frac{d^2\vec{R}}{dt^2} = \frac{q}{mc}\left(\frac{d\vec{R}}{dt} \times \vec{B}\right) \qquad (2.4.26)$$

The original program of trajectory tracing by McCraken et al.[11] in 1962 was modified by Shea et al.[12] 1968 so as to increase the efficiency of the program. In equation 2.4.26 it is noted that if the sign of the charge q and the direction of the particle are both reversed, the equation remains unaltered. Therefore in the numerical computations we assume a negatively charged particle leaving the earth and moving to infinity, which is the same as a positively charged particle coming from infinity to the earth.

We describe here the method of computation used by Shea and Smart[13] in 1975. To integrate the equation 2.4.26, the components of the magnetic field are calculated at a point r,θ,ϕ as given by

$$B_r = -\frac{\partial U(r,\theta,\phi)}{\partial r} \; ; \; B_\theta = -\frac{1}{r}, \; and \; B_\phi = -\frac{1}{r \sin\theta} \cdot \frac{\partial U}{\partial \phi}$$

In equation 2.4.25, the spherical harmonic of degree and order 10 is sufficient to represent the geomagnetic field properly. The Gauss coefficients, g_n^m and h_n^m are quoted in units of nanotesla for a given epoch e.g. 1985 by the International Reference Geomagnetic Field (IRGF), and approved by the IUGG. The secular variations of the constants g and h are given for a period of 5 or 10 years and thus the appropriate values of g and h are used. The geodetic coordinate system, as given in Figure 2.4.12 are used to describe the arrival directions of cosmic ray particles. The procedures to convert the relevant values from the geodetic to the geomagnetic coordinate system and vice versa are adopted from Peddie[14] (1982).

Knowing the values of the components of B at all points it is possible to determine the trajectory of a given particle by the numerical integration of the equation 2.4.26. A forth order Runge-Kutta method is used to solve the second order equation. The problem is divided into two parts:

a. The determination of the initial points, $(r,\theta,\phi)_{initial}$ and the initial velocity of the particle $V_{initial}$.

b. The tracing of the trajectory of the particle from the earth to outside the magnetosphere.

All calculations are carried out for a single nucleon (i.e. proton); but the results are equally applicable to other particles, since the trajectories of two particles with equal momentum to charge ratio i.e. of equal rigidity, are identical.

Using the initial position and the initial velocity, the trajectory is calculated in small steps of time. The gyration time of the particle is divided into small steps, a step size of 1/100th of the gyration time is used for an uniform field in the initial position. This is chosen as a compromise between the large computation time and the errors due to changes in the magnetic field. After every step the position and the velocity of the particle and magnetic field at that point are recalculated. If the current value of the velocity differs from the initial velocity by more than 10^{-5}, the integration is rejected and the trajectory is recomputed with the previous stepsize divided by 2. The gyration time computed for the new position is again divided into 100 small steps. The computation is continued until one of the following conditions occur:

1. If the particle trajectory goes beyond 25 earth radii, then it is considered a cosmic ray particle coming from interplanetary space and the trajectory is allowed.

2. If the distance of the charged particle becomes less than the actual earth's radius, i.e. the trajectory has intersected the earth the trajectory is forbidden.

3. If even after 40,000 steps none of the above conditions are satisfied, the trajectory is considered to have failed and is therefore forbidden.

At each location one begins with a high rigidity, higher than possible cut-off and trajectories are computed with decreasing rigidities in step of $\Delta R = 0.01$ GV, until one reaches the upper cut-off value. Then one comes into the penumbra zone of alternate allowed and forbidden zones. After passing the penumbra zone, one comes to the lower cut-off rigidity (R_L) i.e. Störmer cut-off, below which all particle trajectories are forbidden. The details of trajectory computations and its application to the Spacelab-3 borne cosmic ray experiment Anuradha are given by Singh(15) (1990) and Dutta(16) (1991). In Figure 2.4.8 we show typical illustrations of (a) simple trajectory of a cosmic ray particle arriving from infinity to the earth (allowed trajectory), (b) a complicated trajectory of a forbidden particle. In Figure 2.4.9, we show a typical examples of the results(16) from trajectory computations for cosmic ray particles showing allowed (white) and forbidden (black) zones. The upper cut-off R_U (the main cone R_M), and the lower cut-off R_U (Störmer cut-off, R_s) are shown. The effective cut-off, R_C is given by the relation

$$R_C = R_L + \sum_{R_L}^{R_U} \Delta R \ \ (not \ \ allowed)$$

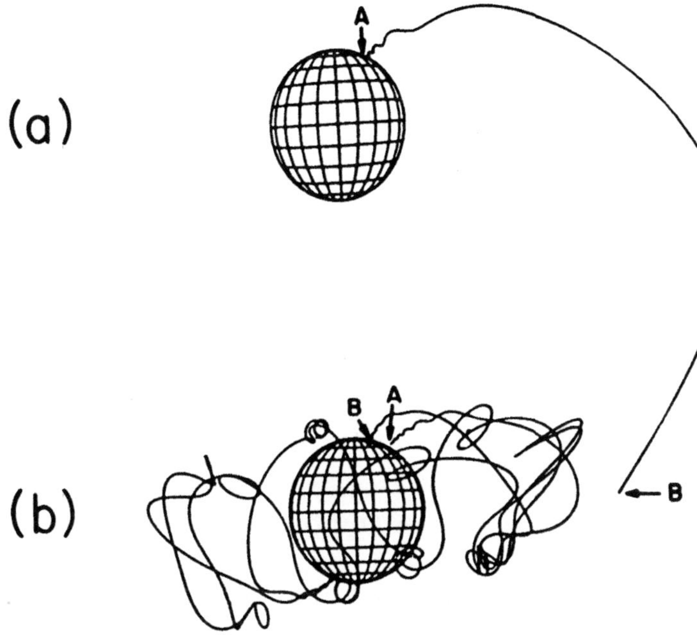

Figure 2.4.8: Trajectory computation results[15] showing a typical (a) allowed trajectory and (b) a complicated trajectory of a forbidden particle

This is shown in Figure 2.4.9.

The computed trajectory results for the vertical cut-off rigidities for the world grid in steps of latitudes of 5° and of longitudes of 15° are given by Shea and Smart[13].

Effective cut-off rigidities R_C determined by the trajectory tracing technique are essential for the understanding of cosmic ray data of several types as follows:

a. R_C values are essential in analyzing the data of the network of cosmic ray neutron monitors.

b. For particles of a given rigidity arriving vertically at a neutron monitor station, one needs its original direction of motion in interplanetary space,

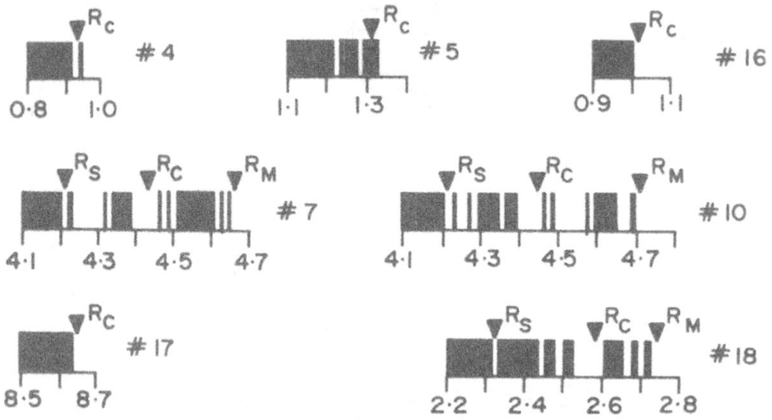

Figure 2.4.9: Typical results of computations of trajectories, showing (a) allowed (white), (b) forbidden (black) zones, and the Main (R_M), the Störmer (R_S) and the effective (R_C) cut-off rigidities

i.e. its asymptotic direction in space.

c. For cosmic ray detectors flown in balloons and space-craft the information on the cut-off rigidities are required for the analysis of the data and interpretation of results.

2.5 The Radiation Belts of the Earth

2.5.1 Early Studies and the Discovery

In 1907 Carl Störmer[(17)] of Sweden published papers in which he showed that there exists some allowed orbits of particles in the dipole field of the earth which are bounded by forbidden regions. This signifies that particles from infinity can not enter this region; also if in some way particles enter into this region, they can not escape to infinity. This represents a zone of trapped particles as shown in Figure 2.5.1 from the drawing of Störmer[(6)]. This idea was however not pursued by him. In order to explain the occurrence of the aurora at $\sim 20^o$ away from the pole. Störmer[17] hypothesized in 1911-12 the existence of a ring current around the earth. Chapman and Ferraro[18] took up the idea and postulated that the main phase of the magnetic storm, when a large decrease of the magnetic field intensity occurred lasting for a day or two, was due to the formation of a ring current located at 6-8 earth radii. Alfven discussed the conditions for the stability of the proposed ring current. In 1956 Singer proposed that a beam of particles from the Sun may enter the geomagnetic field by a perturbation process and some of these may be trapped particles. In 1952-59 Van Allen and co-workers[19] observed large bursts of x-rays during rocket flights in the polar regions indicating the presence of a high flux of electrons.

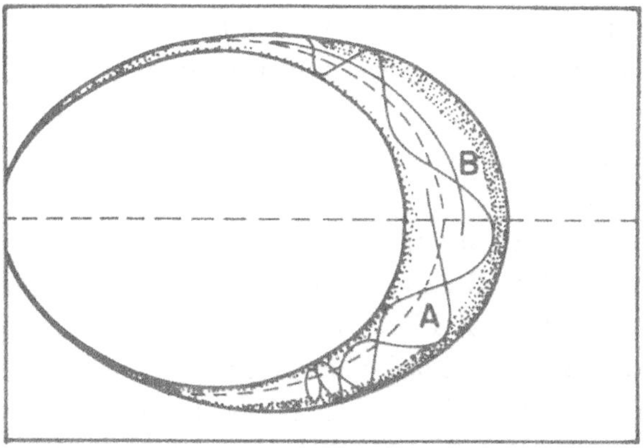

Figure 2.5.1: Störmer's drawing[17] of 1907 of trapped particles

In 1957, the Russian Satellite Sputnik-II first observed a sharp increase in the intensity of particles at 400 to 700 km altitude[20] in Geiger counter, which was certainly due to trapped particles; but it was not interpreted that way. The data for higher altitudes up to the apogee of 1680 km were not received by Soviet ground stations and were not available to Russian scientists. This probably eluded the discovery of the inner radiation belt in Sputnik-II by Soviet scientists. In the first successful US Satellite Explorer-I launched on January 31, 1958, Van Allen, Ludwig, Ray and Mc Ilwain had flown Geiger counters which gave the first evidence[21] for the presence of a high intensity of particles trapped by the geomagnetic field. This, however, came in a dramatic manner. Their Geiger counter recorded an increasing counting rate with altitude, but at the highest altitude of about 2400 km the counting rate dropped to zero. This surprising observation baffled scientists. A few months later, Explorer-III was successfully orbited by the USA on March 26, 1958 and carried the instrument package of the Iowa group. These instruments were identical to that of Explorer-I. On analyzing the data they again found that the counting rate after initial increase dropped to zero at the highest altitude, confirming the Explorer-I results. These observations were then correctly interpreted by the Van Allen's group, after calibration of the counter in the laboratory, as due to the saturation effect of the Geiger counter in the presence of a very high intensity of radiation, of the order of 35,000 counts per second. Van Allen announced the discovery of the high intensity radiation zone above the earth on May 1, 1958 and said that the earth's magnetic field seem to act as an umbrella that kept the radiation particles at about 1000 km above the earth. He visualized them as confined to an reservoir around the earth having a thickness of several hundred to thousand kilometers. When he was asked whether it could be described as radiation "belt" he answered - yes. Thus it came to be known as the radiation belt or Van Allen Belt.

The results of Van Allen and his group(21) created worldwide excitement among the scientists. The Russian scientists launched the Sputnik-III satellite two weeks later on May 15, 1958. The satellite contained several radiation detectors including two scintillation counters. Sputnik-III was launched with a large payload mass of 1325 kg which was more than double of the 510 kg payload of Sputnik-II. Sputnik-III was the first Soviet satellite to have a memory device to store data collected in each orbit and then release it rapidly to the receiving station on ground command. Another innovation by the Russians was to use batteries powered by solar energy. The results from Sputnik-III which was reported by Vernov and Chudakov(22) in 1958 and Vernov et al.(23) in 1959 showed that in addition to the penetrating protons of $>$ 50 MeV in the inner belt there exists a high intensity zone of electrons of energy $>$ 50 keV which came to known as the outer belt. Thus Vernov et al. of the USSR is credited with the discovery of the outer belt.

2.5.2 Observations and General Properties of the Belts

The discovery of the radiation belts triggered a series of space flight experiments in 1958-60 by USSR and USA scientists. In Table 2.5.1 we have summarized some of the important parameters of these experiments. Some highlights of the experimental aspects are discussed here. Explorer-IV was launched successfully by the USA to explore the radiation belts. Van Allen and his group used four types of radiation detectors as follows:

(a) Plastic scintillators shielded by 140 mg·cm^{-2} of Al which recorded protons > 10 MeV, electrons > 650 keV and X-rays > 300 keV,

(b) CsI (Tl) crystal shielded by 200 mg·cm^{-2} of Al and 800 mg·cm^{-2} of nickel and this recorded electrons > 20 keV, proton > 400 keV and X-rays; the total anode current was measured,

(c) Geiger counter, Anton 302 shielded by 1.2 g·cm^{-2} of steel recording electrons > 3 MeV, protons > 30 MeV and X-rays > 10-30 keV,

(d) Geiger counter, Anton 302 shielded by 1.6 g·cm^{-2} of Pb and 1.2 g·cm^{-2} of Fe which recorded electrons > 5 MeV, protons > 40 MeV and X-rays > 80 keV. This experiment provided good information on the inner belt and partial information on the outer belt.

The Sputnik-III satellite of the USSR contained the following intrumentations: (a) Magnetometers, to measure the intensity of the earth magnetic field, (b) Cosmic ray photon counter, (c) Cosmic ray heavy nuclei detector, (d) Cosmic ray intensity recorder and several other instruments. The cosmic ray particle detectors recorded the flux and distribution of energetic electrons and protons in the radiation belts.

The USA satellite Explorer-VI had an extremely elongated elliptical orbit of apogee 2,100 km similar to that of Explorer-IV and thus it could sample a fairly good section of the inner belt and part of the outer belt at high latitudes. Surprisingly good data on the inner and outer belts was provided by USA's spacecraft which was unsuccessful in its primary objective. The Pioneer-III which was aimed at the moon failed to escape the earth's gravity, flew out to about 100,000 km and then fell back to the earth. In this process it traversed through the thicknesses of the inner and outer belts and its radiation detectors gave extremely valuable information on the inner and outer belts, both during the outward and inward trajectories.

The USSR space-probe Lunik-I which flew past the moon gave new data on the inner and outer belts. Lunik-II of the USSR which circled the moon provided further data on the intensity distribution in the outer and inner belts.

Data from all these USA and USSR satellites and space-probes gave detailed information on the radiation belts and showed that there are two radiation zones, the inner and the outer ones separated by an apparent slot. In 1959-60 several US groups used powerful rockets to send detectors such as nuclear emulsions to altitudes of about 1000-1700 km and then retrieved the instrument package from the sea after landing. The analyses of the nuclear emulsion data gave important information on the intensity and energy spectra of the protons in the inner belt. These data had been crucial in establishing the origin of the inner belt as we will discuss later.

General Properties of the Inner and Outer Belts

A. The Location

The locations of the radiation belts in the magnetosphere of the earth are shown in Figure 2.3.1 which shows the meridional cross-section. It is seen that the day side and the night side radiation belts are asymmetrical which is due to pressure of the solar wind on the noon side of the magnetosphere. Outside the trapped radiation belts there are zones of quasi-trapped unstable radiation zones.

B. The Inner Belt

The inner belt consists mostly of energetic protons trapped by the geomagnetic field. If we consider protons > 40 MeV, this typical distribution is shown in Figure 2.5.2 which is an historical diagram on the results of Explorer-IV, Pioneer-III and Pioneer-IV[(24,25)]. In the equatorial region it begins at about 600 km altitude and extends to $\sim$ 10,000 km, the region of maximum intensity is located at $\sim$ 3,600 km altitude i.e. at a geocentric

Table 2.5.1: Early Observations of Trapped Particles: Trapped Protons and Electrons (1957 - 1996)

Vehicle	Launch date and active life	Inclination Apogee(km) Perigee(km)	Detectors	Component measured	Location measured	Ref.
Sputnik-II (USSR)	Nov. 3, '57 (7d)	65° 1700,225	NaI Scin.	Protons >30 MeV	400-700 km	20 26
Explorer-I (USA)	Jan. 31, '58 (~1 yr.)	33° 2500,360	GM counters	High flux of particles	1400-2400 km	21
Sputnik-III (USSR)	May 15, '58	65° 1900,215	2 Scin. counters 2 GM counters	Protons >30 MeV Electrons >30 MeV	Inner and Outer Belts	22, 23
Explorer-IV (USA)	Jul.26, '58 (~2 m)	50° 2,200, 260	2 GM counters 1 Plastic scin. 1 CsI scin.	Protons >30 MeV Electrons >20 keV	Inner and Outer Belts	24
Pioneer-III (USA)	Dec. 6, '58	$1x150^5$ km,0	2 Counters	Protons >60 MeV Electrons >1 MeV	Inner and Outer Belts	25
Lunik-I (USSR)	Jan. 2, '59		GM counters Scin. counters	Protons Electrons	Inner and Outer Belts	22, 23
Explorer-VI (USA)	Jul.26,'59	50° 2,100km	GM, Plastic and CsI scin. Ion counters	Protons >75 MeV Electrons >20 keV, >50 keV, >200 keV	Inner and Outer Belts	25
Lunik-II (USSR)	Sept.12, '59		GM counters Scin.counters	Protons Electrons	Inner and Outer Belts	26
Atlas Rocket (USA)	Feb., July, Aug. '59	~1000 km	GM counters >20 MeV	Protons Electrons >30 keV	Inner and Outer Belts	36
Rocket (USA)	April, May, July '59	~1200 km	Nucl. emulsions >58 MeV	Protons	Inner Belt	34
Rocket (USA)	July 21, '59	~1100 km	Nucl. emulsions > 42 MeV	Protons	Inner Belt	37
Rocket (USA)	Sept.19, '59	~1700 km	Nucl. emulsions >42 MeV	Protons	Inner Belt	38
Rocket (USA)	July 7, '59	~1050km	Mag. spectrometers	Electrons >50 keV	Outer Belt	39

distance of 1.5 earth radii. The peak intensity of protons $J_p > 40$ MeV is $\sim 3{\cdot}10^4$ p.$\cdot$cm$^{-2}\cdot$s^{-1}. The inner belt of protons is fairly stable with time and does not change it characteristics with solar activity.

In Figure 2.5.3, on the left, we show the equal intensity contours of inner belt protons of $E_p > 30$ MeV (Vernov et al.[26] 1967) which has a maximum intensity at 1.5 R_e and it drops off at $R > 2.5$ R_e. Beyond this point the outer belt consisting mainly of electrons of $E_e > 100$ keV occured with a peak intensity at $R \simeq 3.5$ R_e. In this picture the inner and outer belts are separated by a slot which disappears however if we consider low energy protons of $E_p > 0.9$ MeV, as discussed below.

The differential energy spectra of the inner belt protons of $E_p > 10$ MeV were measured by means of recoverable payloads of nuclear emulsion detectors in rocket flights reaching peak altitudes of 1000 to 1700 km. The inner belt protons of $E > 40$ MeV are well accounted for by the Cosmic Ray Albedo Neutron Decay (CRAND) hypothesis (discussed in 2.5.4).

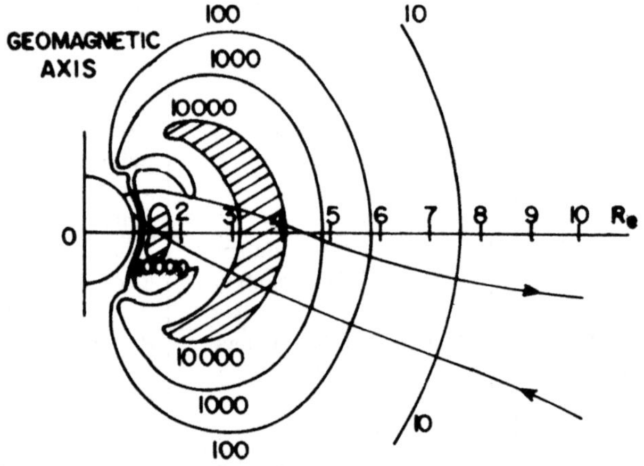

Figure 2.5.2: The inner and outer radiation belts from Van Allen et al.[24,25]

C. Low Energy Proton Belt

When distributions of low energy protons of $E_p > 0.9$ MeV were considered it was found that this intensity started increasing at equatorial altitudes $r > 1000$ km i.e. $R > 1.2$ R_e and reached a peak intensity at $R \simeq 2.8$ R_e i.e. at r $\simeq$ 12,000 km. At the peak zone, the intensity of protons of $E_p > 0.9$ MeV was typically of the order of 10^6 protons/(cm^2·sr·s). Beyond the peak the intensity decreased gradually to $\sim 10^2$ protons/(cm^2·sr·s) at R $\simeq$ 5.5 R_e. This is shown in Figure 2.5.3 (on the right) as measured by the USSR spacecraft 'Electron' in 1964 (Vernov(27) et al.). From the measurements of the trapped proton fluxes of energies $E_p > 0.9$, > 9.0 and > 30 MeV it was shown that their peak positions decreased from $\sim$ R $\simeq$ 2.8 R_e for $E_p > 0.9$ MeV to R $\simeq$ 1.5 R_e for $E_p > 30$ MeV and the peak intensity decreased from $\sim 10^6$ to $\sim 5{\cdot}10^3$ p/(cm^{-2}·sr·sec.) for $E_p > 0.9$ to $E_p > 30$ MeV/N (Vernov(27) 1967). Thus it is seen that low energy trapped protons covered a large region of the magnetosphere of R $\simeq$ 1.2 to 5.5 R_e. The low energy trapped proton flux at R $>$ 2 R_e undergoes both reversible and non-reversible changes following magnetic disturbances.

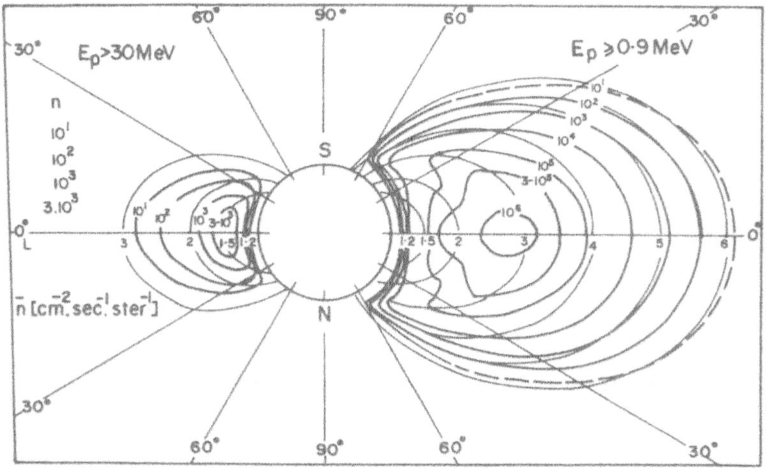

Figure 2.5.3: Intensity distribution of trapped protons of energy $E_p > 30$ MeV (on the left) and $E_p > 0.9$ MeV (on the right) from Vernov(27)

D. Trapped Electrons in the Outer Radiation Belt

In Figure 2.5.4 we show the equal intensity contours of the trapped electrons of energy $E_e > 100$ keV, as measured in 1964 by several authors(28). In the top figure on the noon side of the magnetosphere the highest intensity contour of the 10^7 p/(cm^{-2}·sr·sec) occurred in the magnetic shell of R $\simeq$ 3 to 7 R_E. Beyond this there occurred an unstable zone of quasi-trapped electrons in the magnetic shell of R $\sim$ 9-12 R_e. In the midnight region the trapped electron zone extended from R $\simeq$ 3 to 8 R_E. The unstable zone occurred along the midnight direction from R $\sim$ 8 to 16 R_E.

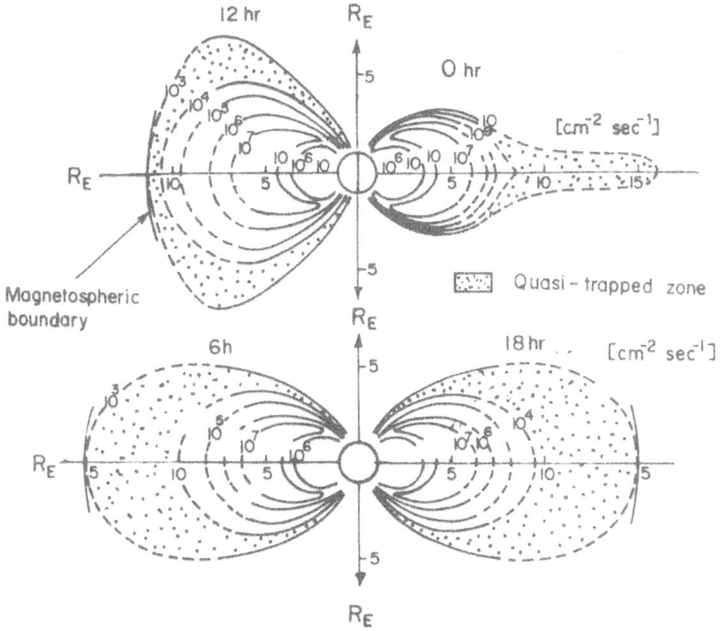

Figure 2.5.4: Intensity distribution of trapped electrons of energy E_e > 100 keV in the noon-midnight plane (upper figure) and dawn-dusk plane (lower figure), from Vernov[28]

The bottom diagram of Figure 2.5.4 shows the distribution of the electron intensity contours in the dawn-dusk plane. In the 6 hr. and 18 hr. directions the contours were symmetrical as seen in the figure and the peaks of the intensity of the electrons of E_e > 100 keV occurred at R $\simeq$ 5 R_E.

A number of satellites and spacecraft had observed the intensities of the trapped electrons in the outer belt undergoing large changes particularly with the arrival of solar plasma and magnetic storms. These observations are important in relation to the origin of the outer belt which will be discussed later.

E. The Unstable Radiation Zone

As shown in Figure 2.5.4 beyond the outer belt there exists quasi-trapped electrons which intensity varies rapidly with time and this region is denoted as an unstable radiation zone. This zone differs from the outer belt electrons in that the intensity of electrons of $E_e > 100$ keV in the unstable zone underwent large changes with variations of the K_p index; with increasing K_p index the average intensity of the unstable radiation zone increased.

F. Heavy Ions in the Radiation Belts

Only a few measurements were made on the flux of trapped ions heavier than protons. This is because of the difficulties in the detection of the low flux of these ions in the presence of a very high intensity of protons and electrons in the radiation belts. In the energy range above 0.6 MeV/nucleon, the ratio of the fluxes of alpha particles to protons in the radiation belt was found to be about $2 \cdot 10^{-4}$. At energy ~ 0.3 MeV/n, the ratio of the intensities of C,N,O nuclei to alpha particles in the outer radiation belt was meausured as about $3 \cdot 10^{-3}$ by Van Allen et al[(29)] in 1970. In 1972 the small flux of CNO nuclei of energies 13-33 MeV/n was measured in the outer belt in the University of Chicago experiment in which an oxygen to carbon ratio of $\sim$ 0.5 was observed.

2.5.3 Dynamics of Charged Particles in Radiation Belts

In this section we discuss the motion of charged particles in the geomagnetic field and the formation of trapped radiation.

A particle of charge q moving with velocity $\vec{v}$ in a magnetic field $\vec{B}$ experiences the Lorenz force $\vec{F}$, where

$$\vec{F} = \frac{d}{dt}(m\gamma\vec{v}) = q\vec{v} \times \vec{B} \tag{2.5.1}$$

Here m = rest mass of the particle, and $\gamma = 1/(1\text{-}\beta^2)^{1/2}$; $\beta = \text{v}/\text{c}$. The particle spirals along the magnetic field in a helical path having the radius of gyration, a. Thus we have

$$\left(m\gamma v_\perp^2\right)/a = (q/c) \cdot v_\perp \cdot B.$$

$$a = \frac{m\gamma v_\perp}{(q/c) \cdot B} = \frac{m\gamma \cdot v}{(q/c) \cdot B} \sin\alpha \tag{2.5.2}$$

Here $v_\perp$ is the component of the velocity of the particle perpendicular to the magnetic field α = pitch angle i.e. the angle between the direction of $\vec{v}$ and $\vec{B}$. Assuming that the magnetic field does not change rapidly in space or time, the motion can be described as a circling motion about a guiding centre which follows the magnetic line of force. The gyrations about the line of force generate a current loop and the magnetic moment, μ, of the loop is given by the current multiplied by the area of the loop:

$$\mu = I \cdot A = I \cdot (\pi a^2) \tag{2.5.3}$$

Since I = (q/c) x frequency of gyration = $(q/c)(v_\perp/2\pi\ a)$,

$$\mu = (\pi a^2) \cdot \frac{q}{c} \cdot \frac{v_\perp}{2\pi a} = \frac{1}{2}\left(\frac{q}{c}\right) \cdot v_\perp \cdot \frac{m\gamma v_\perp}{(q/c) \cdot B}$$

$$= \frac{1}{2}\frac{m\gamma v_\perp^2}{B} = \frac{1}{2} m\gamma v^2 \cdot \frac{\sin^2\alpha}{B} \tag{2.5.4}$$

Alfven showed that in a steady magnetic field the magnetic moment μ is a constant. Therefore this is the first adiabatic invariance. The condition for this invariance is that changes in the magnetic field are small over the gyroradius, a, and during one gyroperiod, T. Therefore when

$$\frac{|\Delta B|}{dt} \ll \frac{1}{a},$$

and

$$\frac{dB}{dt} \ll \frac{B}{T}$$

then,

$$\frac{1}{2} m\gamma \cdot v^2 \cdot \frac{\sin^2\alpha}{B} = constant \tag{2.5.5}$$

Since a static magnetic field does not do work on the particle its velocity or energy is contant. Hence

$$\frac{\sin^2\alpha}{B} = Constant \tag{2.5.6}$$

In the dipole field of the earth if B_o is the magnetic field at a field line in the equatorial plane and α the pitch angle of the particle at that point, as the guiding centre of the particle moves along the line of force towards higher latitude the values of α and B both increases. At a point m, where $\alpha = 90^o$, let $B = B_m$, so that

$$B_m = \frac{B_o}{\sin^2 \alpha_o} \tag{2.5.7}$$

This point m is referred to as the mirror point or turning point, where the direction of motion reverses and the particle travels back to the equator. Then it moves to the other hemisphere where it reaches the mirror point m' and reverses it's direction of motion. Thus the particle oscillates between the two mirror points m and m'. This is the basic motion of a charged particle trapped in the geomagnetic field.

Drift of Trapped Particles

Alfven (1950) showed that, in addition to the above two motions, gyration and oscillation, the trapped particles in the earth's magnetic field undergo another motion which is called 'drift'. This causes the guiding centre of the particles to move along the longitudinal direction, with a drift velocity $\vec{V_d}$, which is given by

$$\vec{V_d} = \frac{\vec{f} \times \vec{B}}{B^2} \tag{2.5.8}$$

where $\vec{f}$ is the small force per unit charge. The directon of the drift velocity $\vec{V_d}$ is perpendicular to both $\vec{f}$ and $\vec{B}$. There are two factors responsible for this force which causes the drift:

(a) Curvature of the lines of force

As the guiding centre follows a curved line of force there is a net centrifugal force acting away from the earth, which is given by

$$f = (m\gamma v_{11}^2)/R_c$$

where R_c is the radius of curvature of the line of force at the point under consideration. Therefore the drift velocity due to this factor is

$$\vec{V}_{d1} = \left(\frac{m\gamma v_{11}^2}{R_c^2}\right) \frac{\vec{R_c} \times \vec{B}}{(q/c) \cdot B^2} \tag{2.5.9}$$

In the static magnetic field with negligible current it can be shown that

$$\vec{R}_c/R_c^2 = -(\vec{\nabla}_\perp \vec{B})/B$$

Thus

$$\vec{V}_{d1} = \frac{m\gamma v_{11}^2}{(q/c)B^3} \cdot \vec{B} \times (\vec{\nabla}_\perp \vec{B}) \tag{2.5.10}$$

Here $\vec{\nabla}_\perp \vec{B}$ is the component of $\vec{\nabla}\vec{B}$ perpendicular to $\vec{B}$.

(b) Gradient of earth's magnetic field:

The magnetic field grows weaker as we go away from the earth. This gives rise to a drift velocity V_{d2} given by

$$\vec{V}_{d2} = \frac{1}{2} m\gamma V_\perp^2 \frac{\vec{B} \times \vec{\nabla}_\perp \vec{B}}{(q/c)B^3} \tag{2.5.11}$$

Thus both the effects work in the same direction and the total drift velocity is given by

$$\vec{V}_d = \vec{V}_{d1} + \vec{V}_{d2} = \frac{\vec{B} \times \vec{\nabla}_\perp \vec{B}}{(q/c)B^3} (m\gamma v_{11}^2 + m\gamma V_\perp^2) \tag{2.5.12}$$

Since in the equatorial plane $\vec{\nabla}_\perp \vec{B}$ is directed towards the centre of the earth, and $\vec{B}$ to the south, $\vec{V}_d$ is directed towards the west for positively charged particles. For negatively changed particles drift is towards the east. Thus the drift of both positively and negatively charged particles produce a westward flowing ring current.

The second adiabatic invariant

The second adiabatic invariant is called the longitudinal invariant and is given by

$$I = \oint m v_{11} dz \tag{2.5.13}$$

where v_{11} is the velocity parallel to $\vec{B}$ and the integration is over a complete north-south cycle. When we consider a trapped particle spiraling around a particular line of force at a certain instant of time i.e. at a certain longitude, the problem of specifying the line of force at a subsequent time i.e. at some other longitude is not simple because of the departure of the field from a dipole and its non-symmetric nature. The conservation of I, given by 2.5.13 allows us to specify the correct line at any longitude. As the magnetic moment is constant, we obtain

$$I = mv \oint \cos \alpha dz = mv \oint \sqrt{1 - \sin^2 \alpha dz} = mv \oint \sqrt{1 - B(z)/B_m dz} \quad (2.5.14)$$

Since v is constant in a static magnetic field, I is determined by the structure of the field only. As μ is constant B_m is also constant. As the intensity of the particles in the inner zone changes very rapidly, say by a factor of 10 for about 3% change in the geocentric distance and the magnetic field deviates from the dipole type by about 3% or more, one finds that the use of a dipole field to describe the intensity changes produces highly discordant results. To overcome this difficulty Mc Ilwain[30] developed a coordinate system called B-L coordinates.

B-L coordinates

McIlwain used the fact that the quantity $I = \oint mv_{11}dz$ is invariant provided the effect of drift and other effects are small during one oscillation. The modified invariant I' = I/mv is used in a dipole field at magnetic latitude λ on a line of force having equatorial radius r_e. He derives the relationship

$$\frac{L^3 B}{M_E} = F\left(\frac{I'^3 B}{M_E}\right) \quad (2.5.15)$$

where the magnetic shell parameter is defined by replacing r_e by L, and I' and B are calculated for the point in question with the use of the **real** field of the earth and F is the function which is calculated with a **dipole** field. A set of values of the function F is given by McIlwain. Given the B and L values of a point in space, one may then determine the equivalent point (r,λ) in a dipole field by using the dipole relations

$$B = \frac{M_E}{r^3} \cdot \left(4 - \frac{3r}{L}\right)^{1/2}, r = L \cos^2 \lambda \quad (2.5.16)$$

Using this method the spatial distributions of intensities of trapped particles are plotted in B-L coordinates and these plots are symmetric and regular.

The third adiabatic invariant

The third adiabatic invariant found by Northrop and Teller is called the flux invariant. This is given by

$$\phi = \oint_S \vec{B} \cdot \vec{n} \cdot ds$$

where the integration is carried over the surface whose boundary line is that traced by one of the mirror points in its drift in longitude. The guiding center of particle moves on a surface of revolution formed by revolving the line of force about the dipole axis. Then ϕ is the flux passing through the open hole at one end of the barrel shaped volume.

Omnidirectional intensities

The understanding of the omnidirectional intensity and its variations with latitude is simplified with the concept of the pitch angle (α) distribution of the unidirectional intensity $J(\alpha_e)$ in the equatorial plane. The omnidirectional intensity at the equatorial plane is then obtained by integration over the sphere of the angular distribution at any point,

$$I_e = 4\pi \int_o^{\pi/2} J(\alpha_e) \sin \alpha_e d\alpha_e$$

By denoting $\cos \alpha_e = \eta$, the above relation is simplified as

$$I_e = 4\phi \int_o^1 J(\eta) d\eta$$

By referring to above it can be shown that knowing the pitch angle distribution in the equatorial plane, one can obtain the angular distribution of the equatorial plane. From equation 2.5.7 it is noted that as the equatorial pitch angle of the particle α_e become smaller, the mirror point takes place at higher B_m value i.e. in the deeper regions of the upper atmosphere. As a result these particles suffer higher coulomb energy loss and may be preferentially removed. Thus there occurs a loss cone in the region of $\cos \alpha_e$ $(= \eta)$ between 1 and an intermediate value, in the pitch angle distribution of particles in the equatorial plane for a given energy at a particular field line. Apart from the loss cone it may be noted that the basic principle is that the differential directional intensity along a line of force is constant which follows from Liouville's theorem (Chapter 6.1).

Lifetime of trapped particles

Trapped Protons: There are four main process by which trapped protons are removed from the trapping region: (a) **Coulomb collisions:** The coulomb collisions of protons with the atoms of the upper atmosphere causes energy loss of the particles and as a result it's spiral orbit shrinks on the line of force; (b) **Nuclear scattering:** Trapped protons may collide with atmospheric oxygen and nitrogen and are elastically scattered. This removes the particles from the trapping region; (c) **Elastic p-p collisions:**Elastic p-p collisions constitute a catastrophic removal process which is important at high energies > 1 GV. (d) **Inelastic p-p:** Inelastic p-p collisions cause the removal of particles and is effective above the pion production threshold. The cross-section, σ, of these processes determine their effectiveness at a given energy. The 1/e lifetime of the particles for the reduction of the intensity to 1/e th of the initial value by nuclear interaction is given by $[\sigma(\mathrm{E})\rho_a \cdot \beta]^{-1}$ where $\sigma(\mathrm{E})$ is the cross-section and ρ_a the mean density of atoms averaged over the orbit. The combined lifetime $\mathrm{T}_{1/e}$ is defined as

$$\frac{1}{T_{1/e}} = \frac{1}{T_{1/e}(energyloss)} + \frac{1}{T_{1/e}(nucl.interaction)}$$

As the mean density $\bar{\rho}_a$ is a variable and somewhat uncertain term, the curves are drawn for the factor $\bar{\rho}_a \cdot \mathrm{T}_{1/e}$ vs energy. The atmospheric density and the effective orbit average density are dependent on atmospheric models and are discussed elsewhere. As an example we note that at the heart of the inner belt at $\mathrm{R}_e = 1.5$, H and H^+ are the main constituents of the upper atmosphere; thus for protons of E > 50 MeV, $\bar{\rho} \cdot \mathrm{T}_{1/e} \approx 5 \cdot 10^{13}$ sec·cm^{-3}. Using $\bar{\rho}_H + \bar{\rho}_{H+} \approx 4.5 \cdot 10^3$ cm^3, we obtain $\mathrm{T}_{1/e} \approx 10^{10}$ sec ≈ 300 years. As the typical omnidirectional intensity of protons > 50 MeV is ≈ $50 \cdot 10^3$ cm^{-2}·sec^{-1}, the density of protons is $\mathrm{J}_p = \mathrm{I}_p/\mathrm{v}$ protons cm^{-3}, where v is the mean velocity $\simeq 1.2 \cdot 10^{10}$ cm·s^{-1}, for $\mathrm{E}_p \simeq 100$ MeV. Hence $\mathrm{J}_p \simeq 4 \cdot 10^{-7}$ cm^{-3}. Thus the injection rate Q_p is estimated as $\mathrm{Q}_p = \mathrm{J}_p/\mathrm{T} \approx 4 \cdot 10^{-7}$ cm$^{-3}/10^{10}$s ≈ $4 \cdot 10^{-17}$ cm^{-3}·s.

Outer Belt

The lifetime of the outer belt electrons is complicated by the fact that these electron fluxes show large variations with time particularly with the magnetic disturbances. Apart from these, the lifetime of these electrons are determined mainly by two factors:

(a) Energy loss

When electrons of energy say > 300 keV lose energy by collisions with ambient matter, the loss rate can be computed from mathematical deduction and the assumed atmospheric model, the details of which are available elsewhere. As an example at R = 4 R_c, the mean lifetime of 300 keV electrons with energy loss only is estimate as ~ 60 years.

(b) Scattering loss

Electrons may undergo no change of energy but large change of angle by 60° or more by scattering processes. Computing the mean lifetime for scattering through 60°, one obtains, for example, $T_s \sim 140$ years.

(c) Non-dipole effects

A particular and pronounced anomaly is found in the South Atlantic near Cape-town where the surface magnetic field is reduced to almost half the value of that at the same magnetic latitude but at distant longitudes. The surface of constant field B at which successive mirror point reflections occur as the particle drifts in longitude is then lowered by ~ 1000 km or so near the anomaly. Because of an increased atmospheric density, the trapped electrons near the loss cone are affected. This additional loss term however is found to be significant for particles at latitudes above 40°.

2.5.4 Origin of the Trapped Radiation Belts

The Inner Belt

Soon after the announcement of Van Allen in 1958 of high intensity trapped radiation at altitudes about 1000 km and above, suggestions were made in 1958-59 by Singer(31), Vernov and Lebedinskii(32) and Hess(33) that these are not particles from solar plasma, but they originated from high energy protons arising from the decay of cosmic ray albedo neutrons. These authors, Freden and White(34), Shoklovsky et al(35) and others calculated with various assumptions and degrees of accuracies, the expected trapped proton flux originating from this process. Following the experimental verification of the trapped protons by Freden and White[33] and the general agreement of the theoretical predictions with observations, this mechanism proved to be a great success. Detailed calculations were made by several investigators. The basic mechanisms of the Cosmic Ray Albedo Neutron Decay (CRAND) theory are briefly as follows. Energetic cosmic ray particles, mainly protons, interact with the atmospheric oxygen and nitrogen nuclei and produce secondary protons and neutrons; the secondaries in turn produce tertiaries and so on. A fraction of the interactions occur sufficiently high in the atmosphere to allow neutrons to escape. The spectrum of the neutrons emitted upwards is estimated as

$$J_n(E)dE = 2E^{-1.8}dE \quad particles/(cm^2 \cdot sec)$$

or 0.4 neutrons $cm^{-2} \cdot s^{-1}$ of energy > 50 MeV. Only very few of the fast neutrons decay within the earth's magnetic field, so that the differential intensity may be taken to be constant along the neutron trajectory. The decay rate is estimated as

$$Q_o(E) = \frac{J_n \Lambda_n}{\gamma \beta c} cm^{-2} \cdot sr^{-1}$$

where the decay constant $\Lambda_n = 9 \cdot 10^{-4}$ s^{-1} and the relativistic time dilatation is taken into account by the factor $\gamma\beta$ c. It can be seen that the distance to the point of decay traveled by the fast neutrons is on the average about one astronomical unit. It is interesting to note that the injection rate of protons near the earth is $\sim 10^{-14}$ $cm^{-2} \cdot s^{-1}$ i.e. for protons > 50 MeV about 1 proton per cm^2 every $3 \cdot 10^6$ years!

The experimentally measured trapped proton fluxes, the energy spectrum and the calculated intensity and spectrum as given by Freden and White are shown in Figure 2.5.5. The calculated line (B), divided by 2.1 and thus normalized at 100 MeV with observations, is shown. Considering various assumptions and uncertainties of the assumed parameters, the agreement within a factor of 2 between the observations and the theory is considered satisfactory for trapped protons of E > 50 MeV. For trapped protons of E $\sim$ 10-50 MeV, the measured fluxes are large and the spectrum is too steep so that CRAND theory can not account for these data (Naugle and Kniffe[38]). It was suggested that solar protons which impinge on the polar

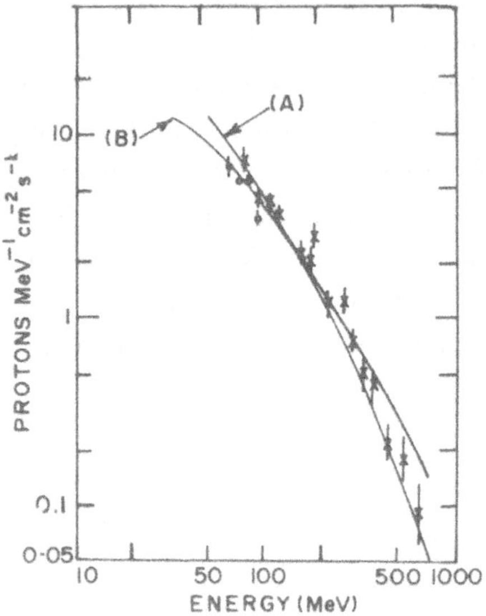

Figure 2.5.5: Observations of trapped protons of $E_p > 50$ MeV in the inner belt[34] (A). The calculated line (B), divided by 2.1 agrees well with observations

cap region following energetic solar particle events undergo spiraling motions in the upper atmosphere and these may generate neutrons. Some of the polar cap neutrons will inject decay protons into the radiation belt. This is an impulsive injection process which is probably superposed on the steady cosmic ray injection process.

Origin of the Outer Belt

The origin of the outer belt electrons has been rather difficult and elusive for a long time. The possible mechanisms for the origin of the electron flux in the outer belt were discussed in literature since its discovery. The oldest suggestion is the hypothesis of the trapping of solar plasma. However the particles in the solar plasma streams have energies of 1-5 keV which are far too small as compared to 30 to > 500 keV of the outer belt electrons. Therefore it was suggested by several authors that there must be some sort of local acceleration process. Radial diffusion of electrons from the boundary region of the belt towards the inner region near the earth, so that the magnetic moment is conserved, may lead to acceleration. This arises from the situation that $\alpha = 90^o$, the magnetic moment of μ of the particle (equation 2.31) is particle energy/magnetic field strength. Therefore the constancy of μ will lead to increased energy due to radiation diffusion. However quantitative calculations lead to several difficulties of this mechanism and this hypothesis was found to be unsuitable as origin process.

The second possible source was the albedo neutron decay mechanism in which electrons from neutron decay are the source of the trapped electrons. This mechanism was quantitatively calculated by several authors and compared with the observations. It was found that the calculation flux is far too small, by a factor 10^3-10^4 or more as compared to the measured flux; hence this hypothesis was found to be unsuitable for the origin of the outer belt electrons.

The complexity in the origin of the outer belt electrons is seen from the early observations in satellites such as Explorer-VI by Winckler's group[(40)] during a magnetic storm on August 16-19, 1959. In a particular orbit passing through the outer belt, the normal intensity is recorded on August 16, whereas in the next orbit on August 17, after the arrival of the magnetic storm, the intensity decreased by a large factor. After the passage of the magnetic storm on the following day on August 18, it was found that the intensity at the same location not only recovered to the original value but showed to be increased by a factor of about two. These observations clearly showed the important role of local acceleration.

Magnetic field reconnection

The electron injection process by this mechanism (Axford[(41)] and Tversky[(42)]) was supported by experimental observations made by two spacecraft simultaneously. It was found that on the midnight side at large distances from the earth there was a collapse of the magnetic field and the trapped electron flux of $E > 50$ keV on the night side increased by several orders of magnitude. Thus it is proposed that at the midnight side at a distance of $L \sim 6$, the collapse or reconnection of the magnetic field leads to a betatron acceleration of the ambient plasma electrons, by a factor of ~ 10, to about 20 keV. Early observations of this mechanism of particle acceleration and injection were made. Recent satellite studies have provided further evidences for this process.

2.5.5 The New Radiation Belt from Trapped Anomalous Cosmic Rays

The discovery and general features of trapped ACR

The discovery of the new radiation belt took place in three stages from the results of orbiting spacecraft. In the Indian cosmic ray experiment conducted in the Skylab-3 mission of NASA, USA in 1973-74, Biswas et al(43) and Biswas and Durgaprasad(44) obtained the first evidence of this new trapped particles originating from the anomalous cosmic rays (Chapter 7.3).

These authors found that while the composition of the anomalous cosmic rays (ACR) detected in the Skylab-3 was the same as those measured in interplanetary space by the Pioneer 10 and 11 and IMP-8 spacecraft, the observed intensity was surprisingly higher, by a factor of 25 or more than that calculated from the interplanetary flux. In order to interpret this strikingly new observations, these authors(44) proposed that the heavy ions such as oxygen, of the anomalous cosmic rays are trapped or quasi-trapped in the geomagnetic field and these were responsible for the excess intensity. They(44) also suggested the trapping mechanism by the 'mirror point stripping' process of Blake and Freisen(45) (discussed later) which has been found to be the correct mechanism.

In the USSR experiments in the COSMOS satellites conducted in 1985-88 Grigorov et al.(46) obtained further evidence for the existence of the trapped heavy ions of oxygen of ACR. They found that the angular distribution of the oxygen ions of ACR recorded in the detectors showed double peaked distributions separated by about 180°. From these they inferred the presence of trapped ACR oxygen ions.

In the Skylab-3 and the COSMOS experiments the authors used solid state nuclear track detectors such as Lexan and CR-39, which have high sensitivities as heavy ion detectors.

In 1993 new evidence was presented from the results of the cosmic ray experiment flown in the US satellite SAMPEX on the presence of the trapped heavy ions of oxygen of ACR by Cummings et al.(47). This experiment identified and located the new radiation belt of trapped heavy ions of anomalous cosmic rays.

In the cosmic ray experiment of the SAMPEX satellite a number of silicon solid state detectors were used which identified nuclear charge, mass and energy of the particles and determined their arrival directions. The spacecraft was a polar orbiting one at an altitude of 500-560 km, and the most dominant component, oxygen particles of ACR of 15-40 MeV were analyzed in addition to other elements of ACR. The Caltech-NASA group(47) observed a high concentration of the oxygen ions in the South Atlantic Anomaly region. On analyzing the arrival latitude and longitude of these particles it was found that these ACR oxygen ions (and some nitrogen and neon ions) are trapped in the L-shell of 2.0. From the trapped L-shells a small fraction of the particles are precipitated near the mirror points close the South Atlantic Anomaly. These are recorded in the detector during the orbits passing through the SAA region.

The new radiation belt is very different from the well known radiation belts in several respects. Its intensity is many orders of magnitudes smaller than that of the inner belt protons. It is composed mainly of He, N, O and Ne reflecting that of the ACR component. At present most of the studies are made on the oxygen ions which are the most abundant heavy component.

Origin of the new belt

The new radiation belt originates from the trapped anomalous cosmic ray (ACR) particles according to the following mechanisms. The ACR ions originate from interstellar neutral atoms which are swept into the solar system and are ionized by solar UV radiation and accelerated in the heliospheric boundary. Thus ACR are composed mainly of He, N, O and Ne ions and are in the singly ionized state (Chapter 7.3). The experimental observations of oxygen and heavier particles in singly ionized state were made in the Spacelab-3(48,49) and Sampex(47) experiments. Singly ionized particles of ACR, e.g. O^{1+} of low energies are incident at high latitudes of $> 50^o$ where a fraction of these ions undergo mirror point reflection in the relatively denser regions of the upper atmosphere. As a result these O^{1+} ions suffer an electron stripping process leading to the productions of O^{2+}, O^{3+}, ... O^{8+} ions(45). These partially stripped ions have magnetic rigidity (total momentum/effective charge) smaller than the cut-off rigidity at that location and these particles, thus, become trapped ions in the geomagnetic field. This process was suggested in 1980 by Biswas and Durgaprasad[44] on the basis of the mirror point stripping mechanism of Blake and Freisen[45]. This is illustrated in the sketch shown in Figure 2.5.6.

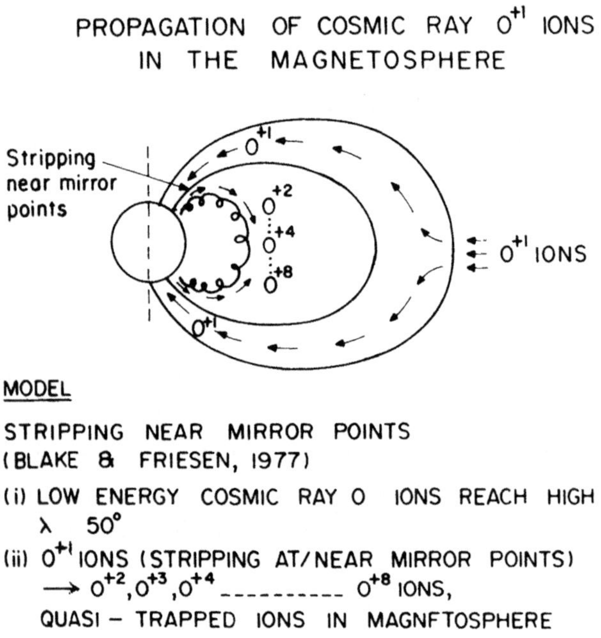

Figure 2.5.6: The sketch shows the model[44] of the origin of trapped oxygen ions of the Anomalous cosmic rays by the mirror point stripping process

The mean lifetime of the trapped ACR oxygen ions of energy $\sim$ 10 MeV is roughly estimated as about 6-12 months assuming an upper air density of $\sim 10^3$ cm^{-3} at R $\approx$ 1.5-2.0 R_e. Further details of the trapped ACR component are given in the works of the Caltech-NASA group[50] and in the review by Biswas[51].

2.6 Heavy Ions in the Magnetosphere of the Earth

Electrostatic deflection spectrometers were designed to identify H, He and heavy ions of low energies and to measure their energies and charge states. The particles were first selected by an electrostatic analyzer according to

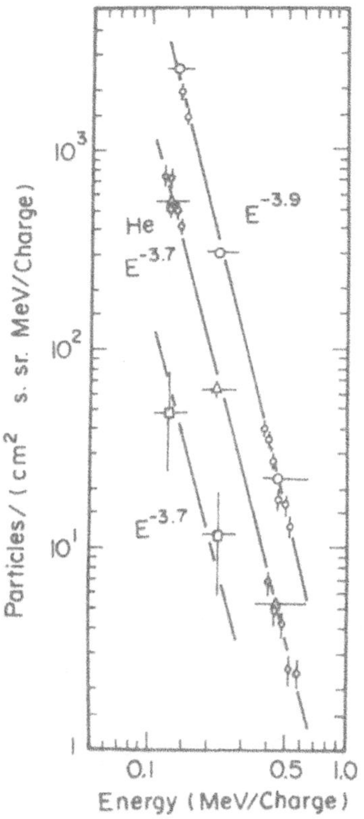

Figure 2.6.1: The differential energy spectra of H, He and Heavy ($Z \geq 6$) ions in the earth's magnetosheath and magnetotail by the IMP-7 satellite[(52)]

their energy per charge and their energies were then measured from the pulse height produced by the energy deposited in the Au-Si solid state detectors. By combining these two sets of data, both energy and charge states were determined(52). This instrument flown on the Interplanetary Monitoring Platform (IMP-7) sampled particles in the earth's magnetosheath and the magnetotail and yielded the fluxes of the particles in the range of 0.1 to 0.5 MeV/charge(52). The differential fluxes of H^+, He^{++} and of CNO nuclei were thus measured in the energy per charge interval of 0.1-0.5 MeV/q. A typical set of data obtained by Fan et al(52) in 1975 is shown in Figure 2.6.1. Most of helium particles were found to be as He^{+2}. Although the instrument could not resolve C^{+5} from C^{+6} and O^{+7} from O^{+8}, most of the carbon and oxygen particles were consistent with fully ionized nuclei as C^{+6} and O^{+8}. The relative abundances of the particles in the 0.12-0.16 MeV/charge range were found to be $H^+ : He^{+2} : C^{+6} : O^{+8} \approx 1 : 10^{-1} : 3.2{\cdot}10^{-3} : 1.3{\cdot}10^{-3}$ where C^{+6} and O^{+8} includes lower charge states of 1 unit.

2.7 Problems on Chapter 2

1. Describe briefly the main regions of the earth's magnetosphere and their principal properties.

2. Calculate the gyroradius of solar wind protons of velocity 300 $km{\cdot}s^{-1}$ in the interplanetary magnetic field of 5 γ at the orbit of the earth and show that it is nearly the same as the thickness of the shock front at the Sun-side of the magnetosphere which is about 600 km.

3. Show that at the magnetosphere boundary at the Sun-side the inward pressure of the solar wind of mean density n and velocity v, is nearly equal to the outward pressure of the magnetic field B. (Use $n = 5\ cm^{-3}$, $v = 300\ km{\cdot}s^{-1}$ and $B = 40\ \gamma$).

4. Explain the main features of the Störmer cone, main cone, penumbra and shadow cone.

5. Using Störmer theory calculate the cutt-off rigidities at a location on the earth at geomagnetic latitude 50° in the (a) vertical direction, (b) at 40° to the east and (c) at 40° to the west in the east-west plane.

6. Derive the first adiabatic invariant condition of a charged particle in a steady magnetic field which states the constancy of the magnetic moment.

7. From the constancy of magnetic moment, derive the conditions for mirror points.

8. What are the two main factors which cause the drift of trapped particles in the longitudinal direction in the earth's magnetic field? Show that both factors cause the drift motions of protons and electrons such that these constitute a ring current flowing from the east to west.

9. What are the processes which determine the lifetime of trapped protons in the inner belt?

10. If we use the peak omni-directional intensity of inner belt protons of E $>$ 50 MeV as $50 \cdot 10^3$ cm^{-2} sec^{-1} and mean lifetime as 300 years, calculate the proton injection rate required to maintain the inner belt (assuming mean velocity of protons of $E_p \simeq 100$ MeV as 0.415 c).

11. Explain the basic mechanisms of the theory of the Cosmic Ray Albedo Neutron Decay (CRAND) of the origin of the inner radiation belt.

CHAPTER 3
THE SUN AND THE HELIOSPHERE

3.1 The Sun and Space Astrophysics

The advent of space age with its new capabilities of satellites and space probes to fly sophisticated instruments in space for a long period of time for studying the Sun and the solar system, opened up many new windows of knowledge on solar physics and space astrophysics. Which led to a series of discoveries which revolutionized our concepts and perceptions of the Sun and the nearby interplanetary medium. This region which is strongly under the influence of the Sun, is called the heliosphere. In this chapter, we shall discuss briefly some of the main areas of the new knowledge of fields and particles relating to the Sun and the heliosphere.

3.1.1 General Features of the Sun

Figure 3.1.1 shows some of the general features of the Sun and the variation of the two most important properties, the temperature and pressure on the Sun as a function of height. The photosphere, the topmost layer of the Sun is visible from the earth and is about 400 km thick. The top of the photosphere is taken as the reference level or zero level height. Below the photosphere lie the unseen layers of the solar interior. Although we speak of the surface of the photosphere of the Sun, there are no sharp boundaries or sharp discontinuities in the gaseous body of the Sun and its atmosphere. At the surface of the photosphere, the density is about 10^{-7} g·cm^{-3} which is roughly equal to the highest vacuum in the laboratory. This falls rapidly with height above the photosphere; at $\sim$ 2500 km height the density falls to 10^{-16} g·cm^{-3} i.e. by 9 orders of magnitude. This region, 0-2200 km in height is called the chromosphere; above this lies a thin region of $\sim$ 200 km in thickness, called the transition region. Beyond this lies the solar corona - the vast region of the solar atmosphere extending to more than 1.5 solar radius. At the coronal heights of about 5000 km and above the density reaches a very low, asymptotic value of about $5{\cdot}10^{-17}$ g·cm^{-3}.

The other remarkable feature of the Sun is its temperature and its dramatic variations. At the top of the photosphere the temperature of the Sun is

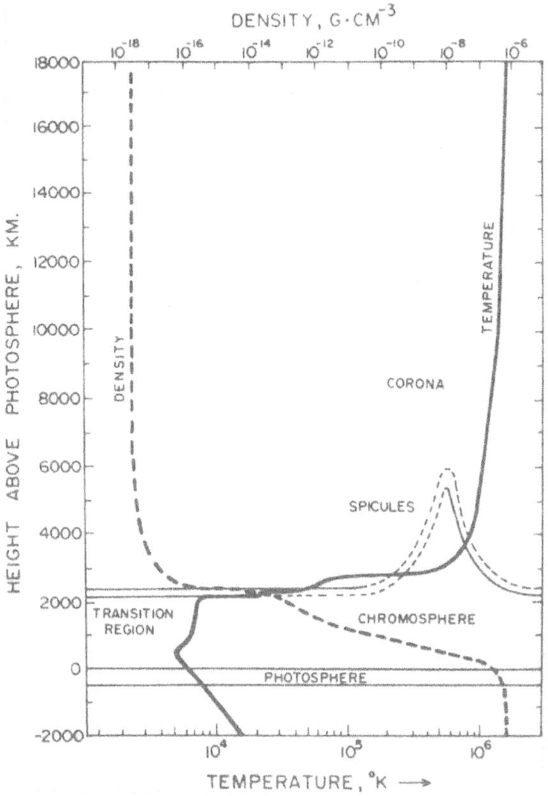

Figure 3.1.1: Variations of pressure and temperature as functions of height above the photosphere in the Sun

about 6000 K. Above the photosphere the temperature continues to fall until a sharp minimum occurs at the lower chromosphere (Figure 3.1.1). Then the temperature in the upper chromosphere begins to rise first slowly and then rapidly till the transition region is reached. Here the temperature sharply rises in several steps until in the lower corona, at about 3000 km, it reaches $\sim$ 500,000 K. Beyond this level, the temperature in the middle and upper corona reaches a staggering value of $\sim 10^6$ K. How and why the Sun maintains its thin atmosphere - the corona - at this very high temperature are some of the major problems in solar physics.

The atmosphere of the Sun above the chromosphere in the corona is highly turbulent. Hot gases shoot up from the top of the chromosphere into the

Table 3.1.1: General Features of the Sun

Age	4.5×10^9 yrs.
Mean Distance from the earth	1.5×10^8 km.
Variation of M.D.	$\pm$ 1.5%
Diameter	1.39×10^6 km.
Volume	1.41×10^{33} cm^3
Mass	1.99×10^{30} kg.
Mean Density	1.41 $g \cdot cm^{-3}$
Surface Density	10^{-9} $g \cdot cm^{-3}$
Central Density	160 $g \cdot cm^{-3}$
Surface Temperature	6000 K
Central Temperature	15×10^6 K
Solar Radiation of the entire sun	3.8×10^{23} kW.
Solar Radiation on the entire earth	8×10^{14} kW.
Solar Radiation at the top of the earth's atmosphere	1.5 $kW \cdot m^{-2}$
Solar Rotation Period	26.8 days
Solar Rotation period at high latitude	30.8 days

corona forming so called "Spicules". They reach heights of ~ 5000 to 15,000 km in the corona. The spicules are thus roughly of the size of the diameter of the earth. These chromospheric spicules can be seen only during the total solar eclipse when the brightness of the solar photosphere is blocked by the disk of the moon. Other spectacular features are the 'Solar Prominences' that can occasionally be seen at the limb of the Sun during a total solar eclipse. These are enormous solar eruptions in which hot solar plasma is shot up and form a flame like appearance which follows the magnetic field lines of the Sun. Some of these giant prominences rise to about 200,000 km or more above the solar surface and some appear in a huge arch-like formation in which the hot solar plasma is guided by the solar magnetic field to come back to the solar surface. These are called Loop prominences of the Sun. In Table 3.1.1 the general properties of the Sun are summarized.

3.1.2 The Solar Energy Source

Below the photosphere lie the unseen layers of the solar interior. Theoretical models and solar surface observations are used to infer about the solar interior. The hot central core of the Sun is believed to have a diameter of about 10^5 km, which is roughly of the size of Jupiter. It is presumed that the central temperature is about $15 \cdot 10^6$ K and the central density is about 160 g·cm^{-3} or 10 times that of ordinary metals. At these extremely high temperatures and densities in the solar core, thermo-nuclear reactions take place. In this process, hydrogen is converted into helium through cyclic processes like the proton-proton cycle, CNO cycle etc. In these processes of nuclear fusion the combined masses of two protons and two neutrons are slightly more than the mass of the helium nucleus formed and the excess mass is converted into energy. For each helium atom formed the energy is emitted in the form of a gamma ray. The gamma rays created normally travel a short distance before being absorbed and re-emitted to be absored again. The energy of gamma rays thus travels outwards through the Sun by a long and tortuous path, first through the radiative interior and then through the convective zone. In the convective zone the outward flow becomes turbulent, creating giant bubbles. When the energy reaches the surface of the Sun - in the photosphere - most of the energy of gamma rays is converted into X-rays, ultra violet rays, visible light and infrared radiation, which are emitted by the solar surface into the solar atmosphere and into interplanetary space. Calculations show that the energy created at the centre of the Sun takes about 50 million years to reach the solar surface!

The Sun emits energy at the rate of about $3.8 \cdot 10^{23}$ kW in all directions. Only a small portion of it - about $8 \cdot 10^{14}$ kW - is received on the surface of the earth. This amount is vastly higher than the full capacity of all power plants of the world together. On each square metre on the earth about 1.5 kW of solar energy is received,this is called the solar constant.

3.1.3 Solar Instruments in Space Craft

Many types of solar telescopes have been flown in a number of space missions. For example in the Skylab mission of NASA, a number of sophisticated solar telescopes were flown in 1973-74 with nine astronauts participating in different phases. These are briefly noted. The Apollo Telescope Mount (ATM) contained six principal instruments which were kept pointed to the Sun with very high accuracy. The X-ray telescopes and an extreme UV telescope observed the full solar disc, the chromosphere and the corona in the wavelengths covering from 2 Å to 200 Å. Grazing incidence mirrors were used for X-ray telescopes. The three UV instruments were the Extreme UV spectroheliograph, the UV spectroheliometer and the UV spectrograph. These studied the Sun, its chromosphere, transition region and lower corona in a wavelength region of 150 Å to 3900 Å. A white light coronograph was used to photograph the corona in visible light with a high resolution. Three occulting discs were employed to block direct sunlight. Two H_α telescopes provided a continuous view of the Sun with monochromator filters that allowed only a narrow band of 0.7 Å centred on the H_α line of 6563 Å. These provided a wealth of new findings on the Sun and its atmosphere. The instruments for measuring hard X-rays and gamma rays from the Sun are discussed later.

3.2 Sunspots and the Solar Cycles

3.2.1 General Features of Sunspots

Galileo was the first who made observations of sunspots with his newly invented telescopes and wrote in 1613: "Having made repeated observations I am at last convinced that the spots are objects close to the surface of the solar globe, where they are continuously being produced and then dissolved, some quickly and some slowly; also that they are carried round the Sun by its rotation, which is completed in a period of one lunar month. This is an occurence of the first importance in itself and still greater in its implications".

Since then, the sunspots have been observed with varous instruments and their various features have been established. The sunspots , as we know now are cooler regions of the photosphere; the typical temperature at the darkest and coolest part of the sunspots is about 3400 K as compared to 6000 K of the photosphere. Hence they appear as dark spots on the photosphere. The sunspots have a central darkest region, called the umbra which is surrounded by filamentary structures called the penumbra. The sizes of sunspots vary from very small, medium and large to very large. The large ones are about the size of the earth and the very large ones are about ten times larger i.e. about size of the Jupiter. These very large sunspots are visible to the naked eye during the period when the Sun is close to horizon. Often the sunspots appear in groups - a group is usually composed of several small, medium and large spots. Spectroscopic observations showed that sunspots are seats of large scale motions so that matter flows outward from the umbra and then along the penumbral filaments.

In the early twentieth century, G. Hale discovered that sunspots are seats of very strong magnetic fields. Using the Zeeman splitting of the iron line with a wavelength of 5941 Å, Hale found that magnetic fields as strong as 3000 Gauss were present in some large sunspots, whose area exceeded several times that of the earth. The discovery of such a strong magnetic field provided an explanation of many observed features of sunspots, e.g. the relatively lower temperature of the sunspots were due to the inhibition of the flow of heat towards the surface by the magnetic field. The penumbral filaments represented the flow of matter along the magnetic lines of force.

Following Alfven we describe the equilibrium of sunspots as due to a strong vertical magnetic field in the spot. Sunspot material is supposed have the same density as material outside the spot at the same level. Thus a lower temperature in the spot causes a reduction in pressure there which is compensated by the magnetic pressure. For a magnetic field of 2000 Gauss in a sunspot the magnetic pressure is $B^2/8\pi$ i.e. $1.6{\cdot}10^5$ dyne$\cdot$cm^{-2}. This permits a significant difference in temperature, as the gas pressure at the photosphere outside the spot is about 10^5 dyne$\cdot$cm^2. In the surface layer the magnetic pressure is too strong and hence it pushes the surrounding material outwards; hence in the equilibrium state the lines of force curve outwards.

3.2.2 Sunspot Cycles

Systematic observations of sunspots for over two hundred years have revealed a number of striking properties of sunspots and their cycles. It is found that the number of sunspots on the solar disc reaches a maximum every 11.2 years. It is to be noted that sunspots occur only in two zones of the solar disc - in the latitude bands of 5° to 40° in the northern and southern hemispheres of the Sun. With the solar rotation, a particular sunspot returns to the same position of the disc after about 27 days which is the synodic rotation period of the Sun as seen from the earth. The mean value of the true rotation period of the Sun, corrected for the motion of the earth is 25.4 days. The sunspots which provide a convenient way of measuring the rotation period of the Sun, show that the solar rotation is nonuniform and a function of latitude; it is about 25.0 day at the equator - increasing to 27.4 days at a solar latitude of 40°.

At the beginning of a sunspot cycle at the solar minimum there are very few or no spots on the solar disc for several days or weeks and the cycle begins with the appearance of a few small spots generally in the mid-latitudes of 30°-40°. As the solar cycle advances, the sunspots increase in number as well as in the complexities during the next 5.5 years or so, till the maximum sunspot number is reached, when there may be as many as 50 to 150 sunspots on the solar disc on a particular day. These large sunspot groups and their complex magnetic field configurations give rise to occasional explosive events of solar flares when the emission of large amounts of electromagnetic radiation, energetic particles and plasma clouds take place. These phenomena will be discussed later. Turning to the sunspot numbers, we note that after the maximum, the number of spots decreases over the following years till the minimum number of sunspots is reached after about 11 years from the beginning of the cycle. It is found that at the sunspot maximum, the mean position of the spots shifts to a lower latitude of about 20° (i.e. 10°-30°) and at the sunspot minimum it reaches the lowest latitude region of 5°-10°. The next cycle begins with sunspots appearing at the mid latitude.

R. Wolf of Zurich Observatory succeeded in standardising the counting procedure of sunspots and sunspot groups and also in collecting the sunspot observations since the time of Galileo. He established the 11.1 year period of the sunspot cycle and introduced the corrected number of spots called the Relative sunspot number, which is also called the Wolf number , is used at present. Another method of using sunspots to determine the level of solar activity is to measure the total area occupied by the sunspots - both umbra and penumbra - as done at Greenwich and other Observatories. The total spot area is expressed in millionth of the Sun's visible hemisphere.

In Figure 3.2.1 we have shown the sunspot cycles during 1610-1976 for which the annual sunspot number (Wolf number) has been plotted for each year since the invention of the telescope. The earlier data for 1610-1730 is less reliable than the data from 1730 onwards. The relationship of the sunspot cycles with several terrestrial and geophysical phenomena are now well established; some of these will be discussed later.

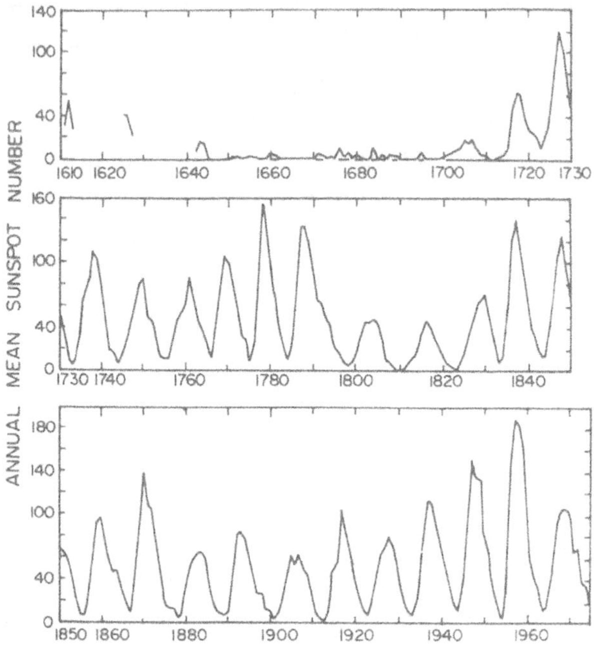

Figure 3.2.1: Sunspot cycles from 1610-1976

3.3 Solar Wind

3.3.1 Early Concepts

The steady corpuscular emission from the Sun, called the solar wind plays a major role in the interplanetary dynamical processes involving plasma and magnetic fields. We first mention some of the early concepts. The idea of corpuscular emission from the Sun was suggested by Birkland of Sweden in 1896 in order to explain the polar aurora. But the phenomena were not understood until the middle of this century and the advent of the space age. In 1931 Chapman and Ferraro(1) proposed the theory of magnetic storms for the emission of corpuscular clouds from the Sun which travel outward with a velocity of 1000-2000 km/sec. In 1950 Alfven(2) showed that ionised gas clouds emitted by the Sun carry 'frozen-in' magnetic fields.

In 1951 an idea of major importance was proposed by Biermann(3) of

Germany who pointed out that comet tails of type I which comprise of molecular ions always point away from the Sun, both in inward and outward journeys. This he showed, needs a very high force which can only be provided by the corpuscular radiation from the Sun. Assuming the explanation as correct, it implied a continuous emission of particles from the Sun. He established that the flux of particles from the Sun should be $\sim 10^9$ particles/cm^2 and the density $\sim$ 50 particles/cm^3. The modern theory of solar wind was formulated by E.N. Parker of the University of Chicago in 1958.

3.3.2 Parker's Theory of the Solar Wind

Parker developed in detail the mathematical theory(4,5) which envisaged the continuous emission of energetic protons and electrons from the Sun; this is basically the outward expansion of the hot ($\sim 10^6$ K) corona into the interplanetary medium. He first considered the hydrodynamical equations for a hot plasma in the gravitational field of the Sun and showed that in the steady state, the corona must expand continuously with the general outflow of matter from the Sun and this is called the solar wind. He showed that because of the very high conductivity of the solar corona, the transport of energy by thermal conduction outwards through the corona can be carried away only by the expansion of the outer corona to a supersonic solar wind. These predictions have been verified in space experiments and this constitutes one of the major milestones in the advancement of space physics.

Parker calculated the expansion velocity i.e. the velocity of the solar wind as a function of distance from the Sun for various values of the coronal temperature. This is shown in Figure 3.3.1 for an isothermal corona with the polytropic index, $\alpha = 1$. It is seen that for the typical temperature of the corona of $0.75 \cdot 10^6$ K, the solar wind velocity at the orbit of the earth is 400 km$\cdot$s^{-1}. Also the velocity of the solar wind is low in the beginning and it becomes higher as the distance from the Sun increases with a decreasing gravitational pull and the velocity attains an asymptotic value at distances beyond about 10^8 km from the Sun. For the high value of the coronal temperature of $4 \cdot 10^6$ K, the solar wind speed at the orbit of the earth becomes 1150 km$\cdot$s^{-1}.

In Parker's theory the isothermal corona with polytrope index $\alpha = 1$

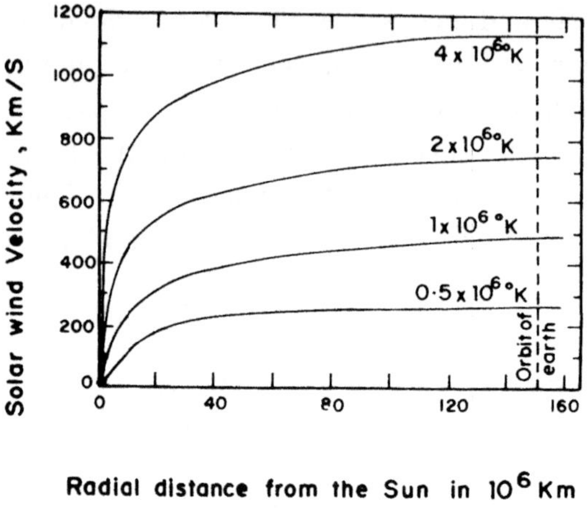

Figure 3.3.1: Solar wind velocity vs distance from the Sun according to Parker's theory[4]

represents an approximate upper limit on the coronal expansion. The heating of the corona is given by the amount by which the polytropic index α is less than the adiabatic value 5/3, so that the temperature of the coronal gas decreases outward as $N^{\alpha-1}$ instead of $N^{2/3}$. For $\alpha = 1$, thermal conduction vanishes so that the temperature is maintained by the dissipation of shock waves. In the actual corona α is somewhat greater than 1.0, probably close to 1.1. The observed temperature gradient of about 3 K/km in the lower corona indicates that coronal heating is sufficient to maintain α close to 1.1. Parker presented models of the solar wind for α = 1.0, 1.1 and 1.25. The effect of increasing α above the isothermal value is discussed by him. The calculated values of solar wind velocities, V_s and solar wind densities, ρ_s at the orbit of the earth are as follows. For $\alpha = 1.1$ and $T_o = 1.77 \cdot 10^6$ K at the reference level r = a, the value of V_s is 400 km·s^{-1} and ρ_s is 40 particles cm^{-3}. These are in good agreement with the range of measured values.

In constructing a mathematical model appropriate to the solar corona, several points are to be noted as pointed out by Parker. The solar corona in quiet times is not uniform in latitude and longitude. The general magnetic field of the Sun is indicated by the coronal streamers; these suggest that the coronal expansion is channelled and directed by these fields. Hence it is only appropriate to consider stationary expansion and the flow of the corona along one given streamline only at a time. The streamline begins from the base of the corona, where the flow velocity is very small, to large distances in space

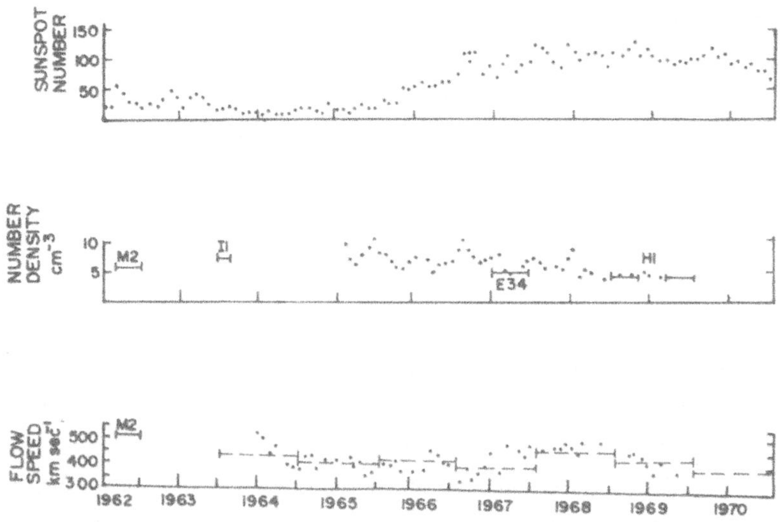

Figure 3.3.2: Solar wind speed and density, and the sunspot number vs year, 1964 to 1970[70]

where the hydrostatic pressure is almost zero. Thus, the flows along adjacent streamlines will depend on the temperature, density and the divergence of the field lines and need not be the same in general.

3.3.3 Observations of the Solar Wind

The first direct measurements of the solar wind were made with the Russian spacecraft Lunik-I and Lunik-II by Shklovskii[6] et al. in January and September 1959 when the Sun was very quiet. They used an ion trap detector to obtain a steady flux of protons of $2{\cdot}10^8$ particles/cm^2. The proton velocities were ≥ 60 km$\cdot$s^{-1} and the solar wind density was < 30 particles$\cdot$cm^{-3}. Soon afterward the M.I.T. group of the USA (Bridge et al.[7]) used an improved plasma probe in NASA's Explorer X satellite in 1960-61 and looking towards the quiet Sun they measured a solar wind speed of 250-400 km$\cdot$s^{-1} (0.3-0.8 keV per proton) and a solar wind density of 7-20 particles$\cdot$cm^{-3}. These yielded the flux of the solar wind as about $5{\cdot}10^8$ protons$\cdot$cm$^{-2}\cdot$sec^{-1}. In 1961, Mustel[8] reported further results from the Russian experiments. The NASA spacecraft Mariner II spent a considerable time outside the magnetosphere in November-December 1962 on its journey to Venus and from these observations Snyder[9] (1963) found that the solar wind varied from day to day and there was no correlation between solar wind velocity and solar activity, (measured by sunspot number or the 10.7 cm radiation from the Sun). In subsequent years systematic measurements were carried out in many spacecraft - specially in the IMP series - and new properties of the solar wind e.g. its composition, ionization states etc. were determined.

A. Bulk Properties of the Solar Wind

A decade of observations of the solar wind established several bulk properties of the solar wind. The measured variations of several properties of solar wind at the orbit of the earth for 1964-70, as reported by Hundhausen[10] in 1975, are shown in Figure 3.3.2 and are compared to the sunspot numbers from 1962 to 1970. These plots show several features of major importance. Firstly the flow speed which varies from day to day, is shown for a 27 day solar rotation period. The yearly average values of flow speed are shown by the dashed lines. It is seen that for 1963 to 1970, the solar wind speed does not depend on the sunspot number and the mean values of the solar wind speed during the period of solar minimum in 1964 and the solar maximum of 1969 are essentially the same. Similarly the density of solar wind and the energy flux are also independent of the solar activity as given by the sunspot numbers from 1965 to 1970. The implications of these observations on the 11 year solar cycle modulation of cosmic rays are discussed later.

B. Composition of the Solar Wind

Measurements of the composition of solar wind particles and their variation with time are important from several respects. (i) Abundances of several ions of different mass numbers and mass to charge ratios provide information on solar wind dynamics in the source region. (ii) Solar surface abundances can be estimated from the solar wind composition. Such information is particularly important for the measurement of the isotopic composition for which spectroscopic data is very limited. (iii) The abundances of noble gases e.g. He, Ne, Ar etc. are very different in the Sun and planetary objects. Hence the composition measurements can be used to trace the solar matter in the planetary system.

H and He: Next to hydrogen, helium is the most abundant element in the Sun and in the solar wind. The He/H ratio in the solar wind is a highly variable quantity, both on short and long time scales. The long-time average of the He/H abundance ratios measured[11] over one solar rotation of 27 days during the period of July 1965 to July 1967 indicate large irregular variations of the ratio in the range 0.2 to 0.65. During this two year period, the long-term average lie between 0.37-0.50. The abundance ratos are found to be weakly correlated to the sunspot numbers during the two year period. On short time scales, the very high He/H ratios of 0.2-0.3 were observed in the solar wind after solar flares. The highly enhanced ratios[12] lasted for about 15 min to 3 hrs.

The abundance of helium in the photosphere is difficult to determine spectroscopically. However, estimates based on solar flare particle abundances, helium lines in prominences and abundances in hot stars or in gaseous nebulae yield a He/H abudance ratio of 0.65 ± 0.15. Therefore it seems that the average He/H abundance ratio of 0.37-0.50 in the solar wind is lower than the photospheric value by a factor of 1.5 to 2. indicating the depletion of helium relative to hydrogen in the solar wind by the above factor.

Oxygen to Fe: Using the energy/charge analyser detector flown in the Vela spacecraft, Bame et al.[13,14] measured the ionic composition of heavy ions of the solar wind e.g. oxygen, silicon and iron (Figure 3.3.3). It is seen that O^{+6} is the dominant component, while silicon and iron occur in various states of ionization.

He, Ne and Ar: Noble gases are studied by a solar wind foil collecting technique used by Geiss[15] in NASA missions and by Boltenkov et al.[16] in Russian space missions. In these experiments specially treated 'clean' metal foils such as of aluminium were exposed to the interplanetary space, such as on the lunar surface, in the Lunar space missions for the available exposure times and then brought back to Earth. Solar wind ions, especially of noble gases, which were implanted in the foil were subjected to vacuum extraction and mass spectrometric analysis. Using high sensitivity mass spectrometers in the laboratory the elemental and isotopic noble gas composition of the solar wind was determined. In Figure 3.3.4, the measured solar wind abundances of some of the major elements are shown and these are compared with the solar system abundances given by Cameron. It is seen that heavy elements

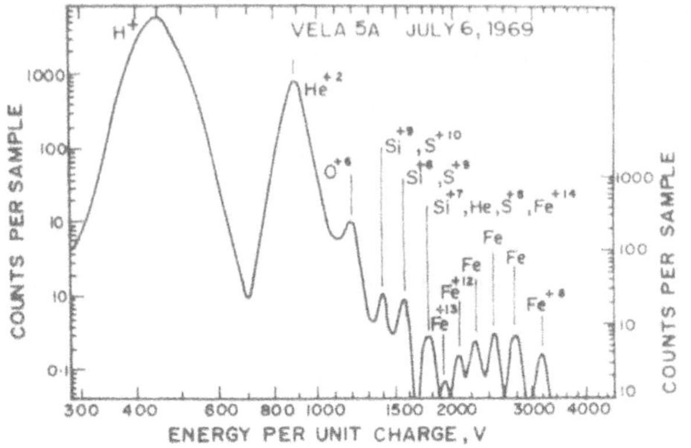

Figure 3.3.3: Ionic composition in the solar wind[14]

upto about iron are not systematicaly depleted in solar wind. This is notable because the He/He ratio is highly variable and He is depleted relative to H in the average solar wind.

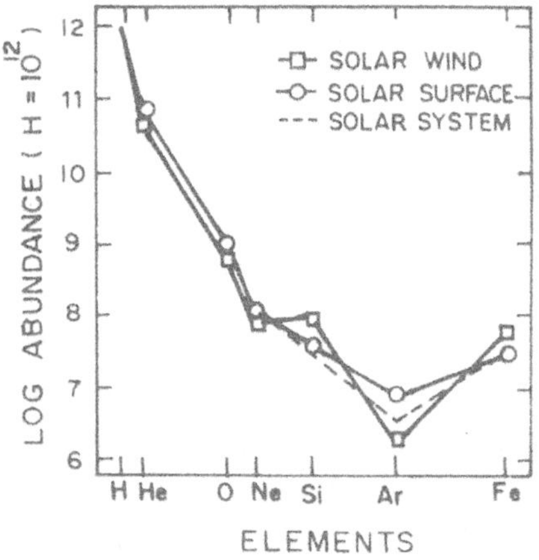

Figure 3.3.4: Abundances of several elements in the solar wind, solar surface and in the solar system

C. Isotopic Abundances

Turning to the isotopic composition of solar wind particles, we show in Table 3.3.1 the summary of results from the review by Geiss[15] including the

Table 3.3.1: Isotopic Abundance Ratios in the Solar Wind*

Ratio	Solar Wind
He^4/H	$0.045 \pm 0.008^{(a)}$
He^3/He^4	$(4.3 \pm 0.3) \times 10^{-4(b)}$
Ne^{20}/He^4	$(1.75 \pm 0.12) \times 10^{-3}$
Ne^{22}/Ne^{20}	0.073 ± 0.002
Ne^{21}/Ne^{22}	0.033 ± 0.004
Ne^{36}/Ne^{20}	0.036 ± 0.011

*Adopted from the review by Geiss (ref. 15).
(a) Highly variable ratio. Long term value averages between 0.037-0.051.
(b) From Apollo 11-16, and Zond 8.

Apollo missions of NASA and Zond mission of the USSR. It is seen that the He^4/He^3 ratio is consistent from mission to mission and the average value of He^4/He^3 was 2350 ± 120 i.e. $He^3/He^4 = (4.3 \pm 0.3) \cdot 10^{-4}$. The He^4/Ne^{20} ratio in the solar wind is systematically lower than the solar system value by a factor of about 1.5. Ne^{20}/Ne^{22} in the solar wind is 13.3 ± 0.3. This will be compared later with solar flare particles and solar system values as these have some important implications.

Extensive studies of the solar wind as well as of solar flare particles entrapped in Lunar surface materials have been made and are discussed in Chapter 4, about the Moon and the Interplanetary medium.

3.4 Solar and Interplanetary Magnetic Fields

3.4.1 Observations of Interplanetary Magnetic Fields

Early conjectures on the solar and interplanetary magnetic fields were based on the observations of cosmic rays during the 1940's. The supposed low energy cut-off of cosmic rays was interpreted as due to the presence of a dipole magnetic field of the Sun which excluded low energy particles from the inner solar system. This explanation was however found to be invalid. In 1955 Davis proposed that solar corpuscular radiation, earlier suggested by Alfven, sweeps out the prevailing field so that the Sun and the inner solar system are embedded in a field-free cavity. This idea was also found not tenable.

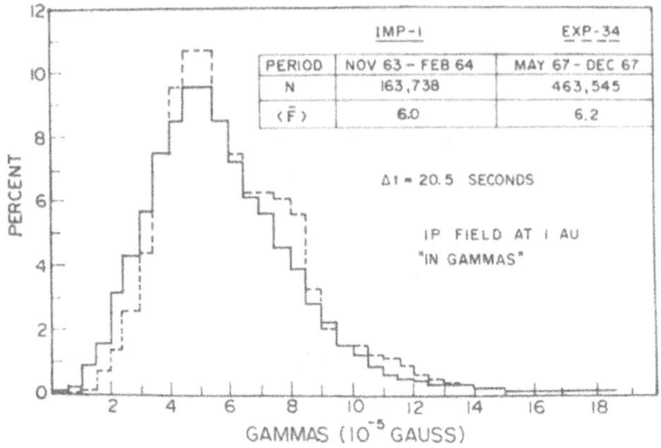

Figure 3.4.1: The histograms of the magnitudes of the interplanetary magnetic field measured by IMP-1 (1963-64) and Explorer-34 (1967)

Magnitude, Direction and Spiral Structure

With the advent of space age magnetometers were flown in satellites (chapter 2.2), and yielded the first remarkable results on interplanetary magnetic fields. These were measured for the first time in a systematic manner by Ness at al.(17) in the IMP-1 satellite of NASA, USA, launched on 27 November 1963. In the very first orbit of IMP-1, excellent data was obtained on geomagnetic and interplanetary fields as given earlier in Figure 2.2.3. It is seen that the interplanetary magnetic fields beyond 16.8 earth radii were very steady having a mean value of $\sim$ 5 gamma (1 gamma $= 10^{-5}$ Gauss). Its direction θ was, close to zero i.e. in the ecliptic plane and the angle ϕ was about 135° from the Sun (for the definition of θ and ϕ, see Figure 2.2.1). Since then extensive measurements were carried out in satellites. The distributions of the magnitudes of the interplanetary magnetic fields measured at the orbit of the earth with IMP-1 by Ness(18) in November 1963 to February 1964 are shown in Figure 3.4.1, together with those from later measurements with Explorer 34 by Fairfield(19) in May 1967 to December 1967. Several important features which come out of these data are as follows: (a) the average magnitude, F, of the interplanetary magnetic field at 1 Astronomical Unit (A.U) from the Sun i.e. at the orbit of the earth is 6 gamma; (b) there is no change in the magnitude of the field from a low level to the increased level of solar activity in 1967; (c) the magnitude is very rarely less than 0.5 gamma, so that a finite magnetic field is always present in the interplanetary medium.

The directional distributions of the magnetic fields measured with the Pioneer 6 satellite during December 1965 - September 1966 by Burlaga and Ness[20] are shown in Figure 3.4.2 (a) and (b). In (a) the directional distributions of the hourly average values in the plane of the ecliptic are shown and those in the perpendicular plane in (b). The striking features are that the magnetic field was directed approximately equally along $\phi = 135^o$ or 315°, which, we shall see presently, correspond to theoretically predicted ones. The field was essentially in the ecliptic plane, having a very small south ward component as shown in (b). Similar patterns were observed in other measurement. For example with Pioneer 7 during August 1966 to February 1967 the ϕ distributions showed the peak along 135° with 33% positive values and along 315° with 58% negative values of the field as shown in Figure 3.4.2. The θ distribution showed a small northerly component.

In addition to the magnetic field direct measurements of the solar wind plasma emanating from the Sun were carried out at the orbit of the earth. It is well established, from the plasma and field measurements, that the expansion of the solar corona into the interplanetary medium extends the solar magnetic field with it, as the plasma is highly conducting. The rotation of the Sun causes the twisting of the solar magnetic field in the classical Archimedean spiral. The geometry of the interplanetary magnetic field at 1 A.U. from the Sun i.e. at the orbit of the earth are shown in Figure 3.4.3(a) and (b). It is seen in the sketch of 3.4.3 (a) of the magnetic field lines in the ecliptic plane that the solar wind is radial from the Sun along the direction OP, where P is the position of the earth. PQ is tangent to the magnetic field line at P and is the direction of the field at P. As the solar wind travels radially along the sun-earth line and the magnetic field line is anchored at the Sun, the rotation of the Sun causes the magnetic field line to be twisted in the form of an Archimedean spiral. At the orbit of the earth at P, the angle ϕ, substended by the magnetic field direction with the Sun earth line, OP, is given by

$$\phi = 90^o + \phi$$

$$= 90^o + tan^{-1}\left(\frac{V_s}{\omega \cdot r}\right) \qquad (3.4.1)$$

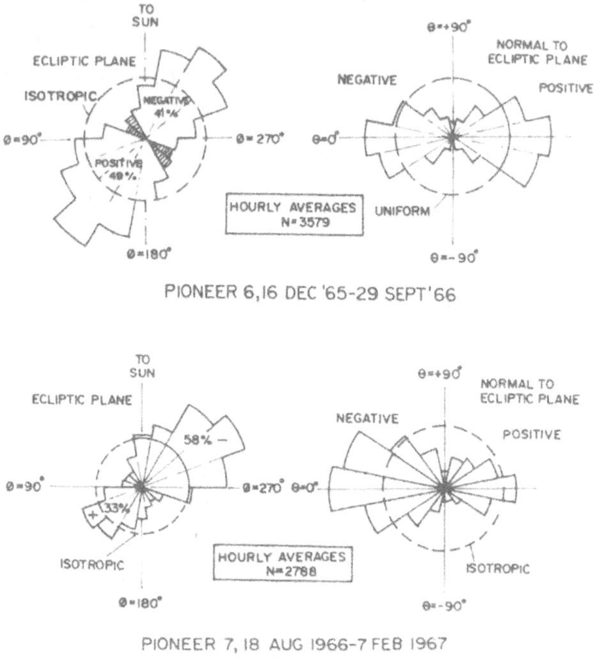

Figure 3.4.2: Directional distributions of the interplanetary magnetic field in Pioneer 6 (upper figure) in the ecliptic plane and normal to the ecliptic plane, and in Pioneer 7 (lower figure)

were V_s is the velocity of the solar wind, typically about 400 km$\cdot$s^{-1}, ω is the angular velocity of the rotation of the Sun at 10°-20° solar latitudes, and r the Sun earth distance ($1.5\cdot10^8$ km). Substituting the numerical values we get $\phi = 90^o + 45^o = 135^o$ for the positive magnetic field i.e. the field directed away from the Sun. The corresponding angle ϕ' is $180^o + 135^o = 315^o$ and is applicable for the negative field i.e. the field directed towards the Sun. Figure 3.4.3 (b) shows the sketch of the formation of the spiral field as solar wind with a velocity of 400 km$\cdot$s^{-1} travels in 5 days from the Sun to the orbit of the earth.

Thus we find that the theoretically predicted magnetic field direction of 135° (or 315°) and the experimentally measured values are in excellent agreement with each other. Hence the theoretical ideas on the solar wind

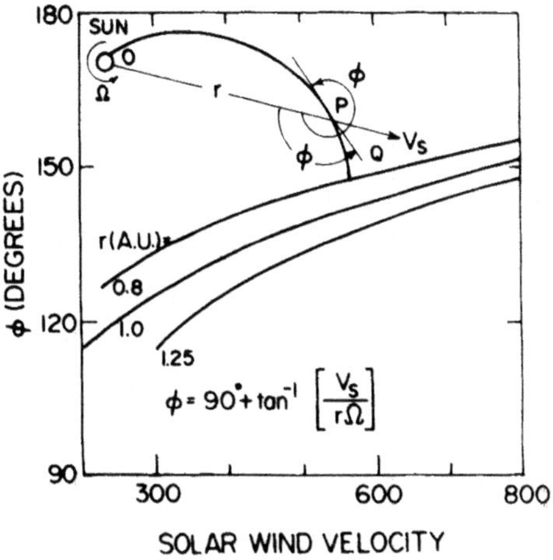

Figure 3.4.3(a): Theoretical streaming angle of 135° of the solar-interplanetary magnetic field at 1 A.U.

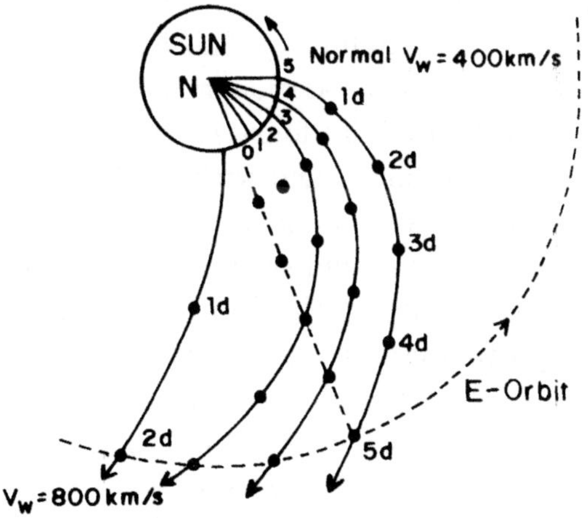

Figure 3.4.3(b): A sketch of the formation of a spiral field as a solar wind with a velocity of 400 km·s^{-1} travels to the earth's orbit in 5 days

and the interplanetary magnetic field are established on a firm foundation.

In the plane normal to the ecliptic the observation of a small component either in southern or northern directions is due to a small angle between the rotational axis and the magnetic axis of the Sun.

3.4.2 General Field of the Sun and Solar Plasma

The striking features of the solar magnetic fields were provided by continuous measurements of the fields and their polarities for several solar rotations. These are shown in Figure 3.4.4 as measured with the IMP-1 satellite by Ness and Wilcox(21). Successive three hour averages of the polarity of the interplanetary magnetic field at 1 A.U. are shown for three solar rotations of 27 days during November 1963 to February 1964. It is seen that the polarity of the field direction repeated itself in a period of 27 days in a remarkable manner. The distortions in the pattern for orbit 1 and 2 of IMP-1 were due to a magnetic storm on December 2, 1963. The orbital period of IMP-1 was 3.9 days. The sector structure of the interplanetary magnetic field was thus established for the first time. During this period of observations when the solar activity was at minimum level, four sector structures were present. The question arises as to the duration of the observed sector structures. Observations by the IMP-2 satellite by Fairfield and Ness(22) in late 1964 indicated the existence of a quasi-stationary four-sector structure observed earlier by IMP-1. A break up of this four-sector pattern was detected in 1964 by Mariner 4 by Coleman et al.(23) and with IMP-3 by Ness and Wilcox(24). The recurrence period was also found to shift from 27 to 28 days. The differential rotation of the Sun would suggest that the source of solar plasma shifted from 10°-15° N latitudes of the Sun for IMP-1 to higher latitudes of 20°-25°. In 1970 only two sector structures were observed by the Marine 10 spacecraft; these two-sector structure were maintained during several solar rotations. In addition to the observation of interplanetary sector structure at 1 A.U., Mariner 10, on its way to Mercury showed that the sector structure extended throughout the interplanetary space from 1 to 0.5 A.U.

The study of the origin of the interplanetary sector structure was carried out by Ness and Wilcox. They used solar photospheric magnetic field measurements, which were grouped into bins with widths of 10° in latitudes and 5.7^o in longitudes which correspond to time intervals of 12 hours.

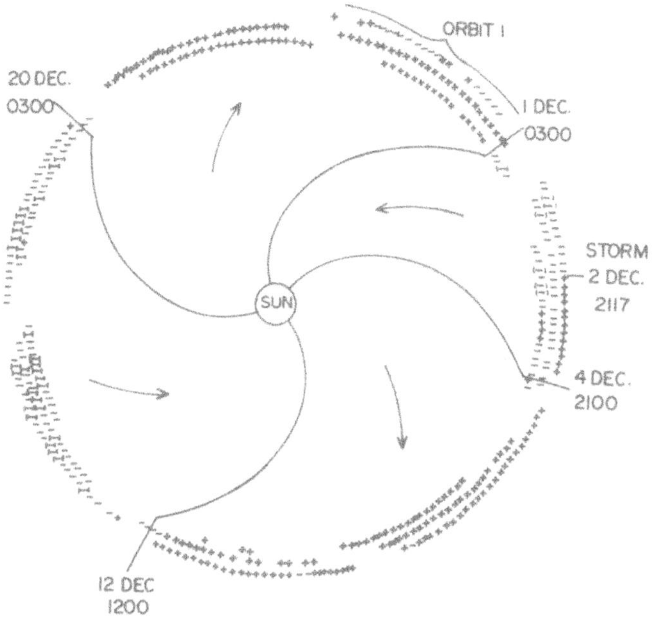

Figure 3.4.4: Sector structures in the polarities of interplanetary magnetic field

Cross-correlations were made between the directions of the photospheric magnetic field and the interplanetary field for solar magnetic fields at three latitudes, 0°, 15° N and 15° S by Ness and Wilcox[25], who found a well defined weak correlation with a time lag of 4.5 ± 0.5 days. These studies conclusively indicate that the interplanetary magnetic field originates from the photospheric field pulled out by the solar wind plasma, 4.5 days being the average travel time of solar plasma to the earth. This agrees very well with the direct measurements of the solar wind in the same satellite.

The observations show the evidence of a quasi stationary pattern in the sector structure of the direction of the magnetic field which forms a co-rotating structure. It should be noted that this co-rotation is not a rigid rotation of the interplanetary medium with the Sun but an apparent co-rotation associated with long lived solar plasma and the magnetic field.

A theoretical model of the interplanetary field direction and the photospheric magnetic field was developed by Schatten[26]. He used the measurements of photospheric field structures and polarities to deduce the closed and open field lines and extended these to interplanetary fields and polarities. Reasonable agreements with observations were obtained.

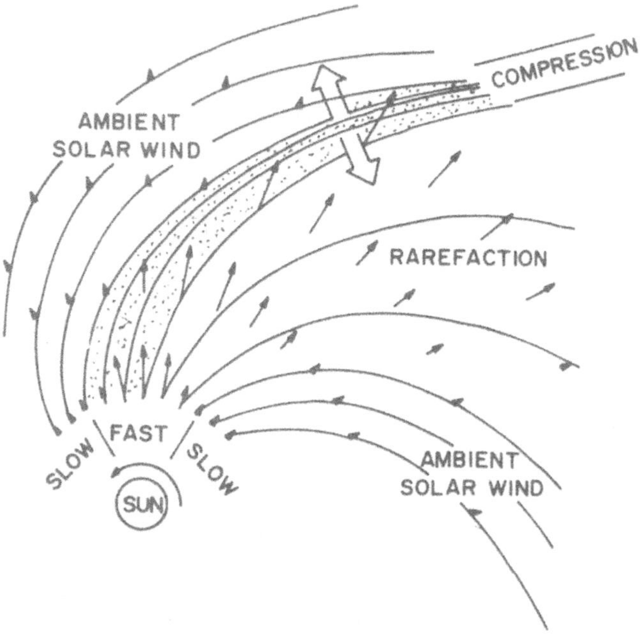

Figure 3.4.5: A sketch of the corotating stream structures formed by the interaction of high speed solar wind streams with slow speed ones resulting in the compression and rarefaction regions of interplanetary magnetic fields originated from the Sun

Interactions of Fast and Slow Wind Streams: Co-rotating Streams

Interplanetary magnetic field structures show effects of the interactions of a high speed solar wind with a slow speed wind as shown[27] in Figure 3.4.6. The high speed wind is emitted by the Sun from a long lived source, such as a coronal hole (3.4.3) extending over a fairly large longitudinal interval. Because of the Sun's rotation, fast plasma overtakes slow plasma ahead of it, which was emitted from a more westerly direction. As a result, the plasma at the interface between the fast and slow streams is compressed. As the magnetic field is "frozen" in the highly conducting plasma (3.4.4), it is also enhanced at the leading front. Temperature is also increased leading to the formation of a pressure wave. Behind the forward shock-front, a rarefaction or reverse shock takes place and the shock front and the compression region together with the enhanced field corotate with the Sun with a period of 27 days. The recurrent, corotating streams with a period of 27 days have been observed at 0.3 A.U., 1 A.U. and beyond with satellites and space probes.

The "recurrent interplanetary streams" having a period of 27 days have been identified as the source of the 27 day variation in cosmic ray intensity, which has been observed for a long time. The observations of recurrent geomagnetic storms are now understood in terms of co-rotating recurrent streams. The 27 day modulation of cosmic ray intensities and its correlation with recurrent solar wind streams have been observed by many investigators. In the observations made by Uicci et al.[(28)] in which daily average solar wind speeds at the earth have been plotted in 27 day solar rotation cycles, it was shown that two long lived and high speed solar wind streams appeared on the 3rd and 15th day i.e. 12 days apart and were present for almost five solar rotations. In the period galactic cosmic ray intensities as measured by a ground level neutron monitor (see Chapter 7.2) were compared. It was shown that the 27 day cosmic ray intensity variations are produced by the recurrent, long lived solar wind streams and the cosmic ray intensity is anti-correlated with co-rotating stream speeds. The high speed streams with enhanced magnetic fields and shock fronts at the earth seem to deflect some of the cosmic ray particles; as a result cosmic ray intensity decreases, whereas at low wind speed days at the earth cosmic ray intensity recovers to nearly normal value. Therefore 27 day variations of cosmic ray intensity are also called the "Co-rotating Forbush decrease" in analogy to Forbush decrease (Chapter 3.5).

Beyond 1 A.U. co-rotating interaction regions of solar wind streams propagate to larger distances. Continuous measurements with the IMP-7 satellite at 1 A.U., simultaneously with the Pioneer-10 spacecraft traveling to Jupiter, showed that high speed solar wind streams measured by IMP-7 on 26-27 August 1973 reached Pioneer-10 at 4.65 A.U., about 10 days later on 6-7 September and the solar wind streams had undergone damping of amplitude and fluctuations. It is generally presumed that the shock front and the interaction regions co-rotating with the Sun travel to about 10 A.U. or so.

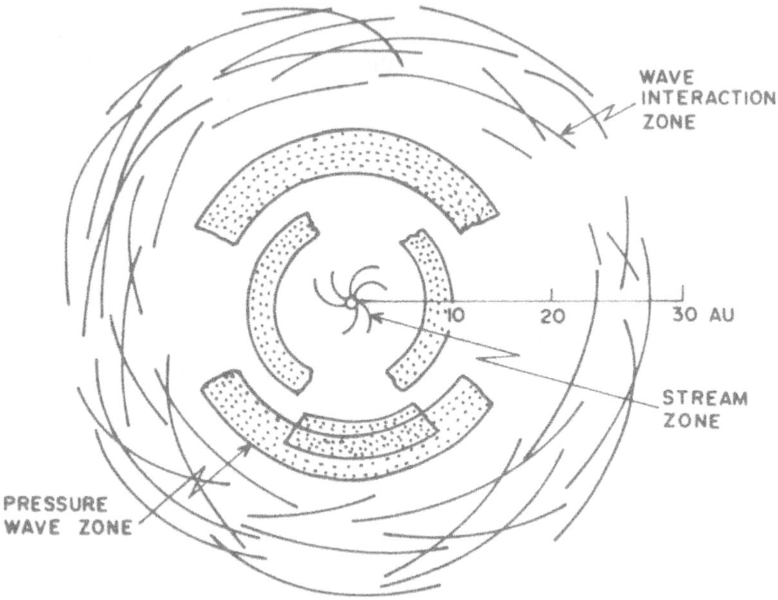

Figure 3.4.6: A sketch showing the evolution of solar wind interactions in the heliosphere which can be described by three zones, (a) the stream zone up to ~ 10 A.U., (b) the pressure wave zone, ~ 10-25 A.U., and (c) the wave interaction zone at > 25 A.U.

The observations of co-rotating streams with a number of spacecraft and theoretical models have given us the following picture of the phenomena in the heliosphere, which is illustrated in Figure 3.4.6. At large radial distances from the Sun the amplitudes of stream speeds diminishes while the magnetic fields are enhanced relative to ambient fields. Further away from the Sun co-rotating streams are eroded and are replaced by co-rotating pressure waves which become dominant feature in the middle regions of the heliosphere. The pressure waves expand and beyond ~ 25 A.U. the pressure wave associated with one stream would interact with those from following streams, so that individual interaction regions can not be distinguished. The merging of the interaction regions would thus result in a "wave interaction zone". This "wave interaction region" is also characterized by irregular magnetic field regions formed by the merging of magnetic field structures. This zone as shown in Figure 3.4.6 forms a kind of shell of irregular magnetic fields through which galactic cosmic rays must travel. Therefore the cosmic ray intensity reaching the earth is reduced as compared to that outside the heliosphere.

This produces the modulation of cosmic rays as discussed in Chapter 7.4.

Thus in the stationary conditions of the normal Sun, the co-rotating streams and magnetic field structures are divided into three regions in the heliosphere: (1) The co-rotating streaming zone near the Sun, say up to 10 A.U.. (2) A pressure wave zone in $\sim$ 10-20 A.U. and (3) a wave interaction zone in the region at $\sim$ 25-30 A.U.. This model was postulated by Burlaga et al.[29] in 1983.

3.4.3 Coronal and Heliospheric Magnetic Fields

Coronal Holes

As discussed in the previous section, we note that a number of re-current events such as 27-day period magnetic storms and 27 day period cosmic ray variations are now understood from the recurrent co-rotating streams formed by the interactions of long lived fast solar wind streams with the slow ones. Now we look into the origin of fast solar wind streams from the corona which remained elusive for a long time.

Measurements made with rocket-borne X-ray telescopes looking at the Sun during 1963-69 revealed[30] a number of major discoveries in solar physics including observations of "coronal holes". These phenomena were studied in great detail with the X-ray telescopes on the Apollo Telescope Mount of the Skylab mission of NASA in 1972-74 and their new properties were established. When an X-ray telescope, like those used in the Skylab mission, sensitive to wavelengths of 2 to 80 Å is pointed to the Sun it does not see the cooler regions of the photosphere and chromosphere, but only the hot corona whose temperature exceeded about 10^6 K, because hot gases at these high temperatures only emit X-rays. This offers a great advantage to astronomers, as with X-ray telescopes they can see the corona without the necessity of blocking out the solar disc with a coronograph. Thus X-ray telescopes can see the corona all over the Sun, and not just at the edges as during a total solar eclipse.

An X-ray photograph of the Sun taken by one of Skylab's X-ray telescopes is shown on Plate 3.4.1. This spectacular photograph illustrates several new features of the solar corona for the first time in great detail. We see bright knots at the edges and on the disc which represent very hot gases trapped in the loops of strong magnetic fields of the active regions of the Sun. Bright arches connect active regions. The X-ray corona is seen as the glowing ring of a very hot X-ray emitting gas which around the Sun. The most energetic hard X-rays come from hot regions of 3 to $5{\cdot}10^6$ K and softer X-rays come from slightly cooler regions of 2 to $3{\cdot}10^6$ K.

We see the new features of the huge dark patch extending from the solar north pole to the equatorial region and further south. These extended regions in the corona are called Coronal Holes where the temperature and density of the gas are unusually low. In these coronal holes the solar magnetic field lines which are normally arched and closed, open out from the Sun in nearly radial directions. The coronal holes are not readily visible in the photosphere or chromosphere and are seen only in X-ray and ultra-violet photographs. Coronal holes which are most striking in X-ray photographs of the Sun were studied in detail in the Skylab mission and several hundred photographs were obtained[(31)]. From these and other observations a number of features of coronal holes were established as follows: (a) Coronal holes are the sources of high speed solar wind particles. Because of the convection of heat from the corona by these high speed solar wind particles, the coronal holes appear as relatively cooler dark patches. (b) In the regions of coronal holes, solar magnetic field lines have open configurations. (c) Coronal holes rotate almost in the same manner as the Sun and the coronal holes persist with changing features over several solar rotations. When observed during three months, it is found that the coronal holes are encroached upon by neighboring coronal arches and are slowly closed, leaving only the persistent hole at the north pole. (d) Coronal holes are enduring features at both poles of the Sun as seen for 9 months in the Skylab mission. (e) The temperature in the coronal hole is about 10^6 K, 2 to 3 times lower than in active regions and about 50% lower than the surrounding 'quiet' areas. (f) The density is about three times lower than in the neighboring corona. (g) Polar plumes originate in coronal holes at the polar region of the Sun and are rooted in bright emission points.

Heliosphere Magnetic Field and the Neutral Sheet

The observation of sector structure of the solar magnetic field as two sector or four sector features and the evolution of the sector boundaries could not be understood until the concept of heliospheric current sheet and neutral

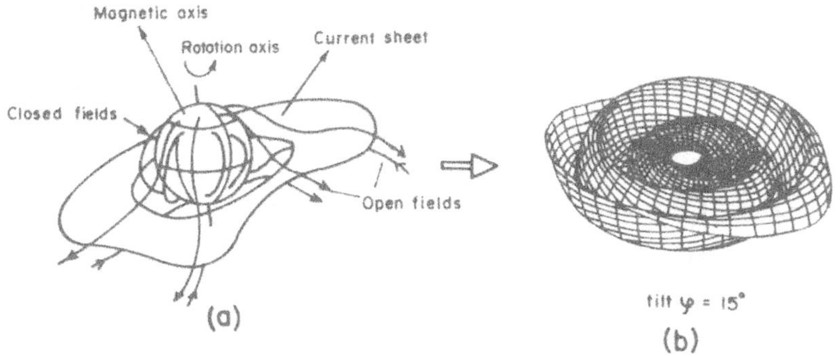

Figure 3.4.7: (a) A sketch of the warped neutral plane and the current sheet which is inclined to the rotational axis of the Sun. (b) A sketch of the wavy neutral sheet in the heliosphere produced by the rotation of the Sun.

plane, analogous to that of the earth's magnetic tail separating regions of opposite magnetic polarity were established. Observational and theoretical studies established these phenomena as follows. As the solar dipole magnetic axis is inclined to the rotational axis of the Sun, the radial flow of solar wind transports the solar magnetic field into interplanetary space and lead to the formation of a neutral sheet as shown in the sketch in Figure 3.4.7(a). The complicated curved surface of the current sheet and neutral plane, instead of a plane surface, arises because the solar magnetic field is not a dipole one, but has a more complicated form. The field lines originating in the north polar region of the Sun lie above or north of the neutral plane current sheet whereas field lines originating from the southern hemisphere lie below the neutral plane. The field lines have their usual spiral structure. As the current sheet and the neutral plane around the Sun, is inclined to the solar equator, the rotation of the Sun generates a wavy structure of the neutral plane as shown in the sketch in Figure 3.4.7(b) for a tilt angle of 15°. These features were described in the review by Smith and Wolf[(32)] in 1979. Burlaga et al. observed that the changes in the configurations of the neutral sheet between a few solar rotations were correlated with the change in sector structure from 2 to 4 sectors.

The wavy neutral sheet configuration in the heliosphere is supported by simultaneous measurements of cosmic ray particle intensities by several spacecraft in interplanetary space. As cosmic ray particles from the galaxy are guided by the large scale magnetic field structures in the heliosphere, it was found that when two or three spacecraft located at nearly the same radial distance from the Sun, but at different latitudes from the ecliptic plane, a latitudinal gradient in the cosmic ray particle intensities. Furthermore it was observed that the sign of the latitudinal gradient changed from positive to negative (or vice versa) when the positions of the spacecraft changed (due to solar rotation) from above the neutral plane to below the neutral plane. The transport of cosmic ray particles in the heliospheric magnetic field and the modulation of its intensities are discussed in Chapter 7.4.

3.4.4 Magnetohydrodynamics and Solar Plasma

Introduction to magnetohydrodynamics

It is now well known that in cosmical physics such as plasma processes in the earth's magnetosphere, in the solar atmosphere and in interplanetary medium etc., electromagnetic processes play a major role. For the understanding of these phenomena classical hydrodynamics had been combined with electrodynamics to form a branch called magnetohydrodynamics. In regions of space such as the interplanetary medium, electromagnetic processes dominate over gravitational forces. As an example, let us consider a hydrogen atom moving in the orbit of the earth with earth's orbital velocity V. The gravitational force of the Sun on the hydrogen atom of mass m, is given by

$$\vec{F}_1 = GM_\odot m\vec{R}/R^3,$$

where G is the gravitational constant, $M_\odot$ the solar mass, R the Sun earth distance. If the atom is ionized, and the interplanetary magnetic field at the earth's orbit is B Gauss, the electromagnetic force on the proton is given by

$$\vec{F}_2 = (e/c) \cdot (\vec{V} \times \vec{B}).$$

On substituting numerical values, we get

$$\vec{F}_2/\vec{F}_1 \approx 10^7$$

which shows the great importance of electromagnetic forces. The forces on electrons and protons being oppositely directed, may however cancel each other in a plasma cloud having equal numbers of electrons and protons, so that the net force on the cloud is zero in first approximation. The motion of a plasma cloud in a magnetic field however causes the separation of ions and electrons and other phenomena will also occur.

Induced electric field in moving matter: It is noted that in the presence of a magnetic field, an electric field is defined in a particular coordinate system. If $\vec{E}$ and $\vec{B}$ are the electric and magnetic fields in a system at rest, we can calculate the fields $\vec{E}$' and $\vec{B}$' in a system moving with velocity $\vec{V}$ in relation to the first, by means of relativistic transformations as follows

$$\vec{E}' = \frac{\vec{E} + (\vec{V}/c) \times \vec{B}}{\sqrt{(1 - V^2/c^2)}} \tag{3.4.2}$$

$$\vec{B}' = \frac{\vec{B} + (\vec{V}/c) \times \vec{E}}{\sqrt{(1 - V^2/c^2)}} \tag{3.4.3}$$

As the velocities of matter considered are much smaller than the velocity of light c, the denominator is close to 1. As electric fields are secondary to the magnetic field and hence are much weaker in cosmic physics, we can write in good approximation

$$\vec{E}' = \vec{E} + (\vec{V}/c) \times \vec{B} \tag{3.4.4}$$

$$\vec{B}' = \vec{B} \tag{3.4.5}$$

Thus magnetic fields are independent of the coordinate system, and there is no difference between the magnetic field in a moving system and that in a 'rest system'. On the contrary, the electric field depends on the co-ordinate system, and it is meaningless to mention an electric field without specifying the coordinate system. Motion of magnetized interplanetary plasma produces electric fields in many cosmic phenomena. The basic ideas of induced electric fields are discussed in this section.

As shown in Figure 3.4.8, let us assume that a slab of cross-section PQRS of conducting matter in a magnetic field $\vec{B}$ in the z-direction, moves with an

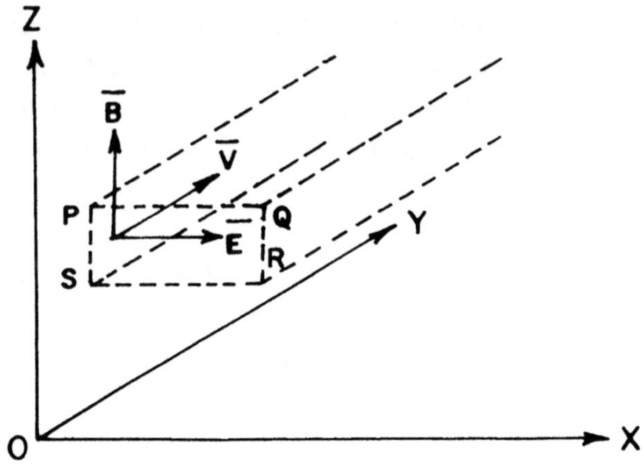

Figure 3.4.8: The coordinate system and the motion of a conducting pillar producing MHD waves

uniform velocity $\vec{V}$ in the y-direction. We define the right handed co-ordinate system in the fixed system, S, of the observer as x, y, z and the co-ordinate system S', moving with the slab as x', y', z'. We denote $\vec{B}$, $\vec{E}$ as quantities measured in the system S, and $\vec{B}$', $\vec{E}$' as measured in system S'. The induced electric field $\vec{E}$' in the system S' moving with the slab is given by

$$\vec{E'} = \vec{E} + (\vec{V}/c) \times \vec{B} \tag{3.4.6}$$

As the slab is assumed to have of infinite conductivity,

$$\vec{E'} = 0$$

Hence,

$$\vec{E} + (\vec{V}/c) \times \vec{B} = 0 \tag{3.4.7}$$

or

$$\vec{E} = -(\vec{V}/c) \times \vec{B} \tag{3.4.8}$$

$$\vec{E} = -(V_o/c) \cdot B \cdot \vec{x}$$

The electric lines of force end in surface charges of density

$$q_s = (v_o/4\pi c) \cdot B_o \ , \ (x = a) \tag{3.4.9}$$

and outside the slab there is no electric field; E = 0 for x > a. In the moving system, S', the magnetic field is the same as in S, whereas the electric field, E' = 0. It may be noted that the magnetic field is almost the same in the moving system and the system at rest. So we may speak of a 'frozen-in' magnetic field which is 'carried' by the plasma as described by Alfven[33].

Various aspects of magnetohydrodynamics and its applications are given in the book by Alfven and Fälthamhar[33].

Magnetohydrodynamic waves: When a medium containing a magnetic field moves, an induced electric field is produced as discussed above. When the medium is of large electrical conductivity and different parts of it move with different velocities electric fields produce currents. These currents interact with the magnetic field producing forces. These forces may in certain conditions generate motion of the medium in the form of wave motions which are called hydromagnetic or magnetohydrodynamic (MHD) waves. Here we shall discuss basic ideas of the phenomena which have many astrophysical applications. It may be noted that the coupling between the electromagnetic and the hydromagnetic processes increases linear with the scale of the phenomena. In the Sun a large number of processes are of a hydromagnetic nature and the same is true for the interplanetary and interstellar medium.

Let us consider an electrically conducting incompressible fluid which is permeated by an uniform magnetic field $\vec{B}$ in the z-direction in the right handed coordinate system (Figure 3.4.8). We assume that at a certain time the whole fluid is at rest except for a pillar, whose cross-section in the z-x plane is given by PQRS and the pillar extends indefinitely in the y-direction. Let the pillar move with velocity $\vec{V}$ in the y-direction.

The motion in the magnetic field produces an induced electric field

$$\vec{E} = (\vec{V}/c) \times \vec{B}$$

in the x-direction. In the conducting medium the electric field produces a current system which is illustrated in Figure 3.4.9(a). In the presence of the magnetic field, the current density $\vec{i}$ produces a volume force given by

$$\vec{F} = (\vec{i}/c) \times \vec{B}$$

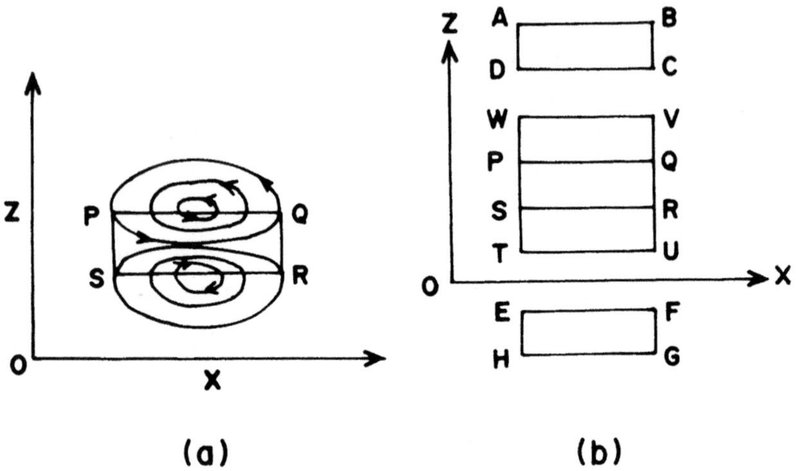

Figure 3.4.9: (a) A sketch showing the current system produced by an induced electric field in the pillar PQRS. (b) The electric fields produce motions in layers above and below PQRS

which is directed to the y-direction i.e. opposite to $\vec{V}$ in the moving pillar. This retards the motion of the pillar. In the adjacent parts of the stationary fluid it is directed parallel to $\vec{V}$. As a result the layer of the fluid initially in motion, PQRS, is retarted and comes to rest and the layers, WVQP above and the RSTU, below, are set into motion in the direction of $\vec{V}$ (Figure 3.4.9(b)). Thus the motion is propagated in the successive layers of the fluid. These wave motions are termed as magnetohydrodynamic waves. Here we see that the induced current system tends to transfer the initial motion to the adjacent regions of the fluid and produces magnetohydrodynamic waves.

The mathematical formulation of the problem is briefly given here. We start with Maxwell's equations,

$$\vec{\nabla} \times \vec{H} = (1/c)(4\pi\vec{i} + \partial\vec{D}/\partial t) \tag{3.4.10}$$

$$\vec{\nabla} \times \vec{E} = -(1/c)\partial\vec{B}/\partial t \tag{3.4.11}$$

$$\vec{\nabla} \cdot \vec{B} = 0 \tag{3.4.12}$$

and

$$\vec{B} = \mu \vec{H} \tag{3.4.13}$$

$$i = \sigma[\vec{E} + (\vec{V}/c) \times \vec{B}] \tag{3.4.14}$$

where μ is the magnetic permeability and σ the electrical conductivity of the fluid. We add to these the hydrodynamic equation of motion

$$\rho \cdot \frac{d\vec{V}}{dt} = \rho \cdot \vec{G} + (\vec{i}/c) \times \vec{B} - \vec{\nabla} p \tag{3.4.15}$$

and the continuity equation

$$-\frac{\partial \rho}{\partial t} = \vec{\nabla} \cdot (\rho \vec{V}) \tag{3.4.16}$$

where ρ is the density, p, the pressure, and G, the gravitational force. For an isotropic plasma we have

$$p = const. \cdot \rho^{\gamma} \tag{3.4.17}$$

and

$$\gamma = c_p/c_\nu \tag{3.4.18}$$

γ being the ratio of specific heats.

Magnetic Field Pressure

In case of a compressible, conducting fluid, let us assume the fluid is in some way compressed in the direction perpendicular to $\vec{B}$, the magnetic field strength is greater in the compressed region. The magnetic field variation in space signifies the existence of a current density as given by the Maxwell equations

$$\vec{\nabla} \times \vec{H} = (4\pi/c)\vec{i}$$

and

$$\vec{H} = \vec{B}/\mu$$

The current density produces a volume force, $\vec{F}$ given by

$$\vec{F} = (\vec{i}/c) \times \vec{B} = (1/4\pi)\vec{\nabla} \times \vec{H} \times \vec{B}.$$

As $\vec{B}$ and $\vec{H}$ have only a z-component, it follows that

$$\vec{F} = -\vec{\nabla}\frac{B^2}{8\pi\mu} \tag{3.4.19}$$

This shows that when the magnetic field is subject to a lateral compression it responds by a restoring force given by 3.4.19. This is equivalent to the magnetic pressure given by

$$p = B^2/8\pi\mu \tag{3.4.20}$$

The Case of Plane Waves: Let us consider the case of plane waves in an incompressible fluid with a homogeneous magnetic field. As the density ρ is constant, we have from 3.4.16

$$\vec{\nabla} \cdot \vec{V} = 0$$

Let the magnetic field be in the z-direction and ρG be neglect-able. The displacement current term $\partial\vec{D}/\partial$t is small in 3.4.10 and can be neglected. The magnetic field is

$$\vec{B} = \vec{B}_o = \vec{b}$$

where the field b is due to current i. In order to study a plane wave propagated in the z-direction (in the direction of $\vec{B}_o$) we assume all vectors are independent of x and y and are dependent on z and t. Thus we have

$$Magnetic\ \ field\ \ B_z = B_o = Const.,\quad B_x = B_y = 0$$

$$Velocity\ \ V_y = V, V_x = V_y = 0$$

$$Current\ \ i_x = i, i_y = i_z = 0$$

$$b_x = const = 0 \tag{3.4.21}$$

Therefore from 3.4.10,

$$(4\pi\mu/c)i_x = (\vec{\nabla} \times \vec{B})_x = \frac{\partial b_z}{\partial y} - \frac{\partial b_y}{\partial z} = \frac{\partial b_y}{\partial z}$$

Therefore

$$i_x = \frac{c}{4\pi\mu} \cdot \frac{\partial b_y}{\partial z} \tag{3.4.22}$$

By substituting 3.4.21 in 3.4.13 and putting $(\vec{\nabla} p)_y = 0$, we have

$$\rho \cdot \frac{\partial V_y}{\partial t} = \frac{1}{c}\vec{i}_x \times \vec{B}_x \tag{3.4.23}$$

substituting i_x from 3.4.22 to 3.4.23, we have

$$\frac{\partial V_y}{\partial t} = \frac{B_o}{r\pi\rho\mu} \cdot \frac{\partial b_y}{\partial z} \tag{3.4.24}$$

$$V_z = 0$$

From 3.4.20,

$$\frac{\partial p}{\partial z} = -\frac{1}{8\pi\mu}\frac{\partial (b_y)^2}{\partial z} \tag{3.4.25}$$

From 3.4.14, we have

$$\vec{E} = \frac{\vec{i}}{\sigma} - (\vec{V}/c) \times \vec{B}$$

or

$$E_x = (i_x/\sigma) - (V_y B_o)/c$$

$$E_y = E_z = 0 \tag{3.4.26}$$

From 3.4.11, we get

$$(\vec{\nabla} \times E)_y = -\frac{1}{c} \cdot \frac{\partial b_y}{\partial t}$$

or

$$\frac{\partial b_y}{\partial t} = -c \cdot \frac{\partial E_x}{\partial z} \tag{3.4.27}$$

From 3.4.26 and 3.4.27, we obtain

$$\frac{\partial^2 b_y}{\partial t^2} = -c \cdot \frac{\partial^2 E_x}{\partial t \cdot \partial z}$$

$$= -c \cdot \frac{\partial^2}{\partial t \cdot \partial z} \cdot \left(\frac{i_x}{\sigma} - \frac{V_y \cdot B_o}{c} \right)$$

$$= \frac{c^2}{4\pi\mu\sigma} \cdot \frac{\partial^2}{\partial t \cdot \partial z} \cdot \left(\frac{\partial b_y}{\partial z} \right) + \frac{\partial^2 V_y}{\partial t \cdot \partial z} \cdot B_o$$

Substituting $\partial V_y/\partial t$ from 3.4.24, we get

$$\frac{\partial^2 b_y}{\partial t^2} = \frac{B_o^2}{4\pi\rho\mu} \frac{\partial^2 b_y}{\partial z^2} + \frac{\partial}{\partial t} \left(\frac{\partial^2 b_y}{\partial z^2} \right)$$

In case of infinite conductivity, we get for $\sigma = \infty$, the simple equation

$$\frac{\partial^2 b_y}{\partial t^2} = \frac{B_o^2}{4\pi\rho\mu} \frac{\partial^2 b_y}{\partial z^2} \qquad (3.4.28)$$

This is the familiar wave equation of magnetohydrodynamic waves which propagate with velocity, V

$$V = \pm \frac{B_o}{(4\pi\mu\sigma)^{1/2}} \qquad (3.4.29)$$

The wave equation 3.4.28 has solutions of the form

$$b_y = A \sin \omega (t - z/V) \qquad (3.4.30)$$

where A is the amplitude and ω the angular frequency. Similarly one can obtain the solution for the wave equation of V_y as

$$V_y = \frac{1}{\sqrt{4\pi\rho\mu}} \cdot A \sin \omega (t - z/V) \qquad (3.4.31)$$

The graphical representations of wave functions are given by Alfven for the induced magnetic field and the velocity Vy as functions of time, t; they also show the wave forms of b_y and V_y as functions of z, at a particular instant of time. The magnetic line of force of the field B_o + b is also shown as a function of z at a given t. The line of force also moves with the same velocity as the material velocity and hence in some cases one may consider, as pointed out by Alfven that the line of force is **'frozen'** in the fluid.

The magnetohydrodynamic waves can also be considered as oscillations of the magnetic lines of force. For a stretched string, the transverse oscillations in the y-direction is given by

$$m\frac{d^2y}{dt^2} = S\frac{d^2y}{dz^2}$$

where m is mass per unit length and S, the tension for unit area and the wave velocity is V = $(S/m)^{1/2}$. In case of magnetic lines of force, the tension per unit area is given by (see Alfven),

$$S = (B^2/4\pi\mu)$$

and the fluid mass per unit volume is, m = ρ. Thus the velocity of wave motion of the magnetic lines of force is given by

$$V = (B/4\pi\mu\rho)^{1/2}$$

which is the same as in equation 3.4.29.

Some of the characteristic quantities of the cosmic plasma are given in Table 3.3.2

Table 3.3.2: Typical quantities of cosmic plasma

Region	$l_{(1)}$ cm	$B_{(2)}$ Gauss	$\rho_{(3)}$ (g/cm^2)	$V_{(4)}$ cm/s
Sunspot	10^9	$2{\cdot}10^3$	10^{-4}	$6{\cdot}10^4$
Solar corona	10^{11}	~ 1	$\sim 10^{-18}$	$\sim 3{\cdot}10^8$
Interplanetary space	10^{13}	10^{-4}	10^{-23}	$9{\cdot}10^6$

(1) l = linear size, (2) magnetic field, (3) density and (4) MHD velocity.

3.5 The Active Sun

3.5.1 Solar Activity

General Features

The solar phenomena we have discussed so far belong to the steady processes in the "quiet Sun". beside these processes there occurs a series of complex phenomena which appear as short lived storm-like events. Many of these storm like emissions or disturbances are connected with sunspots but are not necessarily caused by them. They are collectively called solar activity generated by the 'active' Sun. Among these sporadic emissions from the Sun belong the spectacular phenomena of solar flares in which very large amounts of energies are emitted in optical, ultra violet, X-rays, radio wavelengths and in energetic protons and electrons. Here we shall discuss the basic features of these phenomena of the active Sun.

Major indicators of solar activity are the sunspots. As noted earlier (in 3.2) the number of sunspots visible on the solar disc varies with a solar period of about 11.2 years. The solar activity reaches lowest level at sunspot minimum and highest level during sunspot maximum. However it should be noted that individual solar events of great magnitude do occur even when there are a small number of spots. Sunspots generally appear in groups, centered about two main spots which sometimes grow to a maximum size and increasing complexity in a period of a few days to about a week. Many of these large and complex groups of sunspots are the seats of large events of solar flares and plasma emissions. A large group of spots often persists on the Sun for several weeks and sometimes for a few months i.e. for a few solar rotations, during which in some cases repeated flare events take place. The sunspots are, as noted earlier, the seats of strong magnetic fields. The magnetic energies of large sunspot groups are converted by complex processes into the energies of energetic solar events in which large amount of energies are released in various forms of electromagnetic waves and in energetic electrons and protons and heavier particles. These causes a series of geomagnetic events in the vicinity of the earth e.g. polar cap radio blackout, geomagnetic storms etc.

Plages and Filaments: When we look at the Sun in monochromatic radiation of a strong spectral line such as H_α of hydrogen or H and K-lines of ionized calcium, the layers of the chromosphere are revealed to us, because these lines originate in the chromospheric region of the Sun, in contrast to the continuum spectrum formed by radiation from the photosphere. In photographs of the Sun in the H_α line of hydrogen or the K-line of calcium we see a number of large bright areas which are called plages. The large plages are usually found to surround sunspot groups. They appear to be upward extensions of the facula which are raised regions of the photosphere having higher brightness. In the H_α photographs we see large dark thread like structures which are called filaments. Filaments are relatively cool low density matter sometimes extending to 10,000 km which are seen in the projection of the solar disc. After persisting for several weeks, filaments disappear suddenly, which is sometimes followed by reappearance at the same place.

Prominences: Solar prominences are one of the transient features of the Sun which are seen more frequently near the period of high solar activity near solar maximum. Prominences are large cloud like extensions of gas of the chromosphere into the corona, which are seen, during total solar eclipses in projection of the disc, as a spectacular display of red filaments of enormous size coming out of the chromosphere reaching to great heights of 100,000 km or more and falling back into the Sun. Prominences are routinely seen in solar observatories in the monochromatic light of the H_α line and the K line of calcium. These are associated mainly with the large scale magnetic field of the Sun whose arching lines of force act as pathways for the giant loops of prominences. Quiescent prominences are often seen to occur across the lines of force connecting magnetic regions of opposite polarities. In some cases large masses of gas are thrown out of the Sun into interplanetary space. The temperatures of the prominences seem to be in the range of 8000 to 30,000 K and the particle density is about $2 \cdot 10^{10}$ atoms$\cdot$cm^{-3}. In the surrounding corona the temperature is about $2 \cdot 10^6$ K and the density is about 10^8 cm^{-3}. So prominences are considered as relatively cool and dense gas clouds in the hot and thin corona.

Active Regions and Their Evolution: As large solar flares occur in magnetically active regions and these a chain of geomagnetic disturbances of large magnitude in the vicinity of the earth, it is of great importance to study the active regions and their evolution. These studies were carried out by a number of solar observatories in the world with the goal to understand the occurrence of large solar flares. Here we discuss, as an example, the observations and results of the Mt. Wilson Observatory from a period of 15 years, from 1967 to 1981, which covered two solar maxima, namely the 1968-69 and 1980-81 ones, and the solar minimum of 1974-75[(34)].

Although sunspot groups are important manifestations of the active regions, they themselves are not the active regions, as they represent only a part of the magnetic flux. The active regions are identified as the plage seen in a strong chromospheric line. The magnetic fields of the plage regions are studied using the magnetograms of Mt. Wilson Observatory. The measured parameters are the latitude, central meridian passage (CMP), size in units of 1 or 5 sq. degrees, the magnetic classification and others. Each region is treated as an entity on each disc passage. A total of 5086 regions, with a size $\geq$ 3 deg.^{2}, were analyzed during the 15 year period. The number of active regions having an area greater than A deg.^{2} is given by the expression $N(>A) = 4788 \cdot \exp(-A/175)$. There were about 1000 large active regions with an area > 300 deg.^{2} observed during the observation period. It was found that the average area of active regions during the solar maxima of 1968 and 1979 were about 200-150 sq.deg., whereas the value at the solar minimum was about 100 sq.deg.

The latitude distribution of the 5086 magnetically active regions plotted as a function of time is shown in Figure 3.5.1. Here each point represents an individual active region. This 'butterfly' diagram is similar to that seen for sunspots, but includes many more regions and wider latitude range. It shows that both in the northern and southern hemispheres, the active regions occur at higher latitudes of $\pm$ 10°-30° near solar maximum and these progressively drift to lower latitudes of 5°-10° at solar minimum. At the same time the number of active regions decreases. They reappear at higher latitudes during the next solar maximum.

3.5.2 Solar Flares

General Features

Solar flares, seen in optical, X-ray and radio bursts, are one of the most spectacular phenomena among the transient events in the Sun and are associated with a variety of solar, interplanetary and geophysical processes. In earlier years solar flares were most commonly seen in monochromatic radiation such as the H_α-line of hydrogen. Optical flares are identified as the sudden brightening of the magnetically active plage region in the chromosphere, followed by a slow decline for tens of minutes which may continue for a few hours for very large flare events. In some exceptionally large events, solar flares are also visible in white light.

The British astronomer R.C. Carrington(35) discovered in 1859 the solar flare in white light. While drawing the image of a large sunspot group from the projection of the Sun on a screen he observed a sudden appearance of two intensely bright filaments of white light which lasted for a few

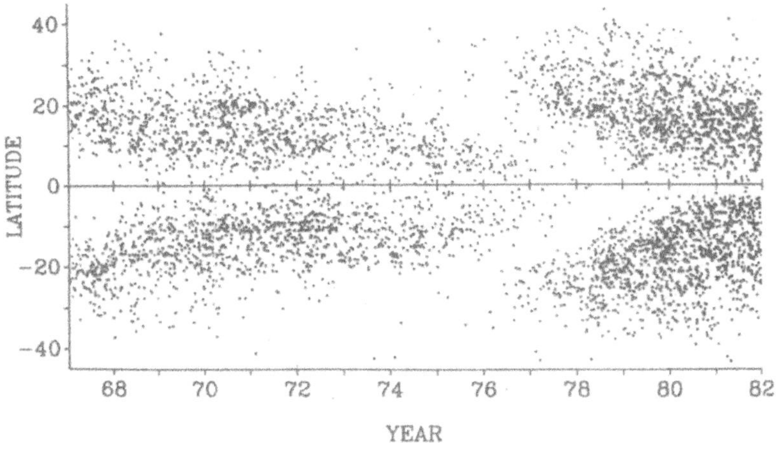

Figure 3.5.1: Latitude distribution as a function of time of 5086 active regions studied for a period of 15 years, 1967-1981 by Mt Wilson laboratory[34]. This is commonly called the 'Butterfly' diagram

minutes. Eighteen hours later the Kew Observatory near London observed the commencement of an intense magnetic storm. This was accompanied by bright aurorae that spread over wide regions of the northern and southern sky, sometimes reaching up to mid latitudes. The fact that the solar flare was associated with these phenomena was found in later years.

In the past solar flares were classified, as seen in H_α, according to their area into five classes: S(sub-flare), 1, 2, 3 and 4, and according to brightness into three classes: F(faint), N(normal) and B(bright). Thus a large and bright optical flare is denoted by 3B. With the advent of satellite-borne detectors and of space stations, monitoring the Sun almost continuously with X-ray detectors, it was observed that X-ray flares with the emission of soft X-ray fluxes in the 1-8Å band are the most common manifestations of solar flares of various types and sometimes there are as many as a few hundred sub-flares in soft X-ray on a day. The optical flare, normally seen in the H_α line reaches its peak intensity in several seconds. The peak is noted as the onset of the impulsive phase and zero time. The intensity then decreases slowly over an hour or more in large events. Solar flares trigger series of solar interplanetary and geophysical disturbances and these reach the earth roughly in the following sequence:

1. Simultaneously with the flare onset there occurs a sudden increase in Cosmic Noise Absorption (SCNA). This is due to solar burst of EUV and soft X-rays which produce increased ionization in the earth's upper atmosphere.

2. Solar radio bursts in meter and decimeter wave bands with the onset of the optical and X-ray flares in large events are followed by radio emissions of various types lasting for about an hour or more in large events. These are described later.

3. Impulsive bursts of hard X-rays of $>$ 10 keV are observed with the onset of the flare and these are followed by the emissions of soft X-rays in the $\sim$ 1-10 keV band lasting for varying periods.

4. High energy protons of $>$ 1 GeV which are sometimes produced in a large solar event, reach the earth with a delay of several minutes and are detected in ground level detectors for relativistic solar particles.

5. Medium energy solar energetic particles of E $>$ 100 MeV arrive at the earth at high latitudes with a delay of a few hours and are studied with instruments flown in balloons and satellites.

6. Low energy solar protons of E $\sim$ 1-10 MeV reach the earth after a delay of several hours and arrive at the polar regions. In case of a large event, they produce intense ionization of the D-layer and cause strong 'Polar Cap Absorption' of cosmic noise i.e. the radio emission from the galaxy. This is called a PCA event which continues for many hours for a large solar event. During the PCA events blackout of radio communication occurs at high latitudes for many hours and sometimes even for a few days.

7. Following a large solar event, a large cloud of plasma is emitted by the Sun. The cloud reaches the orbit of the earth after 1 to 2 days. This causes a large geomagnetic storm when the horizontal component of the earth's magnetic field is rapidly depressed for several hours followed by a slow recovery in one or more days. The arrival of the magnetized cloud at the earth also initiates a depression of the intensity of galactic cosmic rays reaching the earth followed by a slow recovery during several days. This phenomenon was first observed by Scott Forbush in 1937 and is called the Forbush decrease.

The flare associated solar radio emission is discussed in Sec. 3.5.3, followed by the solar X-ray emissions in Sec. 3.5.4. The energetics of solar flares are given in Sec. 3.5.5, followed by the discussion on solar plasma emission in Sec. 3.5.6. As the solar particle emission is a large and complex subject, this is given separately in Chapter 6.

Optical Flares:

Spectroscopic observations in the Fraunhofer spectrum show that the main body of the optical flare occurs in the chromospheric and lower coronal regions somewhere between 0 to 5000 km above the surface of the Sun. When occuring in the limb, the flares may be seen extending to heights of 15,000 km or more. In the main body of the flare, it has electron density of about $3{\cdot}10^{13}$ cm^{-3} and a temperature of about 10^4 K. The density is much higher than the density of the surrounding chromospheric or low coronal matter of about 10^{-10} cm^{-3} or smaller (Figure 3.1.1). Thus in an optical flare a great density is required in a short time in the chromosphere and lower corona. It was found from spectroscopic data that the emitting region of optical flares has an effective thickness in the line of sight of only 10-100 km. This suggests that the optical flare is very inhomogeneous and consists of filaments of high density of unknown size which are separated by much larger non-emitting regions of lower density or higher temperature. These features make solar flares as due to occurrences of a number of complex phenomena, which are not fully understood at present.

Other features which add to the complexity of solar flares are as follows. The relatively cool and dense emitting region of optical flares are accompanied by a very hot, low density plasma emitting thermal X-rays and non-thermal hard X-rays, which are produced by accelerated electrons. These are described later. The properties of solar active regions which produce flares have been studied by several authors using results from increased refinements in the magnetographic measurements. These led to important observations which have been noted by several authors. Flares occur primarily in magnetically complex active regions, where patterns of opposite magnetic polarity mix. The initial flare brightenings or knots, always occur in regions with oppositely directed magnetic fields. These regions tend to occur always closer than about 20" of arc to the neutral line where the vertical component of the magnetic field is zero. The flare brightenings take place on either side of the neutral line and in the regions where the horizontal gradient of the vertical component of the field is large. These flare knots are never located exactly on the neutral line, but on both sides of it. These lead to the two-ribbon optical flares seen in many large events. The morphology of the optical flares establishes the crucial links to the magnetic field structures which act as the energy sources of the accelerated particles and it is now believed that accelerated particles, mainly electrons, lead to the excitation of the hot gas at different heights of the solar chromosphere and corona to generate a variety of optical, radio and X-ray bursts and emissions.

The most remarkable moment of an optical flare is the H_α flash when, within a few minutes, a flare reaches maximum brightness, for example, in the large 3B solar event of November 12, 1960. The intensity then declined in about half an hour followed by a slow decrease during 2 to 3 hours. Coincident with flash phase, large radio outbursts took place as recorded in decimeter and meter wavelengths.

In ultra-violet ($\sim$ 1500-3000Å) wavelengths the flares are not conspicuous whereas in the extreme ultra violet (100-1500Å) the flare bursts have a similar time profile as the hard X-rays and microwave bursts.

The properties of optical flares were studied by many investigators and are discussed in the books by de Jager (1969) and Svestka (1976).

3.5.3 Solar Radio Emission

Early Years

The discovery of solar radio emission in meter wavelengths were made in a dramatic manner in February 1942 during world war II. As the British army was monitoring the hostile German aircraft, with the recently discovered meter wavelength radars, they noticed a strong and variable jamming of the radar. These noise like emissions were traced to the electromagnetic radiation from the Sun by J.S. Hey of the British army. It was also found that a large group of sunspots appeared on the solar disk and a large solar flare occurred on the Sun. This discovery was kept secret until the end of the wear, when the first report on the meter wavelength bursts from the Sun in association with solar flares was published by Appleton and Hey(36) in 1946. At the end of the War, systematic investigations began, notably in the University of Cambridge, UK and the CSIRO Centre in Sydney, Australia, using modified war-time radar receivers and in many other centers of the world leading to new technologies and discoveries.

General Features

Solar radio emissions associated with flares in the centimeter to meter wavelengths are commonly classified as of thermal and non-thermal origin. While the sources of both cases are fast electrons; the thermal emissions are caused by individual electrons which are in thermal equilibrium in the hot gas. Most of the non-thermal radio emissions discussed here are due to coordinated motions of a large number of accelerated electrons. Both thermal and non-thermal solar radio emissions are described in terms of apparent 'black-body' disk temperatures. The actual radiation received is compared to the thermal radiation that would be received from a hypothetical black-body of perfect emissivity which subtends the same solid angle as the actual source. The apparent black-body disc temperature is that required for a black-body so that its radiation is equal to the observed radiation at the given wavelength. As the emissivity and absorptivity of material bodies are less than unity, whereas those of a black-body are equal to unity, the physical temperature of the emitting material of an actual thermal source is somewhat larger than its apparent black-body temperature.

In case of non-thermal sources, the apparent black-body temperatures have no relation to the actual temperature of the body. For example, some non-thermal emission from the Sun may have an apparent black-body temperature momentarily as high as 10^{20} K. Although it is obvious that no part of the solar disk can have such high temperatures, it is convenient to use this fictitious temperature to describe the intensity of non-thermal radiation.

The radio emission from thermal sources are described in terms of the Rayleigh-Jeans approximation of Planck's law for black-body radiation which can be written as

$$b = \frac{2KT_b}{\lambda^2} \tag{3.5.1}$$

where b = 'brightness' of radiation from a black-body at the absolute temperature T_b. λ = the wavelength and K = Boltzmann's constant (= 1.37 erg/K).

The unit of b is erg·cm^{-2} or joules·m^{-2}. The practical unit for b Watts·m^{-2} Hz^{-1}·Sr^{-1} is mostly used.

From equation 3.5.1 it can be seen that if the brightness does not vary as λ^{-2} then the source cannot be thermal whereas the observation of a b varying as λ^{-2} would indicate a thermal nature of the source.

In solar radio-astronomy the received power, w can be specified by equating it to the thermal noise power obtain by replacing the aerial with a resistor at the aerial, or antenna temperature T_a, so that

$$w = KT_a \cdot \Delta f \tag{3.5.2}$$

where K = Boltzmann's constant (in Watts·cm^{-2}·deg^{-1}) and Δf = the receiving band width.

For solar sources with wide band noise like radiation, we define

$$F = S \cdot \Delta f \tag{3.5.3}$$

where F = flux density at the receiving aerial in w·m^{-2}, S = the flux density per unit band width in w·m^{-2}·Hz^{-1} and Δf = the receiving band width.

When the receiving area A is in the direction of the source and the source is small compared to the aerial beam, we have the received power as

$$S \cdot A = KT_a \tag{3.5.4}$$

The received power is determined by the angular distribution of brightness of the source. For each element of the source, the brightness temperature, T_B is defined as the temperature of a black-body for which the brightness of the thermal radiation would be equal to the actual brightness observed. T_B may vary with λ but, like S, it is assumed to be constant over the receiving band width Δf.

Using Rayleigh-Jean's approximation of the black-body radiation equation, we obtain, dw, the power received from a source element subtending a solid angle dΩ as

$$dw = \frac{KT_B \cdot A}{\lambda^2} \cdot \Delta f \cdot d\Omega \tag{3.5.5}$$

and the total power received w, as

$$w = \frac{K \cdot \Delta f}{\lambda^2} \cdot \int_\Omega T_B \cdot A \cdot d\Omega \tag{3.5.6}$$

For an uniformly bright source of angular size much smaller than that of the aerial beam, and when the source of angular size Ω is located on the axis of the beam and has an effective maximum area A_o, we have

$$S = \frac{KT_B \cdot \Omega}{\lambda^2} \tag{3.5.7}$$

and

$$T_a = \frac{A_o T_B \Omega}{\lambda^2} \tag{3.5.8}$$

where S = the flux density received per unit band width (in units of w·m^{-2}·Hz^{-1}) and T_b = the brightness temperature and T_a = the antenna temperature.

In actual practice the received power will be half the total power, because in general only one of the two orthogonal components of polarization of the randomly polarized emission is accepted by the antenna. So the actual flux density of the disc is 2 times that in 3.5.6. The flux density, S, of the quiet

Sun vs λ for meter wavelengths was measured by many authors (see e.g. Kundu and Gergely).

Emission Processes: In solar radio emissions two main processes operate in general. Firstly, an electron moving in a magnetic field, B, with a gyro frequency, f_B

$$f_B = \frac{eB\sqrt{1-\beta^2}}{m_o} \tag{3.5.9}$$

would emit radiation at harmonics of this frequency. For relativistic electrons this radiation is known as synchrotron radiation; for non-relativistic electrons the term gyro-radiation is generally used.

In the case of a gas composed of free electrons, the gas behaves like a dispersive medium for electromagnetic waves having a refractive index n, as given by

$$n^2 = 1 - (f_p/f)^2 \tag{3.5.10}$$

where f is the frequency of e.m. waves and f_p the characteristic frequency of the medium as given by

$$f_p^2 = \frac{N_e e^2}{\pi \epsilon_o m} \tag{3.5.11}$$

where N_e is the electron density and ϵ_o the permeability of free space. When an electron in an ionized gas i.e. plasma is subjected to a linear displacement it behaves as an oscillator with a frequency of f_p as in equation 3.5.11. This is known as the plasma frequency, and the disturbances as plasma oscillations. These are also called Langmuir waves after the discoverer, Langmuir from Russia.

During a total solar eclipse the scattering of sunlight by the free electrons in the corona was used to determine the electron density N_e, at different heights, r, which is given by a model as

$$N_e = 10^8(1.55r^{-6} + 2.99r^{-16})cm^{-3} \tag{3.5.12}$$

Using equations 3.5.11 and 3.5.12, the plasma frequency of f_p as a function of r (the distance from the center of the Sun in units of solar radius) in the solar corona is determined. This is shown in Figure 3.5.2. It is seen that

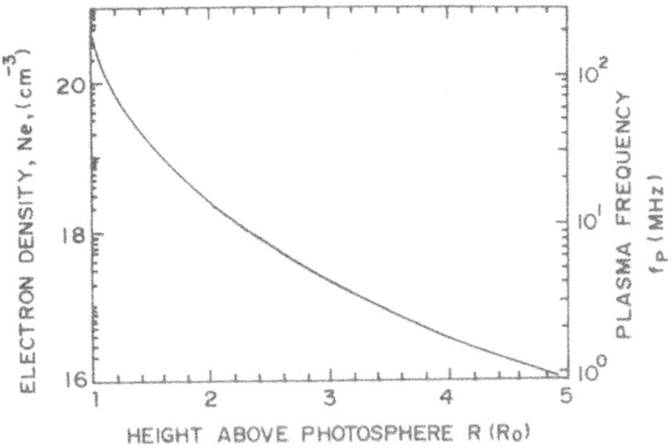

Figure 3.5.2: Variations of electron density and plasma frequency in the corona as a function of height

in the main body of the corona, the plasma frequency is in the range of 30-300 MHz i.e. in wavelengths of 1-10 meter. Therefore the meter wavelength radiation from the Sun reaching us must originate entirely from the corona, and the lower the frequency, the greater is the minimum height of the origin of the radiation. The higher frequency emissions take place from the deeper levels of the corona.

Early studies on this subject are given in the book by Kundu[(37)] (1965).

Types of Radio Emission

Here we briefly describe different types of radio bursts and emissions from the Sun associated with solar flares. Some aspects of the emission mechanisms are briefly noted.

Type I bursts: At meter wavelengths long lasting solar radio emissions, consisting of thousands of short spikes super-imposed on a slowly varying continuum are frequently observed. The first observations of these were made by McCready, Pawsey and Payne-Scott in 1947 and are known as Type I bursts. The bursts occuring in groups of hundreds to thousands are characterized by a band width of a few mega hertz ($\Delta f/f \approx 0.02$) and have a duration of less than a few seconds. The emissions are strongly polarized and they often appear above a background continuum ($\sim$ 50-100 MHz) which is polarized in the same way. Type I bursts begin shortly after onset of a flare and lasts for a long time up to hours and days for large events.

Interpretation: The short time of individual bursts suggests local acceleration of electrons to a few times the thermal energy and from the observed source height at each frequency it is concluded that plasma emission processes are involved. It is generally accepted that these bursts represent some form of fundamental frequency plasma emission due to their high brightness temperature and strong polarization. The long life of a storm seems to suggest that continuing local energy release is taking place which is probably related to the magnetic field reconnections (Chapter 6.5) after new flux extends to the existing fields. Details are given by several authors [see e.g. books by Kundu and Gergely (Ed.), 1980 and McLean and Labrum (Ed.) 1985].

Type II Bursts: Type II bursts accompany solar flares and last for several minutes. These are called 'slow drift' bursts because the spectral features drift in frequency from high to low frequencies at the rate of $\sim$ 200 kHz/s. Early observations of these bursts were made by Wild and McCready(38) in 1950 with the first radio spectrograph near Sydney which recorded the intensities of solar emissions as a continuous function of frequency and time. In the improved instrument operating in 38 to 220 MHz, it was discovered by Wild et. al. in 1954 that Type II bursts show spectral features simultaneously in two frequencies in the ratio of 2:1, indicating the fundamental and the second harmonic frequencies. An example of Type II bursts is shown in Figure 3.5.4.

Type III Bursts: Type III bursts were first identified by the radio spectrograph near Sydney, along with the Type II emissions noted above. It was found that Type III or "fast-drift" bursts are short lived (up to about 10 sec.). Furthermore broad band ($\sim$ 100 MHz) emission of which the frequency of the maximum intensity drifts rapidly to a lower frequency at the rate of $\sim$ 20 MHz/sec which is about 100 times faster than that of Type II bursts was found. Many observations established that Type III fast drift bursts occurred at the very onset of the flare, whereas Type II bursts began a little time later.

Emission Processes of Type II and III Bursts: The first comprehensive theory which accounted for the solar radio emissions of Type II and III was given by Ginzburg and Zheleznyakov in 1956. They suggested the mechanism in which electron plasma waves (Langmuir waves) of frequency f_p are excited by the initiating streams of electrons scattered on ions and electrons of the thermal plasma. The Rayleigh scattering by the ions generate radio waves near the fundamental frequency whereas the scattering on electrons which oscillate at plasma frequency mode produce the second harmonic frequency $2f_p$. This is called combination scattering. Subsequent theories are mostly based on these ideas.

It was established from many studies that Type II bursts were caused by some agency which traveled outwards through the solar atmosphere with velocities of $\sim$ 1000 km·s^{-1}. While Type III bursts were generated by sources traveling upwards with a velocity of $\sim$ 100,000 km·s^{-1}. It was concluded that magnetohydrodynamic shockwaves excited Type II bursts and these shockwaves traveling through the interplanetary medium were responsible for magnetic storms at the earth. The Type III bursts were identified as radiation from plasma oscillations excited by discrete groups of fast electrons. These fast electrons were generated at the onset phase of the solar flare.

Type IV Emissions

Type IV M: In 1957 Boischot and Dennisse of Nancy Observatory, France, identified a distinctive type of meter wavelength continuum emission which was designated as Type IV. These appeared after a large solar flare, often after a Type II burst, and consisted of a smooth continuum of circular polarized radio emission observed in 20 to 400 MHz and their duration was about 30 min at 80 MHz and sometimes as long as $\sim$ 2 hrs. The source of the radiation was found to be moving with a speed of about $\sim 10^3$ ($\sim$ 300-1500) km·s^{-1} through the solar atmosphere. Therefore this type was later classified as "Moving Type IV" (in short IV M), to distinguish it from other types of stationary Type IV events. These are however comparatively rare compared to other varieties of Type IV emissions.

Interpretations of Type IV M: In the beginning Boischot and Denisse suggested these emissions as due to synchrotron radiation of relativistic electrons trapped in the coronal magnetic field, radiating in many harmonics of the gyro frequency. However, observations of circular polarization argued against this as synchrotron radiation is expected to produce linear polarization. From several considerations it is believed that plasma emissions of various types are probably responsible for these radiations. The moving Type IV emissions are related to coronal mass ejections i.e. coronal transients, and coordinated studies in radio and optical bands were made.

The emission processes of Type IV M seem to be rather complex and the results available lead to tentative interpretations. A coherent plasma emission mechanism is needed to explain high brightness temperatures observed at the beginning of bright events. Initial weak polarization favours the second harmonic rather than the fundamental frequency. Later in the event the source brightness decreases to a low value which can be explained by the gyro frequency emission process originally proposed by Boischot. At this later stage high polarization seem to exclude second harmonic emission. Thus a complex hybrid model of second harmonic plasma emission followed by gyro synchrotron emission may be a plausible model (Melrose(39) 1980), although it is not free from some difficulties.

Type IV Flare Continuum: The non-storm continuum emissions in meter wavelengths seen by Boischot were found to extend to continua in decimeter and centimeter regions in the observations of 1961 in the USA (Kundu(40)), in France (Pick(41)) and in Japan (Takakura and Kai(42)). So the term Type IV was applied to all continua emissions during flare events. As not all wavelengths are present in all events various sub-class of Type IV were introduced.

Type IV m: The Type IV meter-wave continuum was divided into two groups. A weakly polarized component which appears early in the event and move outwards (like the original Type IV M events) were called IV mA. The second phase which was stationary and appeared later in the event having circular polarization was denoted by IV mB. Later it was found that not all Type IV mA events moved and some of them were stationary. So IV mA events were further divided into sub-classes of IV mA, (stationary) and IV mA_2 (moving). The Type IV mA_2 is commonly referred to as 'Moving Type IV' (IV M). The later stationary phase IV mB is also called 'Stationary Type IV' (Kundu(37) 1965) or continuum storm (Pick(41) 1961).

Flare Continuum Early (FCE): The term 'Flare Continuum Early' (FCE) was applied by Wild to denote the stationary meter-wave continuum

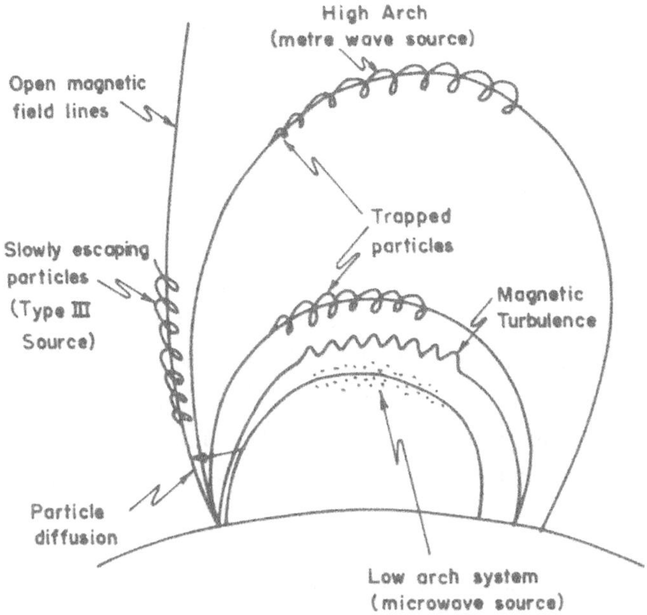

Figure 3.5.3: (a) and (b) Schematic Time vs frequency profiles of different types of bursts and continuum emissions.

which appears early in the flash phase of a flare and to distinguish it from IV mB which appears much later, FCE is always associated with fast drift bursts (Type III emission) and with microwave emission (IV μ). FCE events last for about 3 to 45 minutes (on the average $\sim$ 10 min), and covers the frequency range of $\sim$ 200 to 25 MHz and has a low polarization (Figure 3.5.3a).

FC II Sometimes stationary flare continuum emissions take place immediately after Type II bursts. These are termed as FC II (Figure 3.5.4b). FC II events have low polarization and occurs only a little after the onset, in contrast to IV mB which is also a stationary meter continuum but occurs much later having a strong polarization.

Summarizing the plethora of classes, it may be stated that meter wavelength continua associated with flares can be divided into four classes:

1. FC E - (same as IV mA) Starts early and has a stationary source.

2. FC II - Starts a little later and is stationary.

3. FC M - (same as IV mA_2) Starts early and has a moving source.

4. IV mb - Begins much later and is stationary.

These are illustrated schematically in Figure 3.5.3 (a) and (b).

An important feature of the meter-wave continuum of Type IV is their close association with solar proton events and polar cap absorption which was discovered by Thomson and Maxwell in 1960. Many studies showed association of IV mB and the centimeter to decimeter continuum with solar proton events. Further it was found that the association of solar particle events is very strong for the centimeter wave continuum (IV μ) (Kundu and Haddock[(43)], 1960).

Microwave Bursts and Continuum (IV μ): The onset and flash phase of large solar flares are characterized by emission of microwave and hard X-ray bursts and the beginning of the centimeter wave continuum. A sketch of the time profile of IV μ continuum is given in Figure 3.5.3, showing its relationship with other types of radio emissions. Both centimeter wave emission and hard X-ray bursts are caused by accelerated electrons of $>$ 10 keV. Fast electrons gyrating on magnetic field lines produce centimeter emission and some of these electrons impinge on the dense solar atmosphere and give rise to hard X-rays.

Type V: Type V is a continuum emission which starts during or shortly after a group of Type III burst and lasts for a few minutes only. The emission is strong in the $\sim$ 20-100 MHz range.

Inverted U Bursts: In these the spectrum is initially the same as that for Type III but later the frequency drift stops and then reverses, giving the record the appearance of an inverted U.

Interpretations of Type IV Continuum: Electrons in the Type IV FCE are accelerated in the lower corona and escape to the higher corona producing meter-wave emission. Type IV μ on the other hand are produced

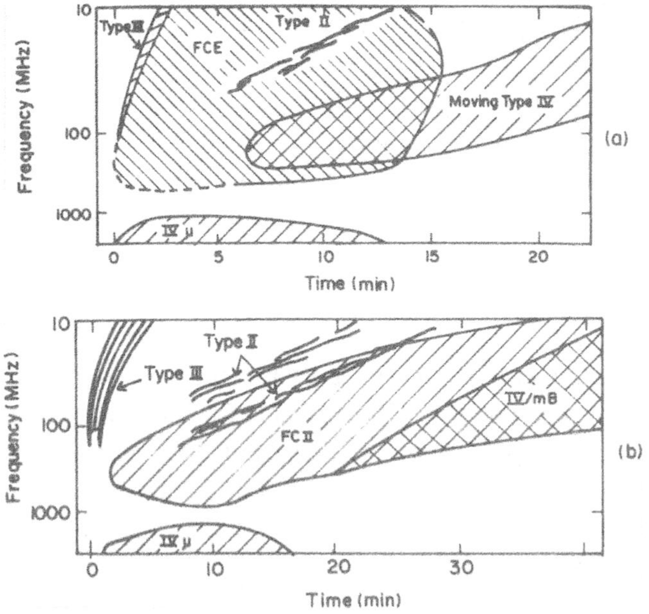

Figure 3.5.4: Schematic model of Type IV flare continuum

by the accelerated electrons which are trapped in closed field lines and emit gyro-synchrotron emission. When the magnetic field structure has a neutral sheet, plasma instabilities sweep magnetic fields to the neutral sheet and the annihilation of the field by magnetic field reconnections lead to an electric field which accelerate electrons to 10-100 keV. Some of these electrons find easy access to open field lines and escape producing Type III; but the majority is confined to the lower coronal magnetic field. The diffusion of the electrons to higher coronal arched field lines in a steady manner lead to the production of stationary Type IV emission lasting for a long period of time (Figure 3.5.4).

Further details on these and various other types of solar radio emission are given in the books referred to in the references.

3.5.4 Solar X-ray Bursts

Early Observations

Early attempts to detect solar X-ray emission associated with solar flares were made during IGY (1956-58) (Chapter 1) when two-stage rockets were kept in readiness with X-ray detectors in the payload. A rocket was launched promptly on observing a large solar flare event and provided an exposure of a few minutes above the atmosphere. Winckler and his associates[45,46] of the University of Minnesota observed the first solar X-ray bursts at balloon altitude in association with a class 2+ flare in March 1958. These hard X-ray bursts of > 50 keV, observed with balloon-borne ionization chambers and Geiger counters, were accompanied by strong radio bursts at 800 MHz. Similar observations were made by them again of the August 11, 1960 solar event.

Solar Flare X-ray studies received great impetus with the advent of satellites with X-ray detectors. In addition to standard instruments e.g. thin window proportional counters, NaI and CsI scintillation counters, new types of detectors with high angular resolutions such as X-ray imaging spectrometers were used in Skylab's solar maximum mission and subsequently. Brief descriptions of these new types of detectors are given in Chapter 6.2.

Soft and Hard X-ray Bursts: Since the 1980's many experimental observations were made with satellite borne detectors. For understanding the observations and the interpretations of hard and soft X-ray emission from the Sun during a solar flare, we shall select one or two well studied events which have been investigated in depth by international teams of solar astrophysicists and these have given us new insight into the basic processes in the flares.

Instruments in the SMM: Detailed studies were made of two homologous flares (i.e. flares connected with each other magnetic field) on November 5, 1980 at 22:26 UT and 22:33 UT with instrumentations aboard the Solar Maximum Mission (SMM). The SMM was a highly successful mission in 1980-82, in which several sophisticated instruments were used to study flare related events simultaneously in the UV, soft and hard X-rays and in ground based optical and radio observations of high resolution. Here we describe the observations of the two flare events with (a) the Hard X-ray Imaging Spectrometer[47] (HXIS) of SRON, Netherlands, and (b) the Hard X-ray Burst Spectrometer[48] of NASA, Goddard Space Flight Center, USA, together with some results from the Ultraviolet Spectrometer and Polarimeter

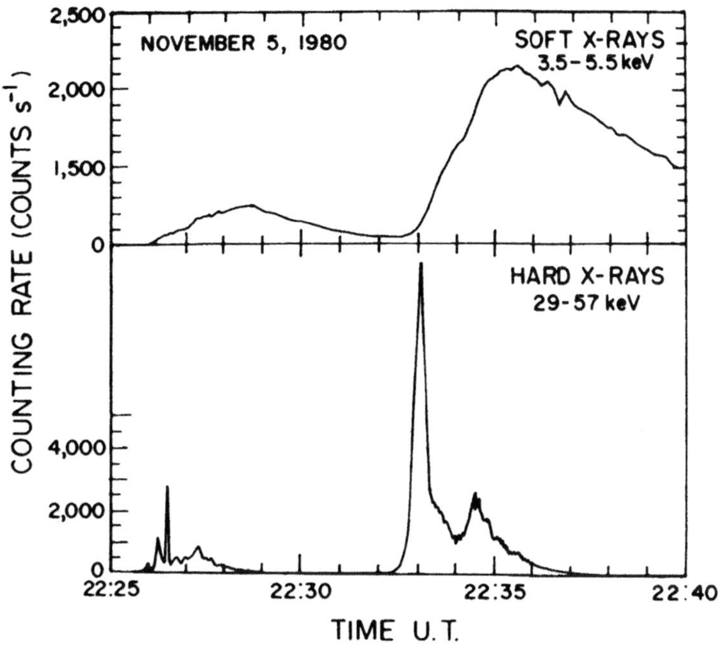

Figure 3.5.5: Time profiles of the soft and hard X-ray emissions of the two solar flare events on November 3, 1980. Measured in the solar maximum mission

(UVSP).

The Hard X-ray Imaging Spectrometer (HXIS) produced images with a *high spatial resolution* of 8"x8", once every 9 sec, in six energy intervals between 3.5 to 30 keV. The accumulations in the counting rates in the band of 3.5 to 16 keV were used to obtain the soft X-ray data, and the counting rates in the 16-30 keV band yielded the hard X-ray results. The Hard X-ray Burst Spectrometer (HXBS) provided spatially integrated fluxes in several intervals in the energy range of 30 to 500 keV once every 128 ms, thereby producing data with a *high time resolution*. The third instrument, UV Spectrometer and Polarimeter (UVSP), obtained high resolution images of parts of the flare at 1371Å in the UV line of OV with a spatial resolution of 10"x10" of arc and a time resolution of $\sim$ 1 s. A 3x3 array with 9 pixels were used for the OV data.

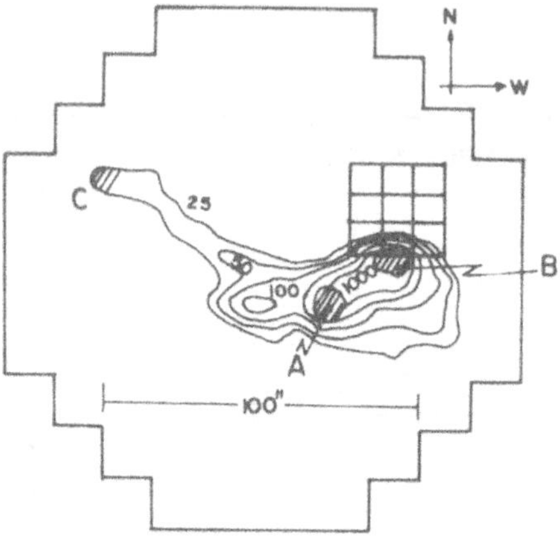

Figure 3.5.6: Results from the high resolution imaging spectrometer of soft and hard X-rays of the SMM satellite for the large solar event on November 5, 1980 (see text and ref. 49 for details)

The fourth instrument was the X-ray polychromator which employed a Bent Crystal Spectrometer measured the blue shift of the Ca XIX line indicating upward velocities in the chromospheric regions during the X-ray peaks.

Observations of Two X-ray Events: In Figure 3.5.5, we show the data of the soft X-rays (3.5.-5.5 keV) and hard X-rays (29-57 keV) during the two solar events on November 5, 1980. The data comes from two instruments, namely the imaging spectrometer (HXIS) and the burst spectrometer (HXBS) noted above. The first event showed impulsive hard X-ray (29-57 keV) bursts within 1 to 2 minutes from 22:26 UT and the soft X-rays (3.5-5.5 keV) showed gradual rise and slow fall in the counting rates during the 5 minutes from the onset. This event was also accompanied by microwave emission at 15 GHz detected by the Very Large Array (VLA). The second event which was the most intense had been studied with much refinements[49] and we shall describe this event in some detail. This event began at 22:32:40, exhibited a very large hard X-ray burst at 22:33:10 UT, followed by another strong burst 1 min later at 22:34:30 UT. The soft X-rays in the 3.5 to 5.5 keV band showed a gradual rise starting at 22:33 UT followed by another sharp rise at 22:34 UT following the second impulsive burst.

In Figure 3.5.6 the results from the high resolution imaging spectrometer

(HXIS) for the second large event are plotted showing the locations of the soft and hard X-ray emissions. The accumulation of the first image began at 22:32:53 UT and each image had an accumulation time of 4.5 sec with a gap of 4.5 sec between images. The contour lines represent the countings in the 3.5-8 keV band at the following levels of counts/pixel of 8"x8": 25, 50, 100, 200 and 400. The peak rate of the soft X-rays was $\sim$ 900 counts/pixel. The shaded areas A, B, C represent the positions of the hard X-ray bursts in 16-30 keV. The outer edges of A and B represent contour levels of 40 counts/pixel per 4.5 sec and that of C 20 counts/pixel. The peak rates at the positions A, B and C were 53 counts/pixel.

The observation of three well resolved patches A, B, C in hard X-rays (16-30 keV) during the impulsive peak of the flare event at 22:53 UT is of great importance. The two patches A and B were separated by $1.6 \cdot 10^4$ km and the third one C was separated from B by $7 \cdot 10^4$ km. Comparing with the H_α images and magnetograms led to the conclusion[(49)] that these bright patches were at the 'foot prints' of two magnetic loops joining A and B and C. It was also argued that simultaneous occurrences of peaks in B and C indicated that most probably fast electrons traveling down the legs of the loops of the magnetic field produced hard X-rays of 16-30 keV by thick target interactions at the foot prints. In the case of the peaks at A and B several arguments lead to similar conclusions. These are further supported by observations of UVSP, which show that OV emissions from foot print B were peaked at 22:53:04 simultaneously with the peaks of hard X-ray bursts at 25-57 and 57-106 keV. Thus it is concluded that fast electron beams existed in the impulsive phase of the flare and these led to a $\geq 10^8$ K thermal source for the hard X-ray emissions. The energy spectrum of the X-ray burst at the time of the first peak at 22:53 UT could be represented by $dN/dE \sim E^{-4}$, whereas in case of the second peak at 22:34:30 UT (Figure 3.5.5) the spectrum was much steeper as given by E^{-7}. Also it was found that the second impulsive burst was located between the bright patches A and B; and this produced further heating of the plasma as seen in increased soft X-ray emission.

Another important feature of hard X-ray bursts is the occurrence of very rapid changes in the intensities on time scales of less than a second. Figure 3.5.7 shows the example of a well studied small flare of May 21, 1984 at 13:26 UT which produced strong hard X-ray bursts in 24-108 keV and 108-219 keV, accompanied by high intensity microwave bursts at 90 and 30 GHz[(50)]. This

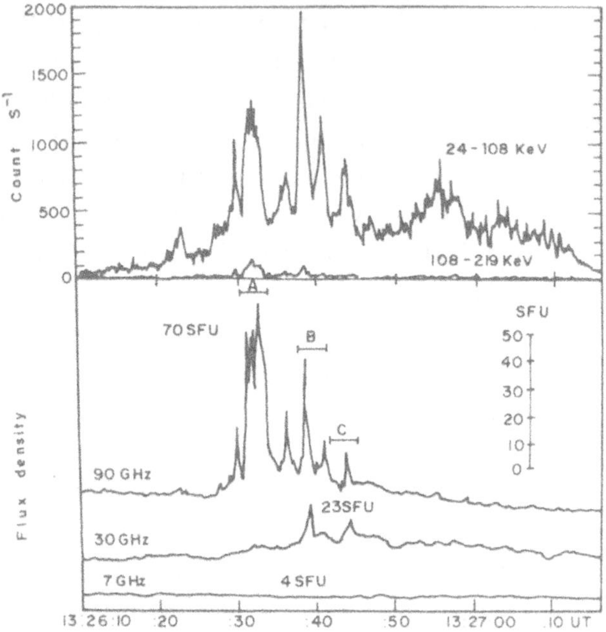

Figure 3.5.7: Time profiles of X-ray bursts and microwave bursts of the solar flare event of May 21, 1984. The 2 sec hard X-ray bursts denoted by "A" were found to be consisting of at least 13 sub-second bursts each with a duration of $\sim$ 0.1 sec (Kaufmann et al.[(50)])

event has provided many clues to the understanding of high energy flares. The remarkable features of the event were that during the 2 sec period, 13:26:30 to 13:26:32 UT, (denoted as "A" in the Figure), a series of complex X-ray bursts and microwave bursts was observed. Further, high time resolution studies showed that these bursts consisted of at least 13 rapid bursts each lasting for $\sim$ 0.1 s with a rise and decay time of $\sim$ 30 ms.

Emission Processes of Hard and Soft X-rays in Flares

In spite of great diversities of these emissions of X-ray bursts, the studies on some of the well-documented events as noted above lead to the following basic mechanisms of hard and soft X-rays as reviewed by de Jager (1986).

1. In the impulsive phase (i.e. when energy is injected into the flare) the energy is liberated by the reconnection between many magnetic flux-threads which exist in the flare region. Each reconnection is responsible for at least one burst. As the feet of the flux threads are rooted in the chromosphere and photosphere, the motions of the feet cause deformations of the flux threads and these lead to their interactions and reconnection. The site of energization of a flare is probably located at about 10,000 km.

2. Dissipation of magnetic energy leads to the instantaneous injection of electrons into the top of coronal loops. These injected electrons are beamed to the foot print areas whereby these chromospheric areas are heated instantaneously to $\sim 7{\cdot}10^7$ K and hard X-ray bursts are produced by thermal bremsstrahlung in a thick target. The hard X-ray complex burst, lasting for say 2 sec should be viewed as a series of injections each with a rise time of 30 msec and declining in the same time. The injected electron beams into the coronal loops produce microwave bursts the gyro-synchrotron radiation in a magnetic field of 1400-2000 G. The general conditions of hard X-ray bursts were applicable to the May 21, 1984 event as follows: (a) Electron temperature, $T_e \simeq 5{\cdot}10^8$ K; (b) Electron density $N_e = 10^{11}$ cm^{-3}, (c) Linear size of emitting area, $l =$ 350 km.

3. By the beam heating a 'well is burnt' in the chromosphere, the bottom of the well being located at about 2000 km above the photosphere. Heated plasma from the well ascends upward with velocities of 150-400 km$\cdot$s^{-1} and spreads over the flaring area. This heated plasma, containing most of the thermal energy of the flare produces a soft X-ray component. The intensity variation of this soft X-ray component thus reflects the evolution of this plasma and these represent the 'gradual phase' of the flare. This picture explains why both the impulsive and gradual phases start at the same time, with the gradual phase continuing longer. The upward motion of the heated chromospheric gas is called the 'chromospheric explosion', whereas the lateral spreading of the gas in the lower corona is called the 'coronal explosion'.

3.5.5 Solar Flare Gamma Rays and Neutrons

It was suggested by several authors that during solar flare events accelerated protons may interact with solar material in the solar atmosphere to produce energetic gamma rays and neutrons (see e.g. Lingenfelter and Ramaty[(52)]). Several investigations were made with instruments in balloons and satellites for studying these aspects (see review by Chupp, 1984). Here we briefly describe one highly successful mission, carried out with the Gamma Ray Spectrometer (GRS) on board of the satellite designated as Solar Maximum Mission (SMM). The GRS discovered the emissions of a number of gamma ray lines and fast neutrons generated from solar surface nuclear reactions during several solar flare events[(53)]. These studies were carried out jointly by the University of New Hampshire, USA, the Max Planck Institute at Garching, Germany and the Naval Research Laboratory, USA.

The Gamma Ray Spectrometer Instrument: The high energy instrument consisted of 7 high resolution NaI (Tl) scintillating crystals each of 7.5 cm in diameter and 7.5 cm high, which yielded pulse heights corresponding to 0.3 to 9 MeV γ-rays, having an energy resolution of $< 7\%$. The gain of each unit was monitored using a ^{60}Co calibration source and were corrected continuously to a common level. In addition the large (25 cm diameter 7.5 cm high) CsI scintillation crystal provided pulses corresponding to 10 to 100 MeV gamma rays. This detector was sensitive to fast neutrons of > 20 MeV, which were identified from the distinctly different pattern of signals. Also the neutron flux was delayed by several minutes with respect to γ-rays because of their flight-time of the sun-earth distance. The GRS instrument was completely surrounded by active anti-coincidence shields, which enabled the system to reject all events from charged particles and to suppress background γ-rays from non-solar directions. Two auxiliary NaI scintillators (8 cm dia, 0.6 cm high) were used to measure the hard X-rays from solar flares in 10-80 and 25-140 keV energy intervals, simultaneously with γ-ray events. These allowed the comparison of time histories of accelerated protons and electrons.

The Gamma Ray Spectrometer in the SMM satellite recorded over 150 solar γ-ray flares of > 0.3 MeV from February 1980 to December 1982. This instrument also recorded several solar neutron events. Here we shall describe two solar flare γ-ray events and two solar neutron events and their salient features.

Table 3.5.1: Important Prompt Gamma Ray Lines

Number	E_γ	Emission Process
1	0.431	^4He (α, n) $^7Be^*$
2	0.478	^4He (α, p) $^7Li^*$
3	0.847	^{56}Fe (p,p') $^{56}Fe^*$
4	1.238	^{56}Fe (p,p') $^{56}Fe^*$
5	1.369	^{24}Mg (p,p') $^{24}Mg^*$
6	1.634	^{20}Ne (p,p') $^{20}Ne^*$
7	1.779	^{28}Si (p,p') $^{28}Si^*$
8	2.313	^{14}N (p,p') $^{14}N^*$
9	4.439	^{12}C (p,p') $^{12}C^*$
10	6.129	^{16}Oe (p,p') $^{16}O^*$

Solar Gamma-ray Flares

Solar γ-ray events are of two types, the prompt and delayed γ-rays. The prompt γ-rays are due to the interaction by accelerated protons in flares with the abundant nuclear species in the solar atmosphere, wherein the excited nuclei return to their ground states by the emission of γ-rays. Some of the major γ-ray lines produced by this de-excitation process are given in Table 3.5.1. As carbon and oxygen are the most abundant nuclei, next to helium, the de-excitation lines in 4 to 6.4 MeV range are the most prominent. From the studies of typical γ-ray flares of June 7, 1980 at 0312 UT and April 27, 1981 at 0807 UT, the following features are noted:

1. The total impulsive phase of the γ-ray flare of 7-6-1980 was about 65 sec, during which γ-ray emission took place in about seven bursts each with a duration of $\sim$ 10 sec. The five strongest bursts were followed by two small bursts. The comparison of the time histories of 4.1 to 6.4 MeV γ-rays and 40-140 keV hard X-rays showed that the onset was almost at the same time for both and the γ-ray and X-ray bursts peaked almost at the same time, although some γ-ray burst peaks were delayed by about 2 sec relative to X-ray burst peaks.

2. The measured γ-ray spectrum of the flare of 27 April 1980 at 0807 UT to 0736 UT, (after subtracting the orbital background and electron

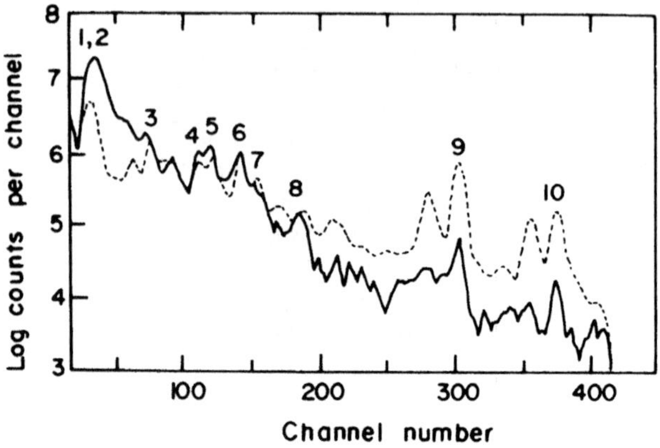

Figure 3.5.8: Solar γ-ray spectra measured by the Gamma Ray Spectrometer in SMM during the solar flare of April 27, 1981. The positions of the peaks numbered 1 to 10 correspond to the γ-ray line numbers given in Table 3.5.1

bremsstrahlung) is shown in Figure 3.5.8. The energies of nuclear de-excitation γ-ray lines shown in Table 3.5.1 and given in serial numbers are indicated in Figure 3.5.8, corresponding to their channel numbers of the pulse height distributions. The full line represents the observations and the dotted line, the model calculations made with some assumptions on the relative abundances and the spectral shapes of the proton beams. The positions of the observed peaks 1 to 10 agree well with those of the calculated peaks. The results and the narrow widths ($\Delta E \leq 180$ keV) confirm these lines as being due to nuclear interactions of accelerated protons and α-particles with ambient heavy nuclei through inelastic scattering, charge exchange and spallation reactions.

3. The other strong γ-ray lines are (a) the 0.511 MeV line from e^+-e^- annihilation and the 2.223 MeV line from the accelerated neutrons capturing protons producing deuterons in $^1H(n,\gamma)^2D$ reactions. Both these lines are of delayed emission types. The e^+ annihilation line of 0.511 MeV was delayed over a time period longer than the production time of the π^+-meson and β^+ emission. The 0.511 MeV line was unresolved from the 0.431 and 0.478 MeV lines denoted by 1 and 2 in Figure 5.3.8. The 2.223 MeV line from the neutron capture by protons forming deuterons was unresolved from the ^{14}N de-excitation line of 2.311

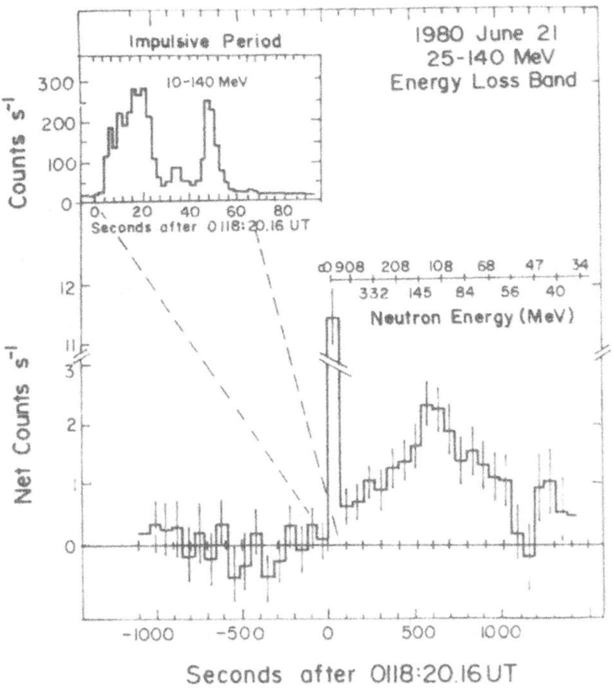

Figure 3.5.9: The solar neutron event following the solar flare at 01:18:20 UT on June 21, 1980 in the GRS experiment of the solar maximum mission. The large excess counting during 0-100 sec after the flare is due to fast neutrons followed by the delayed arrival of $\sim$ 50-300 MeV neutrons

MeV denoted by the peak in the figure.

The γ-ray spectra as shown in Figure 3.5.8, indicate a fairly sharp cut-off at $\sim$ 7 MeV which is characteristic of nuclear γ-ray spectra.

Solar Flare Neutrons

It was suggested by Biermann et.al.[54] in 1951 that high energy neutrons produced in the Sun from nuclear reactions during solar flares could be observed at the earth. Although several attempts were made to observe these solar neutrons, conclusive evidences of the production of fast solar neutrons during solar flares were obtained only in the experiment in the solar maximum mission satellite by the groups of U-NH, MPI and NRL as noted earlier[53].

Here we describe the observations of the two solar neutron events of 21 June 1980 and 3 June 1982. The same instrument (GRS) was used for these studies and the neutron events were identified with their characteristically different signatures. In Figure 3.5.9, we show the excess counting rate vs time, after 01-18-20 UT on 21 June 1980 which was the starting time of the impulsive solar flare event. The counting rates corresponded to the energy loss band of (25 to 140) MeV. The neutron energy scale, shown in the center assumes that the neutrons left the Sun, light travel time 507 sec before 01:18:20 UT. The sharp rise of the excess counting rate at 0-100 sec indicated the arrival of fast solar neutrons reaching the detector followed by the delayed neutron emission of 50-500 MeV peaking at $\sim$ 500 sec. The inset in the figure shows excess counting rates for the energy loss, 10-140 MeV with higher time resolutions for the impulsive phase. The delayed peaks indicated the arrival of bursts of fast neutrons.

In case of the 3 June 1982 solar neutron event measured with the SMM satellite, an additional confirmation was obtained. Within 1 min after the event, the ground level neutron monitor at Jungfraujoch in Switzerland recorded an excess of the counting rate(55). Two other neutron monitors recorded a small increase in the counting phase. The detection of the ground level increase signifies that the solar neutrons with energies $>$ 1 GeV were emitted by the Sun.

In Figure 3.5.9, the delayed neutrons of $>$ 100 MeV are lesser in number due to their energy spectrum at production; out of 50-100 MeV neutrons a large fraction of them decayed in space due to their longer travel time of the Sun-earth distance. Therefore the delayed neutrons showed the peak at an energy of about 100 MeV. Satellite observations have detected following a solar neutron emission event, an excess proton flux of 10-50 MeV which is due to neutron decay in Sun-earth space.

3.5.6 Plasma Emissions of the Active Sun

Early Observations

Earlier in Chapter 3.4.2, we have described solar plasma emissions in quasi-steady conditions of the quiet Sun, which persist for many days or for a few solar rotations. Here we discuss solar plasma emissions of the active Sun which are transient in nature, lasting for only a few days and are associated with impulsive processes in the Sun such as solar flares and prominences.

Scott Forbush found in 1950, that cosmic ray intensities as measured by ground level detectors occasionally undergo a rapid decrease in a few hours followed by a slow recovery over several days. These events called Forbush decreases (Chapter 7.4) were found to be associated with magnetic storms. Chapman and Ferraro ascribed in 1929 the magnetic storms as due to plasma clouds emitted by the Sun and reaching the orbit of the earth. Hence it was natural to attribute the Forbush decreases as due to the interactions of cosmic rays to "magnetized plasma clouds" from the Sun. Alfven showed in 1954 that a plasma beam from the Sun would carry the solar magnetic field with it. But the nature of the magnetic field in the plasma cloud remained unclear, and several models were proposed. In Figure 3.5.10, we show sketches of some of the models proposed on the nature of the magnetic field in the plasma clouds. Morrison suggested in 1956 that the magnetic field is turbulent in the cloud. Gold and Cocconi et.al. proposed that the magnetic field in a cloud is ordered and remains rooted in the Sun. Thus a "magnetic tongue" or a "magnetic bottle" is formed which on reaching the orbit of the earth causes the Forbush decrease of cosmic rays. The existence of shock waves in the solar wind was suggested by Gold in 1955, and Parker showed in 1961 that the interplanetary magnetic field would be compressed and distorted by a shock forming a shell of strong magnetic fields. In Parker's model the magnetic shell would present a diffusive barrier to incoming cosmic ray particles causing the Forbush decrease of cosmic ray particles, somewhat similar to the turbulent magnetic cloud model of Morrison. However in Parker's picture there are no plasma clouds carrying additional magnetic flux.

Radio and Optical Observations of the Coronal Transient

Transient coronal plasma emissions were indicated in the radio burst measurements where flare continuum sources at meter wavelengths were found to travel outward from the Sun. These are denoted as Moving Type IV events as briefly discussed in Chapter 3.5.3. Optical observations of coronal transients were made by the Skylab Apollo Telescope mount where white light coronographs showed many striking events during the nine month period of study. A review of these are given by Gosling[56]. Some of the main properties of coronal transients are - (1) In 1975, the Skylab ATM experiment observed $\sim$ 100 transients during 227 days. After correcting for the unseen portion, the frequency is estimated as (0.5-1.0)/day. (2) Most transients appear as large bubbles rising from the base with velocities of about 200-1100 km/sec.

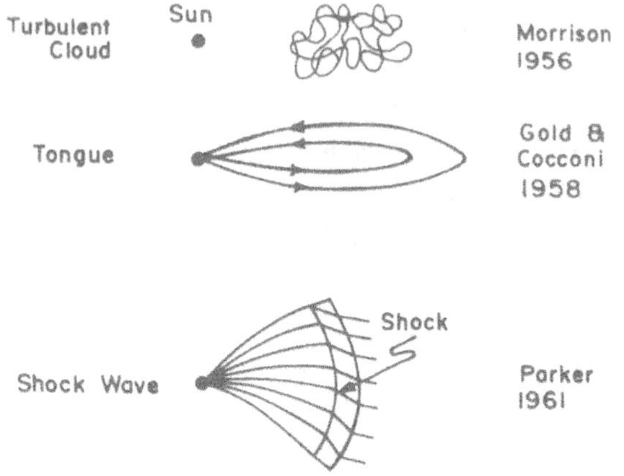

Figure 3.5.10: The sketches of some of the early models of the plasma clouds from the Sun and the magnetic field structures

A shock front is seen in front of the main mass ejecta. (3) About only 10% of coronal transients are associated with solar flares; and $\sim$ 30% are associated with Type II radio bursts. (4) Coronal transients have a strong association with prominence disturbances. (5) The number of coronal transients increases with solar activity, and (6) a typical coronal transient ejects a mass of $\sim 10^{15}$-10^{16} gram of coronal matter and is associated with kinetic energies of $\sim 10^{31}$-10^{32} erg.

Spacecraft Observations of Coronal Mass Ejections

Satellite measurements were made to investigate the magnetic field configuration associated with coronal plasma emissions so as to find out whether the Gold model or the Parker model is applicable. Here we describe the coordinated measurements of several parameters associated with a solar plasma emission event of June 19-21, 1980 seen in the Helios I satellite at 0.5 AU over the west limb of the Sun (Burlaga et.al[57], 1982), as shown in Figure 3.5.11. It is interesting to find that 'in situ' measurements show the evidence of a 'magnetic cloud' with a loop like field line configuration, and a shock front followed by a strong compressed field, which is driven by the magnetized cloud. Thus both the models of the magnetic plasma cloud (Gold model) and the shock front (Parker model) seem to be operative. We describe these with reference to Figure 3.5.11.

1. The dotted lines at times t_1 and t_2 denote the depth of the "shock front" where the total magnetic field B (in a) showed a rapid increase, the directions of the field given by δ and λ showed rapid changes (in b and c). This shock front had velocities much higher than the ambient solar wind (as in d) and a higher matter density, N (in e) and a higher proton temperature T_p (in f).

2. The shock front was followed by a plasma cloud denoted by the dotted lines t_2 and t_3, where the magnetic field was ordered and their direction showed a reversal by about 180° (in b) indicating a loop like structure.

Thus these 'in situ' measurements showed the features which combine all the features of earlier models, such as, a magnetized plasma cloud with an ordered loop like magnetic field configuration, a shock front and a sheath of compressed and turbulent magnetic fields, which is driven by the plasma cloud.

The magnetic cloud described in Figure 3.5.11 was related to a coronal mass ejection seen in white light transient observed by the earth orbiting satellite P-78-1.

Coronal mass ejections and interplanetary shocks were found to be closely related[(56)]. Further studies were made in the ULYSSES mission[(58)] (see 3.6.2) which showed series of coronal mass ejection (CME) events as evidenced by large spikes in solar wind speeds. It was seen that in March 1991, a CME event produced a strong shock, generating an instantaneous solar wind speed of as high as 900 $km \cdot s^{-1}$. It is also shown that between September 1991 and February 1992 there occured a clear sequence of corotating interaction regions (CIR's), a closer look of which indicated the highest density at the upturn of the speed.

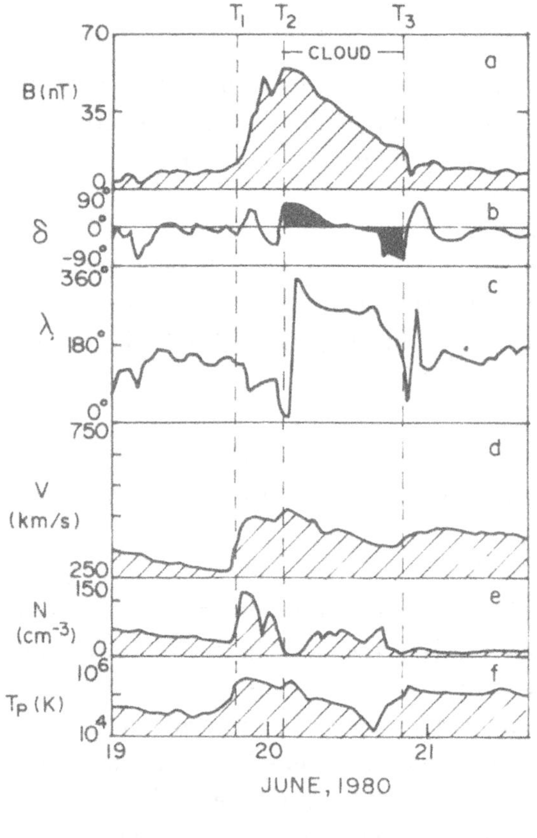

Figure 3.5.11: Co-ordinated measurements of a number of parameters of a magnetic cloud of the June 19-21, 1980 event (see text for explanation) based on the review by Barlaga et al.[57]

3.6 The Outer Heliosphere

3.6.1 Early Concepts

The early concept of the heliosphere, as a domain of space in which the properties of the magnetic field and charged particles are largely influenced by the Sun, emerged in the early 1950's. In these studies the intensity variations of galactic cosmic rays measured at the earth's surface by meson telescopes and neutron monitors played an important role as these showed that the variations originated from the interplanetary dynamical processes (Chapter 6.2.1). In 1955 a tentative idea was that the solar magnetic field pushed out the interplanetary field so as to form a cavity around the Sun. The other idea was that a disordered magnetic field filled up the solar system and acted as a barrier to incoming cosmic rays. The large solar cosmic ray event during the solar flare of February 23, 1956 provided very important data on the intensity variations of the energetic solar particles, lasting for several hours.These were observed by ground level meson telescopes and neutron monitors. The results indicate that the Sun is surrounded by a nearly ordered field in the inner solar system including the earth and a shell of disordered magnetic fields in the outer region of the solar system in which the energetic solar particles were stored for several hours or more. These studies provided the first observational evidence of the presence of the heliosphere.

To our present ideas of the heliosphere, particularly its outer regions was contributed by direct measurements of magnetic field structures and charged particles by a number of satellites and space probes.

3.6.2 Deep Space Probes and General Features of the Heliosphere

The deep space missions with spacecraft of NASA launched in the 1970's have in a highly successful manner provided during the following two decades a wide variety of new information on the magnetic fields, solar wind and energetic particles in the outer heliosphere beyond the orbit of Jupiter. The trajectories of the four spacecraft designed for deep space missions, Pioneer 10 and 11 (P-10 and P-11) and Voyager 1 and 2 (V-1 and V-2) are shown projected on the ecliptic plane with the earth-sun line as the X-axis and the earth as the center in Figure 3.6.1. P-10 and P-11 were launched in 1972 and 1973 respectively and V-1 and V-2 began their space voyages in 1977. P-10 had encountered Jupiter (J) in December 1973 and then traveled in the

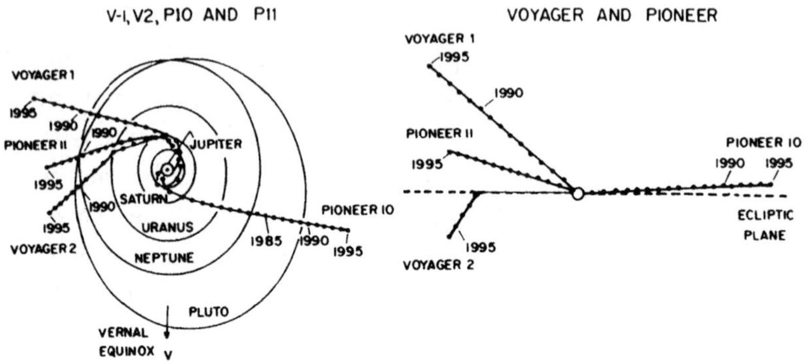

Figure 3.6.1: A composite figure showing the trajectories of the four deep space probes, Pioneer 10 and 11, and Voyager 1 and 2, projected on the ecliptic plane (on the left) and on the normal to the ecliptic plane (on the right) with the earth-sun line as the X-axis and the earth as the center. The dots on the trajectories represent positions one year apart

outer heliosphere to > 50 AU after 1990 in a direction away from the Sun. P-11 has been traveling in interplanetary space > 30 AU since 1990. V-1 encountered Jupiter in 1978, Saturn in 1980 and had its trajectory extended beyond 35 AU since 1990, as shown in the figure. The Voyager-2 made a "GRAND TOUR" of the outer planets, encounting Jupiter in 1979, Saturn in 1981, Uranus in 1986 and Neptune in 1989 at 30 A.U. The orbit of P-10 was close to the ecliptic plane whereas after the S-encounter, the P-11 orbit was at $\sim + 15^o$ latitude in 1979 while the V-1 orbit was at $\sim 35^o$ latitude in 1980. V-2, after the Neptune encounter was at a latitude of $\sim +45^o$ in 1989. These spacecraft have provided information on the solar wind and magnetic field above the ecliptic plane.

The properties of the solar wind in the outer heliosphere[59] acquired by Pioneer 10 and 11 and Voyager 1 and 2 are as follows: (1) The radial variation of solar wind proton number density was measured as $R^{-2.10}$, at 1 to 40 A.U. (2) At higher helio-latitudes the solar wind speed was higher, which was due to coronal hole effects. (3) Some amount of slowing down of the solar wind, by about 30 km/sec, as it propagates to the distant heliosphere was observed. This is ascribed to the pick up of interstellar neutral atoms by the solar wind. (4) Co-rotating interaction regions (CIR) were recurrent structures of solar wind speeds and seen even at 40 A.U. (5) During 1987-93, V-2 data revealed

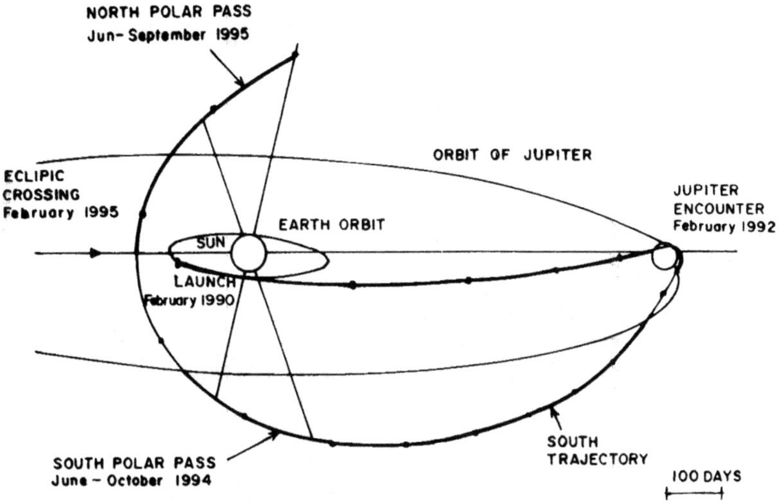

Figure 3.6.2: A sketch of the trajectory of the Ulysses mission (of ESA-NASA) which measured the heliospheric properties in the three dimensions

the new phenomenon of a 1.3 year periodicity in the solar wind speed. IMP-8 results also confirmed this effect. The cause of the unexpected phenomenon is not known at present.

3.6.3 The Heliosphere in 3-dimensions

The highly successful space mission of the European Space Agency (ESA) and NASA - the ULYSSES traversed both the ecliptic and off-ecliptic planes during 1990 and onwards.

The Ulysses spacecraft launched in October 1990 had an in-ecliptic trajectory on its journey to Jupiter. After the Jupiter encounter in February 1992, the spacecraft had a swing-by manoeuvre by Jupiter's gravitational field that propelled it out of the ecliptic plane and sent it back toward the Sun in a trajectory in a plane normal to the ecliptic (Figure 3.6.2). The first high latitude pass of Ulysses took place over the Sun's southern pole from June till October 1994, followed by the ecliptic crossing in February 1995 and the north polar pass from June till September 1995. Thus the joint ESA-NASA project - the Ulysses is the first space mission to circumnavigate the poles of the Sun.

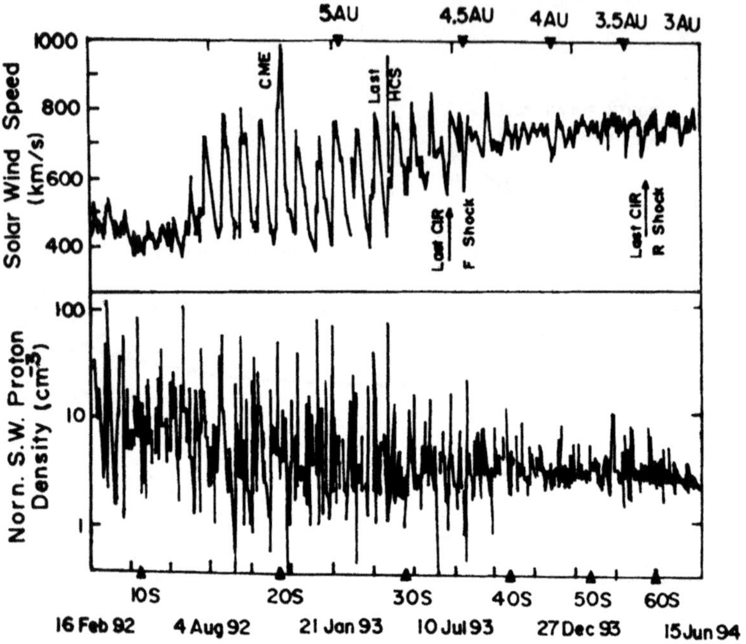

Figure 3.6.3: Ulysses results. The upper trace shows the solar wind speed (6 hr. aver.) vs distance and helio-latitude. The lower trace shows the normalized solar wind density (cm^{-3}) vs solar latitude (after ref. 58).

In 1990-92, just after the solar maximum, Ulysses traveling in the ecliptic plane yielded the following information on the solar wind.

From November 1990 to February 1992, irregular patterns of spikes in solar wind speeds associated with events of coronal mass ejection (CME) were observed. Clear indications of corotating interaction regions (CIR) were seen accompanied by large changes in the density of the solar wind. After February 1992 recurrent high speed streams were present (Figure 3.6.3) which was due to the equator ward extension of the south polar coronal hole. This feature ceased when the spacecraft traveled further south to a latitude of 40° S in August-September 1993. Many co-rotating interaction region (CIR) shocks seen earlier became strange at latitudes of 60° or more. The remarkable feature in the south polar pass was that the solar wind speed became 700-800 $km{\cdot}s^{-1}$, almost the double of the normal value of about 400 $km{\cdot}s^{-1}$. The fluctuations of the solar wind speed and the density were largely reduced. The normalized solar wind proton density became 3 cm^{-3} at $\gtrsim$ 60° S latitude. These striking results[(59)] are shown in Figure 3.6.3.

At 50° latitudes and higher, the magnetic field measurements[(60)] showed a little latitudinal gradient. At 50° S, the observed spiral angle closely agreed with Parker's spiral angle. The variations of the magnetic field indicated the presence of large amplitude Alfven waves.

3.6.4 Heliosphere Boundary Regions

In Figure 3.6.4 we show a sketch of the boundary regions of the heliosphere based on the prevailing concepts. It may be noted that the locations of the boundary regions of the outer heliosphere are largely uncertain as these are dependent on a number of hypothesis on the properties of fields and the solar plasma and their interactions with the interplanetary magnetic field.

The heliosphere is embedded in the interstellar magnetic field and the interactions with the solar wind lead to the formation of two boundary regions. Firstly, the heliopause, as shown, in the figure, separates the interstellar magnetic field from the heliospheric magnetic field in the upstream direction. Here the gas and field pressure of the interstellar medium roughly balances the pressure of the solar plasma and magnetic field. Secondly, the terminal shock front which determines the boundary where the supersonic solar wind becomes subsonic as it travels through the shock front. Inside the terminal shock front, the magnetic field configurations are entirely controlled by the solar processes.

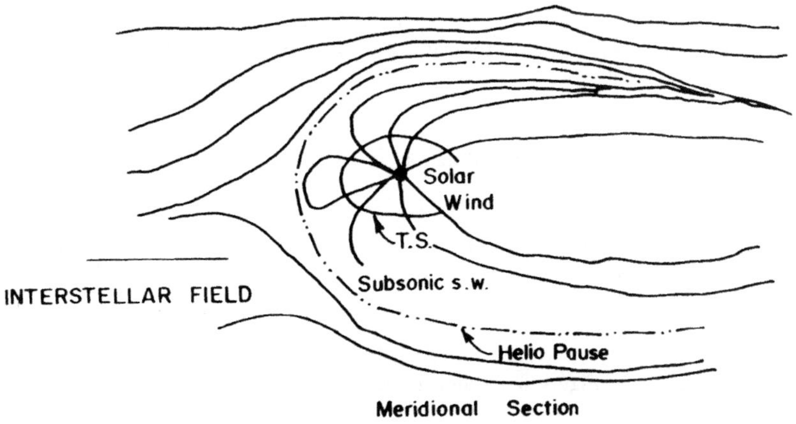

Figure 3.6.4: Schematic diagram of the boundary regions of the heliosphere

In the heliomagnetic tail region the subsonic solar wind and frozen in field probably merge with the interstellar magnetic field forming a wake. It is believed that low energy solar cosmic ray particles probably enter the heliosphere and reach the orbit of the earth through the tail of the heliosphere. However at the present state our knowledge of the outer boundaries and the helio-tail is largely uncertain. The location of the shock front and the heliopause are usually estimated from the energy densities of the fields and particles on both sides of the boundary. In the heliopause and the interplanetary field boundary the pressure from cosmic rays and interstellar magnetic field is estimated as $\sim 7{\cdot}10^{-15}$ dyne/cm^2 for the average interstellar field of $\sim 3{\cdot}10^{-6}$ Gauss. From these and other considerations, the heliospheric shock front and the helio pause boundaries are estimated by different authors as between 50 to 150 A.U.

3.7 Problems on Chapter 3

1. Explain the basic features of Parker's theory of the solar wind. Show the theoretical variation of solar wind speed with the distance from the Sun up to $\sim$ 1 A.U., for $\alpha = 1.1$ and coronal temperature $T \simeq 1{\cdot}10^6$ K.

2. Calculate the Archimedean spiral angle of the solar magnetic field line at the orbit of (a) the earth, and (b) Jupiter. [Use solar wind velocity = 400 km$\cdot$s.$^{-1}$]

3. Explain with a sketch the evolution of the spiral magnetic field as the solar wind with a radial velocity of 400 km$\cdot$s^{-1} travels in 5 days from the Sun to the earth.

4. Explain the formation of a co-rotating interaction region of the solar wind stream in the inner solar system, and their evolution to a pressure wave zone and wave interaction zone in the outer region of the heliosphere.

5. How were 'Coronal Holes' discovered and what are their main features?

6. Consider a conducting fluid in a magnetic field as shown in Figure 3.4.8. If a pillar of the fluid moves with velocity $\vec{V}$ as in the figure, show how the magnetohydrodynamic waves are produced in the conducting fluid.

7. Using notations as in Figure 3.4.8 and using Maxwell and hydrodynamic equations, show that the velocity of MHD waves is given by $V = B/(4\,\pi\mu\sigma)^{1/2}$, where B, μ and σ represent the magnetic field, permeability and conductivity of the fluid respectively.

8. Describe the types and time sequences of occurances of electromagnetic and corpuscular emissions and magnetic disturbances at the earth's orbit which are triggered during a large optical flare on the Sun.

9. Indicate the characteristic features of type II and III solar radio bursts and their association with solar flare events.

10. What are the various types of the meter-wave continuum of Type IV emissions associated with a flare event?

11. Explain by a schematic drawing the locations of emission processes of type IV continuum emissions.

12. Describe the basic mechanisms and various stages and locations of emissions of hard and soft X-rays in solar flare events.

13. What are the basic processes of the two types of prompt and delayed emissions of γ-rays in solar events?

14. Explain how the signatures of solar neutron events are identified in a detector flown in a satellite.

15. What are the different methods of observing Coronal Mass Ejection (CME)? What are the main properties of CME?

16. What are the main features of the solar wind in the outer heliosphere as observed by Pioneer 10, 11 and Voyager 1 and 2 space-craft?

17. Explain the principal properties of the solar wind at large helio-latitudes (λ), particularly at $\lambda > 60^o$, in the polar regions of the Sun observed by the Ulysses mission.

CHAPTER 4
THE MOON AND THE PLANETS

4.1 The Moon

4.1.1 Spacecraft in Lunar Explorations

The exploration of the moon acquired new dimensions with the advent of space age when spacecraft were designed to study the moon, firstly, in the fly-by mode, then as orbiter, and finally in lander missions. About a year after the launch of the first earth satellite, Sputnik I, Soviet Russia launched Luna the I spacecraft in January 1959 which flew by the moon at 6000 km, followed by Luna II in September 1959 which impacted on the moon. In October 1959 Luna III photographed the far side of the moon for the first time. During the 1960's NASA of the USA conducted several missions on the moon such as Ranger, Lunar Orbiter and Surveyer for orbiting, mapping and instrument landing as preparation for manned missions. Soviet Russia conducted several Luna missions. Using the giant Saturn boosters the NASA conducted several Apollo missions to the moon culminating in the historical event of the Apollo-11 mission with the first landing on the moon by US astronauts, Neil Armstrong and Edwin Aldrin on July 19, 1969. Manned Lunar explorations were carried out by NASA in several subsequent missions ending with Apollo 17 in December 1972. Soviet Russia using their Proton rocket boosters for unmanned lunar missions successfully operated a Lunar lander vehicle on the moon in the Venera 7 mission in August 1970; in September 1970 they successfully conducted the Luna 16 mission in which an unmanned automated device scooped out lunar soil samples and returned to the earth.

4.1.2 General Features and Early Studies

The moon having a radius of 1738 km and a mass of 1/81 th of the earth is the largest satellite in the solar system and the earth-moon system may be considered as a double planet similar to Pluto-Charon system discussed later. The orbit of the moon at a mean distance of 384,000 km from the earth has a small eccentricity of 0.055 and is inclined to the ecliptic plane by 5.1^o. The acceleration due to earth's gravity causes tidal bulges in the moon in the direction of the earth and at the anti-earth point. As the rotational axis of the moon is inclined to its orbit by 6.5^o, observers on the earth can see the lunar poles alternately in each month. This is called libration in latitude. The moon makes one orbit around the earth with respect to the stars, in 27.32 days which is the sidereal month; whereas it completes one revolution around the earth with respect to the Sun in 29.53 days i.e. the synodic month. Thus the phases of the moon changes in dates from year to year. In the interval of 19 years both the periods make a complete number of orbits. Hence the phases of the moon and the lunar eclipses are repeated in a cyclic pattern in a period of 19 years.

The relatively large mass of the moon makes the orbital angular momentum of the moon larger than the earth's rotational angular momentum. The mean surface gravity of the moon is 0.165 of the earth's gravity and the escape velocity $(2\ GM/r)^{1/2}$, is about 2.38 $km^{-1} \cdot s^{-1}$

It has been known for a long time that the topography of the moon has two distinct type of terrain. The elevated lunar highlands which are heavily cratered, are called 'terra' (meaning land in Latin), and the low, dark and smooth regions are named as 'mare' basins (meaning sea in Latin). Most of these maria are fairly circular. The very first fly-by of the Soviet Spacecraft Luna-3 showed that the far-side of the moon contains very few mare basin regions as compared to the large number of them at the near-side. The USA spacecraft Lunar Orbiter and Apollo (in orbit) found large concentrations of excess mass, called 'mascon' which coincided with circular maria regions. These suggest occurrences of relatively high density materials in these locations.

The early in situ observations of the moon were made by the US Surveyer-1 lander mission in 1966 in which an unmanned spacecraft landed inside a 100 km crater of a mare basin. Surveyer-2 crashed on the moon, while the follow up missions of Surveyer 3 to 7 successfully conducted scientific experiments and were used to select landing sites for the Apollo missions. The Surveyer missions 3 to 6 landed in a mare basin, while the last one, Surveyer-7 landed in the highlands, on the rim of the crater Tycho. Alpha back-scattering experiment data, telemetered to the earth revealed that the mare basins were mainly of a basaltic nature, having SiO_2, Al_2O_3, CaO, MgO and TiO as main constituents. Other mares are expected to be of similar composition.

For more detailed studies samples of lunar soil returned to the earth were needed. With this in mind the Apollo missions of the USA conducted successfully lunar orbital missions in 1968-69. The USSR carried out Zond missions in Lunar orbital mode successfully during this period. In the historic Apollo-11 mission of NASA in July 1969, US astronauts Neil Armstrong and Edwin Aldrin, landed on the moon and collected lunar rocks and soil of a wide range which were brought back to the earth. While the Apollo-11 mission landed close to the Surveyer-5 landing site, the next Apollo-12 mission successfully landed near Surveyer-3 site. The astronauts visited the the Surveyer-3 spacecraft and removed some parts of it for analysis on earth and a large number of samples of lunar rocks and soil were brought to the earth. The Apollo-13 mission averted near disaster with an exploded oxygen tank of the service module and returned to the earth, without landing on the moon. This was followed by the series of successful flights of Apollo 14, 15, 16 & 17 which brought rock samples from a variety of landing sites. Motor driven Moon-Rover vehicles were used in the last three missions. This phase of manned Lunar missions of NASA ended in December 1972. In September 1970, the USSR flew a spacecraft to the moon with an ingenious device by which the unmanned spacecraft swooped close to the lunar surface and scooped up a samples of lunar soil which was brought back to the earth for analysis. It may be noted that all soil samples of the moon belong to mare basin regions, which are relatively smooth and suitable for landing. Therefore the lunar highlands are poorly studied so far.

4.1.3 Composition of Lunar Rocks and Soil

Before dealing with the composition of lunar rocks and soils, we first discuss briefly the rock types in the moon and their general nature. In the most general way the rocks are classified into three main categories. One, the igneous rocks which are formed by crystallization of molten magma originating from volcanic activities. Most of the lunar rocks are of this igneous type. Two, the metamorphic rocks are those which were once of another type such as igneous, but have undergone substantial reheating and recooling processes. On the moon the most common metamorphic process is meteoritic impact which result in shock heating. A significant portion of lunar rocks fall in this metamorphic class. Three, the sedimentary rocks, as we know on Earth, are formed by particles which are deposited by wind and rain and have been subjected to a low pressure and heating. Such rocks are obviously absent on the moon.

The Lunar surface is dominated by large scale impact craterings of various sizes and these impact basins are filled with basalt type igneous rocks. Recent impact craters sometimes display rays and secondary craters formed by ejecta of the primary impact. Large impact craters sometimes show central peaks which are formed by the fracture of the bedrock which liberate ejecta by a volcanic process. Smaller impact events strike the rubble from large impacts. The impact shocks melt and weld debris fragments which are again fragmented by impacts. Thus the debris layer is being constantly stirred or "gardened". This highly fragmented, reworked debris in finely powdered form on the lunar surface is called "regolith". It may be noted that the lunar gravity is sufficiently strong to prevent the ejected regolith material to escape from the moon. Thus most of the regolith is retained on the moon.

The welding of regolith fragments produces a rock called 'Breccia' which consists of loosely compacted granular fragments. There are seven important major minerals in lunar rocks with abundances $> 1\%$. These are as follows:

(a)	Oxides	:	Ilmenite	$(FeTiO_3)$, abundance	$> 10\%$
			Crystobalite	SiO_2	$\sim (1-10)\%$
			Tridymite	SiO_2	$\sim (1-10)\%$
(b)	Silicates	:	Pyroxine	$(Ca, Fe, Mg)SiO_3$	$> 10\%$
			Feldspar	$CaAl_2Si_2O_8$ / $NaAlSi_3O_8$	$> 10\%$
			Olivine	$(MgFe)_2SiO_4$	$(1-10)\%$

In addition to these a wide range of minerals of (Fe, Ni), non-metal (graphite), sulphides and carbides are present in lunar rocks in small amounts of $< 1\%$.

A few striking differences found in lunar basalts may be noted. The silica abundances are low in lunar basalts as compared to the average terrestrial value. Also lunar rocks have alkali metal abundances which are several times higher than those in the earthen basalts. In addition the titanium abundance is very high which is seen in the large abundance of Ilmenite ($FeTiO_3$), which is sometimes $> 20\%$. The low abundance of silica is reflected in the relatively small content of olivine.

Metallic iron is found in very small amounts in lunar basalts and usually in association with triolite (FeS) with Ni content $< 0.1\%$. In contrast, lunar regoliths and breccias contain metallic iron with a nickel content as high as 30% and Co 1%. So it can be concluded that the regolith metal arises from the debris of asteroidal matter.

Lunar Elemental Abundances

Elemental abundances in lunar basalts and regoliths provide some important clues to the evolution of lunar materials. In regoliths and breccias which were subjected to shock induced metamorphism, observed enhanced abundances of Ni, Ag, Cd, In, Te, Re, Os, Ir and Pt indicate contributions of meteorites. It is estimated that about 2% of meteoritic materials are mixed with lunar regoliths. Several volatile elements such as hydrogen, carbon, nitrogen and rare gases are strongly enhanced. These are indicative of solar wind implantation of ions in the lunar surface materials. Surprisingly, the D/H ratio in lunar materials is 3 to 4 times smaller than that on Earth. The isotopic composition of helium is consistent with its solar wind origin.

In the older geochemical classification of elements, three groups are noted, mainly based on their affinity; these are:

Lithophile (= rock-loving) : Oxides and silicates
Chalcophile (= sulfur-loving) : Sulphides
Siderophile (= metal-loving) : Free metals

In the basalts many elements of the refractory group, (see Chapter 5.4) such as rare earths, actinides U and Th, and Ti, Ta, Ba, Sr, Sc etc. are enhanced by a factor of about 100 relative to C1 chrondritic elements. Ca, Al and Li are enriched by a factor of ~ 10. These abundance enrichments reflect the thermal fraction action processes during lunar evolution. Chalcophile elements e.g. Si, Se, Te, Cu, Zn etc. are depleted by a factor of ~ 10 relative to C1 material, which is due to core forming melting and extraction processes. Pb observed in lunar basalt was formed by radioactive decay, after differentiation and extraction of chalcophiles. Several siderophiles e.g. Ge, Co, As and Sb are depleted by a factor of 10, while the Ag, Cd group

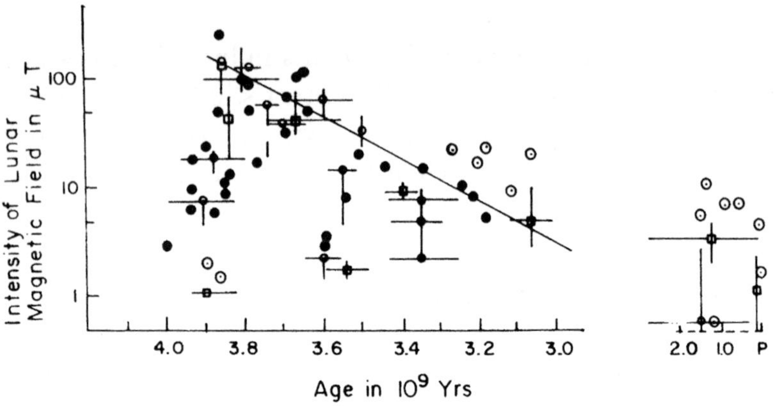

Figure 4.1.1: The intensities of paleo-magnetic fields of moon samples plotted against their age as measured by the radioactivity method. The data points are measured by various methods and these refer to Apollo 11, 12, 15, 16 and 17 mission samples. (Adopted from the review by Runcorn[4]).

by a factor of 100, and Au, Ir, Pt by a factor of 1000. These reflect the fractionation processes.

4.1.4 Magnetic Field of the Moon

Magnetic field measurements on the surface of the moon in the Apollo missions showed very weak field strengths of 6 to 300 γ ($1\gamma = 10^{-5}$ Gauss) which are negligible compared to planetary standards. Thus it is concluded that a cold body like the moon does not have an intrinsic magnetic field. However, laboratory analysis of lunar samples returned by the Apollo missions revealed the surprising result of the presence of a substantial natural remnant magnetization (NRM)[1,2,3]. The implication is that the remnant magnetization of lunar basalts and breccias has been acquired at the time of cooling from a global magnetic field. Such a global field does not exist any more. The absence of the global field now, is understood as the moon is a small body which loses its heat energy more quickly than a planet and the hypothesis that the moon's interior is frozen seems thus to be reasonable.

Detailed measurements of remnant magnetizations of many samples showed surprising results as given in the review by Runcorn[4]. The determinations of paleo-intensities of the magnetic fields when plotted against the age of lunar rocks as determined by the radioactive age method, showed

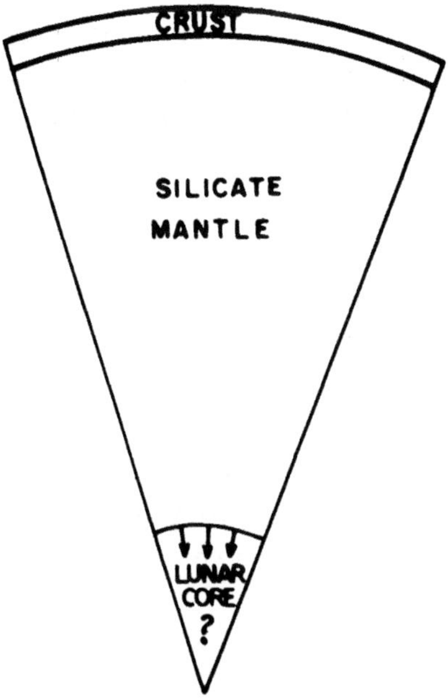

Figure 4.1.2: A sketch of the probable internal structure of the moon.

that the global field with a maximum value of about 100 μT (1 nano Tesla = 1γ) existed about $3.9 \cdot 10^9$ years ago and subsequently disappeared at the present epoch (Figure 4.1.1). At the time of maximum $3.9 \cdot 10^9$ yrs. ago the intensity of the magnetic field of the moon was close to the present magnetic field intensity of the earth. How did the moon acquire such a high intensity and how did it decay to the negligible present value? These striking results which generated attention and controversy, are now generally accepted. However the interpretation of the results involves complex processes and hypothesis which are not fully understood. Possible interpretations are given by Runcorn[(4)] 1994. In summary we note that although not conclusive, many scientists believe the moon possesses a core. Induction by the solar wind of an electrically conducting core is consistent with Apollo magnetometer observations.

4.1.5 Interior of the Moon

In order to probe the lunar interior, a seismic experiment was carried out in the Apollo Program by impacting the burnt-out stage of a Saturn-V booster rocket on the moon's surface. Results from seismic probes indicated that the lunar crust extends to a thickness of about 50 km, in which below 30 km, a large number of cracks and voids were made by many large impacting events. Thus the crust constitutes ~ 8% of its volume. Only 1% of the crust is made of basalt. The mantle of the moon made of silicate material is believed to occupy a large part of its volume. Below the mantle, the core of the moon probably exists. The presence of the lunar core is problematical. Measurements of the electrical conductivity of the moon of magnetic field variations of the solar wind were made. From these data, the upper limit of the radius of the metallic core is placed at about 360 km. This is only 1% of the volume of the moon. In Figure 4.1.2 a tentative model of the lunar interior is shown, which is as yet not established. The thermal structure of the interior of the moon is uncertain, as the thermal conductivities at different levels are not known. With several assumptions, temperatures of 1200 to 1500°C are expected to prevail in the deep interior of the moon, which lead to the conclusion of the moon being a cold object.

4.1.6 Origin of the Moon

The origin of the earth-moon system has been the subject of much discussions and controversy for decades. The earlier hypothesis was that the earth and the moon were a single body that was split apart by rotational instability; then the moon could be the missing part of the earth's crust and upper mantle from the Pacific ocean region. This view is inconsistent with the chemical composition measurements of the lunar surface by spacecraft, which show that the lunar crust is in many respects different from that of the earth.

The other view is that the moon and the earth were both formed out of solar nebular material by an accretion process and developed as a binary system and a double planet. If so the materials of both the bodies should have similar composition. This is not so, as discussed earlier. The lunar basalts have strikingly different composition as compared to those of the earth. Also it is noted that the density of the moon, 3.3 g$\cdot$cm^{-2}, is vastly different from the density of the earth, 5.5 g$\cdot$cm^{-2}. Why does such as large difference exist? As the moon has, practically no core, what happened to the core of the moon, if both were formed at the same time? Therefore this hypothesis is not consistent with many observations.

Following the Apollo and Luna missions, a new hypothesis had been put forward. According to this model in the early period of the earth, a giant planet of the size of Mars collided with the earth. Most of the mass of the impacting body was captured by the earth and the core of the impactor sank into the core of the earth. Its outer parts were mixed with those of the earth and a portion of the impactor splashed out and escaped from the earth forming the moon and a number of smaller bodies forming a ring. It is not known whether this scenario is a valid one or not. There are great diversities in the views of scientists on this subject and the enigma of the origin of the earth-moon system continues.

4.2 Space Missions and Planetary Exploration

4.2.1 Early Planetary Missions

Mariner, Venera, Viking and Vega

In the 1960's and 1970's, a number of planetary missions were initiated by the USA and USSR. The Mariner 4 mission of NASA launched in November 1964 made the first successful fly-by of Mars in July 1965, and Mariner-5 launched in June 1967 did a successful fly-by of Venus at 4000 km range. The first successful atmospheric probe in the atmosphere of Venus was achieved by the Venera-4 mission of the USSR in October, 1967. This was followed by the successful Venera-5 mission in May 1969 with an atmospheric probe for Venus. Shortly afterwards in December 1970, the USSR achieved the distinction of the first successful landing of instruments on the surface of Venus. Planetary explorations continued with vigour in the 1970's with the first successful Mars orbiter in November, 1971 by the Mars-2 mission of the USSR, followed by two more Mars orbiters, Mars-3 and 5 in subsequent years. The exploration of Mars received a big boost with the first landing and surface exploration of Mars by instruments in the Viking-1 mission of NASA in July, 1976. Further success was achieved by Viking-2 in September, 1976. The USSR continued with the successful exploration of Venus by instrument probe landings on the Venusian surface in the Venera 11, 12, 13 and 14 missions during December 1978 to March 1982. With more sophisticated instrument landings on Venus with the Vega-1 and 2 missions, USSR scientists achieved great success in the exploration of Venus.

4.2.2 Deep Space Probes

Pioneer, Voyager, Galileo and Ulysses

For probing deep space and the outer planets and their atmospheres, NASA of the USA launched Pioneer-10 (P-10) and Pioneer-11 (P-11) with complex arrays of instruments in February 1972 and April 1973 respectively. Pioneer-10 made a Jupiter encounter in December 1973 and provided the first excellent in-situ results on the Jovian atmosphere and the magnetosphere. Pioneer-11 has encountered with Jupiter in December 1974 followed by the first historic fly-by of Saturn almost five years later in September 1979.

These historic missions were followed by NASA with the launching of the two deep space probes Voyager-1 and Voyager-2 in August-September 1977. Voyager-1 had a successful observation of Jupiter during its fly-by in March 1979 followed by the Saturn encounter in November 1980. Voyager-2 achieved another historical landmark in planetary exploration as the space probe flew-by Jupiter in July 1979, Saturn in August 1981, Uranus in January 1986 and Neptune in August 1989, thus completing a "GRAND TOUR" of the four outer planets during its space voyage of 12 years.

Planetary explorations are continued in the 1990's with international participation. The Galileo spacecraft of NASA has had its Jupiter encounter and the Ulysses spacecraft of ESA-NASA encountered Jupiter in February 1992. The Hiten spacecraft of Japan launched in January 1990 flew-by the moon and it deployed a small spacecraft 'Hagormo' into lunar orbit. 'Mars Observer' of NASA-USA was launched in September 1992 and Russia is preparing for another launch of their sophisticated space mission in early next century for the exploration of Mars and searching for evidence for life on our neighbouring planet.

In this chapter we shall present briefly some of the highlights of our present state of knowledge on the inner and outer planets, particularly of the physical properties of their surfaces and the atmospheres as derived mainly from spacecraft measurements. This information will be helpful in exploring the present day ideas on the origin and evolution of the solar system as will be briefly discussed in the next Chapter.

4.3 Inner Planets

The four inner planets of the solar system, Mercury, Venus, Earth and Mars are characterized by rocky crusts and large abundances of heavy elements such as iron and nickel in their cores. These inner planets, with the exception of Mercury, have thin to moderate atmospheres. These four planets of moderate size are often called terrestrial planets to distinct them from the giant planets of the outer solar system.

The physical and orbital properties of the four inner planets are given in Table 4.3.1. Here we shall concern ourselves briefly with the two main aspects: the present day knowledge of their surface properties and the composition of the planetary atmospheres as revealed by spacecraft measurements. Here we note that the planetary atmosphere of the inner planets differ greatly from one another, and this is due to several factors. The four main factors are: (a) gradient in the composition of the solar nebula from which the planets were formed by condensation, (b) different escape rates of different gases due to different amounts of heating and gravitational forces, and (c) different distances from the Sun which determine the amount of solar energy influx on the planet together with the (d) different rotational period.

Table 4.3.1: Features of the Inner and Outer Planets

Planet	Radius (km)	Mass (10^{27})g	Density g·cm^{-3}	Density P=0	Rot. Period	Axial Incl.	Dist. A.U.	Orbit Period
Mercury	2,440	0.330	5.43	5.30	58.6 d	0°	0.39	88 d
Venus	6,050	4.870	5.24	4.00	-243 d	177°	0.72	225 d
Earth	6,380	5.970	5.52	4.05	23 h 56 m	23.5^o	1.00	365.25 d
Mars	3,400	0.642	3.93	3.74	24 h 37 m	25.2^o	1.52	687 d
Jupiter	70.850	1899.0	1.36	0.1	9 h 50 m	3.1^o	5.20	11.9 yr
Saturn	60.330	568.4	0.71	0.1	10 h 14 m	26.7^o	9.54	29.5 yr
Uranus	25,400	86.97	1.30	0.3	14 h 42 m	97.9^o	12.50	84.0 yr
Neptune	24,300	102.85	1.8	0.3	18 h 24 m	29.6^o	30.06	165 yr
Pluto	1,150	0.013	2.0	2.0	6.39 d	122.5^o	39.40	248 yr

4.3.1 Mercury

Orbital and Surface Features

Because of the small size of Mercury and its angular proximity to the sun which is always less than 28°, it is never observed in a dark sky and so it is very difficult to make earth-based studies of Mercury. Our knowledge of Mercury comes mainly from the three Mariner-10 fly-by missions of NASA-USA in 1973-75 as no other satellite has visited the planet. Mercury, the inner-most planet, has the shortest orbital period (86 days) among the planets. The eccentricity of the orbit is large, the perihelion and aphelion are 0.307 and 0.467 A.U. As a result the intensity of sunlight varies by a factor of 2.3 and the solar intensity ranges between 4.6 and 10.6 times that at the earth at 1 A.U. (the solar constant is 1.34 kw/m^2). The rotational period of Mercury is 57.3 days which is 2/3 of its orbital period. So Mercury is locked with the sun in a 3:2 spin-orbit resonance, rotating three times on its axis during two orbits around the Sun. As a result there is a most unusual pattern of solar heating on the surface of Mercury. There are two hot spots on the equator on opposite sides. Other regions do not receive overhead sunlight at perihelion.

Because of spin orbit resonance, the temperature on the surface of Mercury varies in a complex manner. At perihelion the day temperature at the subsolar point may be about 700 K, while at aphelion a temperature of about 550 K is expected. Night temperatures would drop to about 100 K. As at the poles the sunlight is always at grazing incidence because of zero axial tilt, the ground temperature is very low.

The Mercurian surface is heavily cratered as photographed from Mariner-10 fly-by missions. The planet has a thick rigid crust of which we have very little direct information on its composition. The rocky planet devoid of any atmosphere appears to be featureless and somewhat similar to lunar highlands. The planet is bombarded by solar wind protons and helium ions. A small trace of hydrogen and helium observed on the surface of Mercury is consistent with the steady state balance between solar wind influx and the escape rate, similar to that on the lunar surface. A small amount of Na and K observed in the atmosphere of Mercury probably originates from sputtering from the surface minerals by energetic particles in space.

Magnetosphere of Mercury

Mariner-10 fly-by data provided by the magnetometer indicated the presence of a bow shock and a weak planetary magnetic field. Strong solar wind at the noon side may press the magnetic field close to the ground level, but it may extend into a small magnetosphere at the night side. The origin of the magnetic field is probably a weak core dynamo. Table 4.3.2 gives the summary of the magetospheric data of the inner planets.

Table 4.3.2: Magnetic Fields of Inner and Outer Planets

Planet	Dipole moment ($G\cdot cm^3$)	Tilt angle (deg)	Dipole equat. field (nT)*	Stagn. dist. (d)+
Mercury	$5x10^{22}$	$+14^o$	330	1.4
Venus	$< 4x10^{21}$	-	-	-
Earth	$8x10^{25}$	$+12^o$	31,000	10.4
Moon	$< 1x10^{19}$	-	-	-
Mars	$< 2x10^{22}$	-	-	-
Jupiter	$1.6x10^{30}$	-10°	428,000	65
Saturn	$4.7x10^{28}$	0°	21,200	20
Uranus	$3.8x10^{27}$	-59°	23,000	20
Neptune	$2.0x10^{27}$	-47°	14,000	26

*1 nano Tesla (nT) = 10^{-5}G.
+d in units of planetary radius.

4.3.2 Venus

Orbit and Thermal Features

The planet Venus, similar in size to earth with a dense and perpetually cloudy atmosphere of CO_2 presents a number of paradoxes. Venus orbits the Sun in a nearly circular path in 224.7 earth days, while it rotates on its axis once in 243 earth days in a retrograde direction i.e. its spin angular momentum is opposite to its orbital angular momentum. It is the only object in the solar system with retrograde spin angular momentum the reason of which is unknown at present. Because of this peculiarity, seen from the surface of Venus the Sun makes one complete circuit in the sky in half a Venus year and the actual length of a day on the surface of Venus is equal to

116.8 earth days.

From measurements of thermal emission by the planet Venus in the radio region of wavelengths $\lambda = 0.2$ to 70 cm, and using the Planck function, the brightness temperature T_B is found to be about 700K at 6 cm. The actual temperature of the emitting body, T is then given by $\varepsilon T^4 = T_B{}^4$, where ε is the emissivity which is probably close to 1. As there is some absorption of radio waves, the actual temperature must be over 700K. In the measured curve of T_B vs λ in cm there is no indication of any absorption at 1.35 cm due to H_2O. Hence it is concluded that the Venus atmosphere contains little or no water vapour and an upper limit of 0.1% (1000 ppm) is placed. In situ measurements by the Venera 11 and 12 landers indicated water abundances ranging from 200 ppm at the cloud base to 25 ppm at the surface. This latter value is about 10^{-5} times less than that on earth.

The Atmosphere and its Constituents

The main constituents of the Venus atmosphere is CO_2 which was identified by infrared spectroscopy of Venus in 1932. Spacecraft measurements with Venera 4, 5 and 6 entry probes of captured samples of the atmosphere, extracted CO_2 and measured the residual pressure. From these the mole-fraction abundances of CO_2 was obtained as 0.965. The other main constituent is N_2 whose abundance is found as 0.035 mole fraction. Other minor constituents in the atmosphere of Venus were found to be SO_2 (150 ppm), Ar (70 ppm), CO (30 ppm) and H_2O (25 ppm) as noted earlier. Rare gases of He and Ne are present in very small amounts of 5 to 12 ppm. The oxygen abundance is puzzling. Spectroscopic measurements from the earth gave a firm upper limit of 1 ppm, whereas the Pioneer-Venus experiments reported a value of 45 ± 25 ppm for O_2. This dichotomy is to be resolved.

Isotopic composition studies made by the Pioneer-Venus mass spectrometer yielded the surprising result that the deuteron abundance i.e. HDO : H_2O was 0.032 ± 0.004 which is about 150 times higher than the terrestrial or meteoritic ratio.

Temperature and Pressure Profiles

Early observations of a high surface temperature implied a massive

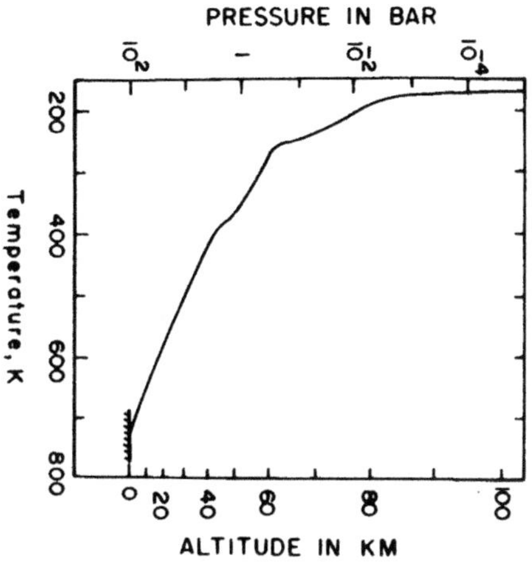

Figure 4.3.1: The temperature and pressure profiles of the atmosphere of Venus as a function of altitude.

atmosphere of Venus. A large number of spacecraft measurements are combined to obtain the mean temperature structure of its atmosphere as shown in Figure 4.3.1. With a surface temperature of about 750 K, the temperature gradient near the surface is about 9 K/km which is close to adiabatic up to about 60 km. Above this it drops faster and above 80 km it is nearly isothermal having a very low temperature of about 180 K. The Venus probes have identified cloud layers which occur between roughly 30 and 70 km altitude and are broadly divided into upper, middle and lower haze with larger particle sizes being more numerous at lower levels. At cloud tops the temperature is about 240 K at 70 km. The surface pressure measured by spacecraft is very high, about 92 bars. Due to its immense atmospheric mass and high infrared opacity, there is practically no variation in day-night surface temperatures. The solar constant at Venus is about 2.62 kw/m^2 which is twice that of the earth, but the absorbed flux is only 0.13 kw/m^2. Only 2.5% of the sunlight reaches the surface.

Observations of cloud tops at UV wavelengths indicate a large retrograde wind speed which is estimated to be as high as 100 m/s relative to the surface.

This high wind speed drops off with lower altitude to about 1 m/s near the surface.

The Magnetosphere of Venus

Spacecraft observations have shown that Venus has no intrinsic magnetic field, similar to that of the moon, its dipole moment being $< 4 \cdot 10^{21}$ Gauss·cm^3.

Spacecraft results have brought many unsolved problems and puzzles on Venus which are yet to be unraveled.

4.3.3 Mars

General Features

The planet Mars orbits the Sun in a slightly elliptic orbit of eccentricity $e = 0.0934$ which gives an aphelion distance of 1.666 A.U. and a perihelion of 1.38 A.U. The intensity of sunlight at the aphelion exceeds that at the perihelion by a factor of 1.45 which leads to a strong annual global variation of temperatures. These seasonal changes of temperature are called eccentricity seasons in order to distinguish these from the axial tilt seasons, similar to those of the earth. During one Martian year of 687 earth days the planet goes through one complete cycle of warming and cooling phases. Mars rotates on its axis with a sidereal period of 24 hrs 37 min and 22 sec which is surprisingly close to that of the earth. The axial inclination of Mars is 25.2^o which is also very similar to that of the earth. The spin axis of Mars precesses about the poles of its orbit with a period of about 50,000 years as compared to the 26,000 years of the earth.

A surprising feature of Mars is its density, 3.93 g·cm^{-3} which is far too small compared to that of the earth and the inner planets which are in the range of 5.2 to 5.5 g·cm^{-3}. The difference is still significant even after correcting the densities to zero pressure. The possible causes of this difference are the subjects of debates among planetary scientists. Some aspects of these causes will be discussed later (in Chapter 5.4). Here we note that the pressure corrected density of Mars is 3.73 g·cm^{-3} which is similar to that of ordinary chrondrites (see Section 5.1). If the Fe:Si ratio is the same for Venus, Earth

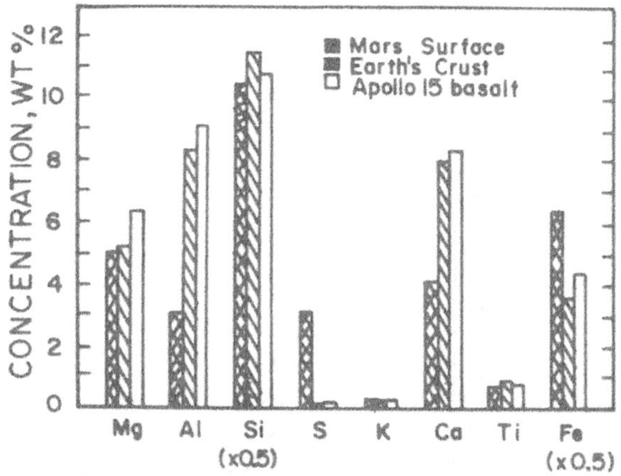

Figure 4.3.2: Composition of some major elements on the surface of Mars and their comparison with lunar and terrestrial basalts. (Fe is plotted as × 0.5).

and Mars, the lower zero pressure density of Mars possibly signifies a high degree of oxidation with nearly all metal being oxidized to sulphides and FeO-silicates.

Surface Composition of Mars

The in-situ observations of the Martian surface by Viking Landers gave detailed information of surface features. X-ray fluorescence studies at the landing sites indicated a fine grained soil composed of high amounts of iron and sulphur, with fair quantities of Al, Mg, Si and Ca. The abundances of K and Ti were very small and the Ca/K ratio was high (Figure 4.3.2). The soil was found to be produced by weathering of igneous rocks, and was composed of roughly 80% iron rich clays, $\sim$ 10% magnesium sulphate, and about 5% each of calcite and higher oxides of iron.

The Atmosphere of Mars

Composition

Spacecraft measurements confirmed the ground based IR results that CO_2 is the most abundance gas in the Martian atmosphere and its concentration

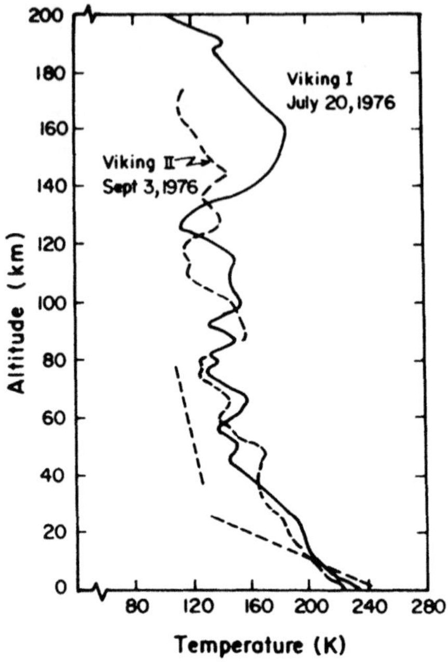

Figure 4.3.3: Temperature in the Martian atmosphere as a function of altitude.

was measured to be 0.953 mole-fraction. A small concentration of N_2 gas of 0.027 mole-fraction was established after several searches. Attempts were made to identify H_2O and O_2 gas in the atmosphere over many years, and many conflicting results were reported. Mariner-9 measurements made during a one year period revealed a very minute and variable amount of water vapour during spring over the north polar cap. Viking orbiters reported barely detectable amounts. Thus it is concluded that the water vapour content is variable in the range of 0 to 0.03%. Other small constituents are CO(0.27%), O_2(0.13%) and ^{40}Ar (0.016). Inert gases of Ne, ^{36}Ar and Kr are present only in trace amounts of 2.5 to 0.3 ppm.

Atmospheric Pressure and Temperatures

Viking Landers 1 and 2 revealed by direct measurements diurnal temperature variations close to the surface of about 190-240 K. The variations of temperature with altitude were determined during the entry periods of V. Lander I (1600 hr. local time, 20 July 1976) and V. Lander 2 (0900 hrs, 3 September 1976). Both detected ground temperatures of about 230-240 K decreasing with wave like structures above 40 km altitude (Figure 4.3.3). The temperature lapse rate was half the adiabatic gradient and the troposphere temperatures at 40-70 km altitudes at day time were within 20 K of the saturation temperature of CO_2 (about 120 K) at that altitude.

Viking Landers 1 and 2 made direct measurements of the surface atmospheric pressure and its variations over one Martian year of 669 earth days. Seasonal variations of atmospheric pressure were found to be from 10.2 to 7.4 mbars at V. Lander 1 site. V. Lander 2 showed similar variations of pressure ($\sim$ 20%) from 9.0 to 6.8 mbar. These large variations are ascribed to the condensation of CO_2 in the winter at the poles and evaporation during summer. Some aspects of the origin and evolution of the planetary atmosphere will be mentioned in Chapter 5.4.

The intrinsic magnetic field of Mars is small or non-existent. The present upper limit of its dipole moment is $< 2{\cdot}10^{22}$ G·cm^3.

Satellites of Mars

Mars has two satellites, Phobos and Deimnos which are rocky and highly irregular in shape with cratered surfaces. These appear to be relatively unaltered samples of small bodies that passed through the mass accretion zone late in the period of planetary formation.

4.4 The Asteroid Belt

4.4.1 General Features and Orbits

The asteroid belt is composed of small solid bodies with masses ranging from a few meters to 1000 km which orbit the Sun at a distance of about 2.2 to 3.3 A.U. The distribution of the orbital semi-major axes (a) of the asteroids is shown in Figure 4.4.1. It is seen that there is a strong concentration of asteroids with a = 2.06 to 3.65, which is called the main belt. In the main belt several peaks and valleys occur indicating several families of asteroids having nearly the same value for their semi-major axis, a. Proceeding from the Sun, we find the group Athens having a $<$ 1.0 A.U., and the next group Apollos with a $>$ 1.0 which crosses the orbit of the earth. The Amors group asteroids approach the earth orbit, but at present do not cross the earth orbit. The Hungarias group lies outside the Mars orbit. The main belt is composed of several families as shown in the figure. Outside the main belt there are three groups of these, the remarkable feature of these groups is that Jupiter exerts very strong influence on them and as a result there is a resonance structure due to perturbations by Jupiter. For example Hildas (3:2) orbit the Sun three times for every two orbits of Jupiter, Thule exhibit the ratio 3:2 and the Trojans 1:1. The Trojans asteroids orbit within the two Lagrangian positions L4 and L4 i.e. where gravitational forces of the Sun and Jupiter cancel each other with the same average period as Jupiter in spite of small perturbations.

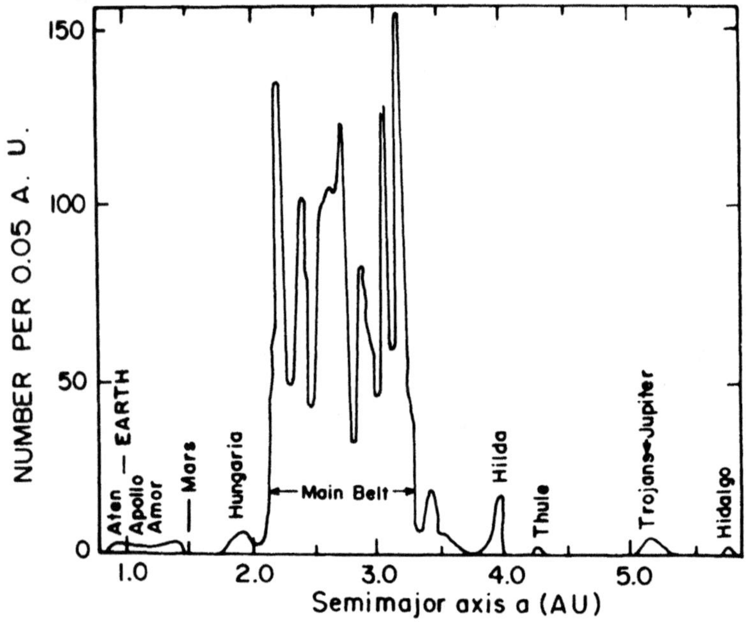

Figure 4.4.1: The number distribution of the semi-major axes of the asteroids.

4.4.2 Masses, Densities and Composition

The rotation periods of the asteroids lie mostly in the range of 7 hr to 30 hrs, the average being close to about 10 hrs, while the extreme values are 2.5 hrs and 50 hrs. The albedos (i.e. the fraction of light reflected by the body) of the asteroids vary over a wide range and the distributions show two clearly separated groups, one with a peak at 0.033, and a minimum at 0.07 and the second with a peak at about 0.15. This probably signifies two different composition and surface properties.

The diameters of several hundred asteroids have been determined and about 30 are known with diameters greater than 200 km. Of these the 7 largest asteroids (larger than 300 km diameter) are Ceres (1020 km), Vesta (550 km), and four others having 440 to 310 km diameters. The mass determination of the three largest asteroids have been made; these are as follows in units of 10^{24} gm: Ceres 1.18, Pallas 0.216, Vesta 0.275. The densities of these three are estimated as, Ceres 2.12, Pallas 2.62 and Vesta 3.16 $g{\cdot}cm^{-2}$ with an uncertainty of 30-40%. The albedo vs color systematic of asteroids is used to investigate the mineralogical composition of asteroids.

The total mass of the asteroid belt is estimated as $2.0{\cdot}10^{25}$g. This is about 27% of the mass of the moon and 0.3% of the mass of the earth.

4.4.3 The Origin and Evolution of Asteroids

The asteroids are remnants of a process that failed to form a planet. The fact that the aggregate mass of asteroids is far too small compared to the earth or Mars can conclusively dispel one popular idea of them being remnants of a broken planet.

In the evolution of asteroids and their families we first consider the inter-asteroidal collisions which are classified into three types, assuming the relative impact velocity of 5 $km{\cdot}s^{-1}$. (A) When the mass of the impactor, m_i, is less than 1/50,000th of the mass of the asteroid m_a, a crater is formed with mass excavated, $m_e \sim 1000\ m_i$. (B) When $m_i => (1/50{,}000){\cdot}m_a$, the impact fragments the entire asteroid, but the fragments do not escape, but fall back to form a rubble piled up in the shape of an ellipsoid. (C) When $m_i > (1/50{,}000){\cdot}m_a$, the asteroid breaks up and fragments move off with escape velocities, the larger fragments forming a family of asteroids, while the dust grains become meteoroid stream of dust. Thus it is presumed that in the initial phase of accretion in the belt, the asteroids were of sizes considerably larger than those present now.

A brief account of the origin and evolution of the asteroid belt is as follows. In Stage 1, at the beginning of the solar system all matter was in the form of dust and gas; these aggregated into loose fractal particles in Stage 2. In Stage 3 they became gravitationally compacted into undifferentiated solid asteroids, of small, medium and large size. In Stage 4 the largest of the

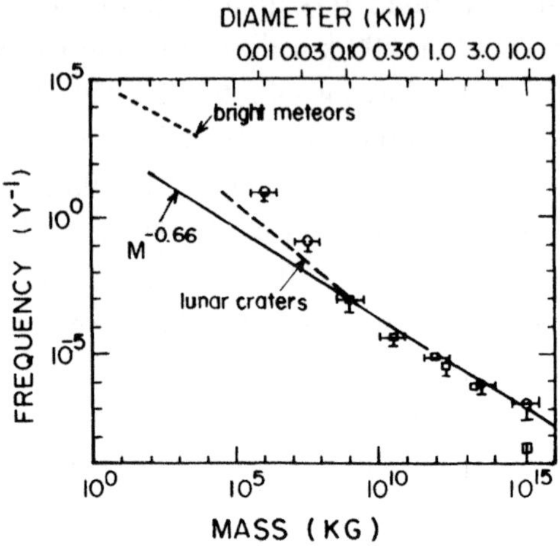

Figure 4.4.2: Observed and estimated rates of impacts on the earth as a function of mass or diameter.

asteroids had their interiors molten and differentiated owing to the decay of long lived isotopes such as U^{235}, $^{26}Al(T_{1/2} = 7.2 \cdot 10^5$ yr). Stage 5 involves the collisional fragmentation resulting in metallic asteroids and mantle asteroids from the large ones, and rocky and undifferentiated asteroids from the break up of small and medium ones. In Stage 6 further fragmentation took place leading to the present day asteroids of metallic, mantle, and undifferentiated rocky types and of meteoroid dusts.

4.4.4 Asteroids and Earth Impacts

As a number of asteroids have earth crossing orbits it is clear that the chance of an asteroid collision is not insignificant (see e.g. Gehrels). Figure 4.4.2 shows the estimated cumulative impact rate (per year) plotted against the mass or diameter of Near Earth Asteroids (NEA)[(5)]. The mean collision velocity can be shown to be about 21 km$\cdot$s^{-1}. In the figure the results of spacewatch of NEA of asteroids with a diameter in the range of 10 to 0.01 km are shown. It is found that the measured cumulative impact rates for a diameter less than 0.1 km, are much higher than the theoretical distribution. Lunar cratering analysis by Shoemaker also indicate higher rates as shown by the dashed line. It is noted that objects of 10 m size hit the earth's atmosphere a few times a year. Most of them burn out in the atmosphere, only the metallic ones come through to make craters. The excess of small objects have been given the name of 'Arjunas' by some authors, after the Pandava prince of the Indian epic 'Mahabharata'.

4.5 Outer Planets

4.5.1 General Features

The outer planets Jupiter, Saturn, Uranus and Neptune are called 'Jovian' planets and are quite different in their intrinsic properties compared to the inner or 'terrestrial' planets. Pluto is also included in this class, called the outer planets. The physical and orbital properties of the Outer Planets are given in Table 4.2.1. The huge planet Jupiter, which is about 300 times more massive than the earth, has an observed density of 1.36 whereas the next most massive planet, Saturn has a density of 0.71. After a correction for self-compression, the intrinsic densities of Jupiter and Saturn are close to 0.1, while those of Uranus and Neptune have values of about 0.3. These reflect the average composition of each planet. A striking feature of the four Jovian planets is that in spite of their enormous masses, they spin about their axis in a much faster rate than the earth and the inner planets, e.g. Jupiters rotation period is only 9 hr 50 min and that of Saturn 10 hr 14 min, and Uranus and Neptune 14.7 and 18.3 hrs respectively. Due to their fast rotation rate their oblatenesses are much larger than the inner planets.

4.5.2 Jupiter

The observed composition

High resolution spectroscopic studies of Jupiter and Saturn have revealed large amounts of information on the composition of their atmospheres; Table 4.5.1 gives a summary of their main constituents. In Jupiter's atmosphere hydrogen and helium are the main gases with mole fractions of 0.82 and 0.18 respectively. Other gases of methane, ammonia and water vapour are present in small amounts. Trace amounts of other gases have been detected, which include ethane (C_2H_6) acetelyne (C_2H_2), ethylene (C_2H_4) and several others. The $^{13}C/^{12}C$ ratio is found to be similar to terrestrial and meteoritic value and the D:H ratio is much lower than in earth's oceans.

Table 4.5.1: Atmospheric Abundances

Component	Abundance in mole fraction	
	Jupiter	Saturn
Hydrogen H_2	0.82	0.94
Helium He	0.18	0.06
Methane CH_4	$1x10^{-3}$	$2x10^{-3}$
Ammonia NH_3	$2x10^{-4}$	$1x10^{-4}$
Water vapour H_2O	$1x10^{-3}$	$1x10^{-3}$

Note: 1. In addition H_2O, PH_3, CO, C_2H_4, C_2H_6, HCN are present in trace amounts with abundances in the range of 10^{-6} to 10^{-10}.

The pressure-temperature profile

The lower atmosphere of Jupiter which is convectively mixed and adiabatic in nature is called the troposphere. The main cloud layer on Jupiter is made of tiny ice crystals of solid ammonia and hence ammonia is partially depleted by condensation in the parts of the atmosphere observable from above. The penetration of visible and IR radiation in the Jovian atmosphere is blocked at a pressure of about 1 bar by dense cloud layers. The cloud top temperature is about 130 K and its pressure about 0.5 bar. The pressure-temperature profile derived for the Jovian atmosphere is shown in Figure 4.5.1. The altitude scale is the depth below the ammonia clouds. The region 0-1000 km represents ideal gas, 1000-4000 km non-ideal gas and the deepest portion 4000-7000 km a dense molecular fluid.

Interior Structure of Jupiter

It can be shown, from mass, radius and density data, that a giant planet like Jupiter must be composed of some material denser than hydrogen. If one assumes a solar-composition type mixture, one gets a too high density. Hence the model of Jovian structure should be able to explain the observed thermal emission from the core and the transport of the heat from the deep interiors to the surface of the planet. From several models the following tentative picture of the structure emerges. The core-mass of > 0.5% of the Jovian mass may be provided by rocky material in solar type composition. The materials forming the rocks and ice in the core would total about 30% of M_J. Above this from $r = 0.11$ to $0.76\ r_j$, there probably exists an extensive liquid and conducting

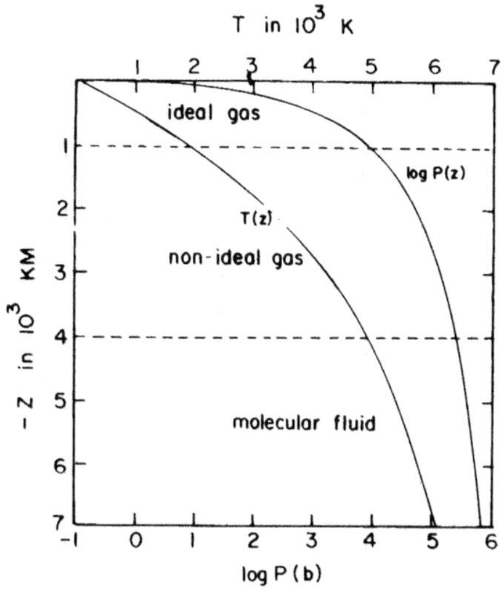

Figure 4.5.1: A sketch of the model of the structure of the deep atmosphere of Jupiter. The ammonia cloud top ($Z = 0$) temperature and pressure are taken as 130 K and 0.5 bar respectively; the altitude scale is below the above level.

metallic mantle of hydrogen. This electrically conducting layer is somewhat similar to that of the earth and is likely the source of Jupiters magnetism. Above the mantle, molecular hydrogen envelope is present (Figure 4.5.2a). There are many enigmas, such as the internal heat source of Jupiter, which are yet to be fully understood.

The Magnetosphere of Jupiter

The most intense magnetic field of all planets exists in Jupiter. The ground based observed non-thermal radio-emission from Jupiter was discovered in 1955 and was later interpreted as due to electrons trapped in the Jovian radiation belt and precipitating in its upper atmosphere. The in-situ measurements of the Jovian magnetosphere by Pioneer 11 and Voyager 1 and 2 provided the detailed picture. The maximum field in the northern hemisphere is 14 Gauss which is in good agreement with the maximum frequency in microwave emission of 42 MHz. The equatorial field lies in the range of 3.3 to 8 Gauss. A summary of the planetary magnetic field is given in Table 4.3.2. In Figure 4.5.3 we show a sketch of the Jovian magnetosphere and its interactions with the solar wind.

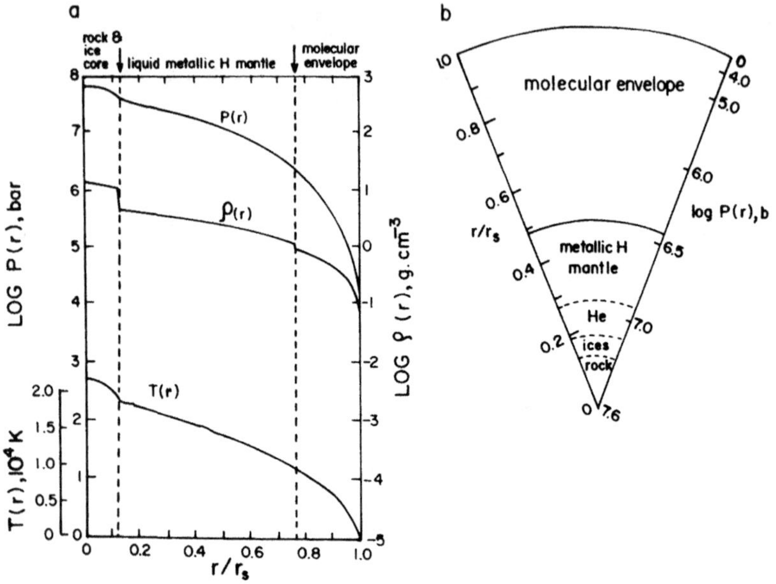

Figure 4.5.2: (a) Interior structure of Jupiter, (b) Interior structure of Saturn.

Jovian Satellite Io

Jupiter has a number of satellites of which the most prominent are the four Galilean satellites, Io, Europa, Ganymede and Callisto. Of these we shall briefly summarize some interesting features of the inner-most satellite Io. The surface-temperature of Io is 140 K. The observed density of Io, 3.5 g$\cdot$cm^{-2} suggests an ice free body of rocky material. Craters of diameter about a few km should have been formed on the surface, but surprisingly these are not seen. Voyager I found volcanic craters and evidence for volcanic eruptions in the recent past with sulphur-compound deposits. So it is concluded that active sulphurous volcanos are present in Io resulting in liquid lava flows which buries other surface features like impact craters.

An interesting feature of the Jupiter-Io system is the observation in the sodium D-line of a very large comet like feature around Io, orbiting tail first around Jupiter (Figure 4.5.4). The source of the sodium cloud is unknown at present.

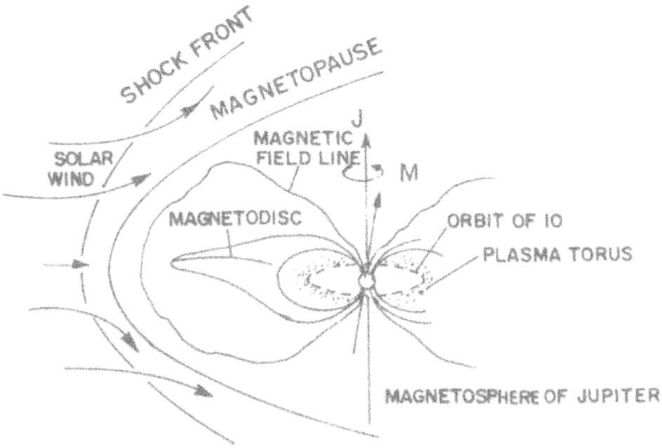

Figure 4.5.3: A sketch of the magnetosphere of Jupiter.

4.5.3 Saturn

General Features and Atmosphere

The physical and orbital properties of Saturn are given in Table 4.3.1 and some important features are discussed here. The second most massive planet of the solar system, Saturn, surrounded by a brilliant system of rings is the most spectacular object in the sky. Its zero pressure density is the same as that of Jupiter indicating similar composition. Its spin rate is very high similar to that of Jupiter; the axial tilt is somewhat larger than that of Jupiter.

The composition of the atmospheres of Jupiter and Saturn studied with high resolution interferometry revealed similar pattern of abundances, as given in Table 4.5.1. Hydrogen and helium are the most abundant gases, with smaller amounts of methane, ammonia and water vapour followed by a number of trace molecules. The cloud top temperature in Saturn is about 120 K and increases adiabatically to 700 K at a depth of about 800 km below the cloud top. In the above range of altitude the pressure varies from 1 bar to about 500 bar. The atmospheric model shown for Jupiter in Figure 4.5.1 can be easily converted to that of Saturn by multiplying the altitude scale by 3 and by doubling the pressure at a given temperature.

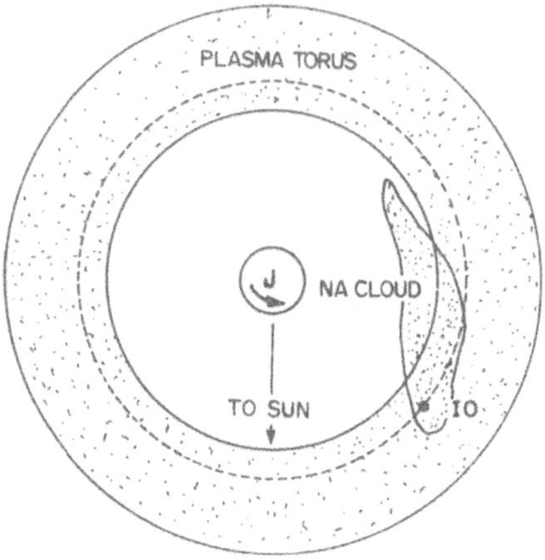

Figure 4.5.4: A sketch of the plasma torus of Jupiter.

Internal Structure of Saturn

From the considerations similar to Jupiter, the model of the Saturn's interior is shown in Figure 4.5.2(b). In case of Saturn the core of rocks and ices and He probably extend to r $\simeq$ 0.25 r_s and the metallic hydrogen mantle is less thicker than that of Jupiter, extending from ~0.25 to 0.5 r_s. The outer envelope of Saturn is much thicker covering r $\simeq$ 0.5 to 1.0 R_s.

The Magnetosphere of Saturn

The strong magnetic field of Saturn, discovered by Pioneer 11 in 1979 was confirmed and further details were obtained by Voyager 1 and 2 in 1980 and 1981. The magnetic field has some striking properties (Table 4.3.2). Its magnetic axis coincides with the rotation axis and the magnetic field can be described mostly by a dipole with much smaller contributions of higher order terms than for the magnetosphere of Jupiter. The magnetosphere of Saturn is shown in the sketch in Figure 4.5.5. The stagnation point is at 20 R_s, compared to 65 R_j for Jupiter. Similar to Jupiter, but unlike Earth, the magnetic field of Saturn is orientated with the north magnetic pole to the north geographic pole. Both almost in coincidence with the tilt angle, with a difference less than 1°. At close distances, < 10 R_s, the magnetic field is closely given by the dipole term whereas at large distances from the planet the magnetic field, as measured by Voyager 1, departs significantly from the dipole value. This indicates the presence of a ring current system in the region of 10-15 R_s.

The energy spectrum of protons in the 50 keV to 50 MeV range trapped in the inner magnetosphere of Saturn is extremely steep, as measured by Voyager, and is similar to Jupiter. These trapped protons most probably originated from solar wind particles. In addition there is a distinct group of trapped protons in the 16-200 MeV range which originate from the decay of cosmic ray albedo neutrons (CRAND) (Chapter 2).

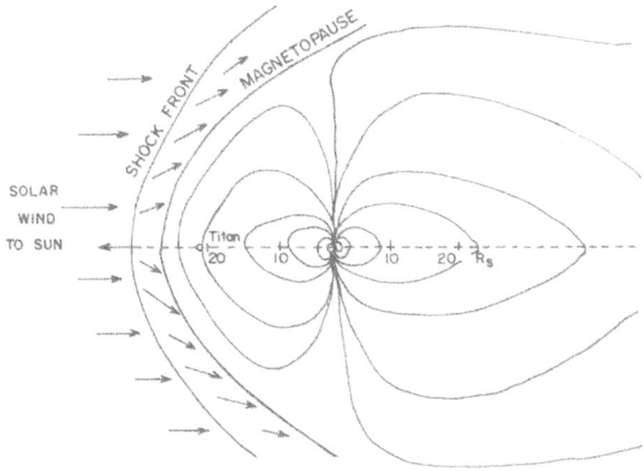

Figure 4.5.5: A sketch of the magnetosphere of Saturn.

The radiation belts of Saturn are strongly perturbated by the satellites and rings of Saturn. Pioneer 11 revealed the striking results shown in Figure 4.5.6 for electrons and protons (80 MeV). It is seen that the A-ring of Saturn completely sweeps off all trapped particles and a dip occurs at the F-ring position. Deep minima in the proton fluxes are shown at the orbits of the minor satellite S-2 and Mimas. As the energetic protons originate from the decay of cosmic ray albedo neutrons, which have a very long replenishment time, these proton slots remain unfilled. Electrons, on the other hand do not show any depression.

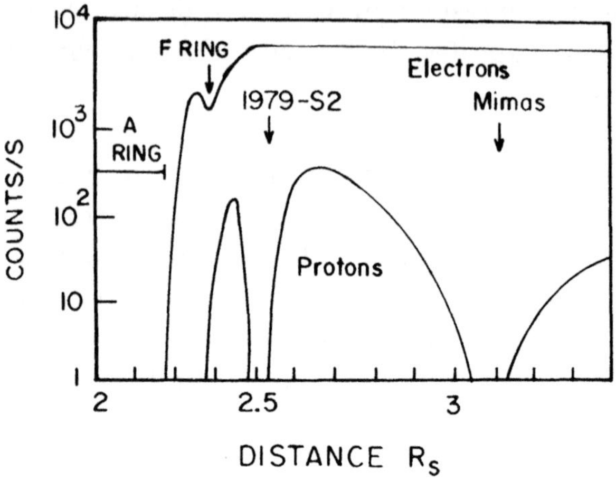

Figure 4.5.6: Results showing the sweeping of radiation belt protons and electrons of Saturn by the rings and the satellites.

Satellites and Rings of Saturn

The fly-by of the Pioneer 10 and Voyager spacecraft with their superior detector systems gave rise to series of new discoveries with spectacular photographs of the rings of Saturn and its many satellites. We can only mention a few highlights of this new features: (a) A very faint D-ring was found inside the A, B, C rings; (b) A satellite Atlas outside the A-ring was discovered; (c) A narrow, faint and stranded F-ring outside A was observed; (d) Further to the outside the very faint G and E rings were discovered. With the new satellites discovered the total number of known satellites became 28. The rings are found to be extremely thin, only 50 to 100 m in thickness. Most of the ring particles have a size of 1 cm to 10 m. The size frequency of the ring particles can be represented by the relation $n(r) = n_o r^{-3}$. So there are 10^9 more particles of 1 cm size, than of 10 m size. As the mass of a 10 m size particle is 10^9 times more, the total mass in each radius is about the same. The small size particles are probably made up of ice-plus-rock combinations and these break up by interactions into smaller particles and small ice crystals. The very bright B-ring is presumably composed of such particles with a diameter comparable to the wavelength of light. Cosmic rays interact with ice and other particles and produce albedo neutrons the decay of which produce a part of the trapped protons mentioned earlier.

Titan is the largest satellite of Saturn and is of interest in several ways, as it is one of the seven very large satellites with a mass of about 10^{26}g in the solar system. Titan has thick and massive atmosphere preventing the observation of its surface. The atmosphere is made up of about 6% methane and 94% nitrogen.

4.5.4 Uranus and Neptune

General Features

Uranus and Neptune, located far from the Sun subtends very small angles at the earth so that we only had rather meager information on them from earth based studies. Voyager 2 made historic fly-by's with Uranus in January 1986 and Neptune in August 1989 which opened new vistas of knowledge of these planets. It may be noted that this spacecraft which reached Jupiter about two years after the launch took nine long years to reach Uranus and twelve years to Neptune. Both Uranus and Neptune have nearly the same radius which is about 2.8 times smaller than Jupiter's radius and they are about 20 times less massive than Jupiter. Both Uranus and Neptune have fairly fast spin rates. However, a very surprising feature is that while the spin axis of Neptune is tilted by about 29.6^o, similar to that of Saturn, Uranus revolves round the Sun with its spin axis titled by 90° to the ecliptic! This unusual orientation of the spin axis can be seen because its five satellites which are coplanar and revolve along its equatorial plane.

Composition

From infrared measurements of these Jovian planets, the composition of the atmospheres of Uranus and Neptune were determined to be mostly of hydrogen and helium. In Uranus the helium abundance is about 0.26 (mole fraction), and in Neptune about 0.30, the rest being hydrogen. These compare well with the solar helium abundance of 0.27. Thus Uranus and Neptune have hydrogen and helium abundances similar to that of the Sun. The other gaseous compound found is methane whose abundance is estimated to be about 2% in their planetary atmospheres. Besides these some other gases were detected only in trace amounts.

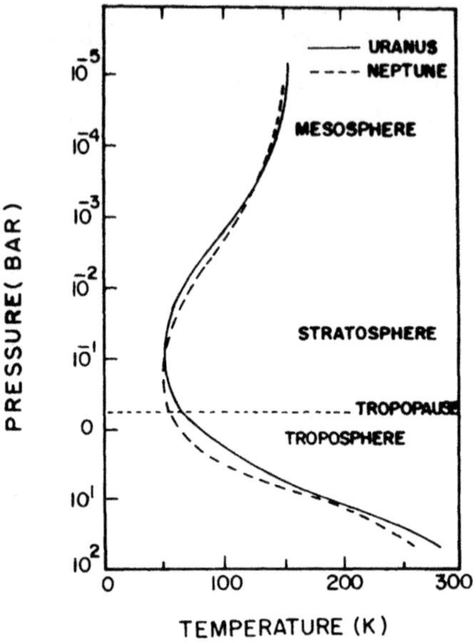

Figure 4.5.7: Temperature and pressure profiles of the atmospheres of Uranus and Neptune.

From infrared and radio brightness measurements models of the pressure temperature profiles of Uranus and Neptune have been made, and the profiles are similar, as shown in Figure 4.5.7. At the surface region the temperature is estimated to be about 250 K and the pressure is about 50 bar. The temperature falls adiabatically to about 65 K at the 1 bar level at the tropopause above which the temperature slowly rises to 100 to 150 K at 10^{-3}-10^{-4} bar pressure level. The effective temperature of Uranus is estimated to be about 57 K and that of Neptune about 59 K. The observations suggest that in Uranus condensations of methane cloud layers take place at about 55 K at ~0.5 bar pressure level and a similar situation occurs in Neptune. The effective temperature of Uranus suggests very little change of temperature between the 42 year long days and nights of Uranus. Voyager 2 revealed in Neptune a giant dark spot and a few smaller spots similar to those in Jupiter. These and other atmospheric phenomena on these Jovian planets are subjects of much discussions.

Internal Structure

If one assumes that the composition of Uranus and Neptune are mainly of the solar composition type, hydrogen and helium are the main constituents with an enrichment of heavy elements. The model of the interior structure has three zones, the rocky core, icy mantle and gaseous envelope.

Magnetospheres of Uranus and Neptune

Voyager 2 discovered a bow shockfront at 23.5 R_U and the magnetosphere of Uranus whose boundary was located at 18 R_U. As the magnetic axes of Jupiter and Saturn are inclined by about 10° and 0° to their spin axes, it was surprising to find Uranus magnetic axis tilted by 60° to the spin axis and the effective dipole displaced from the center by as much as 0.3 R_U. This strange configuration indicates that the dipole is not produced by a metallic core, but by a liquid, electrically conducting mantle. Uranus was found to have a well developed magnetosphere and radiation belt, and a strong dipole moment which is almost one-tenth of that of Saturn (Table 4.3.2). The unusual tilt of the magnetic axis leads to a complicated magnetic field structure. For example the auroral zones are displaced far from the rotational polar regions, and one polar region is almost equatorial. A sketch of the magnetic field configuration of Uranus is shown in Figure 4.5.8. The radiation belts of Uranus are populated by energetic electrons and protons. In their orbits the inner satellites sweep out charged particles of the belt showing depressions in intensity. The presence of a strong magnetosphere and radiation belt in Uranus was unexpected as the planet is not exothermic i.e. it does not emit as much radiation as it receives from the Sun.

Neptune: The discoveries of the magnetic field and the magnetosphere of Neptune are other landmarks of the achievements of Voyager 2. It was found that Neptune possessed a strong intrinsic magnetic field and its dipole axis was tilted by a surprisingly large angle of 47° from the rotational axis. Also the magnetic center was displaced from the rotational center by as much as 0.55 R_N. This produces an unusual configuration of the magnetic field with respect to the rotation axis (Figure 4.5.8). Like Uranus, Neptune is expected to have auroral zone emissions by precipitating electrons from the radiation belt. Because of the unusual tilt of the magnetic axes of Neptune, auroral zones occur far away from the rotational polar region. As Voyager 2 targeted its instruments near the poles, they missed the auroral zone emission of both Uranus and Neptune as these are near the equatorial region.

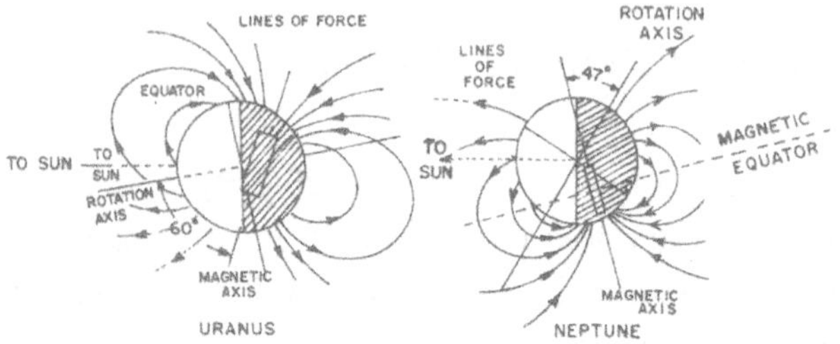

Figure 4.5.8: The magnetic field lines and the rotation axes of Uranus and Neptune.

4.5.5 Pluto

Pluto-Charon Binary System

Pluto is the only planet which has not yet been visited by a spacecraft. So our present knowledge of this planet is rather limited as compared to the others. Pluto was discovered in 1930 and its satellite Charon in 1978. Pluto having a radius 1150 km and Charon of 590 km radius form a kind of double planet which rotates about its common center of gravity with an orbital period of about 6.4 earth days. Charon is at a distance of about 20,000 km from the bary-center and its orbit is inclined to Pluto's orbit by about 90°. Because of the high eccentricity of Pluto's orbit around the Sun, Pluto and Charon cross the orbit of Neptune and spend about 20 years inside before traveling outwards completing an orbit around the Sun in about 250 years.

The mass of the Pluto-Charon system is $1.5 \cdot 10^{25}$g which is far too small compared to the other planets and it is comparable to the masses of satellites of planets in the solar system. Even then this mass is smaller than the seven largest satellites in the solar system, the Moon, the four Jovian satellites, and Titan and Triton. So Pluto naturally falls into the category of a variant of an icy satellite.

Density and Composition

The average density of Pluto-Charon is about 2 g·cm^{-2} which suggests that it is made up of an ice-rock mixture of solar nebula. The atmospheric composition is uncertain, and it is presumed to be similar to that of Triton, with significant amounts of N_2, CO and smaller amounts of CH_4 and NH_3. It is believed that during each orbital winter, when Pluto is beyond Neptune's orbit, condensation of CO, CH_4 takes place in the form of solid ice.

An interesting feature of the orbital periods of Neptune and Pluto is to be noted. It is found that the orbital periods of Neptune and Pluto are locked in a resonance of 3:2 i.e. for every three periods of Neptune, Pluto completes two orbital periods. This is maintained over a long period of time. The unusual features of Pluto have led to the popular suggestion that Pluto and Charon have escaped from the orbit of Neptune. There are objections to this hypothesis. The formation of the Pluto-Charon pair and the Pluto-Neptune resonance are some of the puzzling problems in planetary phenomena which are yet to be unraveled.

4.6 Problems on Chapter 4

1. Explain the formation of lunar regoliths and breccias. What are the main minerals in lunar rocks and soil as (a) Oxides, and (b) Silicates with abundances > 1% ?

2. Describe the principal features of the elemental abundances in the lunar surface and indicate important differences with terrestrial data.

3. What are the main features of the paleomagnetism of lunar rocks and soil?

4. Explain critically the different hypotheses of the origin of the moon.

5. Discuss briefly the four main factors which are responsible for the differences in the formation of planetary atmospheres.

6. What are the main constituents of the atmosphere of Venus and what are the main features of the temperature-pressure profile of the atmosphere of Venus?

7. Describe the results of the Viking Lander missions on the temperature and pressure vs height in the Martian atmosphere. What are the ranges of seasonal variations of pressure at the

Lander sites and what are their interpretations?

8. Explain briefly the six stages of the origin and evolution of the asteroid belt.

9. What are the major features of the magnetic field of Jupiter and Saturn and their radiation belts? How do the rings and satellites of Saturn affect its radiation belt?

10. What are the models of the internal structures of Jupiter and Saturn and what is their physical basis?

11. Explain with sketches the structures of the magnetic fields of Uranus and Neptune. How do their unusual features affect their magnetospheres?

12. Why are Pluto and Charon called a kind of double planet? What are their orbital and physical properties?

CHAPTER 5
COMETS, METEORITES AND THE ORIGIN OF THE SOLAR SYSTEM

5.1 Comets and Cometary Probes

5.1.1 General Properties

Several times during a century a bright comet is visible in the sky and then it is one of the most spectacular objects in the sky. Besides an increasing brightness they also develop long luminous tails while they are in the vicinity of the Sun. Cometary nuclei have a size of a few kilometers and are made of loose aggregates of rock, dust and ice. The astrophysical importance of comets arises from the fact that comets are remnants of the large collection of pristine matter of which the solar system was formed and they remained almost unaltered during the eons. In contrast asteroids are, as discussed earlier, remnants of fragments left over from a process that failed to form a planet.

The British astronomer Edmond Halley was the first astronomer who made a systematic study of the orbit of a comet, namely the bright comet of 1682 and he noted that comets with similar orbits and brightness were seen to pass perihelion 75, 151, 226 and 304 years before 1682, in 1607, 1531, 1456 and 1378 respectively. He concluded that these observations were of the same comet and its orbit had a period of 75, 76, 75 and 78 years; the small changes in the period were due to gravitational perturbations by large planets. In 1705 Halley announced his results and predicted that the comet would return in the winter of 1758-59. Fifty-three years later the comet reappeared exactly as predicted, in December 1758 and it was named after Halley who spent a large part of his lifetime on studying cometary orbits.

At this moment about 700 comets have been recorded, of which the long period (P > 200 years) and short period (P < 200 years) comets occur in a ratio of 6:1. On the average new comets are added at the rate of about 5 per year.

5.1.2 Cometary Orbits

Orbits of short period comets are elliptical with moderate eccentricities between 0.2 to 0.7. They have orbital inclinations of less than 20° to the ecliptic plane. Of 113 short period comets, only 4 have periods of more than 20 years, e.g. Halley 76 yrs, Temple-Tuttle 33.2 yrs. The comet with the shortest period (of 3.3 yrs) is Enke. Most of the short period comets have their perihelion at 1 to 2 A.U. with Enke approaching the Sun as close as 0.33 A.U.

Both long and short period comets are destroyed or lost by two processes. From time to time some of them collide with large planets and are destroyed completely. The most recent example is the collision of Shoemacher-Levy comet fragments with Jupiter in August 1994. The other process is the close encounter of a comet with a large planet whereby the comet trajectory is changed and the comet is ejected out of the solar system. A long period comet with a period of 10^6 years that has been in the present orbit for 10^9 years should have passed through the inner solar system 10^3 times. Its probability of being destroyed by a collision is only 10^{-6}.

Orbit-distribution of Long Period Comets

The majority of long period comets has a period in the range of 10^6-10^8 years. Their semi-major axes are in the range from 10^4 to $2{\cdot}10^5$ A.U.. Long period comets have eccentricities just below unity and their orbital inclinations to the ecliptic are randomly distributed.

5.1.3 Cometary Probes of Halley

The perihelion passage of the Comet Halley in February 1986 triggered new scientific investigations for which specially designed spacecraft were launched. The USSR launched Vega 1 on 15 December 1984 and Vega 2 on 21 December 1984 and they encountered Comet Halley in March 1986, and made the first spectacular in situ measurements of a comet in fly-by mode.

The Vega 1 and 2 spacecraft were transferred to Halley encounter paths by a swing-by in the gravitational field of Venus. The Vega 1 and 2 spacecraft consisted of two modules, an entry probe to study Venus and the fly-by module for Comet Halley and these two flew at the closest distance of 8000-

9000 km to the comet nucleus on March 6 and 9, 1985. The Halley probes were three axis stabilized, with a pointing accuracy of 1° during fly-by. The optical instruments could track the nucleus with an angular accuracy of 5 arc min. The scientific payload of Vega 1 and 2 consisted of a large number of instruments such as Television system, Infrared spectrometer, Dust particle detectors (3 units), Dust mass spectrometer, Neutral gas spectrometer, Plasma energy analyzer, Energetic particle analyzer, Magnetometer and Wave and Plasma analyzer (2 units).

5.1.4 The Nucleus of Halley

The summary of results from Vega 1 and 2 during their closest approach on 6 and 9 March 1985 is as follows. The nucleus of Comet Halley was seen for the first time and was identified as an irregular shaped solid with a dimension of about 16x8x8 km. The nucleus rotates in the direct sense with a period of 53 hrs with an axis which is nearly perpendicular to its orbital plane. The nucleus is covered with a very dark material with an albedo of only 0.04. Jets of dust were seen to emerge at certain times from some parts of the nucleus. The surface of the nucleus which could be seen through the dust layer is probably rough.

One of the most important findings of the Vega missions is the determination of the mass of the nucleus as $M_H = 3{\cdot}10^{17}$g by Sagdeev et al(1). The volume of the nucleus was measured as $5.85{\cdot}10^{17}$ cm^3. Thus the density is estimated to be about 0.5. The effective radius of the nucleus, i.e. the radius of a sphere with the same volume, is obtained as 5.3 km. On comparing with other short period comets it is found that the mass, volume and albedo of Halley's nucleus are close to the average values of short period comets.

The ESA spacecraft GIOTTO launched on 10 July 1985 had a fly-by mission with Halley's nucleus on 13 March 1986 at the closest distance from the nucleus of 2500 km and detailed observations were made with a number of instruments including a TV camera, mass and energy spectrometer, gas and ion analyzer, plasma detector etc. Figure 5.1.1 shows a sketch of the composite image of Halley's nucleus taken with the multi color camera in the GIOTTO spacecraft. With sunlight falling from the left, three sun-lit regions were seen to emit copious amounts of gas and dust in the form of jets. The

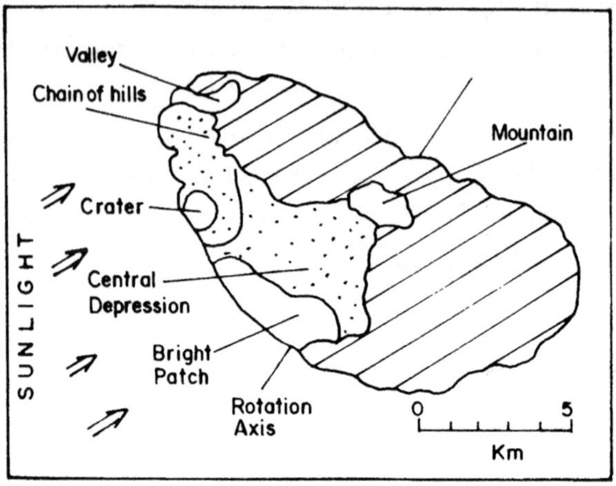

Figure 5.1.1: A sketch of the nucleus of the comet Halley reconstructed from pictures from the GIOTTO spacecraft and showing the principal features. The Sun is on the left.

bulk composition of the gas was measured as $H_2O \leq 80\%$, $NH_3 \geq 10\%$, $CH_4 \leq 7\%$ and $CO_2 \leq 3.5\%$. The expansion velocity of the gas was measured to be 1 $km{\cdot}s^{-1}$. It is estimated that during the perihelion of 1986, the comet lost a mass of $3{\cdot}10^{14}$g, which is about 0.1% of its total mass.

When comets approach the Sun, the surface of the nucleus facing the Sun heats up and the overlaying dust and ice changes to the gaseous phase forming a large cloud of gas and dust around the nucleus called the coma. The emission of jets of dust and gas from a few regions in the nucleus of Halley is coming from cracks or fissures due to solar heating of internal gas and dust. The gas and dust in the coma surrounding the nucleus are pushed away from the nucleus in anti-solar direction by the solar wind and radiation from the Sun and so forming a large tail. The charged ions in the tail form a plasma tail and small dust particles form a dust tail of the comet. Many studies have been made of the composition of the coma and tail of comets and various carbon bearing species such as C^+, CH^+, C_2^+ etc have been found in addition of H_2, H_2O^+, OH^+ etc and their chemistry have been investigated.

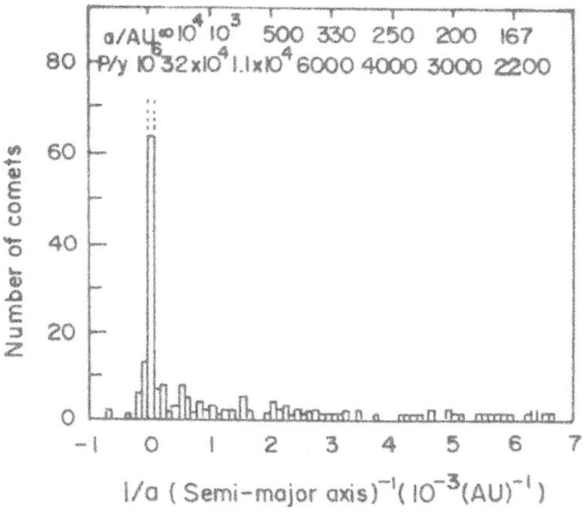

Figure 5.2.1: The number of long period comets plotted against 1/a, where a is the semi-major axis in A.U.

5.2 The Oort Cloud and the Origin of Comets

5.2.1 Long Period Comets

As noted earlier the majority of long period comets have periods in the range of 10^6 to 10^8 years and their major axes are in the range of 10^4 to $2{\cdot}10^5$ A.U.. In order to explore the distribution of the major axes of comets in their entire range of orbital axes (a) ranging from 100 A.U. and up, it is found that a useful plot is the number of comets vs 1/a, the reciprocal of the semi-major axis (a). In this plot, shown in Figure 5.2.1 observational errors are not exaggerated. The significant feature seen in the figure is the strong peak in the distribution of 1/a in the range $0.0 < 1/a < 0.1{\cdot}10^3$ AU^{-1}. The observation of the peak signifies the presence of the Oort cloud as will be discussed later. The idea of plotting frequency vs 1/a is significant also from the fact that 1/a represents the orbital energy of the long period comets.

Here we briefly discuss the orbital velocities and impact velocities of the long period comets in the vicinity of planets, so as to have some idea about their motions in the solar system. The greatest velocity a comet bound to the sun at the orbit of a planet can have is it's escape velocity from the sun at that orbit. At the orbit of the earth the escape velocity can be shown to be 42 km·s^{-1}. When a comet is approaching the earth from behind, the encounter velocity is minimal, and equal to $\vec{V}_{esc}$ minus $\vec{V}_{orb}$. The value of

V_{orb} is $(1/\sqrt{2})\ V_{esc}$. Therefore $V_{imp} = 0.293\ V_{esc}$. In case of a 'head-on' orbit of the comet, the impact velocity is maximal, and can be shown to be $1.707\ V_{esc}$. Therefore at the earth a long period comet may have impact velocities in the range of 12.3 to 71.7 km·s^{-1}. If one takes into account the gravitational pull of the earth, the range of impact velocities of long period comets becomes to be from 16.5 to 72.5 km·s^{-1}.

The strong peak at $1/a \approx 0$ suggests that most long period comets have a $1/a$ value $< 10^{-4}$ i.e. a semi major axis a $> 10^4$ AU. As long period comets have very small velocities near their aphelion, they practically spend all their lifetime at great distances from the Sun, where the temperatures are extremely low. In Figure 5.2.1, it is seen that the frequency of comets with a 1/a between -0.4 to 0.5 is very small. From this and other considerations it can be shown that comets wandering into our solar system from the interstellar medium must be very rare. These observations support the idea of the following origin of long period comets.

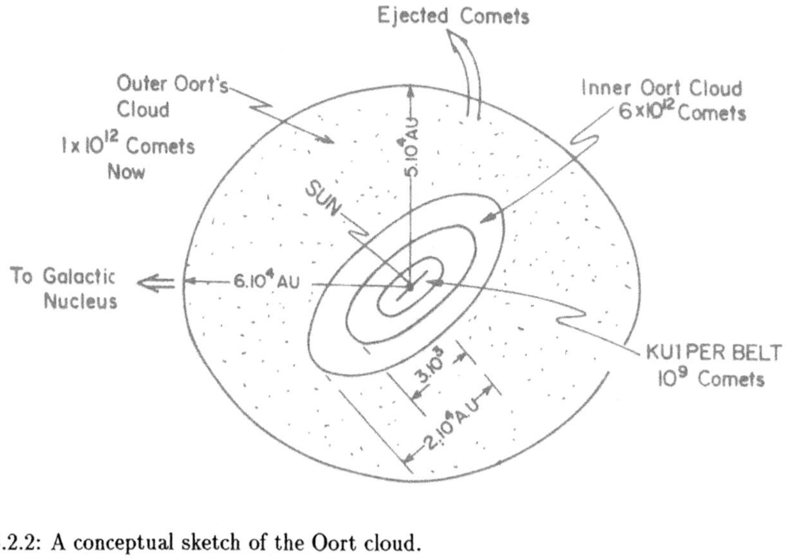

Figure 5.2.2: A conceptual sketch of the Oort cloud.

5.2.2 Oort's Cloud and the Origin of LPC

In 1950 the Dutch astrophysicist Jan Oort[2] first suggested that in the remote periphery of the solar system there is one more component: a giant spherical cloud of long period comets, which has been named as the Oort cloud. A conceptual sketch of the Oort cloud is shown in Figure 5.2.2. It is believed to have a disc shaped inner core with inner and outer radii of 3,000 and 20,000 A.U. respectively[(3)]. This is surrounded by an Oort cloud with the form of a spherical shell with an inner radius of about 20,000 A.U. and an outer radius of about 50,000 A.U.[(3,4)]. Some authors suggested the shape as an oblate spheroid with axes of lengths $12{\cdot}10^4$ and $8{\cdot}10^4$ A.U., the elongated axis being directed towards the galactic center.

From several considerations including the direct observations of comet Halley, the average mass of a new comet is estimated[(5)] as $M \simeq M_H \simeq 3{\cdot}10^{17}$g. The inner Oort cloud contains comets with orbital semi-major axes of about $5{\cdot}10^3$ AU. Perturbation by galactic and stellar objects at the aphelion of the comet would raise the orbital perihelion. In 1990 Weissman[(6)] estimated that the inner cloud contains $6{\cdot}10^{12}$ average comets and the outer cloud has a population of $1.0{\cdot}10^{12}$ comets. These numbers of comets are determined to explain the observed rate of new long period comets. According to Hills[(3)] the number of comets in the core should be 200 times that of that in the halo so that the core can continuously supply comets to the halo to replenish the halo's loss of comets due to passing stars. In this picture the total mass of the inner core becomes quite large, $M_{core} \approx 6{\cdot}10^{31}$ g $\simeq 10^4$ M_E, as compared to the value of 300 M_E derived from Weissman's data, where M_E is the mass of the earth.

The mass of the outer Oort belt having a population of $1{\cdot}10^{12}$ comets is estimated as $6{\cdot}10^{29}$g i.e. $\approx$ 100 M_E. This value is higher than the earlier estimate of about 8 M_E by Weissman, and is due to the higher average mass of a comet determined from the Halley measurements.

The third component of the Oort cloud is the Kuiper belt (Figure 5.2.2) which is believed to be the remnant of the original disc of the solar nebula. It is estimated that there are about 10^8-10^{10} comets in this belt. Their orbits have low inclination and eccentricity and lie just beyond Neptune.

One important consequence of the new estimate of the mass of the Oort cloud as pointed out by the authors[(5)] is that a very substantial part of the total mass in the solar system is located in the Oort cloud in the outer periphery of the solar system. Thus the mass of the Oort cloud is estimated to be about 1/4 of the total mass of the planets. Before the discovery of the Oort cloud, it was noted that the Sun, containing 99.9% of the mass of the solar system carried 2% of the angular momentum, and about 0.1% of the mass and 98% of the angular momentum were to be shared by nine planets and their satellites. With the new mass estimates of the Oort belt, the revised model is as follows. The planetary system contains a total mass of $2.7 \cdot 10^{30}$g i.e. 450 x M_E and an angular momentum of $J_p \simeq 3 \cdot 10^{50}$g·cm^2/s. The planetary system extends to a radius of about 40 A.U., beyond which there is a void up to a distance of about 3000 A.U.. In the region of 3000 A.U. to 20,000 A.U., a giant mass of $6 \cdot 10^{31}$g i.e. 10^4 M_E is located in the form of the inner core of the Oort cloud which contains an angular momentum of about $3 \cdot 10^{52}$g·cm^2/s. The halo of the Oort cloud located at a radial distance between 20,000 to 50,000 AU carries a total mass of 100 M_E and an angular momentum of $3 \cdot 10^{51}$g·cm^2/s. Thus the total angular momentum of the Oort cloud appears to be higher than that of the planetary system by an order of magnitude.

5.2.3 Origin of Comets

The comets are believed to be formed by the collection of planetesimals in the outer regions of the solar system. These planetesimals were produced by the condensation of protosolar nebular matter, a cloud of gas, ice and dust. Many of these planetesimals were used as building blocks for the cores of the outer planets such as Saturn, Uranus and Neptune. During the final stages of accretion, one type of cometary orbits were formed of which the perihelion grew by perturbation by the larger bodies. These led to some comets having parabolic orbits, and these have left the solar system. The outer group of comets having highly eccentric orbits formed the Oort cloud of comets. This huge group of comets had its perihelion in the Saturn, Uranus, Neptune region and an aphelion at 10^4 to 10^5 A.U. from the Sun. Initially they had low inclination, but perturbations by passing stellar bodies randomized their orbital inclination. Perturbation by passing stars would also change the orbital energy of the comets in the Oort cloud. In some cases the perihelion was pushed towards the orbit of Jupiter and thus increasing the probability of them being captured into short period orbits. In case of some others, the eccentricity increased and these comets were ejected from the cloud.

Thus the comet population in the Oort cloud would be periodically stirred by passing stars or giant molecular clouds. The solar system has existed for about $4.5 \cdot 10^9$ years. it is estimated that during this period about $5 \cdot 10^3$ stars have passed within 10^5 A.U. of the Sun. A star of solar mass passing within $5 \cdot 10^4$ A.U. of the Sun at a velocity of 20 $km \cdot s^{-1}$ will form a narrow tunnel through the Oort cloud and will perturb the orbits of comets within about 500 A.U. of their path. It is estimated by Hughes(7) that in the lifetime of the solar system about 3/4 of the comets in the Oort cloud have been ejected and this quantity i.e. about $2 \cdot 10^4$ comets which have been ejected has been replenished from the inner core of the belt.

Further details of comets are given by several authors e.g. by Krishnaswamy in the book on 'The Physics of Comets'.

5.3 Meteorites

5.3.1 General Features

Before the advent of the space age, meteorites were the only source of extra-terrestrial solar system matter available to us. They provided a window to study the physical chemical and geological properties of materials at the time of the formation of the planets. The relationship between the asteroids and meteorites were found over the years and it is now recognized that meteorites originated in the asteroid belt and were thrown out of the belt or from the surface of asteroidal bodies by collisions with other bodies. The studies of meteorites in the laboratories on earth revealed not only their intrinsic physical and chemical properties, but also that they are detectors of corpuscular radiation in the solar system, from the Sun and from galactic cosmic rays of the distance past. These aspects of planetary science will be discussed in Chapters 6 and 7.

Meteorites falling on the surface of the earth have weights varying from a few grams to ten's of tons. As a large meteorite enters the earth's atmosphere at a high speed a portion of its surface material is ablated or vapourised by the high temperature generated by the friction. The residual mass on impacting the surface of the earth forms a large meteoritic crater in the case of a large meteorite and breaks up into fragments during the impact. There are many examples of impact craters on Earth. The famous meteoritic crater in Arizona, 1.3 km in diameter was made by a meteorite with a size of 35 m some 30,000 years ago. Sometimes large objects on entering the deep atmosphere with high speed disintegrate completely with the release of a tremendous energy equivalent to megatons of high explosive. Such an event occurred in 1908 in Tunguska, Siberia and it flattened 2000 km^2 of forest, but did not produce a crater. The Lonar lake of Maharashtra, India is an example of a meteoritic crater formed many years ago. On Earth most of the meteoritic craters of the distant past have been obliterated by the weathering action of wind and rain.

Medium or small size meteorites recovered in different places on Earth are given names according to the names of the locations of their recovery. The first scientific record of the fall of a meteorite was made in 1803 in L'Aigle, France by Biot. In the USA the Weston meteorite in Connecticut of 1807 was investigated by Silliman of Yale. Since then about 10,000 pieces of meteorites from about 3000 separate falls have been recorded and many of these have been kept in museums and are used for scientific studies. Two broad clarifications of these meteorites are made. "The Finds" denote those found on the ground and 'The Falls' are those which have been actually observed to fall. The Dhajala meteorite fall observed in Gujarat, India in recent years has led to the recovery of many fragments which were studied.

Another class of meteorites which originated from impacts on the moon and fell on the earth it called meteorites from the moon. Pieces of small meteorite fragments recovered from lunar regoliths are called meteorites on the moon.

5.3.2 Classes of Meteorites

Meteorites are broadly grouped into three general classes, (1) the iron, (2) the stones and (3) the stony-irons.

The irons are the most spectacular and can be seen in museums. They are tough, corrosion resistant and have a high density of 7.5 $g{\cdot}cm^{-2}$. They are composed mainly of metallic alloys of Fe, Ni and Co, with small traces of other elements. The largest iron meteorite on the earth lies in Hoba in Namibia, Africa, and weights roughly 60 million gm. It probably weighted about 100 million gm prior to the entry. It is so massive that it is almost impossible to move it and so it still lies at the same place where it landed.

Table 5.3.1: Frequencies of Meteorite Types

	Irons		Stones	Iron-Stones
'Falls'	3%		96%	1%
		Chondrites	88%	
		Achondrites	8%	
Finds	42%		52%	6%
		Chondrites	51%	
		Achondrites	1%	

Iron meteorites such as Hoba, are so tough that it is very difficult to cut them. The frequency of iron meteorites amongst the meteorite 'Finds' is about 42% whereas among the Falls it is 3.2%. This reflects the fact that the iron meteorites, can be easily identified compared to other types among the 'Finds'. There are three sub-groups of iron meteorites.

The second class is that of the stone meteorites which are composed mainly of silicate minerals with 0 to 30% metal grains and several percent of sulphides. Some of them have the appearance of a dark stone, while others are hard, igneous rocks, which are formed by recrystallization of silicate melting. The stony meteorites are categorized in diverse types based in their composition, appearance and history of formation. Two main categories of stony meteorites are (a) Achondrites which have an igneous origin, and (2) Chondrites which have never experienced melting. Most chondrites contain small glassy or crystalline spheroidal beads called chondrules with a diameters between 0.1 and a few mm. These are of special importance in planetary science as will be discussed later.

The third group of stony-iron meteorites is composed of nearly equal amounts of silicates and iron group metals. They are less frequent and have three sub-groups. The frequencies of the main groups of meteorites are given in Table 5.3.1.

5.3.3 Orbits

The determination of orbits of meteorites is difficult due to a lack of adequate observational data in most of the cases. Only in a few cases photographic positional information is available. The Pribram stoney meteorite recovered in Czechoslovakia in 1959 was photographically tracked in fortuitous circumstances. Its orbit around the Sun had a semi-major axis of 2.42 A.U. and an eccentricity of 0.67; so its perihelion was 0.79 A.U., just outside the orbit of Venus and its aphelion of 4.1 A.U. was close to the Hilda family (Chapter 4.2), outside the main belt of asteroids. So it is presumed that meteorite Pribram originated from the Hilda family of asteroids. Similarly tracking stations data yielded the orbits of the meteorites, Lost City (USA) in 1970 and Innisfree (Canada) in 1977. The former had an aphelion of 2.35 A.U. in the inner rim of the main belt and the latter ~2.5 A.U. in the heart of the main asteroid belt.

A typical meteorite may enter at say, an angle of 45° to the horizontal and it begins to glow visibly at an altitude of 150 km where it begins to encounter traces of atmosphere. It can be shown that the minimum velocity is 11.2 $km \cdot s^{-1}$, which is the escape velocity of the earth and the maximum velocity of earth entry in a heliocentric orbit is about 73 $km \cdot s^{-1}$. Thus the average entry speed of a meteorite is in the range of 30-40 $km \cdot s^{-1}$. An object entering the earth's atmosphere at a speed of 40 $km \cdot s^{-1}$ has a kinetic energy of $8 \cdot 10^{12}$ $ergs \cdot g^{-1}$, whereas the energy required to heat a gram of typical rock from room temperature to melting point is about $1 \cdot 10^{10}$ erg, and the energy required to crush it to powder form is about one tenth of the above value. Therefore stoney meteorites undergo surface melting and evaporation which is called ablation and some are broken into fragments. Due to ablation a fine dust of glass and oxide grains are left in the upper atmosphere, called trails, which are visible for a few minutes following the atmospheric entry. About 30% of the mass may surive ablation. Stoney meteorites are often broken into numerous fragments spread over a large area. Small fragments are usually vapourised and large ones survive. The meteorite fragments freshly falling on the earth are distinguished from terrestrial rocks by the presence of a thin glassy fusion crust formed by the aerodynamic heating during atmospheric passage. In stony and stony-iron meteorites this fused crust may be chipped ;off to show their glassy surfaces.

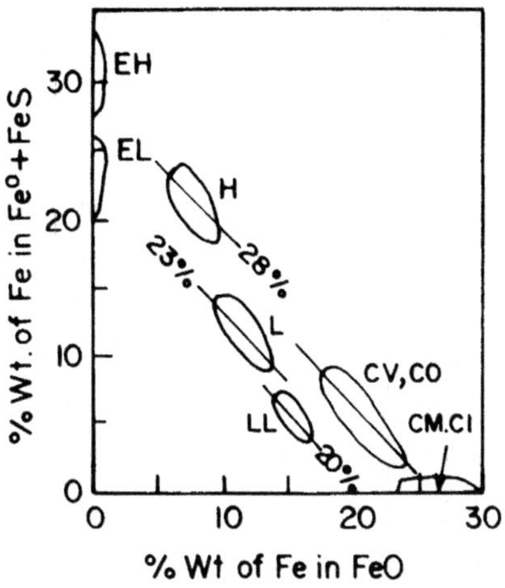

Figure 5.3.1: A modified Urey-Craig diagram showing the variations in the composition of major elements in Chondrite minerals of various types. For notations see text.

5.3.4 Composition

The stony meteorites with highly oxidized constituents and rich volatile matter have low densities of about 2.3, while the metal rich stony ones have higher densities of about 4. In stony meteorites a higher volatile content make these more porous in appearance. The stony-iron meteorites have densities in the range of 5 to 6. Iron meteorites have the highest density of about 7.5, which sometimes may be as high as 7.9.

There exists a vast amount of literature on the composition of meteorites and their interpretations. Here we discuss the highlights and important features. The unmelted stony meteorites - the chondrites - are first discussed briefly. The chondrites are defined as the meteorites that either contain 'chondrules' which are glassy, spheroidal beads with diameters of 0.1 to a few mm, or are similar in structure and composition. They occupy a special position in importance as all chondrites are samples of the original solar nebula which did not undergo melting and differentiation.

The composition of chondrites was first systematized by Harold Urey and Harmon Craig in the 1960's. They used two variables, one, the abundance of reduced Fe, and, two, the abundance of oxidized iron (Fe oxide in silicates and magnetites). They showed that when the oxide content is high, the metal content is low so that the total iron content is remains about the same. The Urey-Craig results are checked and expanded by many authors during the past decades and we show the present representation of the composition of chondrite meteorites in Figure 5.3.1. In this graph we show the weight percentage of Fe in Fe metal and FeS vs the weight percentage of Fe in Fe oxides. This plot called Urey-Craig plot shows clearly that ordinary chondrites fall in three main classes. H represents high iron, L low iron and LL very low iron. The high iron group has low iron oxide content, and the low iron group has high oxide content. The diagonal lines have slopes of -1 and represents a constant total iron content. Chondrites containing $MgSiO_3$ (enstatite) are called E-chondrites. These have a wide range of iron content and are subdivided into two groups, EH (high iron) and EL (low iron). Alternative plots, made by Dodd used (Fe-metal/total silicon) vs Fe in (FeO and FeS)/(total silicon).

Olivine, $(MgFe)_2\ SiO_4$ which is present in all ordinary chondrites but is almost absent in E-chondrites, increases in abundance from H to L to LL. The other major silicate, pyroxene (Ca, Fe, Mg) SiO_3, decreases in importance as olivine increases.

The carbonaceous chondrites (C-chondrites) are an important group of meteorites. Most of this C-chondrites contain easily detectable amounts of organic matter, while free Fe metal is almost absent. The C-chondrites have the highest FeO content and are composed of volatile rich constituents. They have a total iron content at least equal to that of H-chondrites. Based on several considerations C-chondrites are further divided into four groups - named after the type of meteorite. Thus CV stand for the carbonaceous - vigarano type, and CO for the C-Ornans type. These two groups contain metallic iron. The C-M (C-Mighei) and CI (C-Ivuna) groups have a still higher FeO and volatile elements content, while they contain no metal (see Figure 5.3.1). It is noted that the CO type contains more total iron than the CV type. CI chondrites have more total iron content than any C or ordinary chondrites. We also note that in ordinary chondrite almost all FeO occur in the pyroxene and olivine. Thus Fe, Si, O, Mg, Ca and Al are the six major elements in these meteorites. The next major element is sulphur.

Composition of Achondrites: These igneous stony meteorites have undergone at least partial melting at high temperature and subsequently differentiation on solidification. Partial melting of ordinary chondrites will lead to upward separation of low-density crust of silicate melt containing alkali metals, calcium, volatiles and rare earth elements which enter in silicates. This melt solidifies into a crust rich in Feldspar. The other melt of Fe metal sulphide, being much denser than silicates will sink to the center of the parent object. This sinking layer, produces two layers on cooling, triolite on top and an inner core of Fe-Ni metal. The middle region below the crust and above the core is dominated by olivine, which has a high melting temperature and an intermediate density of about 3.6, which are between those of the crust (2.7) and of the core (5.7). Thus rocks containing Feldspar, olivine, triolite and Fe-Ni metal are generated.

Composition of Stony-iron: The stony-iron meteorites fall into groups according to their internal structures. The larger group, the pallasites, have centimeter-size grains of olivine surrounded by a matrix of the metals Fe-Ni. The smaller group, called mesosiderites are composed of metal bodies surrounded by silicate compounds.

Composition of Irons: Iron meteorites have two principal mineral compounds of iron-nickel alloys, kamacite with α-iron and low nickel, and taenite with γ-iron and high nickel. Structurally there are two classes, the hexahedrites containing 4 to 6% NI and secondly the oxtahedrites with 6 to 14% Ni. Other elements such as Ga, Ge and Ir are present in trace quantities.

5.3.5 Isotopic Composition

Studies of the isotopic composition of meteorites provide new clues to the astrophysical aspects of their past history. We shall briefly present the results of oxygen isotopes and their implications.

Isotopic abundances of ^{16}O, ^{17}O and ^{18}O were measured by Clayton and co-workers in a large number of meteorites of different types. Firstly, it is to be noted that in a given body isotopic fractionation takes place during the kinetic processes involved, so that the ratios of ^{17}O/^{16}O and ^{18}O/^{16}O vary in unison, the magnitude of the fractionation being proportional to the mass differences. Thus the plot of ^{17}O/^{16}O vs ^{18}O/^{16}O is made and it is found that it has a slope of 0.5 as given in Figure 5.3.2. A closed isotopic system with an internal process would give fractionations occuring

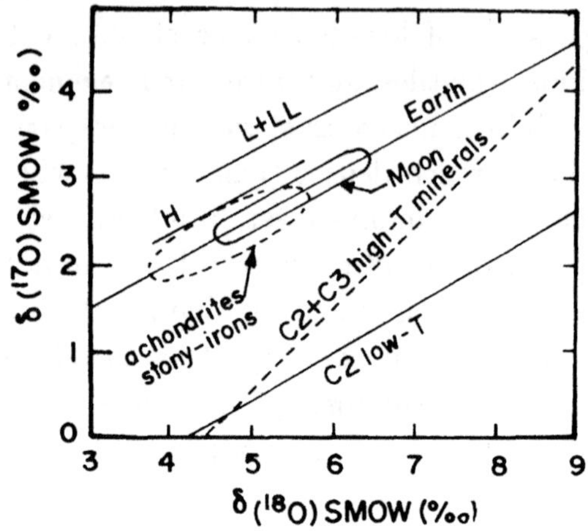

Figure 5.3.2: Oxygen isotopic abundance variations in meteorites. The variations of the ^{18}O abundances are plotted against those of ^{17}O; $\delta(^{18}O)$ SMOW represents the meteoritic $(^{18}O$-$^{16}O)/^{16}O$ ratio relative to the ratio in standard mean ocean water. Isotopic fractionations are in parts per thousand (0/00)

along this line. Two systems that differ in isotopic composition would lie in parallel lines. These ideas are confirmed very well by the results shown in the figure. It is seen that the isotopic fractionations in rocks of the earth and the moon fall on the same line as well as those of achondrites and stony irons. Oxygen in L and LL chondrites definitely shows higher values for the ^{17}O:^{16}O ratios than all other classes. The line for H-chondrites lies a little higher than that of the earth-moon indicating somewhat higher abundance ratios of ^{17}O:^{16}O . The carbonaceous chondrites C2, of low temperature show the lowest abundance ratio of ^{17}O:^{16}O. It may be noted that although achondrite and Earth minerals lie on the same slope, their oxygen isotopic abundances are very different. Hence materials lying on the same line need not have a simple genetic relationship.

The striking features in the results of oxygen isotopic abundances are shown in the materials of C2 and C3 chondrites. As shown in the figure their line has a slope of 1.0. This slope indicates similar ^{17}O:^{18}O ratios, but a variable oxygen-16 content. The simplest explanation is that near the upper end of the line the sample contains isotopically normal solar system material, whereas downward along the line the oxygen-16 content progressively increases, and near the lower end it represents almost pure oxygen-16. Such a rich component of the oxygen-16 isotope signifies the

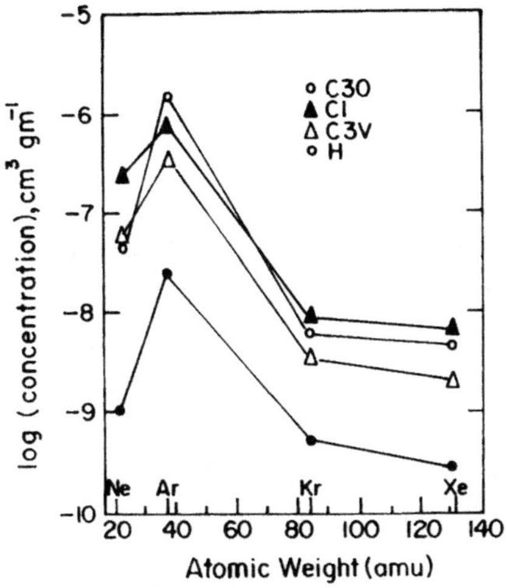

Figure 5.3.3: Abundances of rare gases in chondrite meteorites.

entrapment of oxygen-16 rich material of a supernova explosion. One of the most important implications of the oxygen isotopic data is that the solar system material was not fully vapourised and mixed in the solar nebula. Otherwise all solid materials in planets, satellites, asteroids and meteorites would have shown similar mass dependent fractionation of oxygen isotopes, because they come from a single homogeneous pool of oxygen.

Abundances of carbon and its isotopes, measured in a number of meteorites, show the highest carbon abundance in C-chondrites such as CI and CM. The isotopic composition of ^{13}C:^{12}C is nearly uniform and only slightly higher than the terrestrial value.

Nitrogen abundances are high in CI and CM chondrites and both have a similar isotopic composition, with the nitrogen-15 isotope abundance significantly higher than for terrestrial atmospheric nitrogen. Even higher isotopic ratios of ^{15}O:^{14}N were found in the organic materials in these meteorites. In ordinary chondrites the nitrogen-15 abundance is only marginally higher or similar to that of the earth.

We have little information on the isotopic composition of hydrogen in meteorites because only carbonaceous chondrites contain water, CI have about 20% water by weight. CM's have $\sim$ 10% and CV and CO only 0.3

to 3% water. The isotopic compositions of CI and CM a show somewhat higher ($\sim$ 15%) fraction of the heavier mass as compared to that of the earth.

5.3.6 Rare gases in meteorites

Rare gases in carbonaceous chondrites originate from a number of sources. Some fractions of these gases represent the primordial component which was captured from the solar nebular material at the time of the formation of the meteorites. An example is the rare gas isotope ^{129}Xe which is produced by the radio active decay of the now extinct short-lived nuclei of ^{129}I. Some other isotopes are produced by β-decay e.g. ^{40}Ar from ^{40}K, and α-decay such as ^{4}He from the U and Th radio active chains. Some chondrites were exposed to the solar wind on the surface of their parent and were impregnated with a large amount of rare gases by the solar wind. Another component arises from the binary fission of U and Th, such as Kr and Xe isotopes.

The abundances of rare gases in meteorites are shown in Figure 5.3.3. It is seen that the abundances of all rare gases in C-chondrites are higher than ordinary chondrites (such as H) by a factor of 30 to 200. Ordinary H-chondrites, which have rare gas abundances similar to those of the earth, show a rather large Ar/Ne ratio of about 20, and CO-chondrites have an even higher Ar/Ne ratio of about 40. These are in sharp contrast to the solar or cosmic Ne:Ar ratio of about 30:1 and indicate very strong fractionation with high retention of heavier rare gases in chondrites as compared to lighter gases. This feature is also exhibited in Ar/Kr as well as Kr/Xe ratios in chondrites, as it is seen that the Ar/Kr and Kr/Xe ratios in C-meteorites are about 100 and 1.5, respectively. This is in sharp contrast to the cosmic abundance ratios of Ar/Kr and Kr/Xe which are about 2500 and 10 respectively. Thus it is clear that all rare gas abundances in C-chondrites are strongly fractionated as the retention factor is highly dependent on mass. Even in ordinary chondrites and in the earth the heavy Xe is depleted by a large factor as compared to cosmic abundances, measured with respect to silicon. Because of these variations, two types of average values are used for reference. One is the average carbonaceous chondrite rare gas (AVCC) and the other is the average of terrestrial atmospheric rare gas.

The isotopic composition of these two types of rare gases also differ

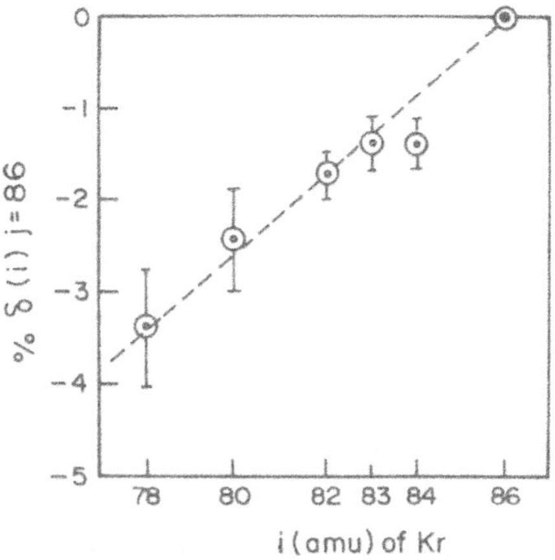

Figure 5.3.4: Isotopic composition of Kr in carbonaceous chondrite meteorites. The earth's atmospheric value of ^{86}Kr is taken as standard.

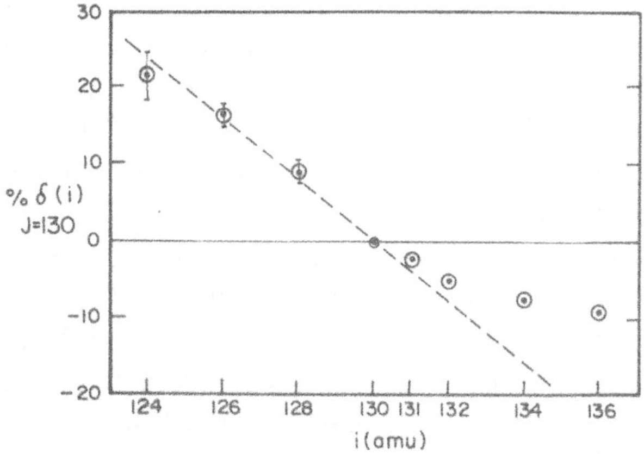

Figure 5.3.5: Abundances of Xe isotopes in chondrite meteorites, normalized to the isotopic abundance of ^{130}Xe in the earth's atmosphere.

significantly. In case of Kr, the isotopic composition, measured with reference to terrestrial ^{86}Kr, shows that abundances of isotopes with mass numbers 78, 80, 82, 83 and 84 are depleted in a linearly mass dependent mode, the lighter mass being depleted more than the heavier one. The data is shown in Figure 5.3.4 as given by Lewis and Prinn (1984). Turning to Xe measurements it is seen in Figure 5.3.5 that for abundances of isotopes of Xe with masses 124 to 136 there occurs a flattening which is due to the presence of fission produced Xe isotopes in these meteorites. But strangely, the depletion of abundance as a function of mass is reversed in Xe - the heavier isotopes being depleted more that the lighter one. The cause of these opposite trends in abundance depletions in Kr and Xe is not known.

Age of Chondrite Meteorites

The ^{4}He in C-meteorites originating from the alpha radio - active decay chains of ^{235}U, ^{238}U and ^{232}Th, provides information on the history of the meteorites. It is found that as alpha particles stop close to the parent nuclei, they are tightly bound in the grains of the crystal. The solar wind helium ions, on the other hand, are implanted close to the surface of the crystals. Therefore mild chemical etching of the surfaces of the grains releases the solar wind gases. The radiogenic helium atoms are released only due to melting at high temperatures. The gas retention ages of chondrites have been measured in many meteorites using the He/(U+Th) ratios. For most of these chondrites, ages are also determined from ^{40}Ar/^{40}K ratios. ^{40}K is a long-lived nucleus which undergoes beta-decay to ^{40}Ar with a half-life of $1.3 \cdot 10^9$ years. It should be noted that ^{40}Ar in chondrites originates only from the beta-decay of ^{40}K as the ^{40}Ar isotope is not present in the argon gas implanted by the solar wind. When the He - (Th,U) age of chondrites are plotted against their Ar-K age, as shown in Figure 5.3.6, interesting features can be seen. Region A is populated a by large number of chondrites with ages of about $4.6 \cdot 10^9$ years. The samples in B indicate diffusive loss of helium, whereas the points below the line indicate mild heating and partial loss of helium and a slightly lower Ar-K age. The other group of samples in region C, indicate that severe reheating by shocks or another process released both He and Ar gas, and thereby the radio-active clock was reset.

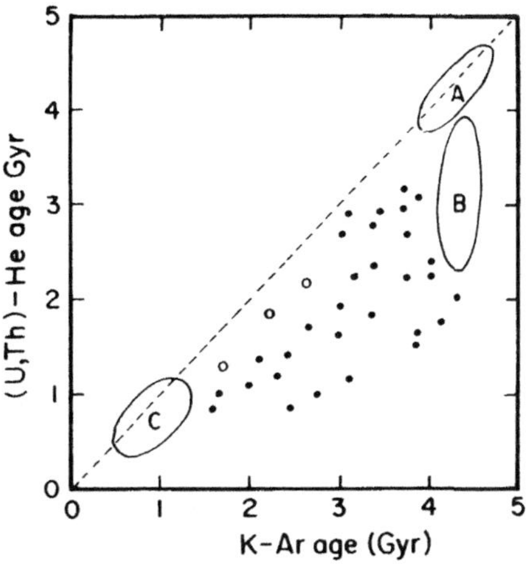

Figure 5.3.6: ^{4}He/(Th,U) age vs. (Ar/K) age of chondrites.

Rare gases in C-meteorites and in Planetary Atmospheres

When we compare the abundances of rare gases in C-meteorites with those in the atmospheres of Earth, Mars and Venus, several striking features are revealed (Figure 5.3.7). It is seen that carbonaceous chondrites have far more amounts of all rare gases per gm of rock, as compared to the atmosphere of the earth. Surprisingly Venus has much more Ne and Ar, as compared to the earth and its Ne and Ar amounts exceed those of C-chondrites. Also in Venus the relative abundances of Ar and Kr are drastically different when compared to those in the earth, Mars and meteorites. The ratio of Ar/Kr in Venus is closer to the solar composition value than that in other bodies. Mars, on the other hand, shows nearly the same relative abundances as in the earth, although the absolute concentration of rare gases in Mars is more than two orders of magnitude smaller than that of the earth, or C-chondrites. Lastly, in the earth and Mars, the Kr/Xe ratio is about 10 whereas in chondrites it is nearly unity. This last anomaly is known as the "missing Xe problem".

This are some of the problems which are not understood at present and are related to the complex mechanisms responsible for the origin and formation of meteorites and planetary atmospheres. For further studies the readers are referred to the books "Meteorites and the Early Solar System" and "Proto stars and Planets" as noted as general texts in the reference.

5.4 Interplanetary Dust

Very small meteoritic particles resulted from the break up of meteorites can survive their entry into the atmosphere. These particles - typically with diameters of 2 to 5 μm, were found to produce impact craters in the metal foils of Skylab's, Long Duration Exposure Facility (LDEF) and in the Space Shuttle orbiter. Terrestrial contaminating materials such as aluminium oxides from the burning of solid fuel booster rockets are identified by chemical analysis. Meteoritic dust particles are identified by their chemical composition by high precision ion probe instruments coupled with a high sensitivity mass spectrometer. In addition to meteoritic dust particles, it was found that a large proportion of particles are of distinctly different composition than those of meteorites. The non-meteoritic particles, numbering over 1500 have been analyzed in the laboratory; they are called Interplanetary Dust Particles (IDP). The pioneering studies of Brownlee and Walker of the USA have led to the identification of these IDP's from the micro-craters in metal foils exposed in space experiments.

These interplanetary dust particles which are identified by their impact craters are found to have a porous and complex physical structure indicating that they were not strongly heated or melted. Meteoritic dust particles are on the other hand black, composite particles. The non-meteoritic interplanetary dust grains are believed to be originating from cometary dust grains.

These suspected cometary dust particles have the distinctive feature that they do not contain Fe_2O_4 and calcium and magnesium sulphates, which are abundant in meteoritic particles. Thus it seems likely that the interplanetary dust particles of cometary origin are weak aggregates of minerals that did not experience the chemical alterations due to heating and partial melting, as the meteorites did. The comet nuclei from which these IDP's may originate are thus believed to be a mixture of low temperature and high temperature materials which have not undergone further chemical alterations.

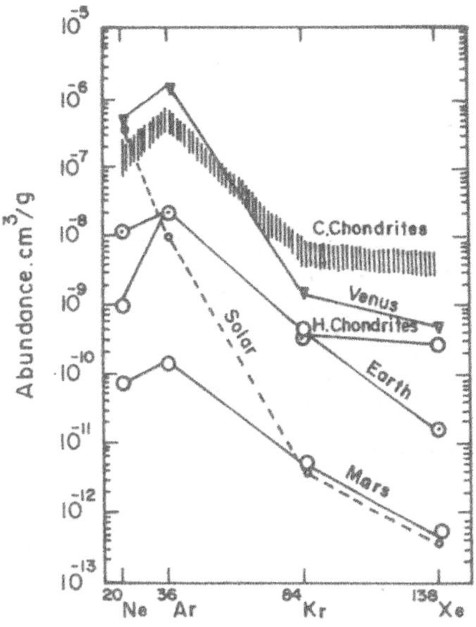

Figure 5.3.7: Comparisons of abundances of Ne, Ar, and Xe in C-chondrites and in the planetary atmospheres of Venus, Earth and Mars; solar abundances are shown for reference.

5.5 The Origin and Early History of the Solar System

5.5.1 General Features of the Solar Nebula

The early history and the origin of the solar system is a long standing problem on which many volumes were written and several theories have emerged over the decades. Here we shall present the broad scenario of the early history of the solar system, starting from the model of the solar nebula.

The collapse of an interstellar gas cloud to form a star - like object such as the Sun produces enormous energy. This energy is released throughout the interior of the nebula and is radiated away from the surface. The high internal temperature and pressure prevent further collapse of the nebula. Rapid radiative loss of internal heat is prevented by the high opacity of the gas to solar radiation. The infrared absorption by the gas molecules and dust contributes to the opacity of the gas.

When a slowly rotating massive molecular cloud collapses to a star size object, what happens to its angular momentum which has to be conserved? In

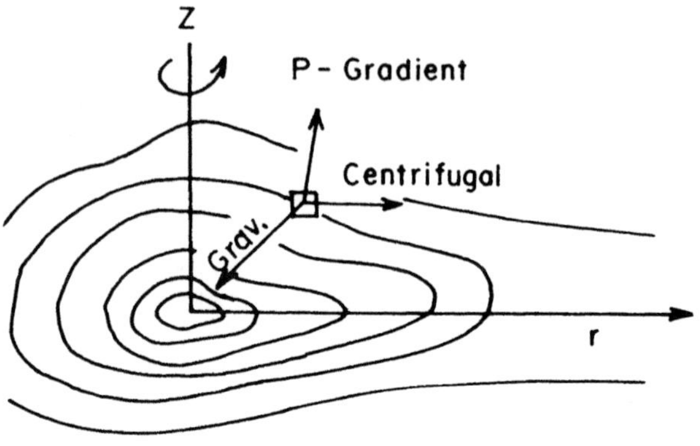

Figure 5.5.1: A sketch showing the three major forces acting on a volume element of the solar nebula.

order that the rotational velocity in the equatorial region of the solar nebula is not unduely high, a significant fraction of the collapsed material forms a thin rotating disc in the equatorial plane of the Sun. Thus the conservation of the angular momentum results in the formation of a flat disk out of the solar nebula.

We may picture the primitive solar nebula as a highly flattened disc, each volume element of which is acted upon by three major forces, e.g. (a) a gravitational force directed towards the center, (b) a centrifugal force due to the angular momentum of the nebula, directed radially outwards, and (c) a pressure gradient directed outwards in the Z-direction. These are shown in the sketch in Figure 5.5.1. In the approximate steady state it is assumed that the pressure gradient balances the Z-component of the gravitational force. The hydrostatic equilibrium equation can be set up and from this the gas pressure P_z as a function of distance Z and r can be calculated as this is given by

$$P_z = P_c exp - [\mu GM/2RTr^3] Z^2$$

were P_c is the pressure in the central plane, μ the mean molecular weight of the solar composition material which is taken as 2.3, and M the mass of the solar nebula, assumed as $\simeq$ 1.1 $M_\odot$. Typically, when one assumes T = 30

K at r = 3 A.U. (i.e. near the asteroid belt, one can estimate Z', where the pressure falls to half its value at the central plane P_c, as 0.2 A.U. Thus Z'/r is very small. A rough sketch (not to scale) of the equal pressure lines in Z-r plane is shown in Figure 5.5.1. This model is used to estimate the lower and upper limits of the mass of the solar nebula as will be discussed below.

5.5.2 Mass and Density of the Solar Nebula

It is now known that solar composition material contains a very large fraction of 98.2% of mass accounted for by the 'permanent' gases H_2, He and Ne, and about 1.4% by condensed compound water, e.g. ice, solid ammonia, methane etc., which are collectively called "ice minerals", and only 0.4% by solid metals, e.g. metal oxides, refractory oxides, silicates etc. collectively called "rocky" minerals. Therefore the minimum mass of solar material required to make the terrestrial planets, Venus, Earth and Mars, is about 100/0.4 times their present mass i.e. a total of about 500 earth masses. This mass is larger than the present total mass of all the Jovian planets. Therefore the solar nebula should have a mass of solar composition type material of at least 1000 earth masses, spread over the region from Mercury to about 40 A.U.. This comes to about $6 \cdot 10^{30}$g which is the $3 \cdot 10^{-3}$th fraction of the present solar mass.

In order to make another order of magnitude estimate how this disk mass was distributed as a function of distance from the Sun, the following procedure may be adopted. For each planet, taking into account its mass and density, the enhancement factor is calculated and hence the mass of the solar composition material of the nebula is obtained. This mass is then distributed in an annular ring with inside and outside radii about half way between the orbits of the planets. The asteroid belt is taken as a proto-planet. The enhancement factors for the inner planets thus, lie in the range 250-350 and 5-20 for the outer planets. Details are given by Lewis on the annular areas, and surface densities of the nebular gas (σ) in the annular areas as a function of distance r from the Sun. It is found that although the solar nebula mass for the outer planets is much larger than the nebula mass for the inner planets, the surface density of matter in the outer rings is much smaller than in the inner rings. The plot of σ roughly falls off as $r^{-1.5}$ to r^{-2}, with truncation of σ to zero at about 0.3 and 80 A.U. The inner edge of truncation probably signifies the infall of matter from the region $r \leq 0.3$ forming the Sun. In this rough model, the total nebular mass becomes $3.5 \cdot 10^{-3}$ $M_{\odot}$, which is the lower limit. The considerations of the T-Tauri phase of the Sun, as discussed later, lead to an upper limit of the nebular mass as 0.1 $M_{\odot}$. So the mean value of the nebular mass is taken as 0.02 $M_{\odot}$.

5.5.3 Formation of the Planetary System

As noted before a slowly rotating molecular cloud has a too high rotational angular momentum to collapse to a sun-like object. So a significant fraction of the collapsing material forms a pressure supported rotational disk, in orbit around the proto-sun. This disk, which has the same composition as the solar nebula, is at a sufficient distance from the central object, cool enough for 1 to 2% of the material to be of solid form, either in condensate form in the disk, or as remnant interstellar grains. The disk is most probably highly turbulent during the infall stage as a result of a mismatch between the angular momentum of the gas falling in the disk and that required to maintain Keplerian rotation. It is assumed in current models that most of the refractory compounds remain as gas or small solid grains until the disk become quiescent.

When the quiescent stage begins the composition is assumed to be solar. The dust particles coalesce due inelastic collisions and gradually settle down towards the middle plane of the disk, and bodies of small, medium and large size are formed. It can be argued that due to a negative radial pressure gradient a small outwardly directed force is produced which acts to reduce effective gravity, so that the gas rotates at slightly less than the Keplerian velocity. Small solid grains rotate with the gas. Large solid bodies orbit with Keplerian velocities and medium sized bodies at intermediate velocities. Thus the large and medium sized, bodies are subjected to a head wind from the gas. This head wind removes some angular momentum of these bodies causing them to spiral inward to the central star. Thus the accretion of bodies into the Sun can take place.

Also the growth of medium meter-sized bodies to large, kilometer size bodies could occur by accretion processes or other mechanisms. Kilometer size bodies are relatively free from loss processes from the disk.

Kilometer-sized and larger bodies, which are called planetesimals, in

Keplerian orbits are subjected to two types of perturbations. These are gravitational interactions and physical collisions. These interactions lead to accretion of planetesimals and the formation of so called 'planetary embryos'. If the random velocities of most planetesimals are less than the escape velocity from the largest bodies, then these planetary embryo's grow rapidly. So evolution of the distribution of sizes of solid bodies is skewed, because larger bodies grow much faster than the smaller ones and this process may be called runaway accretion. Thus planetary embryos absorb most of the slow moving bodies within their gravitational range and the runaway accretion phase ends. Slower growth continues and mutual gravitational perturbations of the planetary bodies lead to increased eccentricity of planetary orbits. When sufficiently massive planets such as those in the Jovian group are formed sufficiently far away from the Sun, they may also eject large mass objects into interstellar space. It is believed by many investigators that Oort cloud comets were planetesimals which were ejected outwards with velocities close to the escape velocity from the Sun and the gravitational perturbations by nearby stars led them to a long-lived, quasi-stable orbit at the outer periphery of the solar system.

These were some of the broad outlines of the physical processes in the formation and the early phases of planetary bodies. It may be noted that there are several uncertainties and complexities in the phenomena which are not yet fully understood and several models and theories have been proposed for these aspects. For detailed discussions of these problems the readers may refer to papers by several authors, e.g. Lissauer, Duncan and others in the Proceedings of the 10th IAP Astrophysics meeting in Paris, in July 1994 and published in the book on "Circumstellar Dust Disks and Planet Formation", (reference 5 of the General Texts).

5.5.4 T-Tauri Phase of the Sun

The dark gas clouds in which star formation takes place show the presence of T-Tauri stars. The characteristics of these stars indicate that T-Tauri stars are in the early stage of their formation. These stars have rapid rotation and strong outflow of gases. The masses of T-Tauri stars are in the range of 0.2 to 3 solar masses. Thus the present T-Tauri stars are believed to be indicative for the early stage of the Sun. The Sun probably entered the T-Tauri phase just after the ignition of its core when its luminosity was high - about 2 to

3 times the present value and this state continued for about 10^7 year till it settled as a main sequence star.

T-Tauri stars are observed to have strong stellar winds with radial velocities of 200-300 km·s^{-1} at 2-3 stellar radii. This is about half the escape velocity of the star at that level. It is assumed that the Sun, in the T-Tauri stage had a similarly high solar wind, and the mass loss of the Sun is estimated to be about 10^{-8} $M_\odot$ per year. One of the main effects of the strong solar wind is its influence on the gas of the stellar nebula. It can be shown that the kinetic energy of the dense solar wind with radial velocities of 200-300 km·s^{-1} at 2-3 solar radii is sufficiently high to blow away the nebular gas. Numerical data indicate that each gm of T-Tauri solar wind carries enough energy for the dissipation of 100 gm of the nebular gas and dust from 1 A.U. to infinity. Thus the solar wind would sweep out efficiently as much as 0.2 solar mass of gas and dust entrained in solids. It is calculated by Ramadurai[(8)] that for a mass loss rate of $5 \cdot 10^{-10}$ $M_\odot$ per year and the mass of the Sun as 1.1 $M_\odot$ and a solar wind velocity of 500 km·s^{-1}, the nebular gas would be swept out in about 3 million years which is well within the T-Tauri phase duration of about 10^7 years.

Another important consequence of the strong solar wind in the T-Tauri phase of the Sun is the radial transport of the angular momentum of the early the Sun. It can be shown that a mass loss of 0.01 solar mass during a period of about 10^7 years would be sufficient to de-spin the Sun considerably and bring it to a typical main sequence phase.

5.5.5 Isotopic Anomalies and the Origin of the Solar System

The origin of the solar system which we have discussed may be called the standard model. In this section we shall discuss some recent observations and suggestions relating to the origin of the solar system. On the basis of several observations it has been suggested by several authors that a nearby supernova may have exploded during the period of the formation of the solar system and injected freshly synthesized atomic nuclei including radio-active nuclei, into the solar nebula. The disturbances produced by the SN explosion may have helped in the collapse of the solar nebula. Although there is at the moment no consensus among the scientists about this collapse hypothesis, supernova injection is confirmed by many observations. We shall briefly discuss the new and interesting observational results which strongly support the supernova injection. The details are given in the review by Schramm.

There are two types of isotopic anomalies measured in carbonaceous chondrites. The first one relating to the stable ^{16}O - isotope was found by Clayton et al.[9]. The other, the radioactive isotope ^{26}Al anomaly was identified by Lee et al. The Allande meteorite which fell in Mexico in 1969 was a massive 2-ton object that provided the first new information on both the isotopes. The stable isotopes of oxygen ^{16}O, ^{17}O and ^{18}O have been widely studied in many meteorites and these isotopes are found to occur in a regular pattern. The striking discovery in Allende, made from mass spectrometric studies, was that there was a component of oxygen which was almost pure ^{16}O with very little ^{17}O and ^{18}O. In some minerals this component amounted to 5% of the oxygen content. This means that when carbonaceous chondrites solidified there were regions with very different isotopic composition. In other words the bulk material of the solar nebula contained primordial solid grains with different isotopic composition which did not completely mix in the nebula. This anomaly does not specify the time scale.

The other isotope, ^{26}Al which decays to ^{26}Mg with a half-life of $7 \cdot 10^5$ years, provides information on the time scale. When we have a sample with a mixture of ^{26}Al and normal ^{27}Al, after $7 \cdot 10^5$ yr half of ^{26}Al will be ^{26}Mg and after $4.6 \cdot 10^9$ years (the age of the solar system) it will have no ^{26}Al but an equal amount of ^{26}Mg. So we shall see an excess of the ^{26}Mg isotope among the normal isotopic composition of 79% ^{24}Mg, 10% ^{25}Mg and 11% ^{26}Mg. In Figure 5.5.2 it is seen that in the Allende meteorite it was found that the higher the $^{27}Al/^{24}Mg$ ratio, the greater is the excess of ^{26}Mg. This shows clearly that ^{26}Al must have been present when the rocks solidified. Otherwise if the decay occurred before final solidification, the resultant ^{24}Mg would mix chemically with all other Mg and the effect will be so diluted that no correlation with the Al/Mg ratio would be present. The observational evidence is that ^{26}Al was alive when the rocks were formed in the early solar system. It also shows that ^{27}Al must have been produced by nuclear reactions within a few half-lives of solidification i.e. within a few million years. This leads to the conclusion that ^{26}Al must have been produced by a supernova. A supernova is considered to be the source of these anomalies because it is well established that all heavy elements like carbon and heavier were synthesized in a supernova. Therefore both the ^{16}O and ^{27}Al anomalies strongly lead to a supernova as their source. Support for the supernova model is also provided by other isotopic anomalies recently discovered, e.g. the ^{107}Pd anomaly with a half-life of $6{\cdot}10^6$ years, and others. Table 5.5.1 shows a summary of extinct radio active nuclei in meteorites which have been confirmed so far. For further details readers may refer to the review by Swindle as given in the book "Proto stars and Planets III", (General Text No. 3).

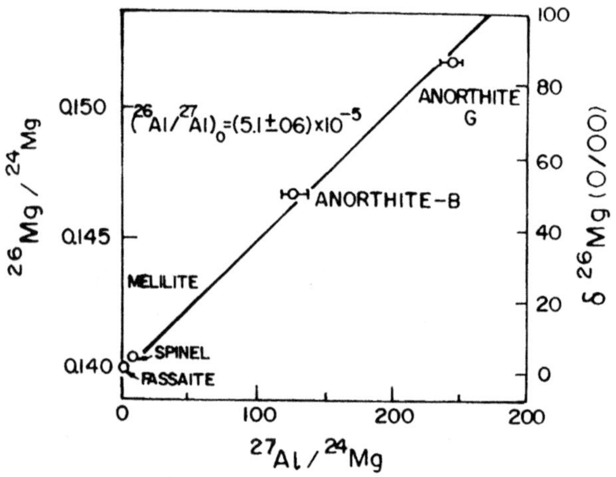

Figure 5.5.2: Results showing the isotopic anomalies of the Al and Mg isotopes. Correlations of ^{26}Mg/^{24}Mg and ^{27}Al/^{24}Mg ratios are shown in the Allende meteorite. See text for details. (Adopted from Swindle).

In summary the scenario of the supernova explosion and the origin of the solar system is given as follows:

The galaxy had been in existence and evolving for about $10{\cdot}10^9$ years prior to the beginning of the solar system. It is assumed that the interstellar medium consisted of well mixed gas and dust. A galactic spiral wave went through the region of gas and dust about $4.7{\cdot}10^9$ years ago; it produced one or more massive stars which evolved rapidly in about $10{\cdot}10^6$ years and blew up ejecting it debris, including I and Pu among other things. The next spiral wave, which came $100{\cdot}10^6$ years later i.e. about $4.6{\cdot}10^9$ years ago, produced a new generation of stars out of the interstellar gas and the debris. Of these the massive ones evolved rapidly and became supernovae. It is then reasonable to assume that one of the first of this new generation of supernovae produced the observed ^{26}Al and ^{107}Pd and helped in the collapse of the solar nebula. It should be noted that the solar nebula had already started to collapse by the time the supernova explosion debris hit it, because it takes about 10^7 years for the solar nebula to collapse from a gaseous mass to solid bodies. The supernova debris containing ^{26}Al, ^{107}Pd along with ^{16}O most probably in the form of grains was added to the outer parts of the condensing solar nebula. Then condensation, mixing, chemical reaction and recrystallization occurred. In large condensed bodies ^{26}Al together with U-Th provided heating by radioactive decay resulting in the melting of the cores.

Table 5.5.1: Extinct Radioactive nuclei in meteorites

Radio Nuclei	$T_{1/2}$ (My)	Daughter	Abundance
^{26}Al	0.7	^{26}Mg	$^{26}Al/^{27}Al = 5{\cdot}10^{-5}$
^{53}Mn	3.7	^{53}Cr	$^{53}Mn/^{55}Mn = 4{\cdot}10^{-5}$
^{107}Pd	6.5	^{107}Ag	$^{107}Pd/^{108}Pd = 2{\cdot}10^{-5}$
^{129}I	16	^{129}Xe	$^{129}I/^{127}I = 1{\cdot}10^{-4}$
^{146}Sm	103	^{142}Nd	$^{146}Sm/^{144}Sm \approx 0.01$
^{244}Pu	82	Fission Xe	$^{244}Pu/^{238}U \simeq 0.005$

5.6 Problems on Chapter 5

1. What are the principal features of the nucleus of the Comet Halley, such as its mass, volume, density etc. and the properties of the emitted gas and dust ?

2. Describe the main features of the inner and outer components of the Oort cloud and their relationships with the origin of comets.

3. What are the distinctive features of the composition of chondrites as given by the modified Urey-Craig representation?

4. Discuss the main features of the results measurements of isotopic abundances of oxygen in several types of chondrites in meteorites and their comparison with those of the earth and the moon. What are their main implications?

5. How are the ages of chondrite meteorites determined and what are their interpretations?

6. How are the interplanetary dust particles detected near the earth and how are these distinguished from meteoritic particles? What is the most likely origin of the interplanetary dust particles?

7. What are planetesimals and how did they evolve in the early phases of the origin of the solar system?

8. What are the main features of the T-Tauri phase of the Sun and how did these affect the evolution of the solar system?

9. What are the basic aspects of the anomalies of the oxygen isotopic abundances in some C-chondrites and what are their significance?

10. State briefly the main stages of the hypothesis of the origin and the evolution of the solar system and the impact of a supernova explosion.

CHAPTER 6
SOLAR ENERGETIC PARTICLES

6.1 General Features and Early Observations

The emission of energetic particles from the Sun is one of the major manifestations of high energy astrophysics and is related to a variety of astrophysical and geophysical phenomena in the heliosphere. An event of solar energetic particles (SEP) is seen as a sudden rise of the counting rate of the background galactic cosmic rays reaching a high value in a few minutes, as measured by ground level or spacecraft borne detectors. The counting rate i.e. the intensity reaches a maximum value which may be 10 to 100 times more than the background level in small and medium events and may be more than 1000 times more in large events. Following the maximum the intensity slowly decreases compared to the background during a few hours in an average event and in a day or so in case of a very large event. A day or two after the optical flare and the solar energetic particle (SEP) event, the plasma cloud emitted by the Sun reaches the earth and causes a magnetic storm and Forbush decrease in the galactic cosmic ray intensity (Chapter 3.5). In contrast to the solar wind and plasma clouds from the Sun which contain protons with energies of 1 to 10 keV, SEP events have protons with energies of 10 MeV to 100 MeV in average events and up to 1 to 10 GeV or more in large events. As these energies overlap the proton energies of galactic cosmic rays, these SEP events are also called solar cosmic ray (SCR) events by some authors.

The discovery of the emission of energetic particles from the Sun was made from observations of the energetic particle events of February 28, 1942 and March 7, 1942 using ionization chambers at ground level, when the intensity increases associated with solar flares were identified by Forbush(1) and others. The sequence of Forbush-type decreases, which was discovered in 1938-39 was also identified. The conclusions were further confirmed in the July 25, 1946 solar flare event.

6.2 Observational Methods and Early Results

6.2.1 Neutron Monitors and Ground Level Events

A new approach for the study of SEP was made with the development of neutron monitors in which fast neutrons produced in the atmosphere by the nuclear interaction of the incident high energy protons (Chapter 7.2). Several neutron monitors were developed by the University of Chicago and several other centers in the USA and Europe during 1948-49. The solar energetic particle event of November 19, 1949 provided the confirmation of the higher sensitivity of neutron monitors for the study of solar energetic particle events. In the early 1950's a series of neutron monitors was constructed by many cosmic ray laboratories of the world and these were operated in a collaborative network for the study of solar events during the International Geophysical Year (IGY) beginning from 1957.

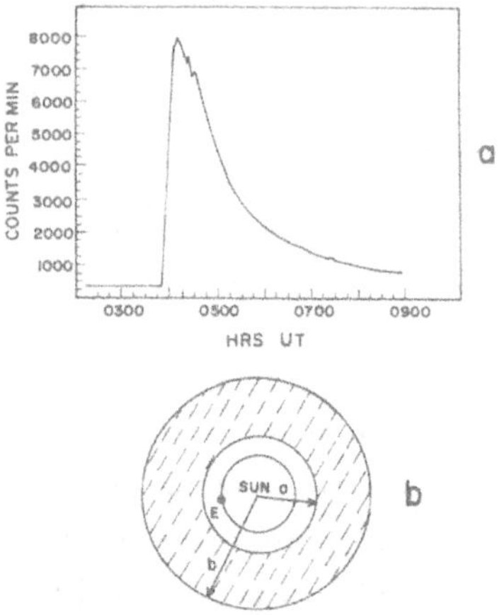

Figure 6.2.1: (a) Counting rate vs time of the neutron monitor of the University of Chicago during the large SEP event of February 23, 1956. (b) A simplified sketch of the assumed disordered magnetic field surrounding the earth and the Sun.

February 23, 1956 Solar Event

This new network of detectors for the IGY got a big boost by the very large solar flare and particle event of February 23, 1956, which was studied by many laboratories. In Figure 6.2.1(a) we show the data obtained by the neutron monitor of the University of Chicago which was analyzed by Meyer, Parker and Simpson[2] in 1956. A class 3 solar H_α flare (located at 23°N, 80°W) began at 0330 UT, reached the maximum intensity at 0342 UT and ended at 0415 UT. The neutron monitor having the background counting rate of 340 counts/min, showed a sharp increase at 0350 hrs UT which rapidly increased to the very high value of about 7800 c/min at 0415 UT. The peak flux $J(> 30$ MeV) was 6200 protons/$cm^2 \cdot s$. It slowly decreased in about 5 hours to near the background level, but remained above the background for another 10 hours or more. This SEP event with a 9000% increase from ground level was one of the largest solar particle event seen and produced an integrated proton intensity of $6.5 \cdot 10^8$ p/cm^2 for $E > 30$ MeV, at the top of the earth's atmosphere.

This event showed that (a) although solar flare associated activities on the Sun ceased 45 minutes after the flare onset, relativistic protons accelerated in the Sun continued to bombard the earth with high intensity for many hours, and (b) in the initial several minutes the energetic protons were reaching the earth from the general direction of the Sun (displaced by the 'garden hose' angle), whereas subsequently solar energetic particles were arriving at the earth almost equally from all directions for many hours. These results led to the major astrophysical conclusion that somewhere beyond 1 A.U. the earth and the Sun is surrounded by a spherical shell of a disordered magnetic field. Accelerated solar protons are stored in the volume of this shell, with inner and outer radii a and b respectively, for a considerable period of time as the protons diffuse through the irregular magnetic field to reach the earth almost isotropically and to escape to the interstellar medium. A simple sketch of the model[2] is given in Figure 6.2.1(b).

As neutron monitors have a much lower threshold (> 0.5 GeV) as compared to earlier ion chambers with a higher threshold (> 5 GeV), from 1956 to 1973, about 21 ground level events (GLE's) of SEP were observed with neutron monitors[3], whereas in the first twenty years 1936-56 only five GLE were detected with ion chambers. The neutron monitor results provide a large volume of information on the propagation of relativistic solar protons in the solar corona and in the interplanetary medium, some highlights of which are discussed later. The summary of 69 GLE's from 1956 to 1972 is given by Smart and Shea[4] in the review of 1985.

6.2.2 Balloon Studies and PCA events

Solar Protons

The direct observations which first showed that the Sun emits non-relativistic protons of $\sim$ 100 MeV more frequently than relativistic protons of GLE, were made in high altitude balloon flight experiments in 1958-59, by the University of Minnesota group of Freier, Ney and Winckler[(5)]. They detected solar protons of $\sim$ 100 MeV with nuclear photographic emulsion stacks and Geiger counters in a balloon flight from Minneapolis made during the March 28, 1958 solar flare event. The flight was made in the main phase of the magnetic storm while studying auroral particles. Anderson had the good fortune to have a balloon at ceiling altitude at Ft Churchill, Canada in August 1958 when a solar proton event occurred and was detected. The USSR group led by Charakhchian[(6)] made series of balloon flight experiments at high latitudes with Geiger counter telescopes during 1958-61 and studied the solar proton event of July 8, 1958 and many other subsequent events.

Polar Cap Absorption (PCA) Events

Polar Cap Absorption (PCA) events were discovered during this period by ionospheric studies in the high latitudes of the polar region and added greatly to the balloon experiments of solar flare protons. Leinbach and Reid of the University of Alaska, following the earlier work of Reid and Collins, observed that low energy solar protons of 1-50 MeV accelerated during large solar flares were incident on the polar region and produced intense ionization of the D-layer at 60-80 km. As a result a receiver with a wide angle antenna measuring cosmic radio noise at 27 MHz or 57 MHz would show strong absorption of cosmic noise signaling the arrival and continuous flux of solar protons. From the ionospheric parameters and the magnitude of the absorption in db, the flux of incident solar protons could be estimated. The absorption in the D-layer in decibels is found to be proportional to the incident flux during day time and the square root of the flux at night time. The instrument is called "riometer" (radio ionospheric opacity meter) and was used to measure 77 PCA events during 1956-1973[(3)].

The PCA events provided a new and powerful technique to monitor low energy, high intensity solar particle events. Using the signaling of a large

PCA event at College, Alaska, a balloon flight at Minneapolis was made in the solar proton event of May 10, 1959, and thus solar protons with a very high intensity were recorded for the first time in nuclear emulsions and Geiger counters[5]. These pioneering studies led to new experiments for the direct measurements on solar energetic particles using balloon and rocket vehicles.

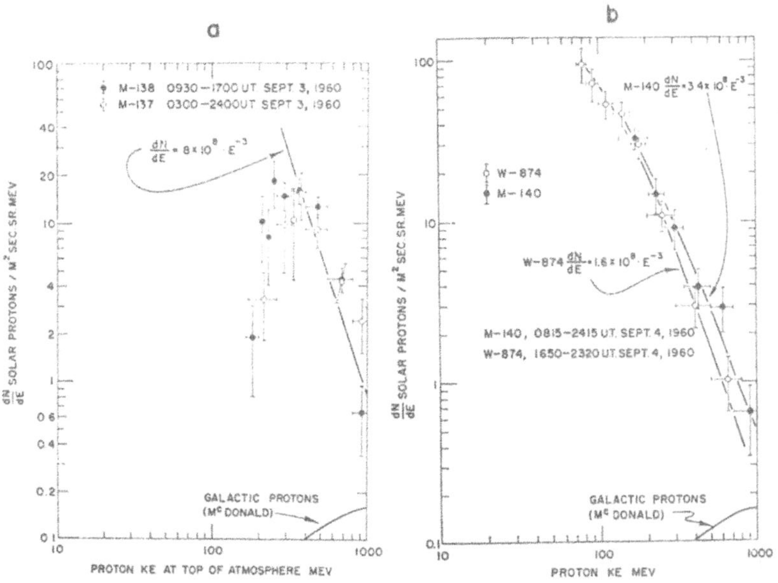

Figure 6.2.2: Solar proton fluxes vs energy in balloon flights from Minneapolis (a) before the commencement of the magnetic storm, and (b) after.

Solar Helium Nuclei

The first detailed observations and measurements of energetic helium nuclei emitted by the Sun were made in balloon flight experiments from Minneapolis during the solar flare event of September 3, 1960 by Biswas, Freier and Stein[8]. A solar flare event of 3^+ magnitude occurred at the east limb (N20, E88) of the solar disc at 0110 UT on September 3, 1960 following which four balloon flights were made during the next 48 hours and solar protons and helium nuclei of fairly high intensities were measured in four nuclear emulsion stacks. As the balloons floated at ceiling altitudes of 4 to 9.5 g·cm^{-2} of residual air, the solar protons with energies of 75 to 1000 MeV and helium nuclei with energies of 75-250 MeV/nucleon could be measured.

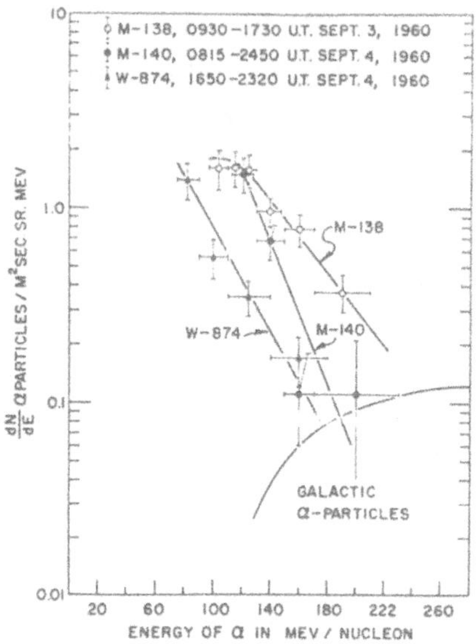

Figure 6.2.3: Fluxes and energy spectra of Solar Helium nuclei before the beginning (M-138) of the storm and after (M-140 and W-874) the magnetic storm.

A large magnetic storm and a Forbush decrease commenced at 0230 UT on September 4, 1960 and continued for more than one day. Interesting effects of solar protons and helium nuclei having the same energy per nucleon but different magnetic rigidities could be seen in these studies, in which the first two balloon flights were made before the arrival of the magnetic storm and the other two balloon flights took place after the commencement of the magnetic storm. In Figure 6.2.2 the differential fluxes and energy spectra of solar protons are shown in (a) before the commencement of the magnetic storm and in (b) after the storm. The fluxes and energy spectra of the helium nuclei before and after the storm are shown in Figure 6.2.3. The following important conclusions are derived:

(a) The solar proton intensity had the maximum value of 15 protons/(m^2·sr·s) at 250 MeV as seen in Figure 6.2.2(a) before the magnetic storm. After the magnetic storm, the Störmer cut-off is almost removed or reduced to a low value so that protons with energies as low as 80 MeV (rigidity 0.6 GV) were reaching the detector with a high intensity of about 100 p/(m^2·sr·s) (Figure 6.2.2(b)). Similar observations were made on the

May 10, 1959 solar proton event by Ney et al.[5].

(b) Solar helium nuclei fluxes were measured (Figure 6.2.3) with significantly large intensities in the two balloon flights before the magnetic storm in the energy range 100-200 MeV/N. Since their rigidities are twice that of protons with the same energies and hence well above the geomagnetic cut-off values, the nuclei had easy access to the detector, whereas solar protons of these energy had a small flux. Thus solar helium could be detected easily with low flux of solar protons. After the magnetic storm when the cut-off was reduced helium nuclei of as low as 75-85 MeV/N could be detected. During this period solar proton flux was also high as seen in Figure 6.2.2(b).

(c) The relative abundances of solar protons and helium nuclei were found to be highly variable due to several factors. Due to different mass to charge ratios of protons and helium nuclei, propagation characteristics are different for these two nuclear species. Therefore helium to proton ratios varied with time as well as event to event, as seen in the SEP events of July 10, 1959, May 4, 1960 and September 3, 1960 events[7]. Above the rigidity 0.8 GV, P/He ratio varied from about 50:1 to 1:1 with smaller ratios being more frequent.

(d) Before the magnetic storm solar protons of 300-1000 MeV had a differential energy spectrum of $dJ/dE \sim K{\cdot}E^{-3}$ (Figure 6.2.2a) with a flat cut-off at $\sim$ 200-300 MeV. This corresponds to magnetic rigidities of 0.6 to 0.9 GV. Thus the actual geomagnetic cut-off at Minneapolis was measured using the steep energy spectrum of solar protons. The measured cut-off R_e = 0.9 GeV was significantly lower than the Störmer R_c = 1.35 GV. Below 0.9 GeV (k.energy $\sim$ 300 MeV) solar protons arrived with partial intensities up to R_c = 0.6 GV (KE $\sim$ 200 MeV). Thus the geomagnetic cut-off at the high latitude at Minneapolis is not sharp and there is a broad energy or rigidity band (0.6-0.9 GV) over which partial transparency occurs. The origin of these effects are not fully understood and is probably due to ring currents, turbulent plasma and magnetic field in the distant regions of the magnetosphere.

6.2.3 Rocket Studies of Solar Energetic Particles (SPICE)

The sounding rocket program of the NASA-Goddard Space Flight Center, Greenbelt, Maryland, USA for the study of solar energetic particles emitted in association with large solar flares was conducted from 1960 onwards.

Beginning with the small solar event of September 3, 1960 by Fichtel and Gauss[8], these rocket studies provided the first new results on several properties of solar energetic particles in the large solar particle events of November 12 and November 15, 1960 and several new discoveries were made in the early 1960's[9,10,11,12]. These rocket experiments were continued with further improvements for the next 15 years or so till early 1970, when studies were made of the extra-ordinarily large solar particle event of August 4, 1972, and a number of new features of SEP were explored. These experiments were code named as SPICE (Solar Particles Intensity and Composition Experiment).

First we present a brief account of the highlights of the experimental methods and the detector system used in SPICE. Solar heavy nuclei of atoms with $Z \geq 6$ are absorbed in the uppermost layers of the atmosphere because of their high ionization loss. Hence these are not accessible at balloon flight altitudes and rocket-borne detectors are needed for their studies. NASA-Goddard space Flight Center used two stage sounding rockets, such as Nike-Cajun and Nike-Apache which could be fixed at short notice and could lift a payload of 40 kg to a peak altitude of $\sim$ 130 km thereby providing about 4 to 6 minutes exposure outside the atmosphere above 60 km altitude out of a total flight time of 600 sec. The rocket flights were conducted from the US Air Force Station at Fort Churchil, Canada, located at high latitude $\lambda = 59^o$ near the polar region, where the geomagnetic cut-off was as low as $\sim$ 1 MeV/N for particles with $A/Z = 2$. Nuclear Emulsion (NE) stacks were used as detectors as these could detect and identify all nuclei with $Z = 2$ to 26 with good charge resolution and tracks of heavy nuclei with $Z > 6$ could easily been seen against the large background of proton tracks. Each NE sheet was circular in shape with a diameter of 20 cm and a thickness of 0.6 mm and 50 NE sheets were assembled in a cylindrical stack with a height of 3 cm. The NE stack detector was surrounded by a cylindrical sheet of stainless steel of thickness with a 100 μm only. This was a critical element, as the shielding must be sufficiently thick so as to protect the delicate NE stack from the heat and pressure change during the flight and reentry of the payload into the atmosphere and at the same time sufficiently thin to allow low energy solar heavy nuclei to reach the detector. The shielding thickness used allowed the measurements of charge and incident energy of P, He and O nuclei greater than 15, 30 and 40 MeV/N respectively. The payload contained in addition, the telemetry sytem which provided the altitude vs time profile

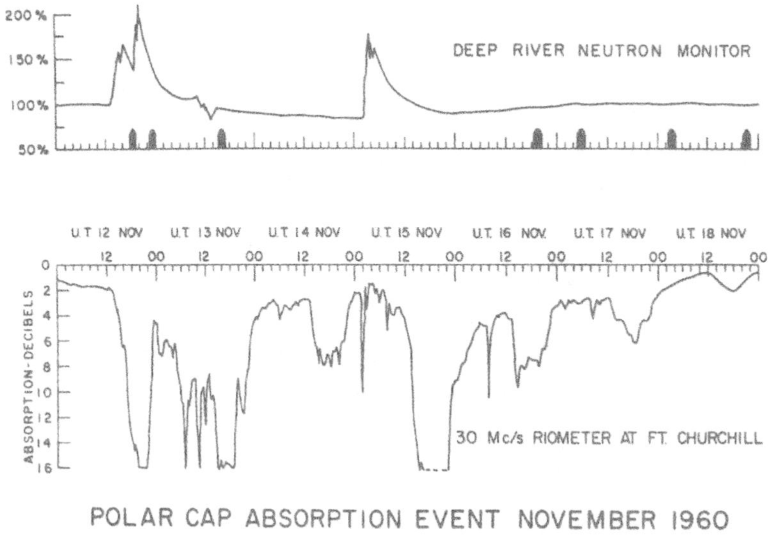

Figure 6.2.4: Polar cap absorption data, neutron monitor counting rates and times of seven rocket flights during the large SEP of November 12 and 15, 1960.

of the payload, and SARAH beacon and a parachute for the safe reentry and recovery of the payload.

The sounding rocket experiment (SPICE) for solar flare particles required elaborate ground support systems. Several payloads with NE detectors were kept in readiness on a 24 hour basis at Fort Churchill, Manitoba, Canada while the Rio-meter continuously monitored for incoming particles. When a large PCA event was found to be in progress, rocket flights were made at a time close to the peak intensity, as in the first SEP event of September 3, 1960. Soon thereafter a series of five rocket flights were made successfully into the solar energetic particle beams during the onset of the extremely large solar particle events of November 12, and November 15, 1960. In Figure 6.2.4 we illustrate the data of very large Polar Cap Absorption (PCA) events following the solar flares of November 12 and 15, 1960 and large increases in the ground level neutron monitor counting rates. The times of seven rocket flights from

Fort. Churchill, Canada are indicated. In five of these flights the payloads were successfully recovered in the arctic region with the help of a microphone system laid on the ground (which detected the shockwaves produced during the reentry of the payload) by triangulation method and with the help of a radio beacon.

In the second generation experiments of 1965 and onwards, the shielding thickness was reduced so that solar oxygen with an energy as low as 20 MeV/N could enter the detector. In the 1970's the techniques were further improved with the following tricks. As soon as the rocket reached 60 km altitude, a motor driven screw quickly raised the nose cone shielding above the detector so that the detector was exposed to free space conditions with little or no shielding matter. On reentry of the payload at about 60 km altitude, the nose cone shielding assembly was brought back to its original position to protect the detector during the reentry phase. In this arrangement ^{16}O nuclei with energies as low as 3 MeV/N could be measured.

Main Results of SPICE

The highly successful rocket flight experiments of the NASA Goddard Space Flight Center which began with the large solar flare events of November 12 and 15, 1960 provided the first detailed picture of the composition of solar energetic particles and the results led to several new discoveries on the multiply charged nuclei emitted by the Sun as given by Biswas et al,[(9–12)] and reviewed by Biswas and Fichtel[(13)]. The measurements were made five times during the two large flare and particle events, and the main results are as follows:

1. The differential energy spectra of solar P, He and C, N, O and Ne were measured with high accuracy in the energy range of 30 to about 200 MeV/N during the five occasions in the two large flare events of November 12 and 15, 1960. Figures 6.2.5(a) and (b) show typical results from two rocket flights at 1951 UT, November 16, 1960 and at 1603 UT, November 13, 1960, plotted in the scales of kinetic energy/nucleon and rigidity respectively. It can be seen that the energy spectra of protons and helium are different indicating that the P/He ratio varies with energy or rigidity, and also from event to event. The fluxes of medium (C, N, O) nuclei multiplied by 60 matches the He fluxes and spectra of He and

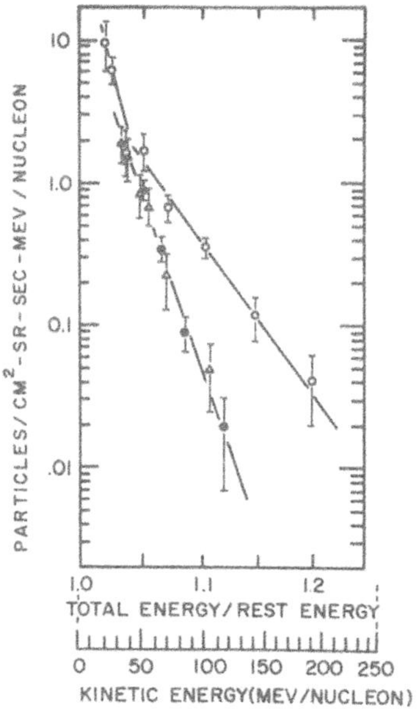

Figure 6.2.5(a): Differential fluxes vs. kinetic energies at 1951 U.T., November 16, 1960 for solar protons (O), He(Δ) and the C,N,O nuclei multiplied by 60 (•).

medium nuclei have the same scales for energy/nucleon and for rigidity.

2. The above features were measured and confirmed five times during the two solar events. The similarity of energy per nucleon spectra of helium and medium nuclei and the same relative abundances of helium to medium were observed six times during three flare events with the mean value of 60 ± 7 in the 42-95 MeV/N interval. This leads to the conclusion that the relative abundances of He to medium represent those in the upper layers of the solar atmosphere. As the nuclei of ^{4}He, ^{12}C, ^{14}N, ^{16}O have the same mass to charge ratios and for the same velocities these have the same magnetic rigidity, their propagation properties are similar. Hence this method of study of solar flare nuclei yields the relative abundances of helium to medium in the Sun[(13)].

3. The relative abundances of He, C, N, O and Ne measured five times during the two flare events show that these are the same within experimental uncertainties and the average values are shown in Table 6.2.1. For comparison spectroscopically measured relative abundances of these elements are shown. It is seen that the relative abundances of C, N, O measured in solar cosmic rays agreed very well with those in the solar photosphere measured spectroscopically. These confirm the ratio of He

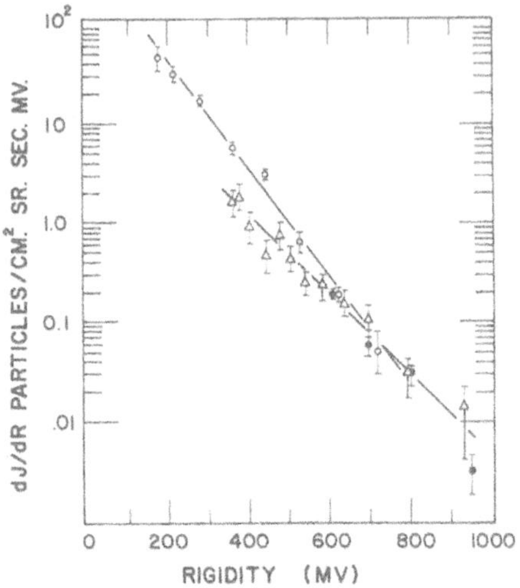

Figure 6.2.5(b): Differential fluxes vs rigidities at 1603 U.T., November 13, 1960 for Solar Protons (O), He(Δ) and the C,N,O nuclei multiplied by 60.($\bullet$)

to medium measured in solar energetic particles of energies > 40 MeV/N as representative of the solar atmosphere. When these measurements were made in the early 1960's, it was not possible to determine the abundances of He and Ne in the solar photosphere spectroscopically. So the abundances of helium (107 $\pm$ 14) and neon (0.16 $\pm$ 0.03) in the Sun relative to oxygen as 1.0 could be obtained for the first time from solar cosmic ray measurements(12,13) (Table 6.2.1).

In later years when it was possible to estimate spectroscopic abundances of helium and neon in the Sun, there was good agreement between the values from solar energetic particles and from spectroscopic measurements, as shown in Table 6.2.1.

4. The next important question was the relative abundance of hydrogen to helium. As shown earlier, this can not be determined from SEP results only. Biswas and Fichtel(13) and Biswas (1968) determined this important quantity by combining the He/O ratio of 107 $\pm$ 14 from solar flare particles with the hydrogen to oxygen ratio of 1690 of the Sun measured spectroscopically, so as to obtain a H/He ratio in the

Sun of 16 ± 2.5 This value is in general agreement with theoretical values assumed for the standard model of the Sun, and yields the mass fractions of hydrogen helium and heavy elements in the Sun as X:Y:Z = 0.79:0.20:0.013. (Biswas 1968, in Coll. on Cosmic rays and Adv. in Astron. and Astrophys., November 1968, TIFR, Bombay, p. 323). Further refinements of the composition results were made in subsequent years as will be discussed later.

5. The upper limit of Be/O, and B/O ratio as 0.02 measured in solar energetic particles as given in Table 6.2.1 can be used to derive some important results. The abundances of Be and B in the solar photosphere is $< 10^{-4}$, relative to oxygen as 1.0, and these are produced by the fragmentation of C, N, O nuclei in SEP during collisions with hydrogen and helium atoms in the solar atmosphere. Using the fragmentation cross-sections available at that time, and using a (Be + B)/(C, N, O) ratio of ≤ 0.02. Biswas and Fichtel[(12,13)] determined an upper limit of the amount of matter traversed in the Sun by the accelerated solar particle of 0.1 g·cm^{-2}. Since the amount of matter traversed in the interplanetary medium is negligible small, this limit of 0.1 g·cm^{-2} of material traversed applies to the solar region. The density of matter at the site of the flare or of the acceleration is poorly known, and is estimated to be between 10^{11} and 10^{13} atoms of hydrogen/cm^3. A particle with an average energy between 10 and 50 MeV/nucleon has a velocity of (0.4 - 1.0) x 10^{10} cm/sec. Thus the acceleration time, T, can be determined using the relation,

$$x = \rho \cdot v \cdot T$$

where x = amount of matter traversed in number of hydrogen atoms/cm^2, ρ= density of the solar material in number of hydrogen atoms/cm^3, v= the velocity of the particles in cm/sec. Thus the acceleration time, T, was determined to be $\leq$ 100 sec by Biswas and Fichtel[(12,13)]. Hence the acceleration time is not more than a few minutes and probably less.

Later experimental results from improved rocket borne detectors are described in Sec. 6.3.

Table 6.2.1: Relative Abundances of Nuclei in SEP and in the Solar Photosphere (with oxygen = 1.0)

Element	Solar Energetic Particles E> 40 MeV/N (a)	Solar Photosphere (b)
He	107 ± 14	110 ± 25 (c)
Li	< 0.02	$< 10^{-5}$
Be	< 0.02	$< 10^{-5}$
B	< 0.02	$< 10^{-5}$
C	0.56 ± 0.06	0.60 ± 0.10
N	0.19 ± 0.03	0.15 ± 0.05
O	1.0	1.0
Ne	0.16 ± 0.03	0.13 ± 0.03 (d)

(a) From the reviews of Biswas and Fichtel[13] and Biswas (1975) (Bull. Astron. Soc. India, 3, 68).
(b) From Pagel, 1973 (Space Sci. Rev., 15, 121) and Lambert, 1967 (Nature, 215, 43).
(c) From chromospheric and prominence data by Hirayama, 1971 (Solar Phys., 19, 384).
(d) Lambert, 1967 (Observatory, 87, 228) from UV data of upper chromosphere and the lower corona.

6.2.4 Satellites and Solar Energetic Particles

Early satellite experiments on solar energetic particles were successfully conducted with Explorer 12 and 14 using scintillation counter telescopes by Frank McDonald and collaborators at the NASA-Goddard Space Flight Center[14]. The energy loss rate, dE/dx vs energy (or range) was used to identify protons and estimate their energy. These telescopes covered the energy range of protons from 3 to 600 MeV and the time history of several solar proton events were measured. These experiments provided continuous measurements of differential fluxes of protons of various energies and their variations with time. Thus time profiles of solar protons from several solar proton events were determined and their relationship with the flares were studied. These clearly showed how the propagation effects the fluxes of protons of 10 to 100 MeV and > 600 MeV. From detailed analysis of the September 28, 1961 solar proton event, these authors plotted for each energy band the arrival time of protons vs their differential fluxes. They found well defined peaks for protons of each energy group. Thus the authors determined for the first time the method of obtaining the energy spectrum of protons at time T = 0 i.e. as these left the Sun. This method was improved and refined in later years to determine the source spectral shape of solar protons.

The details of these and other early measurements are given in the review by Fichtel and McDonald[(15)] in 1967.

Electrons in SEP

The abundances of relativistic electrons in SEP in near earth space are very small due to a very high rate of energy loss by synchrotron radiation in the magnetic field of the Sun. Accelerated electrons are mostly trapped in the solar coronal magnetic field giving rise to several types of solar radio emissions (Chapter 3.5.3). Therefore the flux of flare produced energetic electrons in near earth space is small, and only in a few cases observed. In the July 18, 1961 solar event the flux of electrons with an energy > 100 MeV was measured by Meyer and Vogt[(16)] in 1962 as 0.04 electrons/(cm^2·sr·s) i.e. a few percent of the proton flux in the same kinetic energy region. In the very large solar events of November 12 and 15, 1960, there was no measurable electron flux. In general one finds that the electron to proton ratio above about 10 MeV in SEP is highly variable and less than about 1%.

6.2.5 Space Vehicles and SEP in Deep Space

IMP-7 & 8: With the availability of advanced satellites and space probes in the 1970's, improved detectors with higher charge resolution and better data acquisition systems were flown in interplanetary space by several groups of investigators for studies of different aspects of solar energetic particles and new discoveries were made. NASA-Goddard Space Flight Center group of McDonald and collaborators[(17)] used the highly successful spacecraft IMP-7 and IMP-8 (Interplanetary Monitoring Platform = IMP) and studied eight large solar particle events during the five year period of 1974-78. The Very Low Energy Telescope (VLET) measured the He ions with energy > 1.7 MeV/N, O-ions > 3.2 MeV/N and Fe > 3.6 MeV/N and individual elements up to sulphur were well resolved. The LED telescope operated at somewhat higher energies. Both telescopes operated with a priority data system which allowed solar heavy nuclei of Z > 2 to be sampled adequately during high proton fluxes of large solar particle events. VLET thresholds eliminated proton flux completely. The eight large solar particle events were analyzed to study the systematic differences between solar and non-solar cosmic ray abundances and on compositional variabilities among the events in the low energy range of 6.7 - 15 MeV/N. Some of these are discussed in Section 6.3.

IMP-5 & 7 and OGO-5: New properties of low energy solar cosmic ray nuclei of 5-10 MeV/N were found by the University of Chicago group[18] using the IMP-5, IMP-7 and OGO-5 spacecraft. The IMP-5 telescope provided excellent data for H and He-isotopes, while the Orbital Geophysical Observatory (OGO-5) telescope was designed for high Z nuclei and could measure Fe/He abundance ratios in many flares. The LET instrument of IMP-7 yielded solar heavy nuclei fluxes in the 3 to 50 MeV/N energy region. Using data from these instruments during a number of solar flare events, Mogro-Campero and Simpson[19] found in 1972 that in the low energy solar cosmic rays of 5-10 MeV/N, the abundances of solar cosmic ray heavy nuclei of Z = 6-26 are enhanced relative to solar abundances and the enhancement factors increase with increasing atomic number Z. A new discovery was made by Dietrich and Simpson[20] in 1979 on the isotopic abundances of ^{20}Ne and ^{22}Ne in seven solar flare events from measurements with the IMP-7 satellite. These and other results will be given in the following section.

Voyager 1 and 2: the California Institute of Technology group[21] conducted experiments with the Voyager 1 and 2 spacecraft on solar energetic particles measured in deep space between 1 and 3 AU, during September 1977 and May 1978 (Cook et al.[21] 1979). They used four LET telescopes in each spacecraft with a total geometrical factor of $A\Omega \simeq 3.5$ cm^2·sr. The charge and total energy of a nucleus were determined from double dE/dx vs E analysis. Excellent charge resolution was obtained with a standard deviation of $\sigma_z \simeq$ 0.08 units of charge for carbon and $\simeq$ 0.27 units for Fe. Because of the high collecting power and high charge resolution, gaussian peaks were obtained even for rare elements like Na, Al, Ar, Ca, Cr and Ni. The absence of boron and fluorin events are remarkable and indicate a negligible background. The elemental composition data was obtained for seven large solar events and the data collection period excluded the onset period to avoid propagation effects. The elemental composition and their implications are given in Section 6.3.

ISEE-1 & 3: The Interplanetary Sun-Earth Explorer spacecraft ISEE-3 was launched by NASA on August 12, 1978 and was injected into a roughly elliptical orbit around the sunward libration point of the earth-sun system at about 230 earth radii on November 20, 1978. This orbit is ideal for interplanetary studies since it is permanently in the solar wind and is sufficiently close to the earth to maintain a high bit-rate for data transmission and yet far from the interference with earth's magnetosphere. The detectors consisted of two Very Low Energy Telescope (VLET's) and two High Energy Telescopes (HET's). VLET's had very thin solid-state detectors each with an area of 2 cm^2 and a thickness of 15 μm which allowed measurements of all elements from Z = 2 to 26, with energies above 2 MeV/N. The detectors aboard ISEE-3 measured solar particle intensities in three energy intervals of 2-3, 3.9-6.7 and 6.7-12.4 MeV/N for the major elements, He, O, Fe and others continuously for several days since the flare onset of the September 23, 1978 and November 11, 1978 flares. New results on the variations of the composition ratios of Fe/O and He/O as a function of time were observed. Similar measurements were carried out with the ISEE-1 spacecraft. Solar particle experiments in ISEE-1 were conducted by the Max-Planck Institute in Garching, Germany in collaboration with five US universities of Arizona, New Hampshire, Maryland and Chicago(22) while ISEE-3 experiments were done by the NASA-Goddard Space Flight Center(23).

PROGNOZ-6: Solar cosmic ray studies were made with the PROGNOZ-6 satellite of the USSR in a Soviet-Czechoslovak experiment during the two solar flare events of September 24 and October 12, 1977(24). The instrument comprised of a double dE/dx vs E telescope with three Si-detectors D1, D2 and D3 of thickness 100, 1000 and 1800 μm respectively with an anti-coincidence shield of scintillator D4. The full aperture of the telescope was 38° and the geometrical factor was 0.1 cm^2·sr. Using coincidence conditions D1 D2 $\bar{D}3$ $\bar{D}4$ and D1 D2 D3 $\bar{D}4$, helium to Z = 16 nuclei were identified in the 5 to 20 MeV/N energy interval. In order to obtain more information about rarer particles a four stage priority system was used. The solar particle event of September 24, 1977 occured in a flare behind the limb of the Sun, most probably in the active region at a heliographic longitude of 110^oW. It was accompanied by hard x-ray emission of 30-100 keV and radio emissions of type II, III and IV. Therefore it was typical for a large proton flare with a hard particle spectrum. The flare of October 12, 1977 was a small one of type 1N with heliographic co-ordinates NO4, E02. It was accompanied by hard X-rays, but the particle flux was an order of magnitude smaller. These two solar particle events showed a distinctly different isotopic ratio of $^3He/^4He$. The weak flare of October 12, 1977 showed a very large ratio of

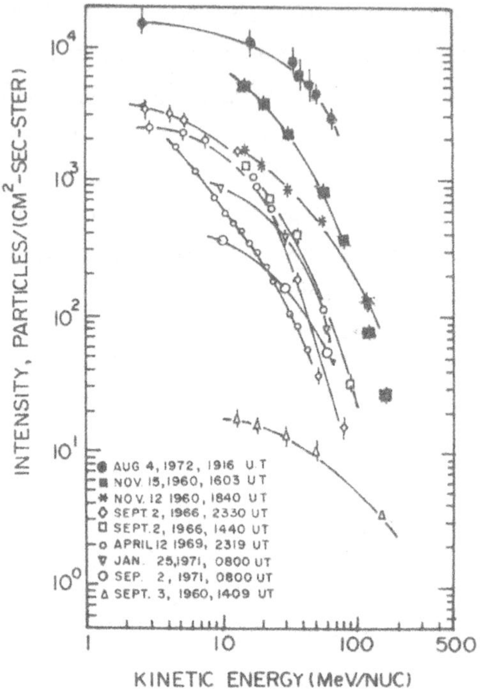

Figure 6.3.1: The plots of Integral fluxes vs energy of solar protons measured in nine large solar events during 1960-72 in the 'SPICE' experiments of the NASA-Goddard Space Flight Center. (From Biswas, unpublished).

^{3}He/^{4}He whereas the ratio was zero in the intense flare. These observations are discussed in the following Section 6.3, with other results.

6.3 Properties of Energetic Solar Particles

6.3.1 Intensity and Energy Spectrum

The measurements of the intensities i.e. the integral fluxes, J (> E) of solar protons and their energy spectra were made during nine large solar events during 1960-72 in the rocket-borne Solar Particle Intensity and Composition Experiment (SPICE) of the NASA-Goddard Space Flight Center. These are shown in Figure 6.3.1. It is seen that the intensities of solar protons range from the barely detectable value of about 3 protons/(cm^2·sr·sec), in the September 3, 1960 event at about 100 MeV, which is close to the galactic cosmic ray background level, to a very high flux level of about 1.5·10^4 protons/(cm^2·sr·sec) for E > 2.5 MeV in the August 4, 1972 event. The

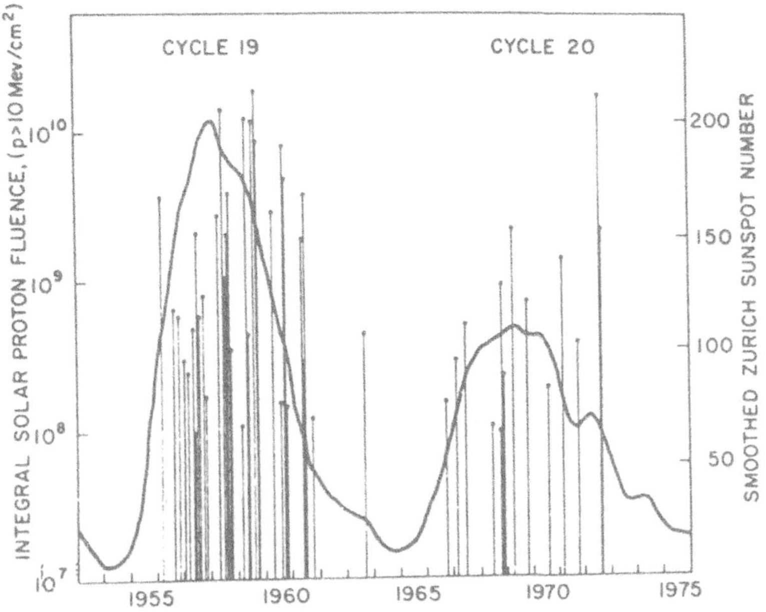

Figure 6.3.2: Fluences of E > 10 MeV for larger SEP events vs time of occurances are shown together with smoothed sunspot numbers, for the two solar cycles 19 and 20. (From ref. 25, courtsey, R.E. Lingenfelter).

solar particle event of August 4, 1972 was one of the largest recorded till then. It may be noted that the total flux of protons emitted in the August 4, 1972 event was larger than of all other previous large SEP events shown in Figure 6.3.1, taken together.

The question that naturally arises is as follows. Is there any corrrelation on annual basis between the occurances of large solar particle events and the solar activity measured by the sun spot number? One study made by Ligenfelter and Hudson[(25)] is shown in Figure 6.3.2, where the fluence of protons of E > 10 MeV from all proton flares of fluences > 10^8 particles/cm^2 are shown in solar cycles 19 and 20 during 1954 to 1975. The fluence (E_p > E) is defined as the integral flux integrated over the duration of the flare and collected over a solid angle of the upper hemisphere. It is seen that while particle events are more frequent during the high level of sun spot number as expected, large particle events do not occur at the time of the maximum sun spot number instead there seems to be a tendency of large particle events to occur a few years before or after the maximum. The reasons for this are not known. Also it is seen that although cycle 19 had a much larger sun spot number at the maximum, as compared to cycle 20, the largest fluence (> 10 MeV) from a single solar proton event in each cycle was the same, about $2 \cdot 10^{10}$ p/cm^2, for the July 14, 1959 event of cycle 19 and the August 8, 1972 event of cycle 20.

While it is clear that the larger the event, the fewer is their number, it is important to determine the number distribution of event sizes. This aspect was studied by Van Hollebeke et al.[(26)] in 1975 from the extensive analysis of measured solar proton events associated with 125 solar flare events between May 1967 to December 1972. He plotted the dN/dI vs I, where dN/dI is the number of events in a given time interval and with an intensity between I and I + dI, and I is the differential intensity of protons with an energy of 20-80 MeV at the time of peak flux. It is found that the size frequency distribution is given by dN/dI α $I^{-\beta}$, where $\beta = 1.15 \pm 0.05$. The flatness of the distribution indicates that the most intense events in a time interval contribute to most of the integrated flux or fluence. It is found that the 47 solar particle events shown in Figure 6.3.2 contributed to $\sim$ 99% of the total particle fluences.

Energy spectrum: The flux and energy spectra of the major components of solar particles such as C, N, O, Ne etc. up to Fe were measured by many investigators in the 1970's using improved detectors, as mentioned earlier. As an example[(27)], we show in Figure 6.3.3 the differential energy spectra of different components in the typical flare event of moderate size of January 24, 1971 of the improved SPICE series. Typically the spectral index, γ, in dJ/dE $\sim E^{-\gamma}$, is an increasing function of energy per nucleon, E. The power law in energy is applicable over a particular energy range. Freier and Webber[(28)] (1965) had shown that later in the event, an exponential rigidity spectrum such as dJ/dR α exp(-R/R_o) gives a better fit over a larger interval where R_o varies from event to event and is usually in the range 200-500 MV.

Another question which is frequently asked is: Is there a correlation between the intensities of SEP as measured by the energy release in MeV range particles and the total energy released in the flare? From some flare events where detailed studies could be made it is found[(29)] that in the three large solar events of August 2, 4 and 7, 1972, the total energies released were

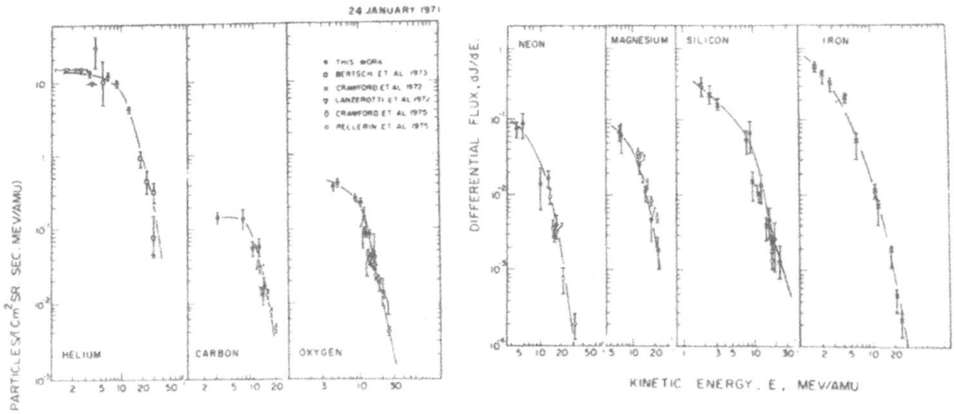

Figure 6.3.3: The flux and energy spectra of the elements He to Fe in the typical solar event of January 24, 1971. (From ref. 27).

nearly the same, ranging from 1 to $5{\cdot}10^{22}$ erg, the energy in escaping SEP of > 10 MeV/N varied from 0.07% to 4% i.e. nearly by two orders of magnitude. Thus the answer is negative. It may be noted that only a fraction of the total number of accelerated SEP, which are connected to the open magnetic field lines are able to escape from the Sun, and this is an important factor in the variabilities of SEP.

6.3.2 Elemental Abundances

The elemental composition of solar hydrogen, helium and heavier nuclei from C to Fe in the 6.7 to 15 MeV/N energy interval was measured by the NASA-GSFC group(17) for eight large solar events during 1974-78, and these are shown in Table 6.3.1. The solar abundances are given which are mainly photospheric with supplementary data. Coronal abundances as available have large uncertainties and hence were not useful. On comparing these low energy solar cosmic ray abundances with other data several new features of abundance enhancements were observed.

The improved SPICE payload had three different detectors, Nuclear Emulsions, CTA plastic and Lexan stack studied by GSFC, Bombay and

Table 6.3.1: Solar Cosmic ray abundances in eight large events in 1974-78 by NASA-GSFC[17].

SOLAR COSMIC RAY ABUNDANCES (6.7-15 MeV/NUC)

	SOLAR ABUNDANCES (a)	4 JULY 1974 0000 UT TO 8 JULY 1974 0000 UT	11 SEPT 1974 1200 UT TO 16 SEPT 1974 0000 UT	19 SEPT 1977 1400 UT TO 21 SEPT 1977 0000 UT	24 SEPT 1977 0800 UT TO 26 SEPT 1977 1200 UT	22 NOV 1977 1200 UT TO 23 NOV 1977 2000 UT	13 FEB 1978 1600 UT TO 15 FEB 1978 1200 UT	29 APR 1978 0400 UT TO 2 MAY 1978 0200 UT	23 SEPT 1978 2000 UT TO 26 SEPT 1978 0000 UT
H/O	1000 ± 250	4600 ± 600	4900 ± 1000	6200 ± 600	5000 ± 1000	21,000 ± 3000	3000 ± 1000	3200 ± 500	2100 ± 500
He/O	110 ± 50	74 ± 10	52 ± 4	80 ± 8	120 ± 15	83 ± 15	48 ± 5	38 ± 5	40 ± 3
Li/O	1.2×10^{-8} (2) (d)	< .004	< .004	< .006	< .02	< .02	< .002	< .002	< .003
Be/O	1.7×10^{-8} (1.6)	< .004	< .004	< .006	< .02	< .02	< .002	< .002	< .003
B/O	2.3×10^{-7} (2)	< .004	< .004	< .006	< .02	< .02	< .002	< .002	< .003
C/O	.56 ± .13	.35 ± .03	.48 ± .04	.53 ± .04	.50 ± .07	.50 ± .08	.46 ± .03	.42 ± .02	.49 ± .03
N/O	.10 ± .04	.10 ± .02	.14 ± .02	.13 ± .02	.12 ± .04	.20 ± .05	.12 ± .01	.13 ± .01	.12 ± .02
O/O	1.00 ± .22	1.00 ± .06	1.00 ± .07	1.00 ± .06	1.00 ± .10	1.00 ± .10	1.00 ± .03	1.00 ± .04	1.00 ± .06
F/O	4.0×10^{-8} (1.5) (d)	< .004	< .004	< .006	< .02	< .02	< .002	< .002	< .003
Ne/O	.15 ± .07	.18 ± .02	.20 ± .03	.10 ± .02	.28 ± .06	.10 ± .03	.12 ± .01	.12 ± .01	.09 ± .01
Na/O	.0024 ± .0004	.021 ± .007	.005 ± .004	.013 ± .007	< .02	.03 ± .02	.008 ± .003	.004 ± .002	.013 ± .005
Mg/O	.049 ± .012	.24 ± .02	.17 ± .02	.19 ± .03	.42 ± .07	.13 ± .04	.20 ± .02	.19 ± .02	.18 ± .02
Al/O	.0037 ± .0009 (d)	.025 ± .007	.008 ± .004	.007 ± .005	.062 ± .026	.03 ± .02	.013 ± .004	.012 ± .003	.010 ± .004
Si/O	.050 ± .013	.14 ± .02	.13 ± .02	.14 ± .03	.16 ± .05	.16 ± .04	.12 ± .01	.14 ± .01	.16 ± .02
P/O	$4.2 \pm .9 \times 10^{-4}$ (d)	< .004	< .004	< .006	< .02	< .02	< .002	< .002	< .005
S/O	.018 ± .005	.044 ± .010	.018 ± .007	.013 ± .007	.083 ± .029	.03 ± .02	.020 ± .005	.025 ± .005	.018 ± .005
Ar/O	3.0×10^{-3} (1.7) (d)	.016 ± .007	< .004	< .006	< .02	.020 ± .014	.0035 ± .0020	< .002	< .003
Ca/O	.0026 ± .0004	.020 ± .008	.006 ± .004	.009 ± .006	.07 ± .03	.04 ± .02	.007 ± .003	.005 ± .002	.006 ± .003
(Ti+Cr)/O	$6.3 \pm 1.9 \times 10^{-4}$	.010 ± .006	< .005	< .01	< .02	< .02	< .005	< .003	< .004
Fe (GROUP)/O	.044 ± .013	.22 ± .03	.06 ± .01	.27 ± .03	1.17 ± .11	.28 ± .06	.11 ± .01	.07 ± .01	.06 ± .01
$^2H/^1H$ (b)	$\sim 1.6 \times 10^{-5}$	$< 2 \times 10^{-3}$	$< 2 \times 10^{-4}$	$< 4 \times 10^{-4}$	$< 8 \times 10^{-5}$	$< 1 \times 10^{-4}$	$< 2 \times 10^{-3}$	$< 3 \times 10^{-3}$	$< 3 \times 10^{-3}$
$^3H/^1H$ (b)	—	$< 5 \times 10^{-4}$	$< 4 \times 10^{-5}$	$< 2 \times 10^{-4}$	$< 8 \times 10^{-5}$	$< 4 \times 10^{-4}$	$< 9 \times 10^{-5}$	$< 4 \times 10^{-4}$	$< 4 \times 10^{-4}$
$^3He/^4He$ (c)	$\sim 1.7 \times 10^{-4}$	< .02	< .012	< .008	< .013	< .013	< .06	< .04	< .06

(a) – BASED ON CAMERON (1973), MEYER AND REEVES (1977) AND MEYER (1978). NUMBERS IN PARENTHESES ARE ERROR FACTORS.
(b) – 8.6 – 11.7 MeV/NUC
(c) – 2.2 – 2.6 MeV/NUC
(d) – UNIVERSAL OR LOCAL INTERSTELLAR MEDIUM VALUES USED.

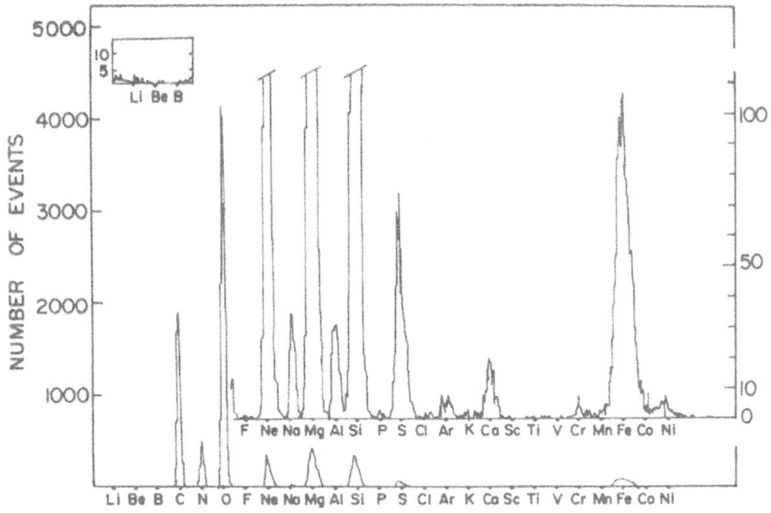

Figure 6.3.4: Elemental abundances of Li to Ni in solar energetic particles of energies 8.7 to 15 MeV/N as measured with Voyager 1 and 2 by the Caltech group (ref. 21).

Berkeley groups respectively. When these studies extended the abundance data to lower energies of ~ 5-10 MeV/N, new features of enhancement of abundances were observed (Bertsch et al.[30], 1973, Nevatia and Biswas[31], 1975 and Crawford et al.[32] 1975). These authors pointed out that in the energy range ~ 5-10 MeV/N He and heavier ions are partially ionized and their charge to mass ratios are different. Hence their acceleration and propagation properties will vary. Therefore their relative abundances will differ from the values at high energies and from the solar values, as given earlier by Biswas and Fichtel[13].

Elemental abundances of the heavy solar ions of helium to nickel emitted in seven major flare events in 1977-78 were measured with a high degree of accuracy with the Voyager 1 and 2 spacecraft by Cook et al.[21], as noted earlier. The element histogram for $Z > 3$ is shown in Figure 6.3.4, for solar energetic particles of 8.7 to 15 MeV/N. The excellent charge resolutions from Li to Fe and Ni were obtained. These composition results yielded abundance enhancements of SEP in much detail.

The abundance measurements of solar energetic particles in the low energy

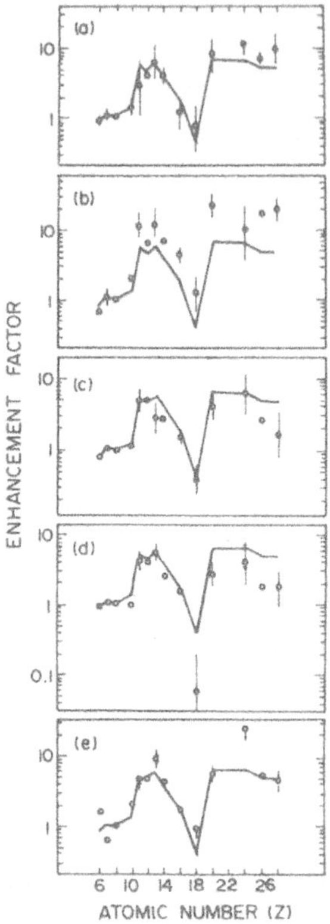

Figure 6.3.5: Abundance enhancement factors vs atomic number Z for four large events, ref. (21).

region of about 5 to 15 MeV/N by several authors, as noted earlier, revealed new features of abundance enhancements. The abundance enhancement factor of an element, i, ξ_i, is defined as

$$\xi_i = \frac{[Abundance\ \ of\ \ an\ \ element,\ \ i,\ \ in\ \ SEP\ \ relative\ \ to\ \ oxygen]}{[Abundance\ \ of,\ \ i,\ \ in\ \ solar\ \ phosphere\ \ relative\ \ to\ \ oxygen]}$$

It is found that there are four specific features of ξ_i, e.g. the enhancements, ξ_i are functions of (a) atomic number Z, (b) energy E, (c) first ionization potential and (d) event to event variation of ξ_i.

Enhancement Factor ξ_i vs Atomic Number, Z:

Early observations[19] suggested a monotonic increase of ξ_i with Z. However detailed studies[17,21] showed that it is not monotonic with increasing Z. The

NASA-GSFC group[17] measured enhancement factors vs Z for 8 major solar events in five years, 1974-78 corresponding to the abundance data shown in Table 6.3.1. It is found that (a) ξ is small and close to 1 for C, N, O and Ne, and (b) ξ is much larger for Na, Mg, Al and Si, (c) Si and Ar are less enhanced than the Mg-group; (d) Ca and Fe show the largest variation with a high degree of enhancements. Results[21] from four large flares in 1977-78 also show the non-monotonic nature of ξ_i vs Z as given in Figure 6.3.5.

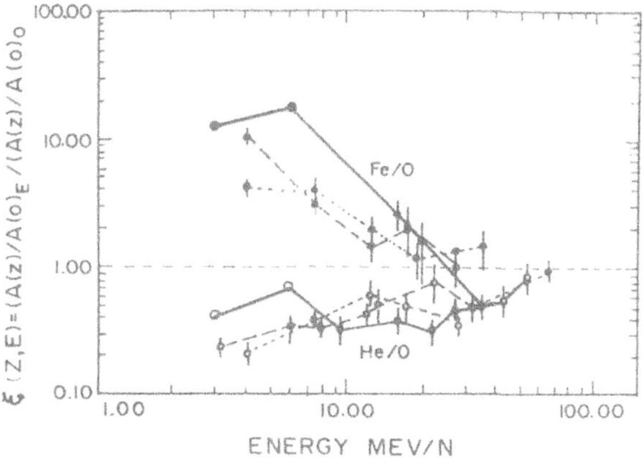

Figure 6.3.6: Abundance factor ξ vs Energy for two elements, Fe and He, with respect to oxygen. See text for details.

Enhancement Factors vs Energy

The enhancement factors ξ vs kin.energy in the 3-60 MeV/N interval for three large flare events of 1971-72 were measured by Biswas et al.[27] for the major elements He, C, Ne, Mg, Si and Fe. It was found that the enhancement ξ increases with decreasing energy. This trend is reversed for He and C. The variation of ξ with energy was small for C and Ne from 3 to 1, fairly large for Mg and Si (about 5 to 1) and largest for Fe by a factor of about 20 to 1. For clarity we show[31] in Figure 6.3.6, enhancement factors ξ_i as a function of energy for two elements, Fe and He. Fe data shows the largest enhancement factor $\xi \sim 20$ at ~ 5 MeV/N which is superposed on event to event variations. In case of He the trend is reversed because we are using He/O ratios and a higher Z element is enhanced relative to lower Z elements, and the O/He ratio would show enhancements at low energies. Hence He/O ratios at low energies show a depletion, ξ being ~ 0.3 at ~ 5 MeV/N and $\xi \simeq 1$ at 60 MeV/N. So the apparent depletion is due to normalization at O. A similar effect is also seen for carbon in a smaller degree. Thus the three large solar events have provided important data on the energy dependence of ξ for He/O in a wide energy range of 5 to 60 MeV/N[31].

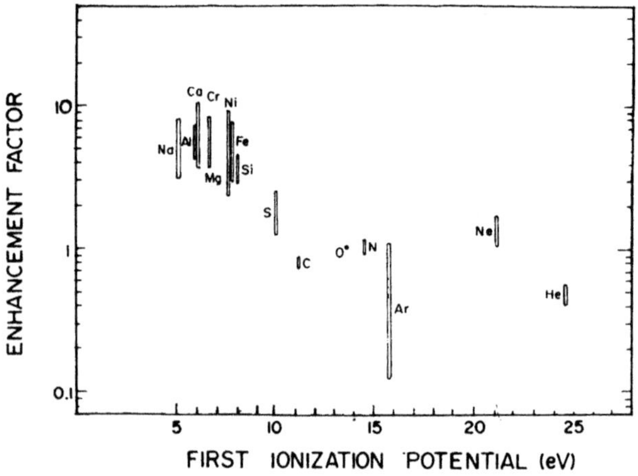

Figure 6.3.7: Abundance enhancement factor vs First Ionization Potential (FIP).

Enhancement, ξ vs First Ionization Potential (FIP)

Using the enhancement factors derived from their measurements, several authors have studied their relationship with the first ionization potential (FIP). For example, we show in Figure 6.3.7 the plot by Cook et al[21]. It is seen that the FIP appears to divide SEP into two groups, with a dividing line at ~ 10 eV: (a) the more volatile elements like, C, O, N, A, Ne and He, having a FIP > 10 eV, are consistent with an enhancement factor of about 1; and (b) for a FIP < 10 eV, the metallic group of elements such as Na, Al, Ca, Mg, Fe, Si etc. are enhanced by a factor of ~ 5. The relative enhancement of the metallic group may be due to injection or acceleration effects or other processes not known.

Implications of SEP abundances and enhancements

1. The almost absence of Li, Be, B, F etc. and also $^2H/^1H < 10^{-4}$ indicate

that the observed solar particles have traversed less than 0.06 $gm \cdot cm^{-2}$ of matter in the Sun.

2. The solar particle energies of E $\sim$ 10 MeV/N are enhanced with respect to photospheric values, the enhancement being larger for high Z elements e.g. Fe. As high Z elements have many more states of ionization than the lower ones these have presumably more channels of injection or acceleration as compared to lower Z elements.

3. Abundance ratios, such as Fe/O, Mg/O etc. as a function of energy show $\xi \approx 1.0$ at the high energy region at $\geq$ 20 MeV/N because at these energies the elements are nearly fully stripped and their charge to mass ratios tend to nearly the same value of about 0.5

4. The SEP abundances and their enhancement factors have been used together with some simple models of initial acceleration to deduce the effective temperature of the accelerating region by Vahia(33), who reported a value of T $\sim 6 \cdot 10^5$ K. This may be considered applicable to the chromospheric region of the flare where the initial injection of heavy ions presumably takes place.

5. The enhancements in the heavy ion composition in solar energetic particles have important implications for observations of solar system materials such as meteorites, lunar rocks etc. which have been bombarded and implanted with solar energetic heavy ions, which also reflect these enhancements. Some aspects of these are discussed in Chapter 6.4.

6. The composition of flare associated solar heavy ions and their enhancements have implications for the comparison of these to the galactic cosmic ray sources composition. It is a point of discussion whether solar type flare stars can be considered as an injection source for the heavy ions of galactic cosmic rays. This aspect is reviewed in Chapter 7.

6.3.3 Isotopic Composition

Hydrogen Isotopes:

^{2}H: In the early observations of the November 12, 1960 flare event, Biswas et al[9] (1962) found a $^2H/^1H$ abundance ratio as less as $3{\cdot}10^{-3}$ in the 25-48 MeV/N interval, while in the July 18, 1961 event Freier and Waddington[34] 1964 found the above ratio to be $< 2{\cdot}10^{-3}$ for $E > 50$ MeV/N. In later studies of several events by Anglin et al.[18] (1975) the $^2H/^1H$ ratio was obtained as $< 10^{-3}$ for $E \sim 10$ MeV/N. The NASA group[17] (McGuire et al., 1979) measured the ratio during the eight large solar events of 1974-78 as between $< 2{\cdot}10^{-3}$ and $< 10^{-4}$ for the 6.7-15 MeV/N energy interval. The solar abundance of the $^2H/^1H$ ratio in the photosphere is about $1.6{\cdot}10^{-5}$. Therefore ^{2}H originates entirely from the interactions of energetic p and He in the solar atmosphere; and the $^2H/^1H$ ratios yield the upper limit of matter traversed by energetic solar particles as 0.6 g$\cdot$cm^{-2} of hydrogen.

^{3}H: The presence of ^{3}H in SEP irradiated material was first found by Firemen et al.[35] in 1961 and Tiles et al.[36] in 1963 by radiochemical studies of material recovered from the Discoverer 17 satellite and from sounding rockets material flown during the large solar event of November 12, 1960. As they did not measure the integrated proton flux, it was deduced from other studies[10,13] as $1{\cdot}10^9$ to $4{\cdot}10^9$ protons/cm^2. Thus the observed $^3H/^1H$ abundance ratio was obtained as $1{\cdot}10^{-3}$ - $4{\cdot}10^{-3}$ above 30 MeV/N[13]. On the basis of radioactive data Lal et al. concluded that more than 80% of the ^{3}H was produced by interactions of incident protons, so that the $^3H/^1H$ ratio in the solar beam was probably $\leq 2{\cdot}10^{-4}$. Similar studies of another event lead to the above ratio as $\leq 10^{-3}$ at < 27 MeV/N[13]. Direct measurements of the November 12, 1960 event by Biswas et al.[9] yielded a $^3H/^1H$ ratio in SEP as $< 10^{-3}$ in 20-45 MeV/N. Freier and Waddinghton[34] reported the upper limit of $6{\cdot}10^{-4}$ for the July 18, 1961 event. Thus the observed ^{3}H concentrations in the materials irradiated with a high integrated dose of solar protons arise from nuclear interactions of SEP protons with the detector materials.

Since 1970 ^{3}H concentrations were measured by radio-chemical methods in more than fifty lunar samples; their concentrations and depth dependences were in general agreement with the calculated values (see Fireman[35], 1980). These have been used to estimate the solar proton flux and energy spectrum above 10 MeV averaged over the past 60 years. This will be discussed in Chapter 6.4.

Helium Isotopes:

In the early observations of 1962, Schaeffer and Zaringer[(36)] reported the presence of ^{3}He by radio-chemical studies in Discoverer 17 satellite material and estimated the ^{3}He/^{4}He ratio as 0.2. It was not known how much of the ^{3}He was due to secondary production, so it was not possible to determine the ^{3}He fraction in the solar beam. However the November 12, 1960 event produced a very large amount of ^{3}H in the material as measured by Fireman. As the ratio of ^{3}He/^{4}He in the solar material is very small ($\sim 1.7{\cdot}10^{-4}$), ^{3}He is produced by interactions between accelerated protons and helium nuclei in the solar atmosphere. As the amount of material transversed by solar protons in the Sun is very small, < 0.1 g$\cdot$cm^{-2}, the ^{3}H (which decays to ^{3}He with a half-life of 12.3 years) and ^{3}He abundances in SEP are estimated to be $\leq 10^{-3}$ relative to ^{1}He. These observations are found to be applicable to large solar flare events. In eight large solar flare events of 1974-1978, the NASA group[(17)] measured an upper limit for the ^{3}He/^{4}He ratio as $\sim 6{\cdot}10^{-3}$.

^{3}He-'Rich' Events:

Now we turn our attention to the striking results of ^{3}He abundances obtained in interplanetary space with the IMP 6 and 7 satellites by the NASA, Chicago and Caltech groups. In 1975, Anglin et al.[(18)] and Garrad et al. found that in several small solar flare events, the ^{3}He/^{4}He ratios were too high, ranging from about 1 to 10 %, in contrast to the normal value of about 0.1% or less. These events were called ^{3}He-'rich' events, in contrast to the others which are called ^{3}He-'poor' events. During the investigations in 1974, the NASA-GSFC group (McDonald et al.[(37)] 1975) found several small flare events with high ratios, one of these having an astonishingly high value of ^{3}He/^{4}He $\simeq 1.5$ in the energy range of 4-18 MeV/N (Figure 6.3.8). The Prognoz-6 of the USSR recorded events with an anomalously high ^{3}He abundance[(24)]. Helios-182 also identified several ^{3}He-rich events in 1974-77[(38)]. The Caltech group, (Hurford et al.[(39)], 1975) reported even higher values of the ratio, namely 6 and 8. The mass histograms of two events measured with the same detector by the NASA-GSFC group[(37)] are shown in Figure 6.3.8. In the large event of November 2, 1969 the ^{3}He abundance was

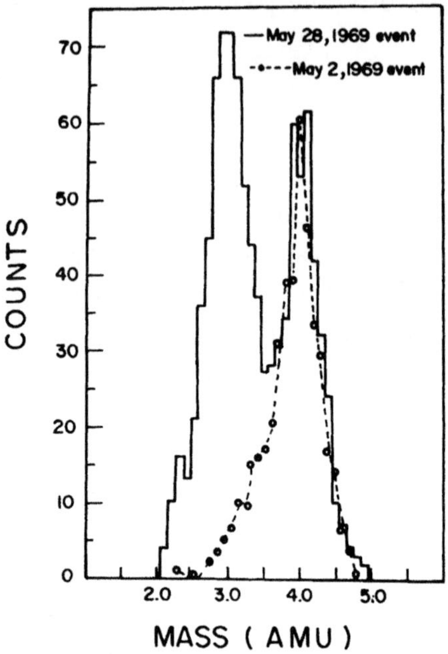

Figure 6.3.8: Mass histogram of ^3He and ^4He. Note the extremely high ^3He in the May 28, 1969 event, in contrast to the very small value of ^3He for the November 2, 1969 event. (Ref. 37).

below detector level, while in the small event of May 28, 1969 the $^3He/^4He$ ratio was as high as 1.5. Similar observations(24) by Prognoz-6 of the USSR are shown in Figure 6.3.9. Such high abundances of ^3He particles in the 3-15 MeV/N energy range among the accelerated solar particles can not be understood in terms of a secondary production of ^3He by nuclear interactions of energetic solar protons with solar surface material, like $^4He(p,d)\ ^3He$, $^4He(p,pn)\ ^3He$ etc. which have thresholds of 23 and 26 MeV respectively, because this would require the material traversed by the solar particles to have a density of about 100 g·cm^{-2} of hydrogen or more, which is clearly not possible. Therefore it is clear that we are dealing with an entirely new type of physical processes in the Sun. Before attempting to interpret these unusual observations, we summarize the important aspects of the available information as follows:

1. Anomalously high ^3He abundances occur in association with very small solar flare events, of type I or with sub-flares and not with large events. Actually there is an inverse correlation of $^3He/^4He$ ratio with flare size.

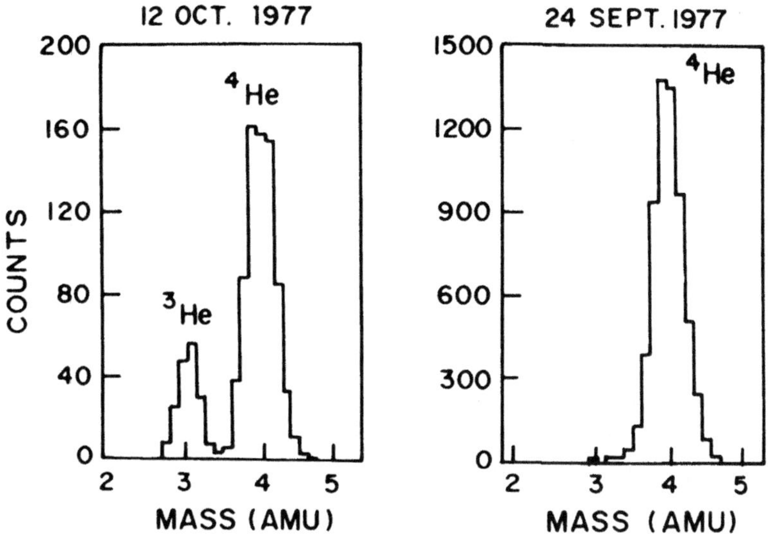

Figure 6.3.9: PROGNOZ-6 Satellite results showing the ^{3}He/^{4}He ratio having the very high value of 0.24 ± 0.02 for the 12 October 1977 event, whereas the ratio was very small, < 0.003 for the 24 September 1977 event. (From Lutsenko and Fischer[(24)]).

2. These events persist over many hours, and sometimes the ^{3}He and ^{4}He data are collected for 10 to 60 hours.

3. When ^{3}He abundances are high, ^{2}H and ^{3}H abundances are small and only upper limits are available.

4. Many ^{3}He-rich events, but not always, are correlated with enhanced abundances of heavy nuclei of SEP in the 5-10 MeV/N energy range, particularly of Fe nuclei.

5. The ^{3}He/H ratio has an anti-correlation with maximum proton flux from $J_p \sim 10$ MeV, up to $J_p \sim 1$ p/(cm^2·sr·S·MeV). For J_p greater than this, ^{3}He/H remains constant. Thus it seems, ^{3}He events are separated into two groups, as given by Ramadurai et al.[(40)] (1984).

A summary of available data on ^{3}He rich events is given by the above authors[(40)]. Several theoretical models have been proposed; these are briefly given as follows: (a) Plasma Processes: Basically these include two stage acceleration processes. In the first stage an excess of ^3He ions relative to ^4He in the Maxwell tail is achieved by invoking the resonant cyclotron frequency of $^3He^{+2}$ with respect to ion-cyclotron waves excited in the medium. Simultaneous enrichment of partially stripped heavy ions, particularly of Fe, is expected to take place. The details are given by Ibragimov and Kocharov(41) (1977) and Fisk(42) (1978); (b) Nuclear reaction model(43). In this model accelerated particles trapped in the flare sites travel through about 2 g·cm^{-2} of solar matter in the closed looped magnetic field structures and produce the majority of the ^3He-rich events by nuclear reactions. The small grammage of < 0.1 g·cm^{-2} of SEP refers to particles escaping along the open field lines and hence the particles in a closed loop may have traveled through a larger amount of matter. The absence of ^2H and ^3H are ascribed to special spallation effects, which are not quite convincing. In this model some events would require 50-100 g·cm^{-2} of matter traversed. These events were considered as a separate class of events. Thus we find that the high ^3He events are paradoxical ones and are not fully understood at present.

C, N, O and Ne, Mg, Si - Groups

The Caltech Heavy Ion Spectrometer Telescope (HIST) of the ISEE-3 spacecraft launched in August 1978 measured the isotopic composition of C, N, O, Ne and Mg nuclei in the energy range covering 6-40 MeV/N during the large solar flare event of September 23, 1978. Using three independent mass measurements, each particle's mass was well resolved and isotopic abundance data of ^{12}C, ^{13}C, ^{14}N, ^{15}N, ^{16}O, ^{18}O, ^{20}Ne, ^{22}Ne, ^{24}Mg, ^{25}Mg and ^{26}Mg were obtained(44) by Mewaldt et al. (1981). More than 10^4 nuclei of $Z > 6$ were measured during this solar event.

The University of Chicago group(45) measured the isotopic composition of Ne, Mg and Si with their multi-element charged particle telescope on the IMP-8 satellite in the energy range of about 30 to 60 MeV/N, (Dietrich and Simpson(45), 1979). An identical back-up instrument was calibrated with heavy ion beams at the Lawrence Berkeley Laboratory, USA and adequate mass resolutions were confirmed. Data from seven solar flares during July 3, 1974 till April 28, 1978 were combined to obtain the isotopic abundances of ^{20}Ne, ^{22}Ne, ^{24}Mg, ^{25}Mg, ^{26}Mg and ^{26}Si.

The results from these two experiments are shown in Table 6.3.2. It is seen that the abundance ratios of $^{13}C/^{12}C$, $^{15}N/^{14}N$, $^{18}O/^{16}O$ in solar energetic particles with 6-45 MeV/N energy are consistent with solar system abundances as given by Cameron 1980. Similarly Mg isotope abundances agree with the solar system values. An interesting situation arises in case of the $^{22}Ne/^{20}Ne$ ratio of 0.13 ± 0.04 in SEP of 10-50 MeV/N. This agrees well with meteoritic or primordial neon, called Neon-A, with the value 0.12. This is used by Cameron as the solar system abundance. Mass spectrometric measurements of neon isotopes in carbonaceous chondrites by many investigators revealed that there are three types of neon: (1) Neon-A or 'planetary' neon; as noted above; (2) Neon-B, with $^{22}Ne/^{20}Ne = 0.08 \pm 0.002$, due to implantation by solar wind, and (3) Neon-C, due to spallation of galactic cosmic rays which produce all three isotopes in roughly equal proportions. The SEP value of 0.13 of $^{22}Ne/^{20}Ne$ signifies that the above ratio represents the solar chromospheric or lower coronal value. In the solar wind some kind of isotopic fractionation probably takes place leading to a solar wind ratio being different from that of the SEP Neon-A value. The type of fractionation taking place in the solar wind is not understood at present.

Table 6.3.2: Isotopic Composition of Carbon to Silicon in SEP

Isotopic Ratio	Energy Interval MeV/N	Observed Ratio*	Solar System (Cameron)
$^{13}C/^{12}C$	6-38	0.010 ± 0.004	0.0112
$^{15}N/^{14}N$	9-42	0.008 ± 0.010	0.0037
$^{17}O/^{16}O$	7-45	< 0.0019	0.00037
$^{18}O/^{16}O$	7-45	0.0018 ± 0.008	0.0020
$^{21}Ne/^{20}Ne$	11-26	< 0.016	0.0030
$^{22}Ne/^{20}Ne$	11-26	0.13 ± 0.04	0.122
$^{22}Ne/^{20}Ne$	30-50	$0.13 \pm 0.04^{+}$	0.122
$^{25}Mg/^{24}Mg$	12-36	0.14 ± 0.03	0.0129
$^{26}Mg/^{24}Mg$	12-36	0.15 ± 0.03	0.142
$^{26}Mg/^{24}Mg$	30-55	$0.22 \pm 0.07^{+}$	0.142
$^{30}Si/^{28}Si$	34-60	$< 0.035^{+}$	0.030

*The results are from the Caltech Group (Mewaldt et al. 1981), except those marked by + which are from the Chicago group (Dietrich and Simpson, 1979).

6.3.4 Ionization States of SEP

The knowledge of ionization states of energetic solar heavy nuclei of 1 to 20 MeV/N is important to understand their acceleration and propagation properties. Ionization states of He, C, O and Fe ions with energies of 0.3 to 2.4 MeV/N associated with ten solar flares were measured by the Ultra Low Energy Z, Energy and Q (charge) sensor of the Max-Planck-Institute/University of Maryland during one year of its operation from September 1978[(46)] and onwards. The atomic number Z of a particle is first identified using the pulse height analyzed signal from the proportional counter i.e. the dE/dx element and the solid state element (E detector) of the dE/dX vs E telescope placed at the exit of the electrostatic deflection system which disperses ions according to their energy per charge. The charge state of a particular element was then determined from the amount of deflection measured by a position detector and the residual energy E. This analysis was done for each incident ion. The instrument was flown in the ISEE-3 spacecraft positioned at about 0.01 AU from the earth at the earth-sun liberation point.

Typical results for the June 6-9, 1979 event show charge state histograms of He of 0.4-0.9 MeV which indicate that the He^{+1} ions are 10% of He^{+2}. Carbon ions of 0.3-2.4 MeV/N show a mean charge state as 5.7, oxygen ions of 0.4-2.4 MeV/N have a mean charge state of 7, and Fe of 0.3-0.9 MeV/N ions have a mean of $Q = 12.5$. The histogram for Fe showed that several ionization states are present, such as $Q = 10$ to 15. There was no evidence for very low charge states such as O^{+1}, Fe^{+3} etc. The He^{+1} fraction varied 8% to 29% in the ten flare events, the even average value being about 15%. The event averaged mean charge state of C, O and Fe were found to be 5.8, 7.1 and 13.5 respectively.

The presence of He^{+1} in SEP with large abundance fraction of 10-15% in all flare events seem to indicate that the original source material of the flare particles belonged to a relatively cooler region of $T_e \leq 10^5$K. The particles are subsequently energized in the second stage of acceleration in other regions of the corona which contain He^{+2}, O^{+6} to O^{+8}, Fe^{+10} to Fe^{+15} etc. The reason why low charge states of ions, such as O^{+1} etc. are not seen is that these have high rigidities and fall in higher energy ranges than the observed narrow energy region. That the solar flare phenomenon involve both 'cold' and 'hot' regions of the chromosphere and lower corona had been pointed out by many

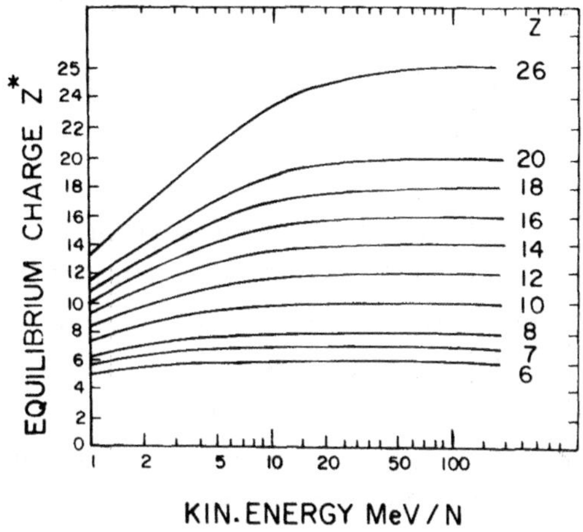

Figure 6.3.10: Calculated equilibrium charges of multiply charged ions passing through a medium.

authors e.g. de Jager and others.

It is useful here to note the equilibrium charges of multiply charged ions of C to Fe with low energies which travel through a medium. The equilibrium charge Z* of an ion with atomic number Z, is given by

$$Z^* = Z\left[1 - exp(-137\beta/Z^{2/3})\right]$$

where βc is the velocity of the particle. This relation is shown in Figure 6.3.10, for Z = 6 to 26. It is seen that at $\sim$ 1 MeV/N, the Z* for C, O and Fe are $\sim$ 5, 6.5 and 13.5. These values represent idealized cases of equilibrium values. In the case of solar particles the situation is much more complicated and for each species at a given energy there are many ionization states each with their distributions. Detailed calculations of the equilibrium charge states at a given coronal temperature taking into various effects of ionization, recombinations, charge exchange etc are given by Jordon[(47)] (1969).

6.4 Coronal and Interplanetary Propagation of SEP

Coronal and Interplanetary propagation of energetic solar particles involve several processes which can be grouped as follows: (1) Solar source of the particles. This includes the acceleration of particles in the lower corona, their source energy spectrum and heliographic location. (2) Coronal storage, coronal and azimuthal transport and release of the particles; and (3) Interplanetary propagation which depends on the large scale structure of the magnetic field, its small scale fluctuations, and the radial and longitudinal dependence of the magnetic field. Observations with ground level detectors and with space-borne instruments during the past three decades have revealed many new aspects of these solar and interplanetary processes which have many complexities and are not fully understood at present. We shall discuss some important features of these phenomena.

6.4.1 Coronal Propagation

Let us consider an idealized and simple case of a flare occuring at a helio longitude θ_F; θ_F is positive west of the solar central meridian and negative at the eastern side. The 'foot print' of the Archimedean spiral of the solar magnetic field which connects the Sun with the earth providing the most favorable propagation path to the earth is located at helio longitude θ_A. Then the helio centric longitudinal distance Φ, from the flare site to the 'fast' of the spiral is given by $\Phi = \theta_A - \theta_F$. Thus solar particles had to diffuse through the coronal magnetic field along the longitudinal distance of Φ to reach the 'front print' and to find the favourable path to the earth. Therefore the delay between the time of flare onset and the arrival of solar protons will be larger for larger values of Φ. Observational data for the delay times are in general in agreement with the picture. The variations in the delay times which are present are due to other complexities. It may be noted that the 'foot print' angle of the solar longitude, θ_A, is given by $\theta_A = (W_s \cdot r)/V_{s.w.}$, where W_s is the solar synodic rotation rate (13.3^o per day), r is 1 A.U. and $V_{s.w}$ is the solar wind velocity which is taken as 400 km/s for the normal case. Thus we obtain $\theta_A = 57.3^o$ or one radian, i.e. when the Sun rotates by one radian, the solar wind plasma travels 1 A.U. during the same time (refer to Figure 3.4.3 in Chapter 3). From studies of over 100 proton flares by Van Hollebeke et al.[48] (1975) the times of maximum intensity of 20-80 MeV protons were plotted against flare longitudinal Φ_F. These plots showed that the time to reach maximum intensity is roughly given as $\sim \Phi^2$ and the maximum delay time of 20-60 min is found for flares at the 'foot print' i.e. for $\Phi \approx 0$.

The effect of coronal transport on energetic solar protons was studied by simultaneous measurements by two or more spacecraft such as Voyager 1 and 2 and Helios 1 and 2 located at different helio longitudes during several solar proton events on September 7-19, 1977. Proton energy spectra were measured for 10-60 MeV/N protons at the time of maximum intensity and assuming a power law for the kinetic energy, $dJ/dE \sim E^{\gamma}$. The spectral index γ was obtained for the September 19, 1977 flare. The γ-values were plotted as a function of flare site longitude so that the spacecraft is connected to 0°. In other words the longitudinal separation Φ of the flare with the 'foot print' longitude of the field line connected to the spacecraft was plotted on the x-axis. It was found that when $\Phi \approx 0$ for Voyager 1 and 2, γ was $\simeq 1.2$ and the spectrum was very hard, while for helios 1 and 2 $\Phi \approx 83^o$, and the spectrum was rather soft with $\gamma = 3.3$. These observations clearly establish that coronal particle release is rigidity dependent. High rigidity particles are released faster from coronal storage into the foot print field lines which connect the spacecraft and thus the spectrum is hard with $\gamma \sim 1.2$. Other flare events also confirm this conclusion of rigidity dependence release of particles from coronal storage. Similar plots showed that the longitudinal dependence of γ as reported by the NASA group[(48)] based on single spacecraft measurements of $\sim$ 100 flares. The September 1977 observations follow the general trend observed earlier.

Source Spectrum

The determination of the source energy spectrum i.e. the proton spectrum emitted by the corona is important for the acceleration process and was derived from the time of maximum intensity by Van Hollebeke et al.[(48)] (1979) and was found to be of the form $dJ/dE \sim E^{-n}$, $n \simeq 3.2$, from 10 to 300 MeV and flattens below 10 MeV. Such power law source spectra were reported by the Prognoz 5 and 6 and Venera 11 and 12 spacecraft[(49)] of USSR-Hungery-France groups[(51)] in 1979. They found source spectra of protons of 70 keV to 300 MeV and electrons of 30 keV to 3 MeV. Below 1 MeV the proton spectrum was much flatter for all three flare events and the exponent n of the power law of the kinetic energy as $dJ/dE \sim E^{-n}$ was equal to 2, 2 and 3.7 in these flare events in a large energy range of $\sim$ 1 to 300 MeV. In the case of electrons, the values of the exponent n were 2.5, 2.5 and 3 for 30 keV to 3 MeV. For $E_p > 1$ MeV, the proton flux in the source spectrum was 25 to 75 times the electron flux for all events of September 24, 1977, November 22, 1977 and September 23, 1978.

6.4.2 Interplanetary Propagation

The interplanetary propagation of solar energetic particles is complicated by several factors as shown in Figure 6.4.1. First, the fast and slow streams of the solar wind deviate the interplanetary propagation paths from the simple Archimedean spiral. Second, the interaction regions formed by the fast and slow wind streams may cause additional effects like a diffusive barrier or acceleration. Third, the large scale magnetic field is embedded with small scale irregularities causing the "scattering" of the particles from their guiding field lines. Some of the observatonal data relating these propagation effects are briefly discussed. When a particle from the Sun spiralling in the magnetic field meets an irregularity, it is displaced to another field line. After further travel another irregularity shifts the particle to another field line. Thus the particle executes a random motion among the field lines as it convects from the Sun to the earth in a diffusive convection mode. This is denoted by 'scattering' of particles. Several studies were made to determine the mean free path of solar protons, λ_{11} which is defined as the average distance traveled by a proton in the plane parallel to the interplanetary field between two scatterings. For solar protons of about 10 MeV, the scattering mean free path is about 0.1 A.U.. This represents the diffusion-convection model of the propagation of cosmic rays and solar particles given by Parker-Fisk-Axford, with a diffusion coefficient $k_{11} = 1/3\ \lambda_{11}$ v. The approximate solution of the diffusion-convection equation with a delta function injection at the Sun is given by

$$U(r,t) \sim t^{-3/2} exp\left[-(r - v^1 \cdot t)^2/4kt\right]$$

where U = particle density (i.e. flux/velocity), $v^1 = 2$ cv with solar wind velocity v and Compton-Getting factor c. Near the earth at early times this reduces to

$$U \sim t^{-3/2}(-r^2/4kt)$$

and at late times the solution gives an exponential time decay with $\tau = k/(v^1)^2$. This is applicable to a diffusion coefficient $k \geq 5{\cdot}10^{20}$ cm^2/sec.

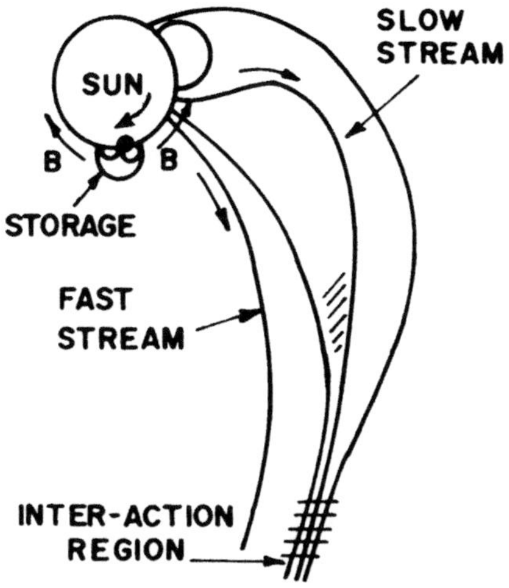

Figure 6.4.1: A schematic diagram of coronal and interplanetary propagation of SEP (adopted from ref. 51).

For 50 MeV protons, with k $\sim 10^{21}$ cm^2/sec, the exponential decay time is about 24 hours. This picture is in agreement with the observations of normal ground level events, with appropriate values for k_{11}, v etc. As an example, the observed and calculated results(49) for the November 22, 1978 event studied well with Prognoz 6 and 7, and Venera 11 and 12, by the Moscow-Budapest-Saclay groups(49) are shown in Figure 6.4.2. In some unusual events solar particle propagations in the interplanetary medium take place through 'scatter-free' paths.

6.4.3 Propagation Effects and Ground Level Events (GLE's)

Propagation effects are studied by ground level events recorded by neutron monitors. The solar flare event of May 7, 1978 provided the largest solar particle event since the November 1960 events till then, and generated relativistic particles with energies greater than 10 GeV which produced strong Ground Level Events (GLE's) recorded by many neutron monitor stations. A summary of which given by Shea et al. (1979) is shown in Figure 6.4.3. The 1N flare occured at 0327 UT at N23, W72 and the proton onset was recorded at 0336 UT with relativistic particles lasting for about two hours. The neutron monitor stations in Europe were at favourable locations and

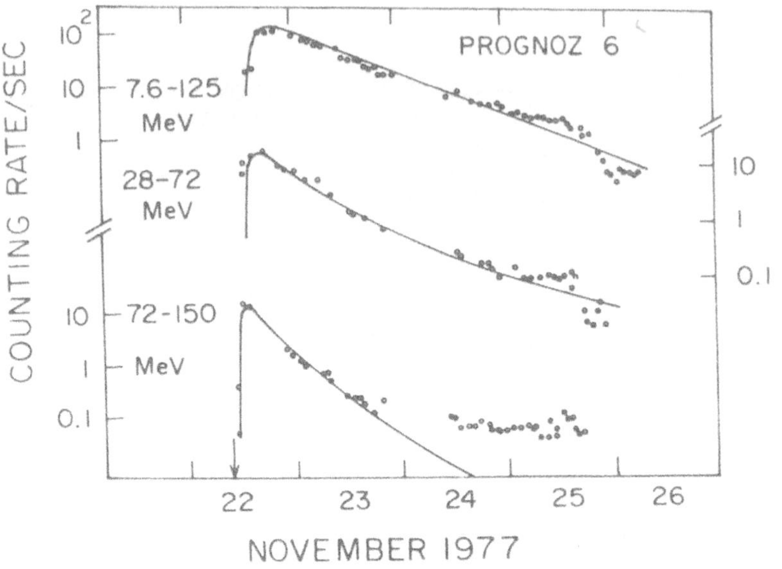

Figure 6.4.2: Interplanetary propagation results of SEP in the November 22 1977 event by Prognoz 6 and 7 and Venera 11 and 12, and calculated values, (adopted from Gombasi et al.[(49)]).

recorded an increase in counting rate as high as 75 to 125%, while non-European stations showed smaller increases. This event was studied by sixteen stations, some of which are shown in Figure 6.4.3. The relative increases and the rate of rise of various stations indicate the complex nature of interplanetary propagation which is not understood with the simplified models.

Details of these aspects are given by Shea et al.[(50)], Webberenz[(51)] and others.

6.4.4 Spacecraft observations

Solar particles and their interplanetary propagation have been measured by several spacecraft including Pioneer 10 and 11, Voyager 1 and 2, IMP-7, 8, Helios 1 and 2 etc. As an example we illustrate these by the observations made with the Ulysses spacecraft during a series of four solar flare events on March 19 to 25, 1991, as shown in Figure. 6.4.4. The Ulysses spacecraft launched on 6 October 1990 followed an in-ecliptic path to Jupiter and was at 2.5 AU in March 1981 and was magnetically connected to W90 on the Sun as shown in the Figure. A 2B flare occured at 0200 at W33^o on 19 March

1991, which showed a small increase in P and He fluxes on 19.3.1991 followed by a nearly exponential decrease till 22/3. The second flare 3B at 2245 hrs on 22/3 at E28^o showed an increase in counting rate at early hours on 23/3. But a very large increase occured at about 1500 hrs on 23/3 which is probably due to the shock front of the previous flare. The increase lasted for a few hours followed by a drop and then a steep rise at $\sim$ 2300 hours which was due to a series of flares at 0429 to 2220 on 23/3 in the same active region at E6^o-14°. The fourth flare was at 2303 UT on 24/3. The major active regions on the Sun were aligned in longitude with Ulysses which observed three shocks as shown in the figure at 1530 on 23/3, 0551 on 25/3 and at $\sim$ 0600 on 26/3. These shocks were produced by plasma emissions associated with the three flares. It may be noted that Flare 2 on 22/3 and Flare 4 on 24/3 produced fast solar neutrons detected at ground level events. The details of SEP events are given by Sanderson et al.(52) of the Netherlands and reviewed by Mandzhavidze of Moscow(53).

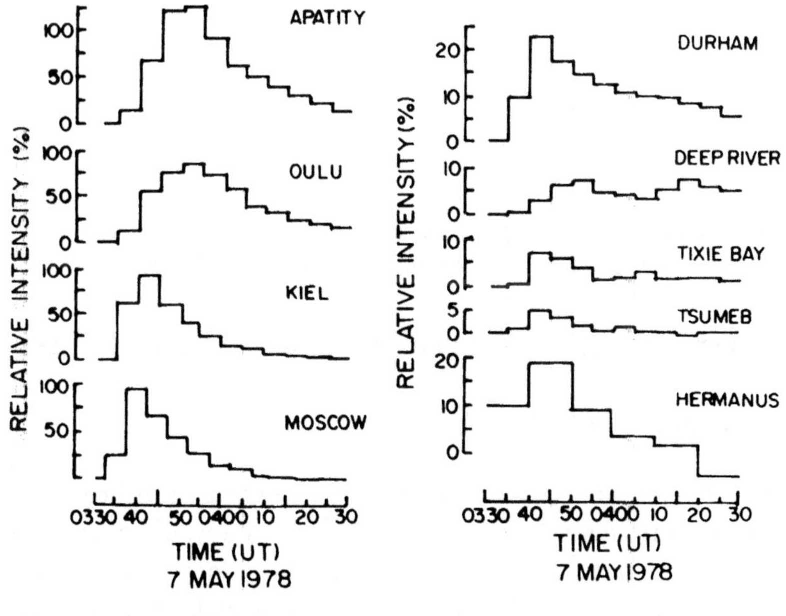

Figure 6.4.3: Relative increases in the GLE of 7 May 1978 as detected by stations in Europe (on the left) and in else where (on the right). (From Shea et al.(50)).

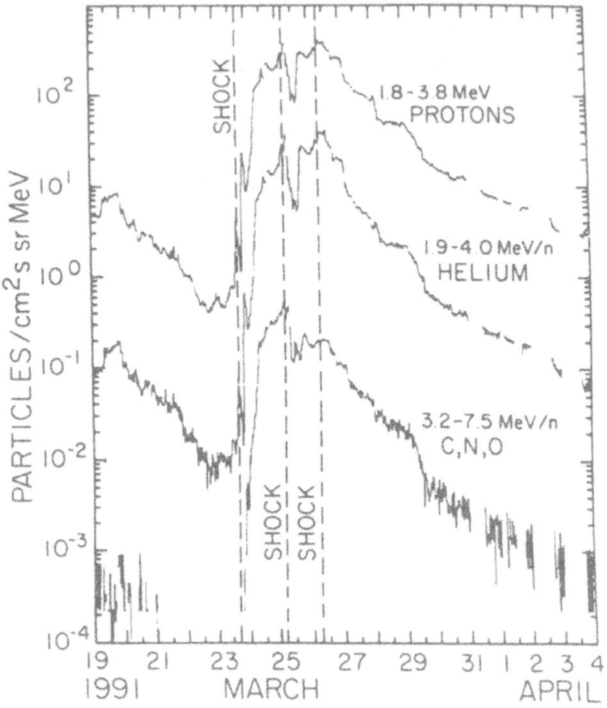

Figure 6.4.4: Protons, Helium and C,N,O nuclei measured in the Ulysses spacecraft during a series of large solar flare events between 19 and 25 March 1991. (From Sanderson et al.[52]).

6.5 Solar Energetic Particles in the Distant Past

6.5.1 Intensities of SEP

Cosmogenic nuclides in meteorites and lunar samples have been studied by many investigators to study intensity variations, if any, of galactic and solar cosmic rays in the past. Flux variations have been looked for by comparing the calculated and measured activities of radio nuclides with different half-lives which originate from the spallation of the target elements in lunar materials and meteorites by exposure to GCR and SEP. Because the flux of SEP decreases rapidly with increasing depth, the activities for SEP-produced nuclides are high at the surface and become small at depths of the order of 5 to 10 cm. Radio nuclei produced by high energy GCR e.g. for long lived ^{10}Be isotope, show a depth-activity profile which is flat near the surface to a depth of about 10 g·cm^2 and decreases at great depths. Thus the large difference in the depth-activity profiles of radio nuclides produced by GCR and SEP allows the two components to be resolved. From the measured equilibrium activities of a cosmogenic radio nuclide with a given half-live, produced by SEP and with known cross-sections from accelerator experiments the average intensity of SEP can be estimated for the past several half-lives. Results from two radio nuclides with different half-lives can be used for the confirmation of the average fluxes. It may be noted that this method is applicable for fluxes of SEP above the thresholds for the reactions.

Lunar rocks are more suitable for studying SEP fluxes in the past than meteorites because of the fixed distance of the moon from the Sun. Also the absence of a magnetic field and an atmosphere on the moon allows all SEP to reach the surface. Apollo missions brought many samples which have been used extensively for SEP and GCR intensity studies.

In Table 6.5.1 we show the results from lunar radioactivity measurements of the solar proton fluxes averaged over various time periods. In lunar samples radio nuclides with half-lives less than about one year e.g. ^{56}Co, ^{54}Mn, ^{35}Ar, were used during 1965-1975, to determine the average solar proton flux. Spacecraft measurements of the solar proton flux during this period were used to estimate the average proton flux from direct studies. Good agreement between the direct measurements and those from lunar samples establishes the validity of the method to study the activity history of the ancient Sun.

The activity measurements of 12.3 yr half-life ^{3}H in lunar rocks were used to determine the solar proton flux in the recent years of solar cycle 19 and to confirm the high average flux level of cycle 19. Similarly short-lived nuclei of ^{22}Na and ^{35}Fe both with a half-live of 2.7 yr were used to determine the average flux of cycle 19 and 20. The nuclide ^{14}C with a half-life of 5730 years did not provide a dependable flux value for the past 10^4 years due to the presence of this radio isotope from other sources. To get an idea of this the thermoluminiscence data whose trapped electrons have a half-life of about $2{\cdot}10^3$ yr were used to estimate the proton flux in the past $5{\cdot}10^3$ yr, and in general this was found to be consistent with other measurements. The radio nuclides ^{26}Al($T_{1/2} = 7.3{\cdot}10^5$ y) and ^{53}Mn($T_{1/2} = 3.7{\cdot}10^6$ y) are

Table 6.5.1: Average solar proton flux over different periods from lunar radio activity measurements

Period	Radio-nuclei	Half-life	Proton Flux, E> 10 MeV	p/(cm^2·s) E > 30 MeV
1965-1975	(a)	(a)	89	28
10^6 yr	^{26}Al	7.3x10^5 yr	70	25
5·10^6 yr	^{53}Mn	3.7x10^6 yr	70	25
5·10^3 yr	(b)	(b)	~ 60	14

(a) Several short lived ($T_{1/2} \leq 1$ yr) radio nuclides e.g. ^{56}Co, ^{54}Mn, ^{37}Ar were used.
(b) From thermoluminiscence data.

suitable for the solar proton flux in the past 10^6 y, because these have good cross-section data for the production from Al, Si and Fe targets for ^{26}Al and ^{53}Mn respectively. Also these reactions have a low threshold of ~ 10 MeV. Several groups have obtained results from these isotopes. The average solar proton flux with an energy > 10 MeV during the past two million years was obtained as 70 protons/(cm^2·s) with an exponential rigidity spectrum J(>R) = K·exp(R/R_o) with mean R_o = 100 MV, as given in Table 6.5.1. These results seem to indicate that the long term average solar proton flux was nearly the same within about 30% as at the present epoch, although short term fluctuations can not be ruled out. The solar proton fluxes of more than 10^7 years ago have not been determined so far, mainly because of difficulties of measuring activities of very-long lived cosmogenic radio nuclides. The details are given in the review by Reedy[54].

6.5.2 Charge Composition

Etched tracks of solar heavy nuclei with atomic number Z ≥ 20 in olivine grains from three carbonaceous chondrite meteorites were analyzed by Goswami et al.[55] to determine their energy spectra and charge composition. These olivine grains received their solar flare particle irradiation in space during the early history of the solar system, about > 4·10^9 years ago, and later formed part of the grains in the chondrites. Track length measurements together with calibration experiments in an accelerator were made to identify the tracks formed by the group of 'very heavy' solar nuclei (VH-nuclei, Z = 20 to 28), and very very heavy, VVH-nuclei (Z ≥ 30). Important results from these studies are as follows: (1) The spectral shape of solar VH-nuclei, time averaged over > $4{\cdot}10^9$ years in the past, was similar to that inferred from Lunar samples and from contemporary observations; (2) Similar to present day observations, ancient solar heavy nuclei of VH and VVH groups with low energies of the 6-10 MeV/N interval had selective enhancements of abundances relative to photospheric values, the enhancement factor, (VVH/VH)SEP/(VVH/VH) of the photosphere was obtained as 2 to 12 for 6-10 MeV/N; (3) It was possible that there were some changes in the ancient solar activity on a time scale of 10^4 years as inferred from variations in the abundance ratios of VVH/VH. Thus from etched track measurements in lunar samples and meteorites, it is established that solar flare activities had persisted for at least the last $4{\cdot}10^9$ years. The details are given by Goswami, Lal and McDougall, and by Crozoz(55).

6.6 Acceleration of Particles in the Sun

Solar energetic particles provide the most direct signature of particle acceleration in the Sun and the understanding of various features of these complex phenomena is one of the basic objectives in high energy solar astrophysics. It is widely believed since the 1960's that solar flares and energetic particles derive their energies from the annihilation of magnetic fields associated with sun spots (e.g. de Jager(56) 1969, Sturrock(58) 1968, Syrovatskii(57) 1969). Many studies made during the past two decades or more revealed some of the basic features of particle acceleration in the Sun which are briefly discussed here. There are several features of particle acceleration the details of which are not fully understood yet.

6.6.1 Energy Release

First let us look at the amount of energy released in the different forms of flare phenomena and in the emission of energetic solar particles. These releases have been estimated in several well studied flare events e.g. in the three flares of August 1972 which were studied by Lin and Hudson(59) in 1976. They showed that (1) mass ejection involves the largest fraction of the energy release amounting to 50 to 85% of the total energy, (2) radiation, H_α, optical, X-rays radio emission, etc. take up about 40 to 15% of the energy and (3) the energies of escaping particles account for only 4 to 0.07% of the total energy release. Also the fraction of energies released in these three channels are highly variable. A very important aspect should be noted. Although the three flares had a total energy release of $1\cdot10^{32}$, $2\cdot10^{32}$ and $4.5\cdot10^{32}$ erg which is typical for large solar flares and are of the same order of magnitude. The amount of energy released through escaping particles was extremely variable and were 0.2%, 4% and 0.07% respectively. Thus we find that there is no relationship between the total energy release in a flare and the energy of the escaping particles. This explains many observations where some large flare events produced little or no particle emission and vice-versa.

Typically total energy releases in flares of different sizes are as follows: Sub-flares: 10^{26}-10^{29} erg, Impulsive flares: 10^{29}-10^{32} erg, and Gradual flares (Large events/Long duration events): 10^{30}-10^{32} erg.

6.6.2 Impulsive Flares

From the studies of the past two cases, it has emerged that solar energetic particle events can be divided into two broad categories, namely Impulsive Events and Gradual Events (Large events).

The Impulsive events are frequent, of short duration, enriched in electrons and heavy ions and these are mostly confined to flare sites. There are no coronal mass ejections from these events. The impulsive events are mostly triggered by energetic electrons accelerated by a magnetic field reconnection mechanism, as will be described later. The examples of impulsive events are shown in Figure 3.5.7 of the Chapter 3 . The impulsive phase is marked by emission of impulsive hard X-ray bursts with energies > 15 keV, and with simultaneous microwave bursts. During the period in which these bursts are emitted, the thermal energy content of the flaring area steadily increases to reach its maximum value during the last hard X-ray burst. The thermal energy is mainly contained in the soft X-ray component of ≤ 10 keV (see the two flare events of November 5, 1980 as given by de Jager[60] in 1987.

The impulsive hard X-ray bursts are emitted from localized small areas on either side of the neutral line of the magnetic field. These areas, called 'foot prints' have a typical diameter of $\sim 10^4$ km (Figure 3.5.6). The fact that two or three foot prints can originate simultaneously indicate that the areas of foot prints are heated by beams mostly of electrons, originating above the foot prints. As a result of this the chromospheric gas in the foot prints is heated up to $7\cdot10^7$ K in the very first instant. The temperature decreases exponentially with an e-folding time of $\sim$ 90 sec. to $2{\cdot}10^7$ K. The beam heating leads to a 'burnt well' in the chromosphere. On average the bottom of the well is located at 2000 km above the photosphere. From there heated plasma rises upward with velocities of 150-400 km/s and spreads over the flaring area. This plasma contains the thermal energy of the flare. Its evolution in time is given by the soft X-ray emission component, and is marked by the gradual phase of the flare.

The period of upward motion of heated chromospheric gas is called the 'chromospheric explosion' and the lateral spreading of the gas in the lower coronal region is called the 'coronal explosion', together these may be called the 'impulsive phase explosion'. The above model, as summarized by de Jager[(60)] is supported by the observations of approximate equality between (a) the energy content of the electron beam, as measured from hard X-rays pulses $>$ 15 keV (because electrons of lower energies do not penetrate into the chromosphere), (b) the total energy of the upward moving hot gas, and (c) the thermal energy of the gradual phase. All three energies for the average flares are of the order of 10^{30}-10^{31} erg. The number of particles involved is $\sim 10^{38}$.

6.6.3 Magnetic Field Reconnection Process

In this model of magnetic field reconnection two oppositely directed magnetic tubes approach each other and merge partly, as shown in Figure 6.6.1. In this process plasma streams into the reconnection area from above and below and is squeezed out laterally as shown in the figure. One may remember, at this stage that , as pointed out by de Jager[(60)], the ideal case is shown. In an actual case, the flux threads are intertwined as in the spaghetti bundle model. Some parameters of the model are as follows.

The ratio b/a = λ is ≤ 1 and also the ratio, η, of the velocity of reconnection V_{rec} to the Alfven velocity is ≤ 1. The interaction time τ can be identified as the burst duration and is given by $\tau \simeq b/V_{rec}$, as a function of b, the particle density and the magnetic field. For average flare bursts, it is found that $\lambda \approx 1$ and $\eta \approx 1$. For some well studied impulsive events, it is estimated that the rise time of bursts is $\approx$ 30 ms and reconnection proceeds over a distance of $\sim$ 1400 km, and in a sheet with a thickness of $\sim$ 70 km.

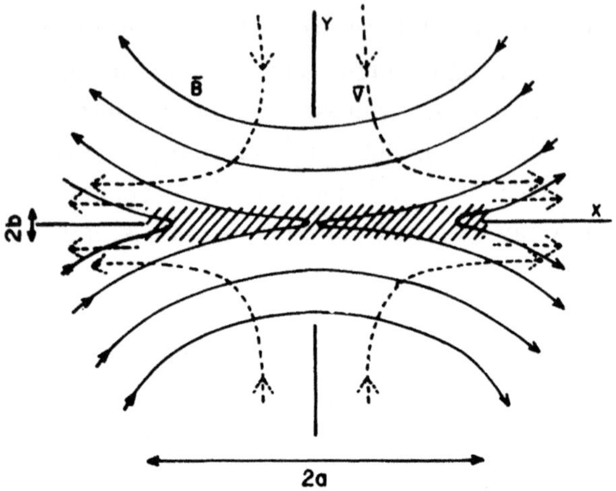

Figure 6.6.1: A sketch of the magnetic field reconnections between oppositely directed flux tubes occurring in an area of 2a x 2b. (From de Jager[(60)]).

Analysis of some impulsive flare events such as the May 21, 1984 event leads to the conclusion that the high magnetic field in some flares may lead to direct acceleration of ions up to 10 to 30 MeV. Ion acoustic turbulence and the high density in the reconnection sheet may lead to differential ion accelerations like for $^3He/^4He$. This may be called the intermediate acceleration phase.

6.6.4 High Energy or Gradual Events

High energy flares or Gradual events have a strong magnetic field $\sim$ 300-2000 gauss, a high electron density $n_e \sim 10^{12}$ cm^{-3} and occupy a large area having linear dimension of (5-10) $\times$ 10^3 km. The total number of particles involved is about 10^{38}. The gradual flares are classified as those where the second stage acceleration takes place and are of two main types, the dynamic type and the confined type. The dynamic type has open field lines and is associated with two ribbon flares and give rise to type II and or type IV radio emission. These flares are long lasting and have an extended phase of hard X-ray emission. Confined flares occur in a low lying closed field configuration which does not produced type IV bursts.

Gradual acceleration events are associated with Coronal Mass Ejection (CME) which lifts off and drives a coronal shock wave. This shock wave is able to accelerate ions efficiently out of the coronal gas. Some of the ions

Table 6.6.1: Impulsive and Gradual (Large) SEP events

	Impulsive Events	Gradual Events
Size	Small	Large
Frequency	~ 1000/year	~ 10/year
Coronal Mass Ejection	No	Yes
Radio Bursts	Type III	Type II, IV
Soft X-ray duration	< 1 hr	> 1 hr
Trapped: Escaping Protons(> 10 MeV)	$> 10:1$	$<10 : 1$
$^3He/^4He$ (at 1.5 MeV/N)	>0.2	<0.2
e/p (10 MeV)	> 0.1	< 0.1

escape the shock and propagate into interplanetary space. In Table 6.6.1 some basic features of impulsive and gradual (high energy) events are listed. The second phase of acceleration is ascribed to the second order Fermi acceleration taking place in the shock wave system associated with the opening of the field lines after the first part of the flare. On the other hand the first phase protons of gamma ray producing protons remain trapped in the Sun, whereas most of the second phase protons escape. Shibata et al.[(61)] in 1994 proposed the model of a CME driven reconnection jet leading to a fast shock front.

In summary, we may state that magnetic field reconnection accelerates particles up to 10 or 100 keV only and leads to small or average impulsive events of hard and soft X-rays. The second stage stochastic Fermi acceleration leads to protons of ~ 100 MeV which in closed loop fields produce solar neutrons and gamma-rays. Further second stage (or third stage) Fermi acceleration in the shock fronts of high energy flares produces relativistic solar protons of 1-10 GeV or more which generate the GLE events. Thus the hierarchy of acceleration processes is responsible for the variety of solar particle events. The mechanisms of acceleration processes are discussed in Chapter 7.5.

The details of solar energetic particle acceleration processes are given by several authors e.g. de Jager[(60)], (1985), Syrovatskii[(57)], (1981), and Ramaty[(62)], 1985.

6.7 Problems on Chapter 6

1. Briefly describe the important aspects of the observations of solar energetic particles in the large event of February 23, 1956 and explain how these led to the major astrophysical concepts of the interplanetary medium around the Sun for the first time.

2. What are the principal new conclusions on the abundances of P, He and C, N, O and Ne nuclei in SEP derived from early sounding rocket studies?

3. Show how from the upper limits of the Be/O and B/O ratios in SEP, the amount of matter traversed by SEP in the Sun and the acceleration time (< 100 sec) are estimated?

4. Is there a correlation between the intensities of SEP and the total energy release in the flares? Explain with examples.

5. Explain the main features of the abundances of $Z = 2$ to 28 nuclei and their enhancements with respect to the solar value, as measured in satellite experiments of IMP and Voyager 1 and 2, and in other studies.

6. Describe the main features of the plot of abundance enhancement factors vs the first ionization potential (as in Figure 6.3.7) and their implications.

7. What are the main properties of the ^{3}He-rich events in SEP? What are the basic ideas of their theoretical models?

8. Describe the main features of the coronal and interplanetary propagation of SEP from the Sun to the earth.

9. Explain how lunar radioactivity measurements yield the solar proton fluxes in different epochs. What are their main implications?

10. How do the impulsive and gradual (or large) SEP events differ from each other? Compare their main properties.

11. Briefly state how the three types of acceleration processes lead to the generation of three stages of solar energetic particles.

Chapter 7
GALACTIC COSMIC RAYS

"It would be one of the romances of science if the study of these remote and prosaic leakages of electricity from well insulated bodies should be the means whereby some of the most fundamental problems of the cosmos came to be investigated ..."

J.J. and G.P. Thompson, 1927

7.1 Early Studies and General Features

7.1.1 Early Years of Cosmic Rays

In the early years of the twentieth century, many scientists were baffled by observations of a slow leakage of electric charge from well insulated bodies. With the discovery of the radioactivity of uranium and radium, it was found that while a part of the loss of charge could be due to ionization by radiation from these substances, it was clear that a residual ionization was present even with a large amount of shielding. The origin of this residual ionization could not be understood. The possibility of extra terrestrial radiation being responsible for the residual ionization was suggested by some and disputed by other scientists. In order to decide on the issue, the German physicist Viktor Hess and two companions, made a manned balloon flight to an altitude of about 5 km and measured the ionization current in the electrometer throughout the flight. He found that at maximum altitude the current was four times larger than at sea level. The same effect was found by Hess for balloon flights at night. From these observations Hess concluded that a very penetrating radiation is entering the atmosphere of the earth from outer space, and the Sun was not the source. Thus Hess discovered a penetrating radiation from outer space which he called "hohenstralung". During the next seven years Köhlhörster made manned balloon flights to still higher altitudes and confirmed Hess's discovery.

In the late 1920's and early 1930's Robert Millikan and his coworkers in the USA made many unmanned balloon flights with more sensitive electrometers developed by them and concluded that radiation with great penetrating power, probably of high energy gamma rays, was incident on Earth, uniformly from all directions with a constant intensity during day and night. Millikan gave it the name "Cosmic Rays" in the context of the gamma ray hypothesis for the radiation. Some of the major milestones in the early years of cosmic rays in the decades of the 1930's and thereafter are as follows:

1929: Bothe and Köhlhörster of Germany developed a new technique for measuring the cosmic ray flux by using pairs of geiger counters one above the other, and showed that, contrary to the current ideas of that time, cosmic rays were composed of charged particles and not γ-rays.

1930-1935: Clay of the USA measured the variation of intensities of cosmic rays with latitude, and this 'latitude effect' conclusively showed the particle nature of cosmic rays. It was then thought by many scientists that cosmic rays consisted of fast electrons. During this time Johnson of the USA used Geiger-Müller counter telescopes to determine that the intensities of cosmic rays are different in eastern and western directions. As the intensity was higher from the western direction, the primary cosmic radiation must be mainly composed of positively charged particles i.e. protons, as determined from the direction of deflection by the magnetic field of the earth. Thus the "East-West Effect" of cosmic rays excluded electrons as primary cosmic rays. The positively charged corpuscular nature of primary cosmic rays was thus established.

1932: Direct evidence of the existence of fast charged particles in cosmic rays was obtained by Skobelzyn of Russia during his investigations of cosmic rays with a cloud chamber in a strong magnetic field.

1932: Bruno Rossi of the USA detected showers of cosmic ray particles by observing two or more particles in coincidence with three Geiger counters placed in a triangle. Hard and soft components of cosmic rays were identified by Rossi and coworkers.

1930-32: The development of a counter controlled cloud chamber by P.M.S. Blackett and G.P.S. Occhialini of the University of Manchester, UK, brought clear cut observations of 'Cosmic ray showers' of charged particles and of the identity of the 'soft' and 'hard' components of cosmic rays at sea level.

1932: Using a cloud chamber in a strong magnetic field C.D. Anderson of the USA discovered positrons (e^+) in cosmic rays.

1933: Vallarta of Mexico made further developments of the geomagnetic theory of propagation of cosmic rays in the earth's magnetic field extending the early works of Störmer.

1937: H.J. Bhabha and W. Heitler while at Cambridge, UK, developed the cascade theory of cosmic rays which explained the production of a large number of particles in a cosmic ray shower due to successive interactions cascading in the atmosphere.

1947: C.F. Powell and his collaborators of Bristol University, UK, discovered π-mesons and π-μ-e decay processes on newly developed nuclear photographic plates exposed for several months to cosmic rays at the high altitude station of Jungfraujoch in the Alps.

1948-53: During this period heavy unstable particles like K-mesons and hypersons were detected by Rochester and Bulter in the cloud chamber in the UK. The large variety of particles of intermediate mass such as τ-meson, K^+,K^- etc and of heavier mass hyperons such as Λ^o, Σ^o etc. was discovered in nuclear emulsions flown in balloons by scientists of the UK, Europe and the USA.

1948-50: The groups of E.P. Ney of the University of Minnesota, USA, and of B. Peters and H. Bradt of the University of Rochester, USA, flew compact cloud chambers and nuclear photographic emulsion plates to high altitudes of about 80,000 ft by means of the newly developed plastic balloons, called the 'Skyhook', developed by US Navy, and they discovered the presence of heavy nuclei in cosmic rays. It was found that primary cosmic rays incident on the top of the atmosphere were composed of, not only protons, but also heavier nuclei of He, Li, Be, B, C, N, O etc. up to Fe.

The exploration of cosmic rays with satellites orbiting the earth began with the launching of Sputnik by the USSR in 1957 and of Explorer by the USA in 1958. These and subsequent satellites enabled instruments to be exposed to cosmic rays for a long period of time completely outside earth's atmosphere.

A historical account of cosmic rays is given in the book on cosmic rays by Jannosy.

7.1.2 General Features of Cosmic Rays

With the unfolding of the new horizons in cosmic ray studies, it was realized that cosmic rays opened to us a new window to the universe. Our knowledge of the universe comes from signals received by us in different forms. The information received by us in the form of visible light constitutes the branch of optical astronomy, in radio waves the radio-astronomy. Cosmic ray particles moving with speed close to velocity of light bring to us the information in the form of high energy particles of interplanetary and interstellar space, from the galaxy and beyond and this opened the new vistas of high energy astrophysics.

Cosmic rays coming from the galaxy, called galactic cosmic rays consist of about 90% protons 9% helium and 1% heavier nuclei of Li to Fe; and also in very small proportions, heavy nuclei of practically all elements of the periodic table up to uranium. Electrons constitute only about 1% of the proton flux. Cosmic gamma rays coming from discrete sources and the diffuse background have been studied and these form the subject of gamma ray astronomy. Unlike electromagnetic radiations such as X-rays and γ-rays which travel in rectilinear paths from the source to the earth, cosmic ray particles travel in tortuous paths in the galactic magnetic field in the interstellar medium. Thereby cosmic ray particles reaching us from the galaxy completely lose their memory of the direction of the source from which they originated and of the time when they left the source. Thus cosmic ray particles reach us almost uniformly from all directions and with an almost constant intensity at all times. These features render the study of the origin of cosmic rays doubly difficult due to the absence of information on the source direction and the time of origin. These features, however, make the study of cosmic rays an exciting branch of high energy astrophysics in the following respects. First the cosmic ray particles bring to us the only sample of matter from distant stars, the galaxy and probably from beyond. There is no other way of obtaining these samples of matter from outside the solar system other than by cosmic rays. Therefore studies of the composition of cosmic rays provide us with clues about the composition of the sources of their origin. Secondly, because of the long interstellar travel in the galaxy, the primary cosmic ray nuclei of cosmic ray sources, such as nuclei of H, He, C, O ... Fe, etc, undergo nuclear interactions with interstellar atoms whereby several types of nuclei, which were not present in the source are produced by fragmentation of the primary cosmic ray nuclei. Thus nuclei of ^{3}He, Li, Be, B, N, F, sub-iron (Z = 21-25) etc are produced by fragmentation processes in the cosmic ray beams reaching the earth. Thus if we can measure the relative proportions of these 'interstellar secondaries' such as ^{3}He/^{4}He, B/C, (Z = 21-25)/Fe, and their production cross-sections in laboratory accelerators, we can determine the mean amount of matter traversed by cosmic rays in the galaxy before reaching the earth. From various measurements it is found that cosmic ray particles had traversed in their tortuous path in the galaxy the mean amount of about 7 g·cm^{-2} of hydrogen in the interstellar medium. If we assume a mean density of interstellar hydrogen of 0.03 atoms/cm^3, we obtain a mean life time for cosmic rays reaching us of about 10 million years.

The other fascinating aspect of cosmic rays is their enormous span of energies. Cosmic ray particles reaching us have energies ranging from about 1 MeV (i.e. 10^6 eV) to $\sim 10^{20}$ eV which covers about 10^{14} decades of energy. The lowest band of energy (1 to 100 MeV/N) belongs mostly to one class of cosmic ray particles of local interstellar origin called Anomalous Cosmic Rays (ACR); the low energy Galactic Cosmic Rays (GCR) cover the energy range of about 10 MeV/N to 10 GeV/N. In this energy band the cosmic ray particles are most well studied, as these energies are not too difficult to measure and these particles are deflected by Earth's magnetic field and the earth magnetic field can be used as a momentum analyzer for these particles with energies up to about 50 GeV/N. The high energy cosmic rays of 10 GeV/N to 10^{14} eV, are more difficult to measure, particularly at the higher energy range of 10^{12}-10^{14} eV. Only recently direct measurement of these high energy particles could be made with very elaborate techniques. In the ultra high energy cosmic rays covering an energy range of 10^{14} to 10^{20} eV the flux of cosmic ray particles is too small for direct measurement by detectors of permissible size in a space craft, even with a collecting time of a year or more. These cosmic rays are studied by the Extensive Air Shower method. On entering the atmosphere a cosmic ray particle of ultra high energy, initiates a series of interactions with atoms of the air and produces a cascade shower of particles which is spread over a large area on the surface of the earth. By measuring the density distribution and energies of this shower of particles, the energy of the primary and its nature are deduced. The energies of ultra high energy cosmic rays are larger than the highest energies available in accelerators. Therefore ultra high energy cosmic ray interactions provide new information to high energy physics. The production of high energy and ultra high energy cosmic rays in the astrophysical scenario's is one of the challenging problems in high energy astrophysics. Although several mechanisms of cosmic ray acceleration have been developed, the problems of the acceleration of high energy cosmic rays still remains unsolved. As the sources of cosmic rays, several possible sources such as supernova and their remnants, pulsars have been suggested, there are several aspects of the cosmic ray origin which are still obscure and elusive.

There are several areas of astrophysics, physics and geophysics where cosmic rays play an important role. As cosmic ray particles travels through the heliospheric plasma and magnetic field, their intensities undergo different types of small and large variations. Studies of those cosmic ray variations provide clues to the magnetic field and plasma configurations in the solar system. Energetic cosmic ray particles interact with Earth's atmospheric nuclei and also with surface materials of the moon in which radio active nuclides with different life times are formed. Studies of these long lived radio active nuclides provide new information on the early history of the solar system, and of planetary geophysics. Radioactive nuclei produced in the earth's atmosphere yield important information on atmospheric and stratospheric circulations. The well known isotope, ^{14}C having an half life of 5400 yrs is produced by cosmic ray protons interacting with atmospheric nitrogen, and has been widely used in radio carbon dating and in archaeology.

7.1.3 Definitions and Nomenclatures.

Energy, Momentum and Rigidity of Particles:

In cosmic ray physics the following definitions and nomenclatures are generally used. Let W_o = rest energy of the nucleon = M_oC^2 = 938 MeV, v = its velocity, β = v/c, c being the velocity of light, E = kinetic energy in MeV, P = momentum in MeV/c, W = Total energy (= M_oC^2 + E), and $\gamma = 1/\sqrt{1-\beta^2}$, the Lorenz factor, then

$$W = \frac{M_oC^2}{\sqrt{1-\beta^2}} = M_oC^2 \cdot \gamma \tag{7.1.1}$$

$$E = W - M_oC^2 = M_oC^2\Big[\frac{1}{\sqrt{1-\beta^2}} - 1\Big] = M_oC^2(\gamma - 1) \qquad (7.1.2)$$

$$Pc = \frac{M_o v \cdot c}{\sqrt{1-\beta^2}} = \frac{M_oC^2 \cdot v/c}{\sqrt{1-\beta^2}}$$

$$Pc = M_oC^2 \cdot \beta \cdot \gamma \qquad (7.1.3)$$

The total energy W is related to the particle momentum P by the relativistic equation

$$W = (P^2C^2 + M_o^2C^4)^{1/2} \qquad (7.1.4)$$

For protons and heavier nuclei of cosmic rays of mass number A and atomic number Z, the kinetic energy is expressed as energy per nucleon, i.e. E/N. Thus protons, alpha particles and heavier nuclei having the same energy per nucleon have the same velocity.

Example: For protons and alpha particles, both having kinetic energy 1000 MeV/nucleon have the same velocity, β, which is calculated by

$$E = M_oC^2(\gamma - 1)$$

or γ - 1 = 1000/938 = 1.066 or $\gamma = 2.066$ or $\sqrt{1-\beta^2} = 1/2.066$ or 1-β^2 = 0.2342. Therefore $\beta = 0.875$.

The momentum per nucleon P in MeV/c is

$$P = \frac{M_o\beta c}{\sqrt{1-\beta^2}} = M_oC^2\beta\gamma/c$$

$$= 938 \times 0.875 \times 2.066/c$$

Therefore

$$P = 1695 MeV/c, or \quad Pc = 1695 \quad MeV$$

The Rigidity of a proton of kinetic energy 1000 MeV is given by

$$R_p = \frac{P \cdot c}{Z} = 1695 MV.$$

and the Rigidity of a ^{4}He nuclei is given by

$$R_{He} = \frac{P \cdot A \cdot c \cdot}{Z} = P \cdot c \cdot \frac{A}{Z} = 1695 \times \frac{4}{2} = 3390MV$$

Thus for the same energy per nucleon i.e. same velocity the rigidity of multiply charged nuclei such as He, C, N, O, etc. having A/Z = 2, is twice that of protons.

As shown earlier the relation between the Rigidity of a particle and its curvature in a magnetic field is given by

$$R_{eV} = 300H\rho \tag{7.1.5}$$

where the Rigidity is in eV, the magnetic field H is in gauss and the curvature ρ is in cm.

The Flux or the Intensity of cosmic ray particles is measured as follows. Let $J_i(>E)$ be the flux of particles of type i with an energy > E MeV/N incident on an unit area per unit solid angle, per unit time. If N is the number of particles incident on an area A from the upper hemisphere in time t, then

$$J_i(> E) = N/A \cdot \Omega \cdot t$$

$$Ad\Omega = A \cdot \cos\theta(r \cdot d\theta) \cdot (r \cdot \sin\theta \cdot d\phi)$$

$$= Ar^2 \cos\theta \sin\theta d\theta d\phi$$

$$\int Ad\Omega = A \cdot \int_o^{\pi/2} \cos\theta d(\sin\theta) \cdot \int_o^{2\pi} d\phi \quad for \quad r = 1$$

$$= A \cdot \int_o^{\pi/2} \frac{\sin^2\theta}{2} \cdot 2\pi$$

$$= A \cdot \frac{1}{2} \cdot 2\pi = A\pi$$

Therefore

$$J_i(> E) = \frac{N}{A \cdot \pi \cdot t} \quad particles/(cm^2 \cdot sr \cdot sec) \tag{7.1.6}$$

$J_i(>E)$ is defined as the Integral Flux or Integral Intensity of particles of type i.

Similarly, $J_i(E)dE$ is defined as the differential flux, which is the number of particles of energy between E and E+dE incident per unit area, unit solid angle and unit time.

$$J_i(E)dE = \frac{N(E)dE}{A \cdot \pi \cdot t} \quad particles/(cm^2 \cdot sr \cdot sec/MeV/N) \tag{7.1.7}$$

Cosmic ray particles exhibit a power law spectrum for the total energy per nucleon, over a certain energy interval, i.e.

$$J_i > (m_o C^2 + E) = K \cdot (m_o C^2 + E)^{-n} \quad or \quad J_i > (W) = K \cdot W^{-n} \tag{7.1.8}$$

Then the differential energy spectrum is given by

$$J_i(m_o C^2 + E)dE = \frac{K}{n}(m_o C^2 + E)^{(n+1)} \tag{7.1.9}$$

The power law index of the integral energy spectrum is about 1.6, and for the differential spectrum about 2.6.

7.2 Interactions of Cosmic Rays with Matter

7.2.1 Ionization loss of charged particles in matter

When a fast charged particle such as a proton or a heavier nucleus passes through matter it interacts electromagnetically with nearby atomic electrons causing excitation or ionization of the atoms. At lower velocities nuclear collisions and capture and loss of electrons become important. The energy loss of the incident particle in an individual collision is rather small and its energy is gradually reduced by numerous small losses until it can no longer ionize. The distance it has traversed is called the range. Statistical fluctuations in the number of collisions lead to fluctuations in the energy loss per unit path length and in its range.

The primary ionization refers to the initial ionization by primary particles, whereas total ionization includes the secondary ionization produced by the knock on electrons, called δ-rays produced by a primary of charge Z, and velocity β c, per unit length given by $N_\delta = K \cdot Z^2/\beta^2$, where K is a constant which can be determined experimentally. Specific ionization is defined as the energy loss by the primary per unit path length.

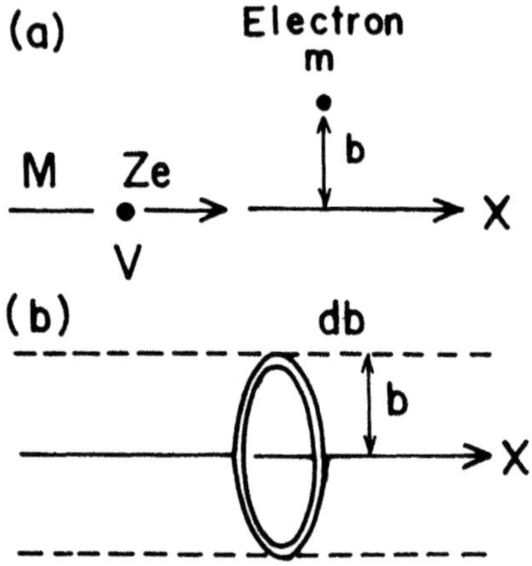

Figure 7.2.1: A sketch showing the interaction of a moving particle with an electron in a medium

Bohr's formula:

Let us consider an electron of mass m at a distance b from the path of an incident particle of mass M and charge Ze moving with velocity v in the x-direction (Figure 7.2.1a). Let the electron be free and at rest and its displacement during the collision small. The momentum acquired by the electron during the collision is first calculated and then the energy acquired is determined. As the particle passes the force the electron changes its direction. It is seen that the parallel component of the force $F_{||}$ vanishes due to symmetry and we have the perpendicular component $F_\perp$ only. We consider a cylinder with its axis along x, and radius b (Figure 7.2.1b). The electric field intensity $\bar{\epsilon}$ of the charge Ze leads to the electric flux on a part of the cylinder surface as $\int \epsilon_\perp \cdot 2\pi$ b dx = 4π Ze by Gauss's theorem. Therefore we have

$$\int_{-\infty}^{+\infty} \epsilon_\perp dx = \frac{2Ze}{b} \tag{7.2.1}$$

The variation of $\epsilon_\perp$ at the electron with time is the same as that of keeping Ze fixed and observing a point moving with velocity v along the surface of the cylinder. Hence

$$\int_{-\infty}^{+\infty} \epsilon_{\perp}(t)dt = \int_{-\infty}^{+\infty} \epsilon_{\perp}(x) \cdot \frac{dx}{v} = \frac{1}{v} \cdot \int_{-\infty}^{+\infty} \epsilon_{\perp}(x)dx = \frac{2Ze}{vb}$$

Therefore, the impulse,

$$I_{\perp} = \int_{-\infty}^{+\infty} \epsilon_{\perp} \cdot e \cdot dt = \frac{2Ze^2}{vb} = p \tag{7.2.2}$$

where p is the momentum acquired by the electron. The energy received by the electron is

$$\frac{p^2}{2m} = \frac{2Z^2e^4}{mv^2b^2} \tag{7.2.3}$$

If n is the number of electrons per cm^2, the energy lost in collisions with electrons in the volume of the cylinder between b and b + db is

$$dE(b) = 2\pi b,\ \ db \cdot \frac{2nZ^2e^4}{mv^2 \cdot b^2} = \frac{4\pi nZ^2 \cdot e^4}{mv^2} \cdot \frac{db}{b}$$

The total energy change in all shells is given by

$$-\frac{dE}{dx} = \frac{4\pi n \cdot Z^2e^4}{mv^2} log_e \frac{b_{max}}{b_{min}} \ \ erg \cdot cm^3 \tag{7.2.4}$$

The b_{max}: Relativistically the electric field of the incident particle is contracted in the direction of motion and the field is increased by a factor $1/(1-\beta^2)^{1/2}$, $\beta = v/c$. This sharpens the pulse of force on the electron which is given by $(b/v)(1-\beta^2)^{1/2}$. The b_{max} is chosen so that if $1/\tau$, (where τ is the impulse time) is much less than the vibration frequency ν of the electron in an atom, then the electron absorbs no energy. Thus we set

$$b_{max} = \frac{v}{\bar{\nu}(1-\beta^2)^{1/2}} \tag{7.2.5}$$

where $\bar{\nu}$ is the average frequency of the electrons in the material.

The b_{min} is derived from the maximum energy that can be imparted to the electron. The classically maximum velocity of an electron for a head on collision is less than, 2v, and its energy is less than $(1/2)m(2v)^2$. Therefore, we have, classically,

$$b_{min} \approx Ze^2/mv^2$$

Quantum mechanically we find that if λ is the deBroglie wavelength of the electron as seen by the incident particle, only b > λ is meaningful. Therefore we use

$$b_{min} = \lambda = \frac{\hbar(1-\beta^2)^{1/2}}{mv} \tag{7.2.6}$$

Substituting b_{max} and b_{min} in 7.2.6 we get

$$-\frac{dE}{dx} = \frac{4\pi n Z^2 e^4}{mv^2} \cdot ln \frac{mv^2}{\hbar\bar{\nu}(1-\beta^2)} \tag{7.2.7}$$

More accurate calculations from quantum mechanical derivations by Bethe give the following Bohr formula for heavy particles such as protons and ions

$$-\frac{dE}{dx} = \frac{4\pi n Z^2 e^4}{mv^2}\left[ln \frac{2mv^2}{I(1-\beta^2)} - \beta^2\right] \; erg \cdot cm \tag{7.2.8}$$

where I is the mean ionization potential.

A slightly different form for the energy loss by incident relativistic electrons of energy E is given by

$$-\left(\frac{dE}{dx}\right)_{electron} = \frac{2\pi n e^4}{mc} ln \frac{E}{2mc^2 I} \tag{7.2.9}$$

The following features of the ionization loss formula (7.2.8) are noted.

1. The rate of energy loss is nearly proportional to $1/v^2$, so that slow particles lose much more energy than the fast one. The energy loss rate is proportional to Z^2.

2. The rate of energy loss decreases with increasing primary energy, and it reaches a minimum value at energies where the particle becomes relativistic, i.e. at $\gamma \approx 1$ to 2.

3. At extremely relativistic energies, the energy loss rate increases slowly and logarithmically with energy. In dense medium, dE/dx reaches a plateau value due to polarization of the medium.

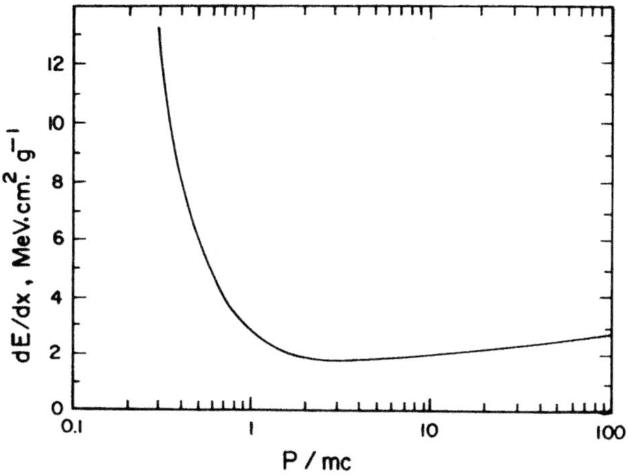

Figure 7.2.2: A sketch showing the variation of the energy loss rate with energy

In Figure 7.2.2 we show a sketch of the variation of the energy loss rate of a proton with velocity p/mc.

For numerical computation it is convenient to use $-\frac{1}{\rho}\frac{dE}{dx}$, where ρ is the density of the medium, so that we use energy loss per gm$\cdot$cm^{-2} of the medium. Expressing the energy loss in MeV, it is found that in an air proton the energy loss rate at 1 GeV is about 2.3 MeV per gm$\cdot$cm^{-2}, which is nearly its minimum value.

As dE/dx = Const$\cdot Z^2$ f(v), two particles having the same velocity will have an ionization loss proportional to the square of their nuclear charge Z^2,

$$\frac{dE_1/dx}{dE_2/dx} = \frac{Z_1^2}{Z_2^2}$$

The energy loss formula can be integrated to give the range of the charged particle,

$$\int_R^o -dx = \int_o^{E_o} \frac{dE}{(-dE/dx)}$$

Therefore

$$R(M/Z) = Z^2 \int f(v)dE = \frac{M}{Z^2} f(v) \tag{7.2.10}$$

Thus it is seen that two particles which have the same velocity have their ranges proportional to M/Z^2. Therefore for the same energy per nucleon, the ranges of the two particles are

$$\frac{R_1}{R_2} = \frac{M_1/Z_1^2}{M_2/Z_2^2} \qquad (7.2.11)$$

The ranges of protons and other particles in different media such as Al, Fe, Cu, etc and nuclear emulsions are best determined by the calibration with experiments with accelerators.

The rate of energy loss by protons, -dE/dx vs. (or momentum) as given by equation 7.2.10 is shown in Figure 7.2.2. It is found that dE/dx decreases with increasing energy of the particle and the minimum ionization loss rate is reached at about 2 GeV beyond which there is a slow increase which reaches a plateau value at energies of 50-100 GeV. The relativistic rise in the ionization loss rate has been verified in cosmic rays by measuring the specific ionization in cloud chamber tracks in a magnetic field and in nuclear emulsions by measuring grain densities as a function of kinetic energy. It is found that the relativistic rise in the energy loss rate is about 7% in case of nuclear emulsions; it varies between 5-15% depending on the medium.

The minimum ionization loss rate is experimentally found by calibrations in accelerators and is given by Barkas as 1.6 MeV/(gm·cm^{-2}) for nuclear emulsions. The dE/dx is measured in terms of (dE/dx) min. Knowing the dE/dx vs E curve for protons, the ionization loss rate vs energy/nucleon curves can be computed for nuclei of different elements with M, and Z such as He, C, O, etc, using the relation 7.2.10. Range vs Energy curves of various nuclei of M, Z are thus determined from the proton calibration curve. These are used for the identification of cosmic ray nuclei.

7.2.2 Inelastic Interactions of Cosmic Ray Particles

Interactions of High Energy Protons with Heavy Target Nuclei

Many studies have shown that high energy cosmic ray protons interacting with target nuclei with a mass number A $\sim$ 10 to 60 or more, produce secondaries which consist of (a) meson showers in a narrow angle in the forward direction, (b) high energy secondary protons and neutrons in the forward hemisphere and (c) evaporation particles of protons, neutrons, alpha particles with low energies which stop within a short distance and often (d) the residual of the target nucleus, called the spallation product. Total inelastic cross-sections of the above processes are required for interpretations of the observations of cosmic ray interactions in the atmosphere with nitrogen and oxygen nuclei and with other target materials of cosmic ray experiments. These cross-sections are needed for the study of the spallation products, both radio-active and stable ones, produced in the interactions with the earth's atmosphere and with the surface of the moon. The number of measurements of these cross-sections are at present sufficiently large so that semi-empirical equations can be constructed which fit fairly good to the experimental data. As more experimental results are available improvements in the semi-empirical formulae are made.

The total inelastic cross-section of the proton-nucleus (A_t) interactions are needed to calculate the propagation of cosmic rays in the atmosphere. As the inverse reaction, nucleus-proton has the same cross-section, nucleus-proton cross-sections are important to determine the fraction of primary cosmic ray nuclei of a given type which survive after collisions with interstellar matter which is mainly hydrogen. The σ_i for p-A_t collisions are constant for $E > 3$ GeV and we shall consider the relativistic σ values now. The simplest fit to the total inelastic cross-section σ_i for p-A_t collisions is given by $\sigma_i =$ const. $\cdot A_t^{2/3}$, which is more accurately expressed as

$$\sigma_i = 45 \cdot A_t^{0.7} mb \tag{7.2.12a}$$

This equation gives σ_i for targets with mass number A $\sim$ 10 to $\sim$ 240 i.e. lithium to uranium with an accuracy of $\leq 2\%$. Empirically additional terms are used for higher accuracy. The energy dependence of the σ_i is given empirically as

$$\sigma_i(E) = \sigma_i(H \cdot E)\Big[1 - 0.62 exp(-E/200)\sin(10.9E^{0.28})\Big] \tag{7.2.12b}$$

where E is expressed in MeV/N. This equation gives a good fit for carbon-proton collisions in the energy range of 20 to 1000 MeV/N.

Charge Changing or Fragmentation Cross-Sections:

Cosmic ray nuclei heavier than helium have undergone interactions in the interstellar medium, mainly with hydrogen in the which dominant process is the partial break up of the incident nucleus, leading to the primary fragment traveling in the forward direction with almost the same energy/N as the primary nucleus. It has been observed extensively in nuclear emulsions exposed to relativistic cosmic rays that heavy nuclei colliding with C, N and O nuclei of the emulsions are most frequently only partially broken up leading to a fragment of the original particle traveling in the same direction with the same energy per nucleon. For example an energetic Fe nucleus track in an emulsion when followed for some distance is often found to have a fragmentation product of, say, chromium (Z = 22) and the secondary chromium nucleus after some distance may interact and produce a low charge fragment say lithium. These are called the fragmentation interactions of cosmic ray heavy nuclei. In the case of the cosmic ray primaries carbon and oxygen nuclei which are next to helium the most abundant nuclei, it is found they often produce by fragmentation collisions, nuclei of lithium, beryllium and boron. In fact, as shown later, it is found that the Li, Be, B nuclei which constitute about 30% of the C and O nuclei in the primary cosmic rays, originate entirely from the interstellar collisions and fragmentations of C, O and heavier nuclei. Therefore it is of great importance to measure the appropriate cross-sections in order to determine the major astrophysical aspects of the primary cosmic rays. More details of these aspects are given in the section on interstellar propagation of cosmic rays (Chapter 7.6).

Charge changing or partial cross-sections were measured by the Orsay group of Raisbeck and Yiou, the Berkeley group of Heckman et al., the New Hampshire group of Webber et al. and the Minnesota group of Waddington et al.; and semi-empirical cross-section formulae were developed by the Naval Research Laboratory group of Silberberg et al. Studies made by using projectiles of Fe and lighter nuclei of energies $\leq$ 2 GeV/N in accelerators,the total interaction cross-sections, σ_{total} are measured for various target elements. σ_{total} can be represented by

$$\sigma_{total} = 10\pi\, r_o^2\left(A_T^{1/3} + A_B^{1/3} - \delta\right)^2 \qquad (7.2.12c)$$

where $r_o = 1.35 \cdot 10^{-13}$ cm

A_T = Mass number of the Target
A_B = Mass number of the Beam
δ = Overlap term

The overlap term δ is either constant or a slowly varying function of target and beam mass, and is given by

$$\delta = C \cdot \left(A_T^{1/3} + A_B^{1/3}\right)$$

Other forms of δ are also used.

When N_o particles are incident on a target of thickness x cm, the number of transmitted particles, N is given by

$$N = N_o\, e^{-x/L} = N_o(1 - x/L) for \;\; x \ll \;\; L \qquad (7.2.13)$$

Where L = interaction mean free path in cm = $1/n\sigma$, n is the number of atoms/cm^3 of the target.

If ρ is the density of the target material in g·cm^3, we get

$$L\rho = \lambda = \rho/n\sigma$$

when λ is the mean free path in g·cm^2.

It is convenient to express mean free paths in g·cm^2 of the material because in this form of λ (in g·cm^2) the attenuations of different materials are approximately the same.

From equation 7.2.13, we get the fraction of particles interacting in the target material as $\Delta N/N_o = (N_o - N)/N_o = \rho \cdot x/\lambda$ i.e. $\Delta N/N_o = x/\lambda$, where x is in g·cm^2 and λ in g·cm^2.

λ or σ for a composite medium such as nuclear emulsions, composed of gelatine (H, C, N, O) and silver bromide (Ag, Br), are calculated firstly by determining the λ_i or σ_i for individual elements i and then calculating the weighted average $\bar{\sigma}_{av} = \Sigma\, w_i\, \sigma_i / \Sigma\, w_i$ where w_i is the weight factor for element i.

Important Cross-Sections for Cosmic Ray Studies:

For cosmic ray experiments and propagation studies the important beam elements, (in addition to p and alpha particles) are ^{12}C, ^{16}O, ^{20}Ne, ^{22}Ne,

^{24}Mg, ^{28}Si and ^{56}Fe. The target elements include common materials like C, Al, Fe, Pb etc and N and O for air nuclei. For interstellar propagation calculations, the most important target element is hydrogen, followed by helium as their abundances are about 95:5 in the ISM. Of these the hydrogen target is the most difficult to obtain experimentally. Therefore a commonly used procedure is to determine σ_{int} for a plastic compound with a composition like CH_2 and for carbon (as graphite). Then by a difference method σ_t for H is determined. The beam energies commonly used are 650, 800 and $\sim$ 1000 MeV/N.

The C, O and heavier nuclei beams are used for the partial cross-sections of Li, Be, B and F secondaries for cosmic ray pathlength computations. Of these Be/(C+O) and B/(C+O) are some of the most useful 'secondary' to primary ratios in cosmic ray studies. The primary beam of Fe and secondaries Sc, Ti, V, Cr and Mn are important as these Fe-secondaries to Fe ratios are sensitive indicators of interstellar propagation. These are discussed later (Chapter 7.4.2).

In Table 7.2.1 we show the total interaction and partial cross-sections for C and O beams. Elemental and isotopic cross-sections are given, as determined experimentally and some by semi-empirical methods. For further details the readers may refer to the studies of 1973 by Silberberg and Tsao(1), Silberberg, R., Tsao, C.H. and Letaw, J.R.,(2) and Tsao(2) of 1983, Webber et al.(3) of 1982, Peron(4) of 1976 and others referred therein.

7.2.3 Interactions of Gamma Rays with Matter

Gamma rays interact with matter in several discrete elementary processes by which the photon is absorbed completely or scattered with or without a decrease in energy. Thus if I_o is the number of photons incident on a material of thickness x, the number of undisturbed photons passing through the material, I(x), then is

$$I(x) = I_o e^{-Kx} \tag{7.2.14}$$

where K is the absorption coefficient of the radiation for the material. The unit of K is reciprocal to that of x i.e. cm^{-1} when x is in cm.

The mass absorption coefficient μ is defined by the relation: $I(\xi) = I_o$

$e^{-\mu\xi}$, where ξ is the thickness of the medium expressed in g·cm^{-2}, and ξ = x·ρ where ρ is the density of the medium. The unit of μ is the reciprocal of that of ξ, i.e. cm^2·g^{-1}. It can be seen that $I/I_o = e^{-\sigma N x}$, where σ is the absorption cross-section in cm^2, N is the number of atoms per cm^3. Therefore,

$$\mu \cdot \xi = \sigma \cdot N \cdot x, \quad or \quad \mu \cdot (x \cdot \rho) = \sigma \cdot N \cdot x$$

Thus we have the relation between the mass absorption coefficient μ, and σ, which is

Table 7.2.1: Cross-Sections (in Millibarns) for Collisions of Various "Heavy Nuclei" with Hydrogen (E ≃ 2.3 GeV)

Product	^{11}B	^{12}C	^{14}N	^{16}O	^{20}Ne	^{24}Mg	^{28}Si	^{56}Fe
^{6}Li.....	14	**15***	11	**14**	13	13	13	30
^{7}Li.....	20	**13**	13	**14**	13	13	13	20
^{7}Be.....	7	**10**	**12**	**11**	10	**10**	**10**	**8.5**
^{9}Be.....	5.5	6	3	**3.7**	3	3	3	5
^{10}Be....	**14**	3.5	1.9	**1.0**	1.9	1.9	1.9	4
^{10}B.....	26	16	14	**12**	11	10	9	9
^{11}B.....	...	**51**	25	**25**	18	15	12	9
^{12}C.....	...	...	26	24	18	13	10	7
^{13}C.....	...	...	25	20	14	10	8	5
^{14}N.....	...	...	...	26	18	13	10	6
^{15}N.....	...	...	...	**50**	23	17	13	6
^{16}O.....	...	...	...	...	24	18	13	6
^{17}O.....	...	...	...	...	25	19	14	6
^{18}O.....	...	...	...	...	23	16	**12**	6
^{19}F.....	...	...	...	...	45	19	14	7
^{20,21,22}Ne....	...	...	...	...	...	**69**	**55**	**21**
^{23}Na....	...	...	...	...	...	51	23	9
^{24,25,26}Na....	...	...	...	...	...	...	77	25
^{27}Al....	...	...	...	...	...	...	52	10
σ_T.....	195†	205	235	260	315	355	400	676

* Bold values refer to cross-sections based on experimental information.
†σ_T is the total inelastic cross-section.

$$\mu = (\sigma N)/\rho \ \ cm^2 \cdot g^{-1} = \sigma \cdot No/A \tag{7.2.15}$$

where N_o = Avogadro No and A = Atomic mass no.

(1)**Rayleigh Scattering:** When the momentum transfer to the bound atomic electron is small so that the atom remains in its ground state, the gamma rays are scattered with unchanged energy or frequency, with a definite phase relation between the incoming and scattered wave. The intensity of the radiation scattered by an atom is determined by adding the amplitudes of the radiation coherently by each of the electrons. The Rayleigh scattering is well known in the X-ray region and is a basic process for Bragg interference in the spectroscopy of very low energy gamma rays.

(2)**Photo-electric Effect:** A photon whose energy is less than the ionization energy of the atom, is usually absorbed by the atom when its energy corresponds to the difference of the energies of two bound states of the electrons in the atom, $E_i - E_f = h\nu$. When the photon energy exceeds the ionization potential, the absorption spectrum is continuous. The kinetic energy of the ejected electron, is then $T = h\nu - I$, where I is the ionization potential of the shell in which the electron originally belonged. The photo-electric effect is large when the photon energy is close to the ionization potential.

As the frequency ν of the incident photons is increased with a small value, they are able to excite electrons in the optical orbits of a few eV binding energy. As ν is increased further electrons in the deeper orbits absorb the energy. When ν reaches the binding energy of a particular shell, the absorption increases abruptly due to absorption say, by the K-shell electrons, which is called the K-absorption edge. The L-absorption edges are three due to three sub-shells. A sketch of photo-electric absorption as a function of ν is shown in Figure 7.2.3.

The photo-electric cross-section for the absorption of a K-shell electron has been calculated quantum mechanically and is given by

$$\sigma_{ph}(perelectron) = \frac{128\pi}{3} \cdot \frac{e^2}{mc} \cdot \frac{\nu_k^3}{\nu^4} \frac{e^{4\varepsilon cot^{-1}\varepsilon}}{1 - e^{-2\pi\varepsilon}} cm^2 \tag{7.2.16}$$

where $\varepsilon = (\nu_k/\nu\text{-}\nu_k)^{1/2}$, and ν_k is the frequency of the K-absorption edge. In the case of ν near ν_k, the following simplified relation is used

$$\sigma_{ph}(perelectron) = \frac{6.31 \times 10^{18}}{Z^2} \left(\frac{\nu_k}{\nu}\right)^{8/3} \tag{7.2.17}$$

As an example we note that for 0.5 MeV gamma rays in Pb, the photo-electric absorption of the K-shell can be shown (from equation 7.2.17) to be responsible for 80% of the absorption of photons. It may also be noted that the sharpness of the absorption edge of the K-shell of different elements can be used for measuring the energy of low energy gamma rays. For example, K-absorption edges of Al, Cu, Sn, and Pb are of energies 0.005, 0.02, 0.07 and 0.19 in units of mc^2, respectively.

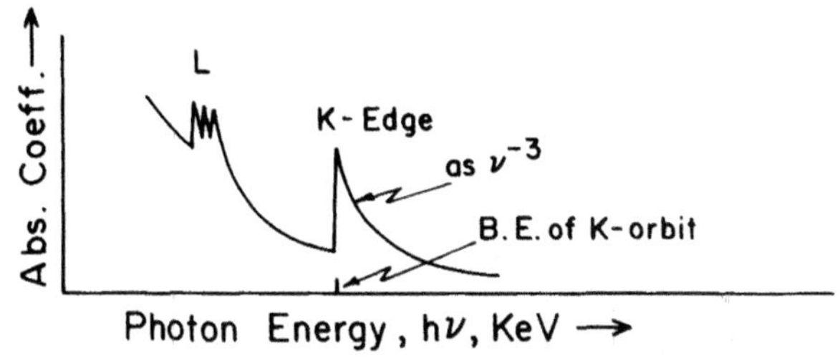

Figure 7.2.3: A sketch of the photo-electric absorption coefficient vs. photon energy, hν

(3)**Thomson Scattering:** In the low energy region of $h\nu \ll mc^2$ the scattering cross-section is given by Thomson scattering where the incident wave is reradiated by the free electrons in different directions with the same wavelength. The Thomson scattering cross-section per electron is given by

$$\sigma_T = \frac{8\pi}{3} \cdot \left(\frac{e^2}{mc^2}\right)^2 \cdot \frac{8\pi}{3} r_e^2 = 0.66 \times 10^{-24} cm^2 \qquad (7.2.18)$$

(4)**Compton Scattering:** For photons with energies beyond the K-shell absorption edge of the medium, the predominant absorption process is Compton scattering. It is not a true absorption since this process sends the incident energy in a reduced magnitude in a different direction. In this process incident photon hν collides with an electron which is considered free as its binding energy is small compared to hν. The scattered photon suffers a change of energy which can be calculated by applying the conservation of relativistic energy and relativistic momentum as follows.

In Figure 7.2.4, let E(= hν) be the energy of the incident photon, E'(= hν') the energy of photon after the collision, θ the angle at which it is scattered, E", the energy acquired by the electron, and p" the electron momentum. The conservation of energy gives

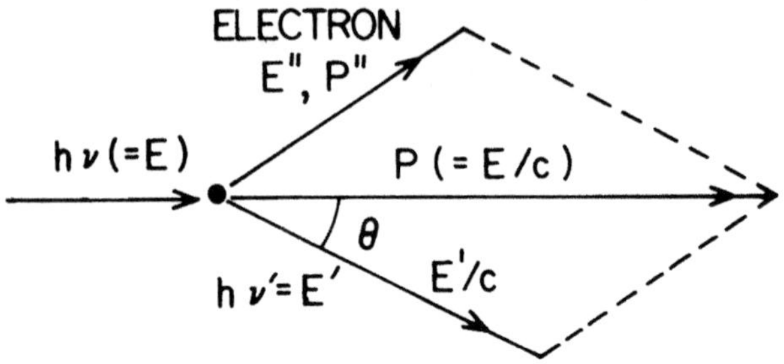

Figure 7.2.4: A sketch of Compton scattering of an incident photon hν by an electron

$$E - E' = E" \tag{7.2.19}$$

and the conservation of momentum gives

$$(p")^2 = \left(\frac{E}{c}\right)^2 + \left(\frac{E'}{c}\right)^2 - \frac{2EE'}{c^2}\cos\theta \tag{7.2.20}$$

Expressing p" in terms of E" and m, we get

$$(E")^2 + 2mc^2E" = E^2 + (E')^2 - 2EE'\cos\theta \tag{7.2.21}$$

By substituting E" = E-E', and simplifying we get

$$E' = \frac{E \cdot mc^2}{mc^2 + E(1 - \cos\theta)} \tag{7.2.22}$$

This relation 7.2.22, called the Compton equation, gives the energy E' of the scattered photon in terms of the incident energy E of the photon and the angle of scattering. We note that for $\theta = 0$, i.e. in the forward direction E' is maximal (= E) and it is minimal for $\theta = 180^o$ where E' = $mc^2/2$, when E $\gg mc^2$.

The cross-section for Compton scattering is given by the Klein-Nishina formula as given by Heitler:

$$\sigma_c = 2\pi r_e^2\Big[\frac{1+\alpha}{\alpha^2}\Big\{\frac{2(1+\alpha)}{1+2\alpha} - \frac{1}{\alpha}ln(1+2\alpha)\Big\}$$

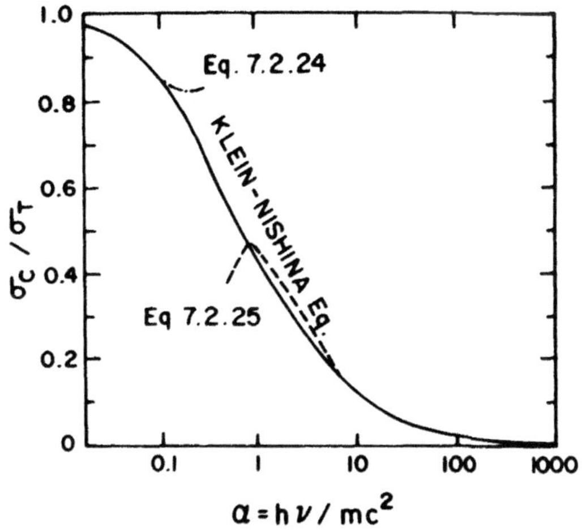

Figure 7.2.5: Klein-Nishina cross-sections for Compton scattering, σ_c, in units of σ_r

$$+\frac{1}{2\alpha}ln(1+2\alpha)-\frac{1+3\alpha}{(1+2\alpha)^2}\Big] \tag{7.2.23}$$

where $\alpha = \mathrm{h}\nu/\mathrm{mc}^2$.

Approximate expressions for σ_c for low energies and and high energies are as follows:

$$Low\ \ energies,\ \ \alpha \ll 1,\ \ \sigma_c = \sigma_T\Big\{1 - 2\alpha + \frac{26}{5}\alpha^2 + ...\Big\} \tag{7.2.24}$$

$$High\ \ energies\ \ \alpha \gg 1,\ \ \sigma_c = \sigma_T\frac{8}{3}\frac{1}{\alpha}(ln2\alpha + \frac{1}{2}) \tag{7.2.25}$$

Figure 7.2.5 shows the plots of the Klein-Nishina equations 7.2.23 to 7.2.25.

Radiation Loss of Fast Electrons: Bremsstrahlung

This topic is included here, with gamma rays, as calculations of this section will be used.

We consider a fast electron with velocity v passing a nucleus Ze, and we calculate the energy emitted by the electron in the form of photons, as it is accelerated by the field of the nucleus. We first consider the rest system of the electron in which the nucleus Ze moves with a velocity of almost c and the electric field $\bar{\varepsilon}$ of the nucleus is contracted and its magnetic field is perpendicular to $\bar{E}$. Therefore the electron sees the moving nucleus like a plane electromagnetic wave. The waves of photons representing the nucleus undergoes Compton scattering by the electron. These scattered photons, when viewed from the rest system of the nucleus i.e. in the laboratory system appear as photons emitted by the incident electron. Detailed calculations are given by several authors e.g. Heitler, which show that a spectrum of frequencies, ν, of photons is emitted and the total average energy loss per path length dx for all frequencies is given by

$$(\bar{d}E)_{rad} = -dx \int_{\nu_o}^{\nu_{max}} h\nu_o N \sum(\nu_o) d\nu_o \qquad (7.2.26)$$

where $\sum(\nu_o)d\nu_o$ is the cross-section per nucleus for scattering at frequency ν^o in the lab system. Substituting the cross-section, and integrating, we get

$$(\bar{d}E)_{rad} = 4Z^2 \frac{N \cdot E}{137} r_e^2 ln \frac{183}{Z^{1/3}} dx \qquad (7.2.27)$$

where $E = h\nu^o_{max}$ = energy of the electron, and N the number of atoms$\cdot$ cm^{-3}.

It should be noted that in radiation losses an electron loses an appreciable fraction of its energy in the form of a single quanta, so the actual energy loss may differ significantly from the average value given in 7.2.27. Since (dE/dx) by radiation is proportional to the energy E of the electron, it is a very important loss factor for high energy cosmic ray electrons.

Radiation Length

We define the radiation length l_r as

$$l_r = 1/\frac{4Z^2 \cdot N}{137} \cdot r_e^2 ln \frac{183}{Z^{1/3}} \qquad (7.2.28)$$

then

$$(\bar{d}E)_{rad} = -\frac{Edx}{l_r} \; or \; E = E_o e^{-\frac{x}{l_r}} \qquad (7.2.29)$$

For $x = l_r$, 1/e th of the energy is lost on the average by radiation.

$$l_r \approx \frac{C}{Z^2 \cdot N} \; where\ C\ is\ nearly\ constant.$$

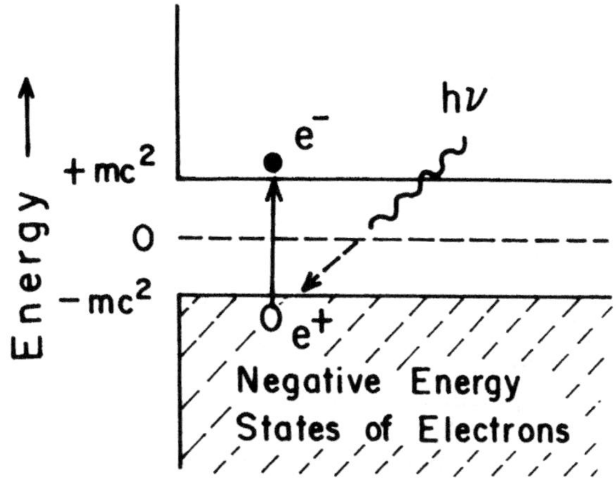

Figure 7.2.6: A schematic diagram of pair production by an energetic photon

The radiation loss per g·cm^{-2} is greater for a medium of heavy elements as $l_r \propto 1/Z\rho$. In air at NTP, the radiation length is 330 m and the thickness of the atmosphere is equivalent to 8 km of NTP air or about 20 l_r. The values of l_r for Al and Pb are 9.7 cm and 0.517 cm respectively. The energies of an electron for which $(dE/dx)_{rad} = (dE/dx)_{ion}$ in air, Al and Pb are 120, 52 and 7.0 MeV respectively.

Energy Loss of Gamma Rays by PAIR PRODUCTION

The important process by which very high energy gamma rays lose energy is by e^+ e^- pair production. Qualitatively the process is described as follows. In the relativistic theory of the electron an electron has an energy of $\pm\sqrt{p^2c^2 + m^2c^4}$, which permits negative energy states. Dirac theory states that in vacuum all negative energy states are filled up uniformly. A photon of sufficiently high energy may lift an electron in the negative energy state to the positive energy state (Figure 7.2.6). The energy threshold for the photon is 2mc^2. Physically it means that we shall see an electron of negative charge and with positive energy $\geq +$ mc^2, and also a void in the sea of negative energy states of the electron. The absence of a negative energy state appears to us as a positive energy state of an electron with a positive charge. Thus we get a pair of electron and positron produced ($h\nu \rightarrow e^+ + e^-$). Conservation of energy and momentum state that the process can not take place in vacuum, and it can only take place close to a nucleus or an electron which can take

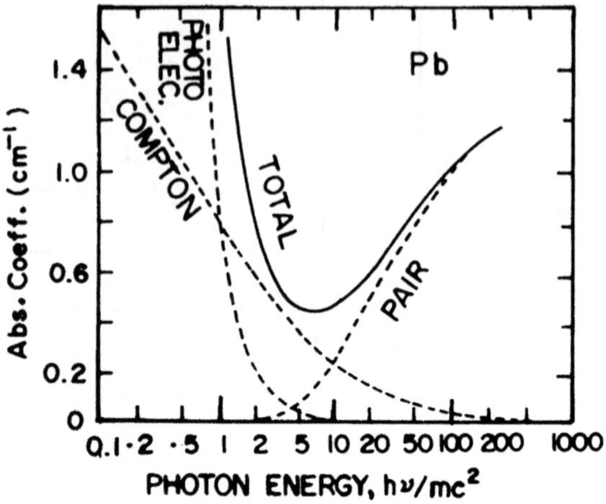

Figure 7.2.7: Absorption of gamma rays by photo electric, Compton and pair production processes

away the residual momentum.

The cross-section for pair production near a nucleus is given by the relation (Heitler):

$$\sigma_{pair} = \frac{Z^2}{137} r_e^2 \left(\frac{28}{9} ln \frac{183}{Z^{1/3}} - \frac{2}{27} \right) cm^2 \tag{7.2.30}$$

for h$\nu \gg$ mc^2.

The mean free path for pair production is defined as

$$l_p = \frac{1}{N\sigma_{pair}} = \frac{1}{\frac{28}{9}\frac{Z^2 N}{137} \cdot r_e^2 ln \frac{183}{Z^{1/3}}} \tag{7.2.31}$$

So that a decrease in the intensity of gamma rays is given by

$$I = I_o e^{-x/l_p}$$

On comparing with the radiation length of an electron as in equation 7.2.28, we get

$$l_r = \frac{7}{9} l_p \tag{7.2.32}$$

The energy loss processes for gamma rays traversing a medium are summarized in Figure 7.2.7 which is given for Pb. It is seen that the total absorption co-efficient, as the sum of the three processes, passes through a minimum value at a specific energy for a medium. In Table 7.2.2 we show the features of the total absorption coefficient in commonly used media of Al Cu and Pb.

Table 7.2.2: Properties of media for the absorption of gamma rays

Material	Energy for minimum total absorption coeff. (MeV)	Thickness ξ for 1/e reduction in intensity ($g{\cdot}cm^2$)
Aluminium	35	47 at 35 MeV
Copper	10	31 at 10 MeV
Lead	3.5	25 at 3.5 MeV

7.2.4 Electron Photon Cascades

The main features of the cascade development in air can be described as follows.

1. Photons of energy $> 10^8$ eV (mostly from π^o decays) produce electron-positron pairs, the cross-section for this process is independent of energy. Less energetic photons produce soft electrons by Compton scattering.

2. Electrons and positrons above 10^8 eV produce low and high energy photons by the bremsstrahlung process the cross-section of which is nearly independent of energy.

3. The energetic photons produce next generation of electron-positron pairs and the multiplying process continues, leading to a cascades of particle which reaches a maximum. Thus the incident energy is divided among many particles and quanta.

4. When the energies of e^+e^- become less than 10^8 eV their ionization loss rate is larger than the radiation loss rate; so they are slowly brought to rest by ionization. Thus the cascade particles gradually decrease in number.

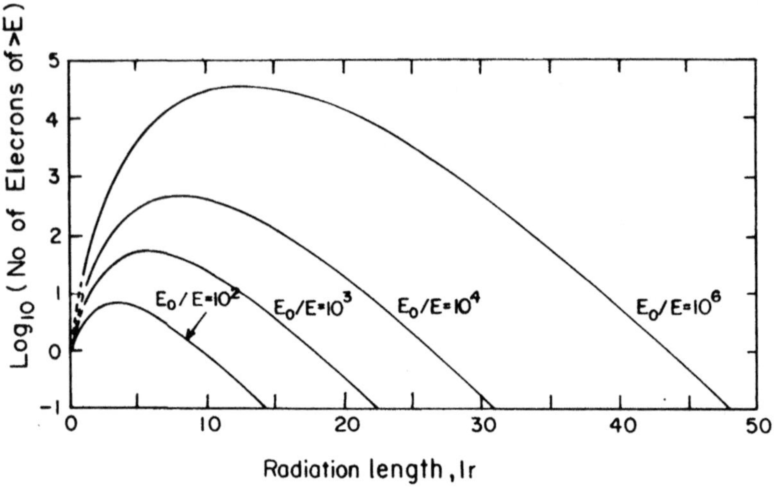

Figure 7.2.8: Log of average number of electrons with energy > E, produced by an incident electron of energy E_o, as a function of matter traversed in radiation length units

5. Because of momentum conservation, the high energy photons and electrons travel mostly in the forward direction with only small deviations from the shower axis.

As a rough approximation we note the process as follows. As mentioned earlier it is convenient to use the "radiation length" (equation 7.2.28) in an absorber as the unit of length. Let N = 2 be the number of particles (electrons and photons) to start with. This number is doubled after each traversal of radiation length and the average energy of the particles is correspondingly reduced. Thus after traversal of p radiation lengths, the initial energy E_o produces a cascade containing N particles, where $N = 2^p$ particles, of energy $E = E_o/N$. The multiplication continuous till $E \leq \epsilon_o \approx 10^8$ eV. At the maximum of the cascade development the number of particles is $N_{max} = E_o/\epsilon_o$ photons and electrons and the maximum occurs after the traversal of

$$p_o = \frac{ln(E_o/\epsilon_o)}{ln2} \quad radiation \;\; lengths.$$

The number of particles with an energy greater than $E(> \epsilon_o)$ reaches its maximum earlier at $p_o(E) = \ln(E_o/E)/\ln 2$. Beyond p_o the number of particles decreases as the energy fed into the electron photon component is lost by ionization.

The complete solution of the cascade showers involves complex processes which were first developed by Bhabha and Heitler[(5)] in 1937 (see Heitler for details) and also independently by Carlson and Oppenheimer[(6)] in 1937. The results of the above calculations are shown in Figure 7.2.8. From this we obtain the number of electrons with energy > E, produced by a primary photon of energy E_o which reaches a maximum at a depth p(E) given by

$$p(E) = 1.01\left[ln(E_o/E) - 1/2\right] \tag{7.2.33}$$

and the maximum number is given by

$$N_{max}(E) = \frac{0.137}{\left[ln(E_o/E) - 0.18)\right]^{1/2}} \cdot \frac{E_o}{E} \tag{7.2.34}$$

It may be noted that in the above original derivations certain approximations were adopted. Also it is noted that in individual cascades the number is likely to fluctuate significantly about the average value given. These large fluctuations occur due to the statistical nature of the radiation loss process and therefore initial variations are greatly magnified. These problems were treated in subsequent studies.

Cosmic Ray interactions in the atmosphere

The generic relationships between the main components of cosmic rays produced in the atmosphere by the high energy protons (and heavy nuclei) are shown in Figure 7.2.9. The soft component of the electron-photon cascade originates mainly from the decay of π^o produced in meson showers; μ^+ mesons decaying to $e^{\pm}$, and producing knock electrons, also feed to the soft component. Secondary fast nucleons produce further interactions and give rise to a nucleonic cascade. The μ-meson component arising from the decay of π mesons reaches the surface and constitutes the 'hard' component which can traverse 10 to 15 cm of Pb. High energy μ-mesons penetrate deep into Earth's surface. At the surface of the earth μ-mesons constitute about 80% and electrons about 20% of the particles.

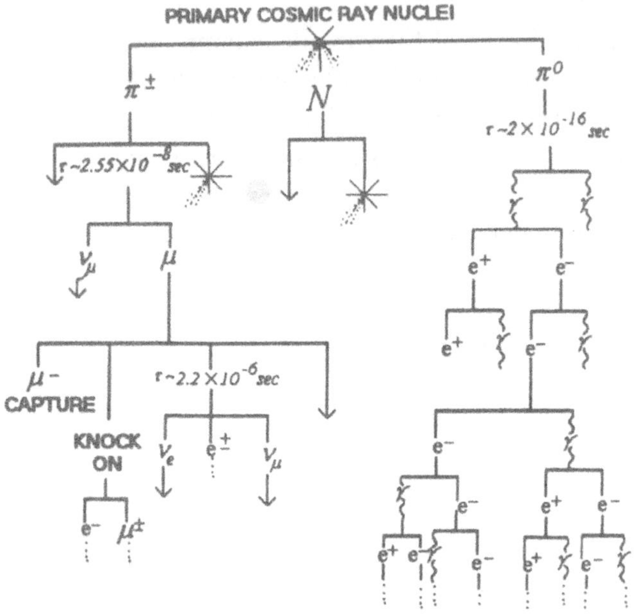

Figure 7.2.9: A schematic diagram showing generic relationships between main components of cosmic rays in the earth's atmosphere

7.3 Methods of Studies of Cosmic Ray Nuclei

7.3.1 Detectors of Low Energy Cosmic Rays ~ 10-1000 MeV

Nuclear Emulsions: In the early years of cosmic ray studies in the 1950's and 1960's nuclear emulsion stacks were widely used in balloon experiments. Energetic cosmic ray nuclei while passing through nuclear photographic emulsions suffer ionization loss and the energy released to electrons is used to dissociate the AgBr compounds into Ag^+ ions along the trajectory of the particle. The chemical processing of the nuclear emulsions reduces the Ag into metallic silver grains along the particle trajectory which are conveniently visible under optical microscopes with high magnification. The number of silver grains per unit length (i.e. grains/100 μ), called the grain density is measured which represents the ionization loss, dE/dx, of the particle. The range of the particle track in the stack can be accurately measured. Combining these and using calibration tracks, the mass M, atomic number Z and velocity v i.e. energy/nucleon are determined (equations 7.2.10 and

7.2.11). In case of heavy primaries of cosmic rays such as C, O, Fe etc almost continuous and thick tracks of silver grains are formed which are accompanied by small curved tracks of electrons, called δ-rays. The number of δ-rays per 100 μ (N_δ) is used to measure the Z of a fast particle track ($\beta \approx 1$) by the relation $N_\delta = aZ^2 + b$, where a and b are constants, determined by calibration. Nuclear emulsion techniques have been a powerful tool in the discoveries of many aspects of cosmic rays such as the elemental abundances, energy spectra and are used by many investigators.

Several refinements and new features of nuclear emulsion techniques are given in the early works e.g. of Shapiro, Stiller and their group(7) at the Naval Research Laboratory, Washington, Waddington(8) of the University of Minnesota, Minneapolis, and Biswas, Peters and collaborators(9) of TIFR, Bombay. A detailed account of the nuclear emulsion technique is given in the book by Powell, Fowler and Perkins of the University of Bristol, UK.

Solid State Nuclear Track Detectors (SSNTD)

In the 1960's and 1970's it was discovered by Fleisher, Price and Walker, then at General Electric Research Laboratory, USA that several types of dielectric substances such as Cellulose nitrate (CN), Lexan Polycarbonates etc. are efficient threshold detectors of energetic heavy nuclei of cosmic rays and other sources (General text, ref. 6). Each dielectric detector was found to have a specific threshold ionization loss rate below which no particle was recorded even if this has a high intensity. Because of these properties the SSNTD's have been widely used in balloon and satellite borne experiments for the study of cosmic ray heavy nuclei. Energetic heavy nuclei in passing through these detectors e.g. CN, Lexan, CR-39 (allyl digycol carbonate) etc produce radiation damages in the medium whereby molecular chemical bonds are broken along the trajectory of the particles. This damage trails with a diameter of about 50 to 100 Å can be seen only with electron or ultra-microscopes. To render these latent tracks easily visible, the exposed plastic materials in the form of sheets with a thickness of about 0.25 mm or so are etched in strong alkaline solutions of 6.25 normal NaOH or KOH for 10-20 hrs or more. The chemical solution strongly attacks the portions of the material where the damaged trails of the incident particle were formed and thus these are revealed as etched cones on the upper and lower surfaces of the detector sheets. The cone length (L) or track etch rate ($\Delta L/\Delta t$) is found to be a measure of $\partial E/\partial x$ of the particle. The end point of the particle where it comes to rest is also measured accurately as the range of the particle in the stack of detector sheets. Thus from the measurements of cone lengths (or track etch rates) vs residual range, the mass, atomic number and velocity (i.e. energy per nucleon) of the incident particle can be determined. Details of solid state track detectors are given in the book by Fleischer, Price and Walker.

Lexan in Skylab: Large area Lexan track detectors were used by Price et al.[10] in the Skylab mission of NASA, USA, for the study of ultra heavy nuclei of cosmic rays in 1973-74. In a small stack of Lexan exposed on the outside of Skylab for 73 days, Biswas et al.[11] of the TIFR-PRL group of India observed anomalous cosmic rays in near earth space for the first time in 1974-75.

ANURADHA in Spacelab-3: Using both active and passive elements of CR detectors in the Anuradha Cosmic ray experiments in the Space shuttle Spacelab-3 mission of NASA, Biswas et al.[12] of the TIFR-PRL-ISRO team of India, determined the arrival direction and arrival time of each of the incident anomalous and galactic heavy cosmic ray particles with a good accuracy of about 10 sec. For this purpose they used a fixed thin stack of single sheet CR-39 with a thickness of 250 μm and a diameter of 40 cm, below which a thick stack of about 40 CR-39 sheets in cylindrical form was placed. The bottom thick stack was slowly rotated in steps of 40 sec of arc once each 10 sec so that one full revolution takes place in 90 hrs. Tracks are scanned in annular strips in the fixed top sheet and in the top sheet of the bottom stack. By matching the track segments in the two sheets, the arrival time is determined. By using Spacelab's position and orientation at this time together with the track direction parameters in the stack, the arrival location, latitude and longitude and arrival direction θ, ϕ of the incident cosmic ray particles could be determined. These parameters were then used to compute the trajectories of the particles in the geomagnetic field and hence, their threshold rigidity. By combining the threshold rigidity and the momentum of the particle, the ionization state of the cosmic ray particle could be determined (Z^* = Total momentum/Rigidity; (Ch. 7.1.3)). Such measurements could be successfully made for the first time in the Anuradha experiment with facilities provided by the Space Shuttle - Spacelab-3 mission as summarized by Biswas et al.[12]. Details are given by Singh[13] and Dutta[14]. A sketch of the 'Anuradha' experiment detector assembly is shown in Figure 7.3.1.

Lexan and CR-39 in Space Missions: Several passive Lexan stacks were exposed in Cosmos satellites of the USSR[(15)] in 1985-88 for studies on anomalous cosmic rays.

LDEF: In 1985-90, large area Lexan stacks were exposed to cosmic rays in the Long Duration Exposure Facility (LDEF) of NASA, for ultra heavy Cosmic Ray ($Z > 60$) studies by the Dublin and ESA groups[(16)]. Sixteen experiment trays of LDEF contained 47 pressurized cylinders with 1 atmosphere of dry air and each cylinder contained four stacks of lexan each with a size of 20 cm x 26 cm and a thickness of ~ 4.7 g·cm^2. Thus about 15 m^2 of Lexan was exposed in the LDEF with an orbital inclination of 23° for almost 5 years. The successful exposure provided a large amount of data for $Z > 60$ nuclei of cosmic rays which was more than three times the combined data previously available.

ELECTRONIC dE/dx vs E DETECTORS

Scintillation Counters: Some transparent materials produce visible light when a charged particle passes through them, and this light can be detected with photo multiplier tubes (PMT's) to produce an electrical pulse. Commonly available plastics are widely used as scintillators and as dE/dx counters because of their low loss, easy availability in various shapes, and of their low atomic number. Crystals of NaI and CaI with some suitable impurities provide good light output as scintillators. These are relatively more expensive. Because of the high atomic number and hence large stopping power these were widely used as total energy detectors of particles stopping in these detectors.

Solid State Detectors: Semiconductors such as Si, doped with suitable elements e.g. Li, are excellent particle detectors in which the electrical pulse output is linear with the energy loss rate and therefore they are widely used. The basic idea in this type of detector system is to use a few, thin Si-detectors as dE/dx elements and these are followed by a thick solid state or scintillation crystal which measures the total kinetic energy of the particle. These dE/dx vs E elements for a counter telescope are surrounded by anti-coincidence counts so that only particles arriving the telescope from the given

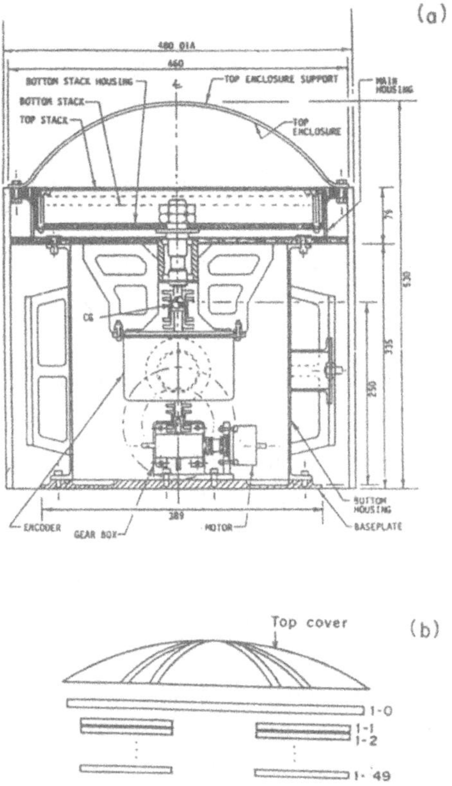

Figure 7.3.1: (a) A sketch of the ANURADHA cosmic ray instrument in Spacelab-3, (b) Schematic diagram of fixed stack (1-0) and rotating stack 1-1 to 1-49

solid angle are recorded for analysis. Using the calibration from accelerators, the dE/dx vs E signals are unfolded to determine the mass, nuclear charge and velocity i.e. energy per nucleon of the incident cosmic ray particle. Many studies with this basic method were made by a number of investigators with various improvements and modifications, such as by Simpson et al. of the University of Chicago in the IMP Satellites, McDonald and collaborators of NASA-Goddard Space Flight in the Pioneer-10 and 11 Spacecraft, Stone and team members of Caltech in the Voyager 1 and 2 Spacecraft and by many other groups. A sketch of the dE/dx vs E system of the Mass Spectrometer Telescope of the Caltech-NASA/GSFC groups flown in 1992 in the SAMPEX[34] satellite is shown in Figure 7.3.2. M1 to D2 are thin silicon detectors with increasing thickness; D3 to D7 are composed of thicker silicon wafers with anti-coincidence rings, G.

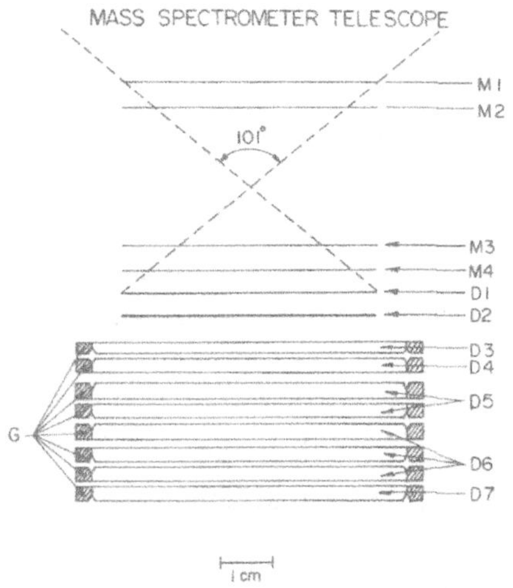

Figure 7.3.2: A sketch of dE/dx vs E system of a cosmic ray mass spectrometer telescope[34].

7.3.2 Detectors of High Energy (∼ 1-1000 GeV) Cosmic Rays

Cerenkov-Scintillation Detectors: A transparent medium having refractive index n, will emit Cerenkov radiation of optical photons when a particle of charge Ze traverses the medium with velocity, βc such that $\beta n > 1$. The light-yield is given by

$$N(C) = 0.0458Z^2\Big(\frac{1}{\lambda_2} - \frac{1}{\lambda_1}\Big)\Big(1 - \frac{1}{\beta^2 n^2}\Big) \;\; photons \cdot cm^{-1} \qquad (7.3.1)$$

of wave lengths between λ_1 and λ_2. This is maximal for $\beta \approx 1$ and zero for βn = 1. This is the reverse of dE/dx detectors. The light output is generally small, typically $\sim$ 200 Z^2 photons·cm^{-1} for Lucite, and is collected by photomultiplier tubes (PMT's). As the Cerenkov photons are emitted in the forward direction at an angle θ, so that $\cos\theta = 1/\beta$n, they provide directional information of the particle. It is seen from relation 7.3.1 that Cerenkov radiation provides important information on the velocity or energy/nucleon of the particle when $\beta \lesssim 1$/n.

For example in glass with n = 1.5, the threshold β of the particle $\beta_{th} \approx$ 1/n = 0.67 which corresponds to protons and helium nuclei with a threshold energy of 300 MeV/n. In plastics with n = 1.3, the threshold energy for P and He is about 500 MeV/n. Therefore Cerenkov detectors in combination with dE/dx plastic scintillators have been widely used in the early years in the 1950's and 1960's for the study of P, He and heavier nuclei in the energy range of $\sim$ 300 to 1000 MeV/n, as discussed in 7.2.1. In the near relativistic region many types of Cerenkov radiators with varying thresholds were developed in the 1970's and 1980's and these were successfully used for the elemental and isotopic composition study of cosmic ray nuclei by several groups e.g. Webber and collaborators(17) of the New Hampshire University.

For relativistic particles of $\sim$ 10 to 1000 GeV/n, the energy measurements are made with air or gas Cerenkov detectors which have small refractive indices and hence higher threshold energies. A gas Cerenkov detector assembly has a large size, as large path lengths are required to have a significant amount of light output, and hence they are of heavy weight. In case of gas Cerenkov detectors the Lorenz factor for the threshold value of the particle is given by:

$$\gamma_{th} = 1/\Big(1 - \beta_{th}^2\Big)^{1/2} = 1/\Big(1 - \frac{1}{n^2}\Big)^{1/2}$$

$$= \frac{n}{(n^2 - 1)^{1/2}} \simeq \frac{n}{[2(n-1)]^{1/2}}$$

$$or \;\; \gamma_{th} = \Big[2(n-1)\Big]^{-1/2} \qquad (7.3.2)$$

as n $\approx$ 1. For air at NTP, the refractive index is 1.0002918, hence γ_{th} (air) $\simeq$ 41 i.e. the kinetic energy threshold is about 38 GeV/n. Thus with suitably designed gas Cerenkov-Scintillator telescopes it is possible to measure the fluxes and energies of protons, helium and heavier nuclei in cosmic rays in the energy range of about 20 to 200 GeV/n. Several versions of Gas Cerenkov-Scintillator assemblies were flown in balloons, including double gas-Cerenkov assemblies.

An advanced version of the gas Cerenkov detector system was flown in balloons in 1991-95, by the University of Chicago group[18]. The basic elements of the instrument are as follows: (1) Plastic scintillators S_1 and S_2 on the top and the bottom form a coincidence telescope which provides the traggering of the instrument, and the measure of the nuclear charge of the incident particle from its dE/dx. The measurement of the time of flight between S_1 and S_2 allows one to remove upward moving particles. (2) The top and bottom hodoscopes which consist of four layers of multiwire proportional chambers, one pair being along the x-direction and the other in the y-direction, give the particle trajectory and also additional measurements of the ionization loss of the particle. (3) Below the top hodoscope the position-sensitive photon detectors are located. These are specially designed multiwire proportional chambers sensitive to UV radiation from Cerenkov light. These provide imaging points of Cerenkov radiation photons reflected from the spherical mirror placed at the bottom of the gas chamber so as to give the Cerenkov ring a radius which yields the Lorenz factor γ of the incident particle. (4) The instrument employed a 3m long and 1.5 m^2 base area gas chamber filled with C_4F_{10} gas which gives a threshold $\gamma = 16$. This instrument can measure the flux and energy of primaries with $\gamma \approx$ 20-200.

HEAO-3 Detectors: In the High Energy Astrophysics Observatory of NASA, the French-Danish joint experiment[19] used three solid state Cerenkov counter systems for studying cosmic ray nuclei of Be to Fe of about 1 to 20 GeV/N. In another HEAO-3 experiment the St. Louis-Caltech- Minneapolis team[20] of the USA used Ion chambers and Cerenkov counter assemblies together with x and y layers of hodoscopes to measure the elemental abundances of ultra-heavy cosmic ray nuclei of $40 \leq Z \leq 80$.

Magnet Spectrometer:

Superconducting magnet spectrometers have been developed by the groups of the USSR(21) and the USA(22) for the study of proton-antiproton ratios in the energy range 2-10 GeV. The basic features of the instrument used for balloon experiments by the US groups are: (1) A superconducting magnet coil with 8 layers of multiwire proportional counters placed above and below the magnetic field, which determine the trajectory of the particles. (2) A pair of Scintillators above and below the magnet and a thick cerenkov detector form the counter telescope, and provide the triggering system and also measure dE/dx and the nuclear charge of the incoming particle. The Cerenkov counter is used to reject upward moving particles. The magnet spectrometer studies enabled the first measurements of antiproton fluxes in the 2-5 GeV energy range in the primary cosmic rays in balloon experiments by Bogomolov et al.(21) of Leningrad, USSR in 1979 and by Golden et al.(22) of the USA.

Cosmic Ray Nuclei Detector in Spacelab-2:

The University of Chicago group exposed a massive instrument in the Spacelab-2 mission of NASA in 1985 which measured cosmic ray heavy nuclei of C to Fe in the energy range of 50-1000 GeV/N with good statistical accuracy. The main components of this instrument are as follows:

1. Two identical gas cerenkov counters C_1 and C_2 are hemispherical in shape and fit in the top and bottom portions of the instrument. The gas chambers are filled with a N_2-CO_2 mixture at one atmosphere pressure which gives a threshold Lorenz factor of $\gamma \approx 40$. This provides energy measurements in the range of $\sim$ 40-150 GeV/N.

2. Two plastic scintillators T_1 and T_2 in light integration boxes serve as the main triggering and charge measuring devices. The flight-time measurements between T_1 and T_2 ensures that only particles from the top are selected.

3. Six layers of Transition Radiation Detectors (TRD) are placed between the two scintillators. Each layer consists of a poly-olyfin fiber radiator and a multiwire proportional chamber (MWPC). TR-detectors are based on the observation that when a very energetic charged particle passes

through two media with different dielectric constants, electromagnetic radiation like soft X-ray is emitted. In the present arrangement, this is detected by MWPC and provides the primary energy from a threshold of about 500 GeV/N to the observable high energy of $\sim$ 1000 GeV/N. Below 500 GeV/N, dE/dx signals with a relativistic rise are used to measure energies in the 200-500 GeV/N region. This instrument with a geometrical factor of 5 $m^2 \cdot sr$, was flown in Spacelab-2 for 8 days and provided a significant number of events of $Z > 3$ of $E > 100$ GeV. The flux of $Z > 3$ of $E > 100$ GeV/N is about 10^3 per $m^2 \cdot sr \cdot day$.

Calorimeters for High Energy Cosmic Rays:

The measurements of the flux and energies of the cosmic ray nuclei P, He and $Z \geq 6$ were made by several groups in the 30-300 GeV/n energy region using a calorimetric method in balloon experiments. The basic apparatus used by Ormes, Yodh and collaborators[(23)] of the NASA Goddard Space Flight Center and the University of Maryland, consisted of the following elements. (1) Plastic scintillators S_1 and S_2 and a cerenkov detector constitute the charge identifying module. (2) A set of seven multiwire proportional counters (MWPC) determine the trajectory of the incident particle and its accompanying track if any. These allow to separate the incident particles from the upward moving secondaries i.e. albedo from the calorimeter. (3) An ionization calorimeter which is composed of five layers of iron absorber, each with a thickness of half a proton interaction length. The incident particle generates a nuclear electromagnetic cascade which is sampled by scintillator counters placed below each absorber. (4) A set of three layers of crossed x and y scintillator hodoscopes located in the calorimeter determines the trajectories of the cascade and identify multiple hadrons. This also provides additional energy measurements. Such an instrument with a 24 hour long balloon flight may provide proton and helium data up to about $5 \cdot 10^{12}$ eV. The instrument is a complex one and is very massive so that special balloon flights are needed.

High Energy Cosmic Ray Electron Detectors:

For the measurements of the flux and energy of primary cosmic ray electrons in the energy range of 5-200 GeV, two kinds of instruments have been employed. (A) The emulsion chamber method of the Japan-US collaboration by Nishimura et al[24], employed layers of nuclear emulsions and X-ray films which are inserted in graduated intervals in a stack of Pb plates and electromagnetic cascades are sampled. Showers with E $\geq$ 500 GeV produce spots on X-ray films which are readily visible to the naked eye and are used as templates to identify the showers on the emulsions. Cascades with lower energies are found by area scan of the emulsion plates. Relatively small area of emulsions provide sufficient numbers of events of E $\sim$ 100 GeV. Each 40x50 cm^2 of emulsions consisted of 100 μ thick Fuji ET 7B emulsion on both sides of a plastic sheet. A balloon flight of about 30 hrs at Palestine, USA provided a typical exposure factor of 2 m^2·sr·day. A total of four balloon flights in 1976 to 1980 excluding earlier flights yielded a total exposure factor of about $6 \cdot 10^5$ m^2·sr·sec. The electron flux and the energy spectrum were thus measured in the 50-1000 GeV region. (B) In the second method the measurements of cosmic ray electrons were made in balloon flight experiments by the University of Chicago group[25] using a transition radiation-cum-shower detector telescope which covers the 5-300 GeV energy range. The top plastic scintillator T1 identifies singly ionized particles. This is followed by a series of 6 polythene foam radiators and multiwire proportional counters for transition radiation detectors. Protons produce only ionization signals in MWPC's, whereas electrons because of their large Lorenz factor produce transition radiation X-rays, and their signal is higher than the ionization signal by a factor two. Lower parts of the detector consist of shower detectors of one CsI plate followed by eight plastic scintillators sandwiched between 8 Pb plates each of 2.5 radiation length thick, providing a total of 18.5 radiation lengths. Cascade developments are measured to determine the electron energy.

7.3.3 Detectors of Ultra High Energy Cosmic Rays (E > 1 TeV)

The JACEE Detector:

The direct measurement of ultra high energy cosmic rays is one of the most difficult problems of cosmic ray studies. This arises from two factors: First there is the extremely low intensity at these energies so that a very large exposure factor (area x time x solid angle) is required. Secondly there are the difficulties of reliable measurements of these high energies. To overcome these problem nine research groups of Japan and the USA had jointly undertaken the JACEE collaboration program[26], in which a large complex instrument was built and this repeatedly flown in balloon flights of long duration since 1979. This Japanese-American Co-operative Emulsion Experiment had two basic configurations, one of which follows. Balloon flights were made with one, three or four chambers, each of size 40x50 cm, and 12-25 cm thick weighing about 120 kg. The assembly is designed to optimize the tracing of particles produced in the interaction in the target. Typically, this assembly consisted of the following sections: (1) Charge determination module of 100 acrylic sheets (800 μm) double coated with nuclear emulsion layers of 100, 200, 400 μm, $\sim$ 20 sheets of CR-39 plastic track detectors; (2) The target module included 40 acrylic sheets, 20 Pb plates of 1 mm and 2.5 mm thickness, spacers and 40 X-ray films; (3) The spacer module consisted of honey-combs of light materials, and (4) The Emulsion calorimeter module consisted of Pb sheets, emulsion plates, and X-ray films. The target module had a thickness of 0.2 interaction mean free paths for protons, and the calorimeter of 0.2 proton interaction m.f.p. and 7 radiation lengths.

Interactions in the target or in the calorimeter produce cascade showers of nuclear active particles, electron-proton showers which cause visible dark spots on X-ray film when the track density is high i.e. with a threshold energy of $\sim$ 700 GeV for electromagnetic cascades. Following the mapping of detected events in X-ray films, the cascades are easily observed in adjacent emulsion plates with microscope. Up stream locations of the cascades are determined and the cascade is traced up to the interaction vertex. Above the interaction vertex the primary charge is measured, and protons, helium and C, N, O and heavier nuclei are identified in emulsions. The energy measurements are made by counting cascade electrons in nuclear emulsions and by a densitometer of dark spots on X-ray films. When the interaction takes place in the target module, the cascades are well separated in the calorimeter, and the electrons are counted within a small radius at various depths. The transition curves are then compared to the calculated ones to assign the individual cascade energy, and the total gamma ray energy $\sum E_\gamma$ is determined. Extensive simulations of nucleonic and electromagnetic cascades in the multi media of the emulsion chamber are made to complete the analysis including the factor of energy going to gamma rays. Following these the primary energy is determined.

Between 1979 and 1988, nine successful long duration balloon flights were made with as longest exposure times 120 and 150 hours in 1987 and 1988 respectively from Alice Spring. Following these, the JACEE instrument was flown in circumpolar balloon flights over Antarctica for 200 and 300 hours in 1993 and 1994 respectively. These provided the very large total exposure factor of the JACEE experiment.

7.3.4 Detection of Ultra High Energy Cosmic Rays ($E > 10^{14}$ eV) by Extensive Air Showers

Air Shower Method

Beyond $E > 10^{14}$ eV the cosmic ray flux is so small that direct measurements are not possible and the Extensive Air Shower (EAS) method is used. The development of EAS in the atmosphere is as follows. When an incident cosmic ray particle, say a proton of $E > 10^{14}$ eV interacts with an air nucleus in the upper atmosphere, it produces: (a) several fast neutrons and protons which produce further interactions in the atmosphere and contribute to successive generations of particles, (b) $\pi^{\pm}$ mesons some of which further interacts and a fraction decays and (c) $e^{\pm}$-photon cascades from the π^{o} decays. These are briefly discussed in 7.2.4 and sketch is shown in Figure 7.2.9.

Thus for experimental studies of ultra high energy at sea level and mountain altitudes, the following components are available:

(i) The soft component of $e^{\pm}$ and gamma rays, (ii) the penetrating component of muons, (iii) the hadronic component of nucleons, pions, kaons etc, (iv) Cerenkov light in the optical band by the ultra relativistic electrons, (v) radio emission from 1 to a few 100 MHz, from coherent emission by electrons.

The energy radiated in radio emission is only about 10^{-4} times less than that in Cerenkov radiation. So the latter is extensively used in air shower arrays. Observationally the number spectrum is determined from the relative frequency of showers of different sizes i.e. number of electrons and the effective areas for a 100% efficiency of detection. For example we note that in the energy range of 10^{14}-10^{17} eV, it is possible to measure electron densities at various locations and thus determine their lateral distribution using an array of about 20 scintillation detectors, each with an area of about 1 m^2 spread over a circle with a diameter of about 400 m. From these data core locations and shower sizes for individual showers are determined.

For primary energies of 10^{18}-10^{19} eV which produce very large showers it is necessary to have effective areas of several km^2. Such a large array needs a very large number of detectors so that the spacings between them are not too large. In these large arrays, water Cerkenkov counters are used as energy flow detectors. In the 1970's and 1980's, several large arrays provided many new results on the very high energy air showers and these include the Haverah Park array in England, Giant Air Shower Array in Sydney, the Large Yakutsk Array of the USSR.

The Haverah Park Air shower array had a central 500 m array of four detectors each of area 34 m^2 made up of 15 water Cerenkov detectors with a water depth of 120 cm. The outer assemblies are located at six locations on the circumference with a radius of about 2 km from the center. Each of the outer detectors had four water Cerenkov counters each with an area of 13.5 m^2. The total area of the array was 11 km^2.

The Giant Air Shower array of Sydney group located at Narrabari, Australia has 34 stations and at each station there are two scintillators each of area 6 m^2 and with a separation of 50 m, and these are buried under 1.6 m thickness of rock. The 34 stations were located in a rectangular area of 8 km x 5 km, the effective area being 34 km^2.

The Yakutsk array of the USSR had 43 stations at each of which there were a particle detector of 4 m^2 area and a Cerenkov light receiver of 176 cm^2 area. In the central region of the array there are one 12 m^2 muon detector, a 6 m^2 neutron detector, and a 2 m^2 GM counter array. These are located in a rectangular area of about 8 km x 4 km.

In recent years more complex arrays have been built and others are in progress. These include the Akino (Japan), Baskan (Russia), CASA-MIA (Chicago-Michigan) arrays and the 'Fly's Eye' High resolution unit. The latter makes coincident measurements of atmospheric fluorescence light generated by extremely high energy extensive air shower.

The above discussions provide the broad outlines of the Extensive air shower method. The main results from the EAS studies are briefly discussed in Chapter 7.4; the details are given in the General texts and references.

7.4 Properties of Cosmic Rays and their Implications

7.4.1 Anomalous Cosmic Rays (ACR)

In 1973-74 a new kind of cosmic ray was discovered in detectors flown in interplanetary space by McDonald (of NASA) et al.,[(27)] Hovestadt (of MPI, Munich) et al.[(28)] and Gacia-Munoz (of Chicago) et al.[(29)], who found in the 5-100 MeV/N energy range a new component of non-solar cosmic rays of unusual composition that was named Anomalous Cosmic Rays (ACR). The ACR particles were seen in near earth space for the first time in 1973-74 in the Skylab-3 mission of NASA by Biswas et al.[(11,30)]. Since then the ACR components have been extensively studied in interplanetary space and inside the earth's magnetosphere in the last two decades. The main properties of ACR are as follows:

Energy Spectra and Composition: The energy spectra of ACR components measured during the period close to the solar minimum of 1972-75 and of 1992-93 are shown in Figure 7.4.1(a) and (b) respectively. It is seen in (a) as given by Klecker et al.[(31)] that as Galactic Cosmic Rays (GCR) He and O nuclei decrease in flux below 100 MeV/N, there is a sharp rise of the anomalous cosmic ray flux of oxygen in decreasing energy range of 50 to 3 MeV/N; the helium fluxes show a broad maximum in $\sim$ 3-20 MeV/N. In $\sim$ 1 to 3 MeV/N there existed a quiet time solar component, which we shall not discuss here. It was found that He, O, N, Ne and Ar constituted the main components of ACR with a weak component of carbon. The ACR energy spectra in the 1992-93 solar minimum measured by Mewaldt[(32)] et al. are shown in Figure 7.4.1(b). In the following solar minimum of 1993-95, ACR spectra with C, N, O and Ne were measured inside the magnetosphere by the SAMPEX satellite in the 3-20 MeV/N energy range by Klecker et al.[(33)] and in the 15 to $\sim$ 100 MeV/N energy interval by Mewaldt et al.[(34)].

The elemental abundances of ACR measured in 1972-75 are shown in Table 7.4.1(a), (normalized to carbon = 1.0 and also to O = 100). It may be noted

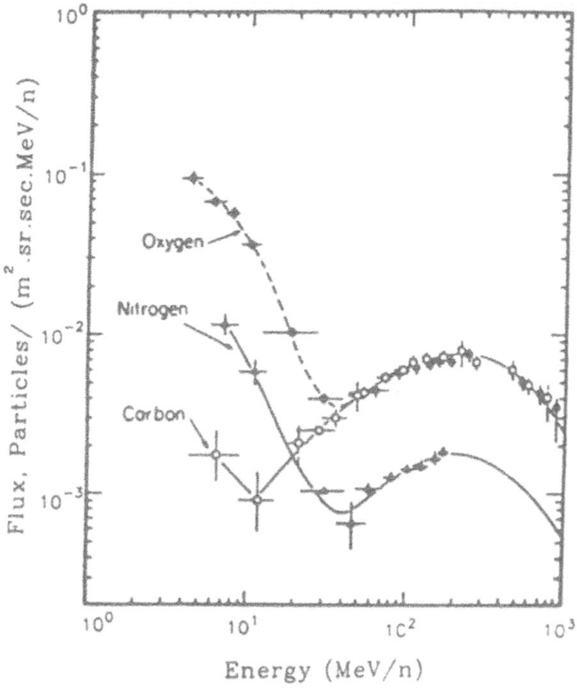

Figure 7.4.1(a): Anomalous Cosmic Ray Oxygen and carbon near the solar min in 1973-75 (Klecker et al. 1977, from ref. 11)

that as the energy spectra of the components are different, small variations in the energy range may cause a significant change in relative abundances, which is particularly applicable to carbon. Only a small number of carbon ions were detected, hence uncertainties are large. So the carbon abundance in the table should be treated as an upper limit. The statistical uncertainties are not shown for clarity and they range from 20 to 35%. It is seen that next to helium, oxygen is the most abundant element in ACR followed by nitrogen and neon.

More detailed measurements of ACR were made in the solar minimum periods of 1987[(35)] and 1992-93[(34)], and these are shown in Table 7.4.1(b). There is good agreement between the 1987 and 1992-93 data as well as with earlier data for O, N and Ne. The carbon abundance is now more reliably determined. The errors in Table 7.4.1(b) are not shown for clarity. These range from about 15% for strong elements e.g. O to 35% for weak elements

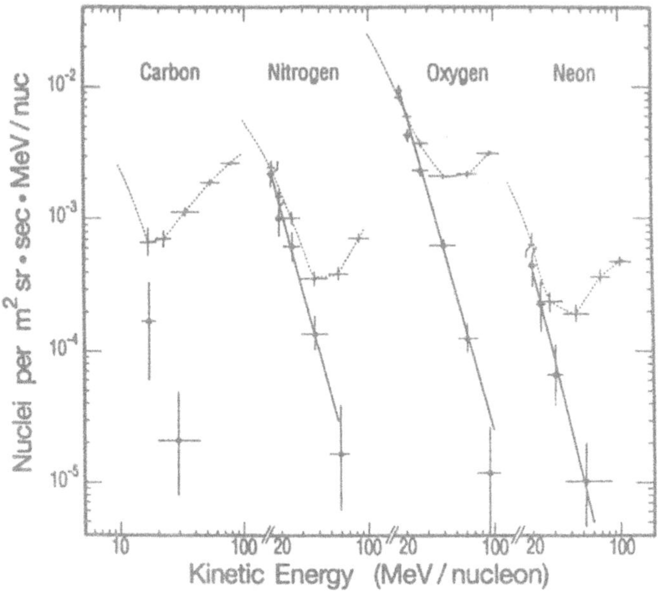

Figure 7.4.1(b): Anomalous Cosmic Ray C, N, O and Ne near the solar min. of 1992-93 as measured by Mewaldt et al. (1966) (ref.34)

like C.

The question whether there is an anomalous proton flux has been ambiguous in earlier studies. This problem was resolved with the 1994 data of Voyager-1 and it is established that there exists a small but finite flux of ACR protons(36) detectable at the minimum solar activity period and the ACR P/He flux ratio is about 40 $\pm$ 4%.

Isotopic Abundances of ACR: The isotopic composition of ACR was measured for N, O and Ne by the Voyager(37) and in SAMPEX(38) spacecraft and is: ^{15}N/N = $<$ 0.023, ^{18}O/O $<$ 0.0034 and ^{22}Ne/Ne = 0.077. The ACR ^{22}Ne/Ne ratios are in agreement with solar wind "Neon B" and SEP values of 0.076; whereas the galactic cosmic ray value, 0.40, is much higher.

Ionization States of ACR: Soon after the discovery of ACR several theories were proposed on the origin of ACR; some of these suggested unusual stellar objects as the source, e.g. by Hoyle and Clayton(39) and others(40) and these theories predicted all ionization states for ACR such as O^{+2},

Table 7.4.1: Composition of Anomalous Cosmic Rays

(a) In 1972-75:
[Relative to C = 1.0, and within bracket with the based O = 100]

E(MeV/N)	8-20 Skylab[11]	9-30 P-10[27]	11-24 IMP-8[31]
C	1.0(21)	1.0(22)	1.0(18)
N	0.97(21)	0.96(22)	1.5(27)
O	4.67(=100)	4.35(=100)	5.56(=100)
Ne	0.39(8)	0.30(7)	0.13(2)

(b) In 1987, 1992-93 and 1994:
[Relative to O = 100 for C to A]

E(MeV/N)	1987 Voyager-2[35] 16-30	1992-93 SAMPEX[34] > 17	1994 V-1[36] 10-300
P	-	-	40
He	-	-	100
C	2.0	1.4	
N	19.4	19	
O	100	100	
Ne	4.8	6	
Ar			

O^{+4}, O^{+8} etc. The other theory by Fisk et al.[41] proposed the origin of ACR from neutral atoms of the local interstellar medium accelerated in the heliosphere, whereby it was predicted that ACR ions should be singly ionized, such as, N^{+1}, O^{+1} etc. Therefore it was realized that the determination of ionization states of ACR is of crucial importance for deciding its origin. The measurements made in the interplanetary medium could not distinguish between the particles of different ionization state. The Skylab cosmic ray experiment of the Indian group[11] suggested that ACR particles must be of a partially ionized state. The Spacelab-3 cosmic ray experiment 'Anuradha' of the TIFR-PRL groups of India was designed to measure the ionization states of ACR ions by using the earth's magnetic field as a rigidity filter and this measured successfully for the first time from direct observations that ACR ions of N, O and Ne are singly ionized[42–44]. Two methods were used. In the orbit averaged flux method[45], we use the flux and energy spectrum of ACR oxygen in IPS and calculate the expected fluxes and energy

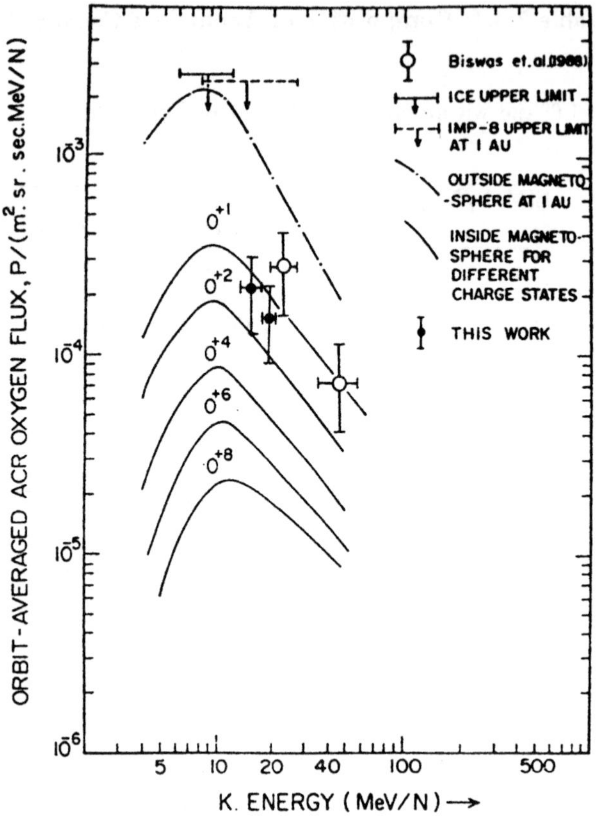

Figure 7.4.2: Orbit average flux of ACR oxygen and its ionization state (ref. 45)

spectra inside the magnetosphere at the orbit of Spacelab-3 using geomagnetic transmission factors, for different assumed ionization states, Z^*, as O^{+1}, O^{+2}, O^{+4} ... etc O^{+8}. Using measured values Z^* is determined (Figure 7.4.2). In the second method, ionization states, Z^*, of each ACR particle were measured directly using arrival time and cut-off rigidities, as described in 7.3.1. Table 7.4.2 shows the results of a sample of tracks of ACR ions of N, O, Ne. The energy per nucleon of the particles are accurately known from their ranges and their arrival time given their arrival locations and directions in space. Thus from trajectory computations the cut-off rigidities, R_c were determined by a complex series of computations details of which are given elsewhere[42,43]. Thus ionization states Z^* were determined from the relation, $Z^* = (A \cdot pc)/R_c$ where A is the mass number pc, the momentum per nucleon in GeV and R_c, the rigidity in GV (see 7.1.3). It is found by both methods that ACR N,O and Ne are singly ionized.

Table 7.4.2: Ionization States, Z* of N, O, and Ne of ACR

Sr.No.	Element	Energy Mev/N	Cut-off Rigidity(GV)	Upper limit of Z*
1	N	14.8	1.38	1
2	N	14.6	0.67	3
3	O	16.4	0.57	5
4	O	17.9	2.17	1
5	O	17.5	1.87	2
6	O	16.1	1.82	1
7	O	18.0	1.35	2
8	O	15.1	1.88	1
9	O	15.5	2.25	1
10	O	19.3	1.57	1
11	Ne	17.3	2.34	1
12	Ne	17.9	3.10	1
13	Ne	24.3	2.73	1
14	Ne	20.7	2.39	1

The method of using the earth's magnetic field as a rigidity filter as described for the Spacelab-3 experiment conducted in 1985 was used in the 1990's by the USSR and USA groups to make more detailed measurements of the ionization states of ACR. The first method of orbit average flux(45) was used in the USSR-USA collaboration experiment(46) in which the data from a number of 'COSMOS' flights inside the magnetosphere were combined with the fluxes of ACR oxygen in the interplanetary space measured by IMP-8, to determine the ionization state of ACR oxygen, $Z^* = +1$. The second method of direct measurement of Z^* by arrival location, arrival direction and the cut-off rigidity determinations was used in the Heavy Ion Large Telescope in the SAMPEX satellite by Klecker et al.(47) who found that in the 8-16 MeV/N energy range practically all ACR ions are singly ionized, and O^{2+} ions, if any, is less than 10% as shown in Figure 7.4.3. The ACR oxygen fluxes are plotted against the invariant latitude, Λ, which is determined from the relation $R_c = 0.11 + 15.5 \cos^4 \Lambda$, where R_c in GV being obtained from trajectory computations(48).

In the Mass Spectrometer Telescope experiment in SAMPEX Mewaldt et al(49) measured ACR oxygen ions in the 20-55 MeV/n energy interval in addition to the 16-20 MeV/N range, and found that for $E > 20$ MeV/N a

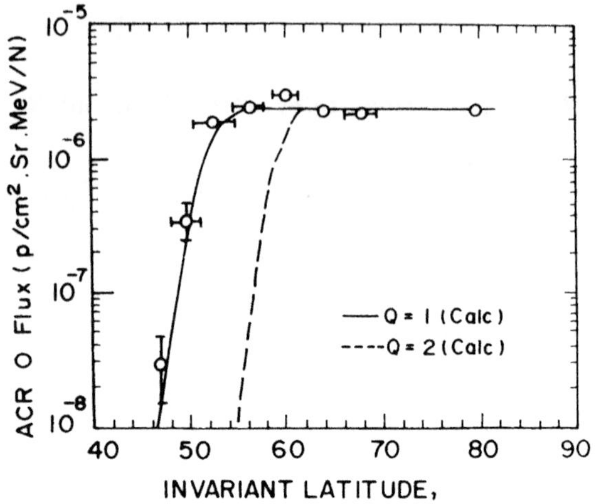

Figure 7.4.3: Ionization state of ACR oxygen from SAMPEX (ref. Klecker et al.[(47)])

number of these ions has ionization charge O^{2+} and O^{3+} and higher. This is believed to be due to a stripping process during their acceleration phase to higher energies.

Spatial Distribution: As the fluxes of ACR oxygen ions were simultaneously measured at different radial (r) and latitude (θ) positions by several satellites, it was found that these ions have strong positive radial gradients. If f_1 and f_2 are fluxes measured at the locations r_1, θ_1, and r_2, θ_2 respectively we have the equation

$$ln(f_1/f_2) = G_r(r_1 - r_2) + G_\theta(\theta_1 - \theta_2) \tag{7.4.1}$$

neglecting azimuthal variation which is small. G_r is the radial gradient and G_θ the latitudinal gradient. Using three spacecraft at different r and nearly the same latitude e.g. IMP-8, V-2 and P-10, the radial gradient G_r, was found to be as high as 14% per A.U. in the inner heliosphere at 1 to 20 AU, whereas it was about 3% per A.U. in the outer heliosphere between 20-40 A.U. by Cummings et al.[(50)]. This can be expressed as $G_r \propto 1/r^n$, where $n = 0.96$. The latitude gradient G_θ was found to be -5.6% per degree i.e. 10.8% per A.U., a value close to G_r. The sign of G_θ depends on the tilt angle of the neutral plane of the Sun and the direction of the solar magnetic field polarity. The details of variations of G_r and G_θ for ACR oxygen in different solar cycles are given by Cummings et al.[(51)].

Origin of ACR: The observations of singly ionized states of ACR N, O, Ne ions by the groups of India[42–45], Russia[46], USA[34,46] and Europe[47] established the origin of ACR from the interstellar neutrals as proposed by Fisk et al.[41] This is shown in Figure 7.4.4. The model involves the following four stages. The interstellar neutrals such as He, O, N, Ne etc are drawn into the solar system by its motion through LISM; these atoms are then photo ionized by the solar UV in the solar system; the ionized atoms are then coupled to the solar wind and are swept to the boundary region of the heliosphere where these are accelerated from 1 keV/N to $\sim$ 100 MeV/N; and finally the accelerated ions propagate into the solar system. The ACR ions thus provide us with a new sample of matter from the very local interstellar medium. Using correction factors for ionization by solar UV and acceleration efficiency the abundances of neutral atoms in the local interstellar medium were determined by Cummings et al.[51] which are given in Table 7.4.3.

Further details of ACR properties and implications are given in the reviews by Biswas et al.[44], Klecker[48] and Mewaldt et al.

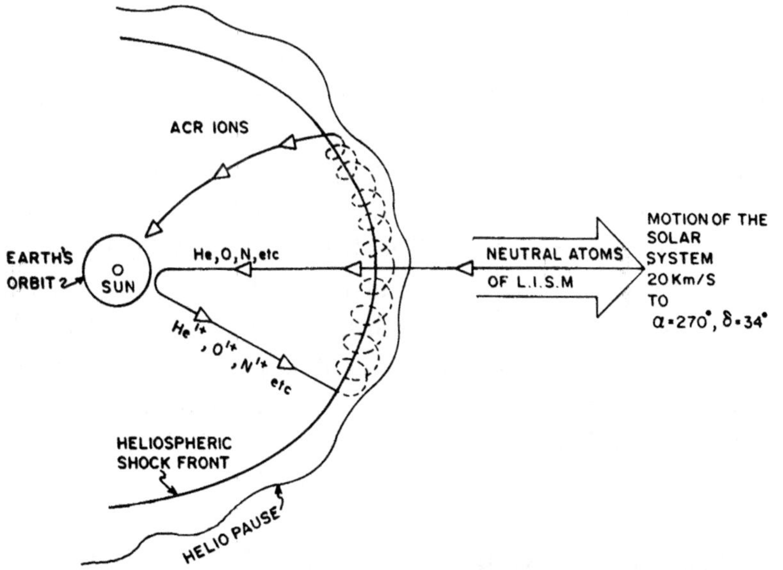

Figure 7.4.4: A sketch of the model of Fisk et al. of the origin of ACR from interstellar neutrals

Table 7.4.3: ACR Dervied Neutral Gas Abundance in LISM

Element	ACR-derived LISM abundance
He	114
C	0.5
N	25
O	125
Ne	15
A	0.6

7.4.2 Low Energy Galactic Cosmic Rays (~ 10 MeV/N-100 GeV/N)

7.4.2.1 Protons and Helium Nuclei: Energy Spectrum and Relative Abundances

The knowledge of the energy spectra of protons and helium nuclei and their relative abundances forms the basis of our understanding of their acceleration in cosmic ray sources and their propagation in the interstellar medium. These abundances have been measured in the past decades with increasing refinements and accuracies by many investigators. In Figure 7.4.5(a) and (b) are shown the differential energy spectra of P and He measured during the solar minimum of 1987 by Ormes et al.(52), together with some other data. Above 10 GV rigidity, the rigidity spectra of protons and helium nuclei are nearly the same, their power law indices being 2.74 ± 0.02 for protons and 2.68 ± 0.03 for helium nuclei. P/He ratios as function of kinetic energy per nucleon increase from about ~ 5 at 100 MeV/N to ~ 20 at 10 GeV/N and become nearly flat; this arises due to the fact that for the same energy per nucleon P and He have different rigidities below ~ 10 GeV/N. Below 1 GeV/N the fluxes of P and He vary significantly with the changing level of solar activity due to the solar modulation with an ~ 11 year period. The calculated curves in Figure 7.4.5 represent different fluxes corresponding to different modulation parameters, ϕ; "a" represents local interstellar spectra and "b" to "g" represent $\phi = 200$, 400, 500, 600, 800 and 1000 MV respectively. The meaning of the ϕ parameter is given in chapter 7.5.

7.4.2.2 Lithium to Nickel Nuclei

The differential energy spectra of some of the principal elements of GCR e.g. H, He, C, O, Fe measured near a solar minimum in the energy range ~ 0.1 to

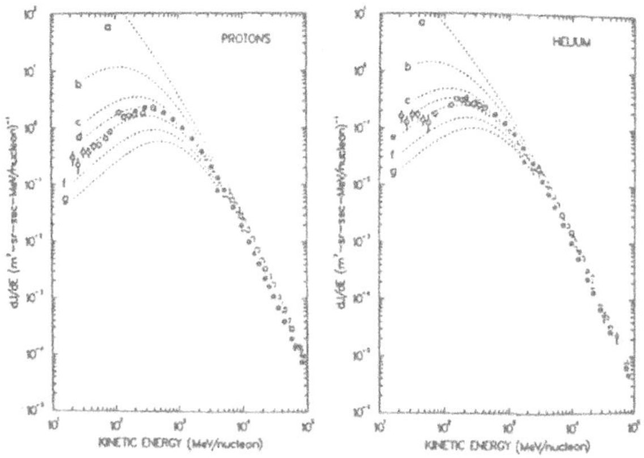

Figure 7.4.5: P and He spectra in 1987; calculated curves for various modulation levels are shown

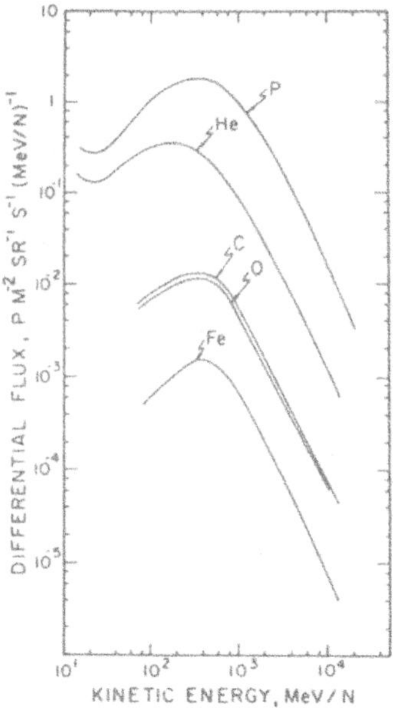

Figure 7.4.6: P, He, C, O and Fe at a solar minimum

20 GeV/N are shown in Figure 7.4.6. Below energy less than ~ 1 GeV/n the differential fluxes vary with solar activity levels similar to solar modulation parameters for He shown in Figure 7.4.5(b). As He, C, O etc have same mass to charge ratios these are modulated in a similar manner and their spectral shapes are nearly the same. For E > 1 GeV/N, the differential flux of the i-th component can be roughly expressed as a power law in total energy per nucleon, given by

$$(dJ/dE)_i = C_i/(m_o C^2 + E_i)^\gamma \tag{7.4.2}$$

where C_i is constant for the element i, and $\gamma \simeq 2.7$. The relative abundances of elements in the GCR from H to Ni are shown in Table 7.4.4. The composition of elements in the two energy ranges are essentially the same with only minor variations. Two major aspects of the composition are as follows. Firstly the major elements H, He, C, O, Ne, Mg, Si, Ca and Fe represent the elements present in the cosmic ray sources. Secondly elements like Li, Be, B, N, F, Sc to Mn are not present in cosmic ray sources and these are produced by the interactions of heavier primaries with interstellar matter, mostly hydrogen, whereby these secondaries are produced by fragmentation processes. The fragmentation cross-sections are determined from laboratory measurements for some and by semi-empirical calculations for others as discussed earlier. Using these the amount of interstellar hydrogen traversed by the GCR primaries is estimated as ~ 7 g·cm^{-2} of hydrogen. Using this mean pathlength value and the interstellar propagation model, (C. 7.6), the composition of cosmic ray primaries at the source region is determined. These source abundances are shown in column 4 of Table 7.4.4.

7.4.2.3 Sources Composition

The ratios of abundances in cosmic ray sources (CRS) to those in the solar system (compiled by Cameron), as function of atomic number are shown in Figure 7.4.7(a) and have important astrophysical significances. It is seen that from Ne up to Ni the ratios CRS/SS are close to 1, whereas there is progressive decrease in the ratios from Ne to H. Thus H, He , C and O abundances in cosmic ray sources are depleted relative to Mg-Si and Fe-Ni abundances, when normalization is done at Fe = 1. If we normalize as oxygen = 1, H and He are depleted and Mg-Si and Fe-Ni are enhanced, relative to solar system abundances. These characteristic features of cosmic ray source abundances are attributed to the preferential acceleration effects

Table 7.4.4: Relative abundances of elements in Galactic Cosmic rays and the solar system*

Z	Nucleus	Energy 100-300 MeV/N (a)	Energy >850 MeV/N (b)	Cosmic ray source (c)	Solar system (d)
1	H	11500±1700	-	$4.7x10^4$	$1.5x10^5$
2	He	4020±575	-	2430	$1.03x10^4$
3	Li	22±4	16.7±0.7	-	$2.3x10^{-4}$
4	Be	11±2	9.7±0.5	-	$3.8x10^{-6}$
5	B	30±5	30.5±0.9	-	$1.6x10^{-6}$
6	C	115±11	114.0±2.0	94±2	54.9
7	N	25±3	27.5±0.9	10±2	17.4
8	O	100	100	100	100
9	F	2.1±0.6	1.3±0.2	-	$1.1x10^{-2}$
10	Ne	15±2	17.8±0.7	15±2	16
11	Na	3.4±0.8	2.9±0.3	0.8±0.4	0.3
12	Mg	20±2	20.6±0.7	22±1	4.9
13	Al	3.9±0.7	2.6±0.3	2±1	0.4
14	Si	14±2	13.6±0.6	19±3	4.7
15	P	1.0±0.2	0.45±0.11	$0.2^{+0.4}_{-0.2}$	$4.5x10^{-2}$
16	S	2.7±0.3	3.5±0.3	2.8±0.6	2.3
17	Cl	0.4±0.2	0.56±0.12	-	$2.65x10^{-2}$
18	Ar	1.2±0.3	1.3±0.2	0.7±0.5	0.55
19	K	0.8±0.5	1.2±0.2	0.4±02	$1.95x10^{-2}$
20	Ca	2.1±0.5	2.6±0.25	2.1±0.7	0.3
21	Sc	0.4±0.2	0.4±0.11	-	$1.6x10^{-4}$
22	Ti	2.2±0.6	1.4±0.2	-	$1.29x10^{-2}$
23	V	0.9±0.5	0.5±0.12	-	$1.2x10^{-3}$
24	Cr	2.8±1.8	1.07±0.18	0.3±0.3	$5.9x10^{-2}$
25	Mn	-	0.85±0.16	$0.2^{+0.4}_{-0.2}$	$4.3x10^{-2}$
26	Fe	12.9±1.8	10.5±0.5	19±3	3.9
27	Co	-	0.39±0.10	-	$1.0x10^{-2}$
28	Ni	-	0.44±0.11	0.8±0.2	0.22

*From the review by Biswas et al. ref. 11.
(a) Cartwright et al. (1971)
(b) Webber et al. (1972)
(c) Shapiro et al. (1973)
(d) Cameron (1973)

which depend on the mass, charge and the first ionization potential of the nuclei. In Figure 7.4.4(b) we show the ratios of abundances of CRS/SS as a function of the first ionization potential (FIP) of the element. It is seen that CRS/SS ratios are in general well organized with FIP and an important role is played in FIP by the elements of GCR somewhat similar as with the solar energetic particles, but with some important differences. In the above systematics H and He, the most abundant elements do not fit in, and these elements with depleted abundances represent a separate problem of GCR sources. It may be noted that the H/He ratio of $\sim$ 15 in GCR is similar to those in normal astronomical objects.

7.4.2.4 Isotopic Abundances

H and He isotopes: Cosmic ray ^{2}H and ^{3}He isotopes are of secondary origin and therefore these are probes of the propagation of cosmic rays in the galaxy. In the 0.2 to 3.0 GeV/N energy range ^{2}H/^{1}H ratios are in the range of 0.02-0.04, which is consistent with a propagation pathlength of $\Lambda_c \approx$ 8 g·cm^{-2} (see Chapter 7.6). ^{3}He/^{4}He ratios have been measured by many groups in the energy range of 0.2-3.0 GeV/n and the ratios slowly increase from $\sim$ 0.10 to 0.15. This fits well with the standard leaky box model (see Chapter 7.6), as shown later in Figure 7.4.10.

Be and B isotopes and cosmic ray life time

As Be and B are produced by the fragmentation of heavier cosmic ray nuclei in the interstellar medium, their isotopic composition provides important information on cosmic ray propagation in the galaxy. ^{9}Be is the stable isotope and ^{10}Be is the long lived radio active isotope with a half life of 1.5·10^{6} yr. Therefore if the travel time of cosmic ray nuclei in the galaxy (T) was much less than 10^{6} yrs, very few ^{10}Be would decay and the ^{10}Be/^{9}Be ratio should be 0.54, as determined by fragmentation cross-sections. If on the other hand, the mean life time of cosmic rays, T, is $\gg 2 \cdot 10^{6}$ yrs, most of the ^{10}Be would decay and ^{10}Be/^{9}Be ratio would be small, ≤ 0.1. The measured ratios of ^{10}Be/^{9}Be by several authors are shown in Figure 7.4.8, which shows that most of the ratios lie in the range of 0.05 to 0.2. These indicate that most of the ^{10}Be have decayed and the mean life time of cosmic rays is about 10·10^{6} yrs. B/C ratios yield the mean path length, x, as $\sim$ 7 g·cm^{-2}. Thus x is the matter traversed in time T, we have

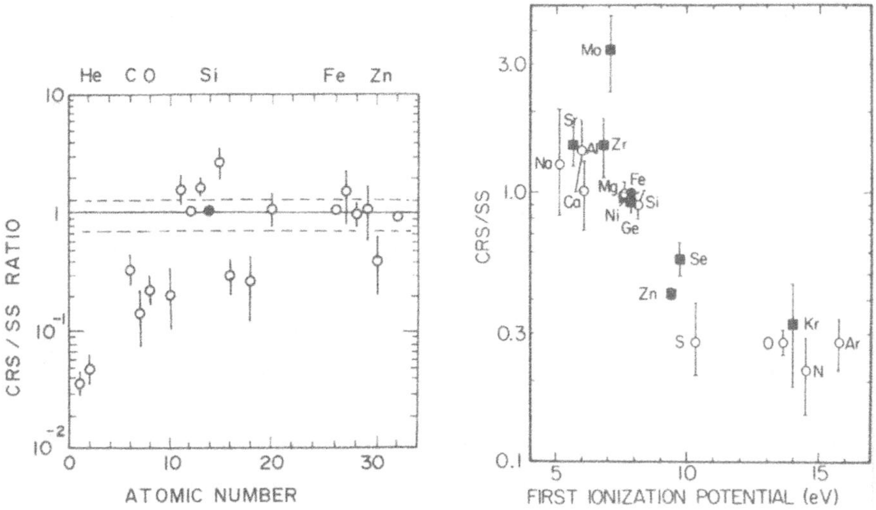

Figure 7.4.7(a): cosmic ray sources (CRS)/solar system (SS) abundances vs. Z.
Figure 7.4.7(b): CRS/SS vs First Ionization Potential

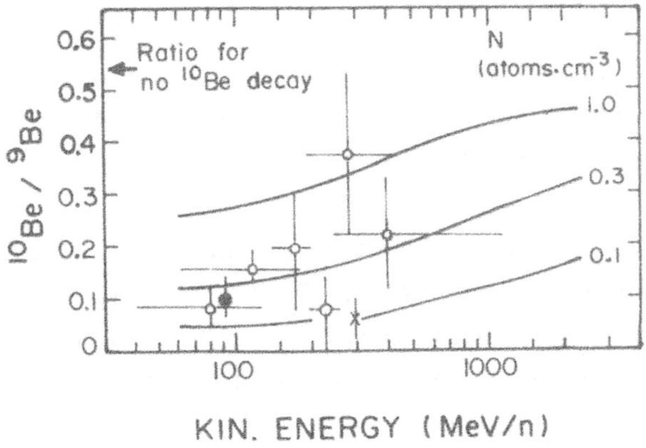

Figure 7.4.8: $^{10}Be/^{9}Be$ ratios vs kinetic energy with calculated lines assuming different densities of the ISM gas

$$x = \rho\beta cT \tag{7.4.3}$$

where ρ = mean density of matter and βc is the velocity of the particles. If we substitute $x \approx 7$ g·cm^{-2}, $T = 10^7$ yr and $\beta c \leq 3 \cdot 10^{10}$ cm·s^{-1}, ρ is obtained as ≤ 0.1 atoms·cm^{-3}. Thus the mean density of matter in the galaxy where cosmic rays spend most of their time is about 0.1 atoms·cm^{-3}.

Voyager 1 and 2 yielded isotopic composition results during 1986-88 at $\sim$ 100 MeV/N as ^{7}Li/Li = 0.46, ^{9}Be/B = 0.44, ^{10}Be/Be = 0.51 and ^{10}B/B = 0.32 (Ferrando et al.[(53)]). The ^{10}Be/B results fit with a matter density of $\sim$ 0.3 hydrogen atoms·cm^{-3} and the energy dependent mean escape path length is in agreement with the B/C results (see Chapter 7.6).

Carbon to Nickel: The isotopic composition of C to Fe measured by Voyager 1 and 2[(54,55)] and of Fe-Ni by Ulysses[(56)] are given in Table 7.4.5 for important isotopes. The salient features are as follows: (1) The measurement of a ^{13}C/^{12}C ratio of 0.05 extrapolated to the CR source after propagation in the interstellar medium gives a value of 0.001, which about 3 times lower than the solar system value of the ratio, (2) ^{15}N is entirely of secondary origin and the ^{14}N/^{16}O ratio in the source after correcting for propagation effect is small (0.05), which is about 3 times lower than solar system value of 0.12, (3) the ^{18}O/^{16}O ratio in the cosmic ray source region is 3.7 times higher than the solar system-value, (4) The most striking difference occurs for the ^{22}Ne/^{20}Ne ratio. The measured value of 0.61, after propagation effects leads to a CR source ratio of 0.45. This is 4 times higher than the solar system (solar photospheric) value of 0.12.

The large differences in the cases of the above four isotopic ratios, particularly of the large deficiency of ^{14}N and large excess of ^{22}Ne, in cosmic ray sources are puzzling and the reasons are not clearly known. The possible explanation of the large excess of ^{22}Ne could be the contribution of the He-burning Wolf-Rayet component.

The principal isotopes of Mg, Si, Fe and Ni have nearly the same abundances in cosmic ray sources as in the solar system, as seen in Table 7.4.5

These isotopic abundance measurements provide crucial tests for the

Table 7.4.5: Isotopic abundances of C to Ni in GCR[a]

	Meas. Ratio	Source Ratio	CRS/SS		Meas. Ratio	Source Ratio	CRS/SS
$^{13}C/^{12}C$	0.05	0.001	0.3	$^{54}Mn/^{55}Mn$	0.45	-	-
$^{15}N/^{14}N$	1.10	0.005	-	$^{54}Fe/^{56}Fe$	0.11	0.89	1.4
$^{14}N/^{16}O$	0.12	0.05	0.37	$^{55}Fe/^{56}Fe$	0.055	0.017	-
$^{18}O/^{16}O$	0.016	0.002	3.7	$^{57}Fe/^{56}Fe$	0.040	0.038	1.6
$^{22}Ne/^{20}Ne$	0.61	0.45	5.0	$^{58}Fe/^{56}Fe$	0.003	0.001	<0.8
$^{26}Mg/^{24}Mg$	0.23	0.02	1.26	$^{59}Ni/^{58}Ni$	0.034	0.016	-
$^{30}Si/^{28}Si$	0.09	0.03	1.40	$^{60}Ni/^{58}Ni$	0.51	0.49	1.3

(a) for details of C to Mn, see ref. 54 and 55; and of Fe-Ni ref. 56.

cosmic ray source matter.

7.4.2.5 Secondary to Primary Ratios

Secondary nuclei i.e. those nuclei which are almost absent in cosmic ray sources and in most astronomical objects, but are produced by the interaction and fragmentation of primary nuclei in the interstellar medium, act as major indicators of interstellar propagation characteristics of primary cosmic rays. Of these, three ratios have been well studied by many investigators, namely B/C, (Sc-Cr)/Fe [3] and He/^{4}He. These ratios vs energy are shown in Figures 7.4.9(a) and (b) and 7.4.10. The B/C ratios are fitted with the line corresponding to the empirical model of the 'leaky box' with exponential distribution of path lengths with parameters of the mean escape path length $\lambda_e = 14.8\ \beta$ gm·cm^{-2} for rigidities R < 4.0 GV and $\lambda_e = 33.6\ \beta\ R^{-0.6}$ gm·cm^{-2} (see Chapter 7.6). The propagation problems have been discussed by many authors and are reviewed by Garcia-Munoz et al[57] (1987). Here we note that the parameters derived from B/C ratios, give a good fit for the ^{3}He/^{4}He ratios as shown in Figure 7.4.10. However, it is found that the (Sc-Cr)/Fe ratios shown in Figure 7.4.10(b) do not fit the line derived from the parameters of the B/C ratios. For better agreement some authors use 'truncated' exponential path length distributions. Here we note that light nuclei, B, C are sensitive to long path lengths whereas sub-iron (Sc-Cr) and Fe nuclei having a smaller interaction mean free path (λ_{int} of Fe in hydrogen $= 2.5$ g·cm^{-2}) are sensitive to shorter path lengths. Thus attempts are being made[58], using new cross-section data and new energy loss computations, to obtain the propagation models which satisfy both B/C and (Sc-Cr)/Fe ratios.

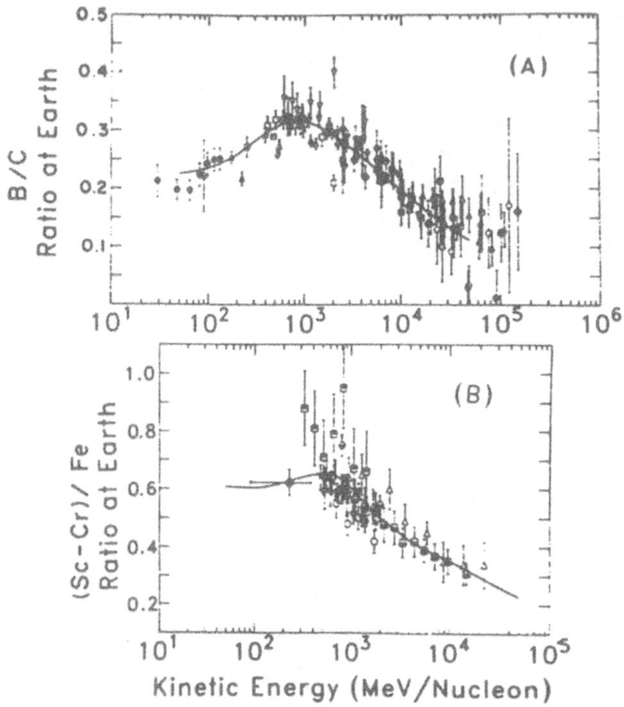

Figure 7.4.9: (A) B/C ratios vs energy; (B) (Sc-Cr)/Fe vs energy, as reviewed by Garcia-Munoz (ref. 57)

This problem is not yet fully resolved.

7.4.2.6 Radio-active Nuclei

Secondary Radio-active Nuclei provide information on the propagation time of cosmic rays in the galaxy. This includes ^{10}Be($T_{1/2} = 1.5{\cdot}10^6$ yr),^{26}Al($T_{1/2} = 7.4{\cdot}10^5$ yrs), ^{36}Cl($T_{1/2} = 3{\cdot}10^5$ yrs) and ^{50}Mn($T_{1/2} = 2{\cdot}10^6$ yr). The ^{10}Be results, acting as a clock of cosmic ray life time are discussed earlier.

K-Capturing Radioactive Nuclei are used to determine the time elapsed between nucleo-synthesis at the source and the time of its acceleration; these nuclei and their K-capture half lives are: ^{56}Ni (6.10d), ^{57}Co(270d) and ^{59}Ni($8{\cdot}10^4$ yr). The measured ^{59}Ni/^{60}Ni abundance ratio is

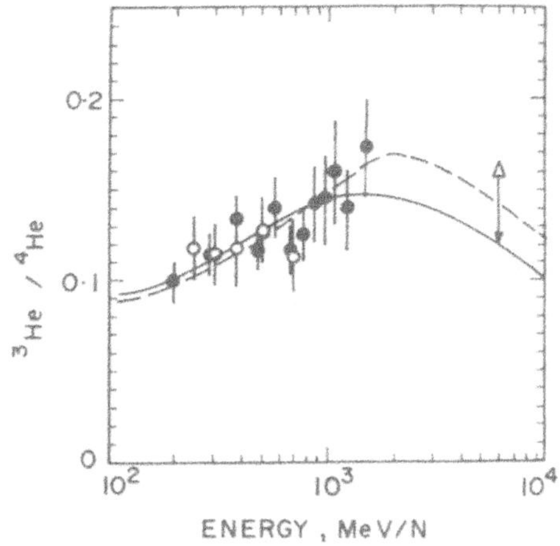

Figure 7.4.10: ^{3}He/^{4}He vs energy.

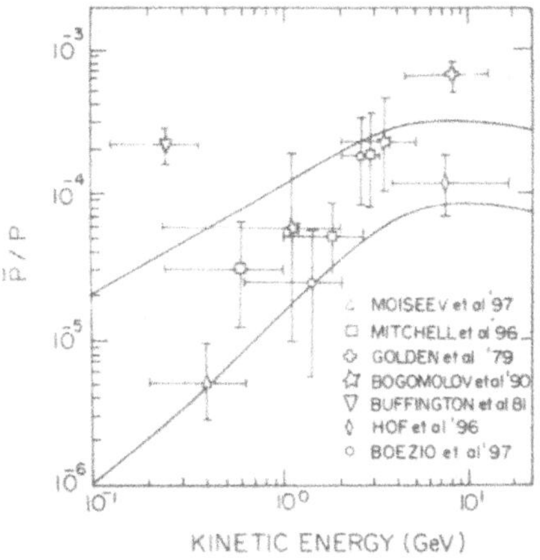

Figure 7.4.11: $\bar{p}$/p vs energy (review by Stephens(60)).

very small and negligible(57) (see Table 7.4.5) which suggests that ^{59}Ni has disappeared in cosmic ray sources, indicating a significant time delay between ^{59}Ni synthesis and acceleration.

Long-lived Radio active Nuclei such as trans-Bismuth radio nuclei e.g. Th, U, Np, Pu etc can be used to determine the time elapsed between nucleo-synthesis and the time of observation and can give major clues to their origin. Little information on these nuclei is available at present.

7.4.2.7 Anti-Particles in Cosmic Rays

Antiprotons in cosmic rays are produced from nuclear interactions of high energy cosmic ray nuclei, mostly protons with interstellar gas. As the anti-proton flux at the top of the atmosphere is rather small, one $\bar{p}$ particle for 10^3-10^4 protons, only a small number of $\bar{p}$ events were detected in the early studies in the 1979 and in the 1980's. These observations presented some major anomalies and discrepancies. Firstly it was found that the observed $\bar{p}$/p ratios in the 1-10 GeV energy range were apparently 3-5 times higher than the calculated values from the standard 'leaky box' model of cosmic ray propagation in the interstellar medium which has been confirmed with many cosmic ray studies of heavy nuclei. Secondly, one measured $\bar{p}$/p ratio at 0.3 GeV was more than 10 times higher than the calculated value(59). The validity of this low energy ratio was questioned by many observers. It was suggested that if the large excesses of $\bar{p}$/p ratios were correct, it would be of major cosmological importance in having an additional source of anti-protons. Further measurements carried out in 1992 by the NASA-CalTech groups and others in the Isotope, Anti-matter Experiment (IMAX) partially clarified the anomalies. These and earlier $\bar{p}/p$ ratios are reviewed by Stephens[(60)] and shown in Figure 7.4.11, which yield the following conclusions. Firstly the earlier high ratio of $\bar{p}/p$ at 0.3 GeV by the Berkeley group is not correct, and secondly, the revised calculated values, the limits of which are shown taking into account interstellar proton and anti-proton spectra and their solar modulation in 1992, agree fairly well with measured $\bar{p}/p$ ratios. Thus large parts of the anomalies are resolved. Further studies are needed to confirm these findings.

Search for anti-helium: An anti-helium nucleus, if it exists in cosmic rays would be detectable as a slow helium nucleus coming to rest and then producing a nuclear disintegration with large energy release, which can be easily distinguished in a nuclear emulsion detector. Several studies made yielded no anti-nuclei event, and an upper limit for the anti-He to He ratio is given as $\sim 10^{-3}$-10^{-4}.

Positrons: Cosmic ray electrons may either be produced in primary acceleration sites or can be of secondary origin from nuclear interactions of cosmic rays in the interstellar medium. The secondary component is composed of roughly equal parts of e^+ and e^-. In the 1-10 GeV range, the measured $e^+/(e^++e^-)$ ratio is $< 10\%$ indicating a dominant electron component in this region. Above 10 GeV an unexpected rise in the ratio was observed by several investigators. This excess has been attributed either to depletion of the primary source of electrons with high energies or to the appearance of a new source of $e^\pm$ pairs. A recent experiment[(61)] however does not show evidence for the increase in the ratio above 10 GeV (see Tarle et al.[(61)] and Muller et al.[(62)] for recent and earlier results and interpretations).

7.4.2.8 Cosmic Ray Electrons

Cosmic ray electrons consist of e^- and e^+ fractions. Their combined energy spectrum and flux have been measured in balloon flights and range from $\sim$ 5 to 1000 GeV. In Figure 7.4.12 a compilation of the data by Müller[(62)] is shown. It may be noted that the fluxes are multiplied by E^3, so that steepening of the spectrum beyond about 30 GeV is clearly discernible. This steepening about $\sim$ 30 GeV is generally taken as evidence for energy loss of electrons by synchrotron radiation and inverse Compton processes (see Chapter 8) in the interstellar medium. We note the following features of the cosmic ray electrons. It is assumed that the majority of the cosmic ray electrons are of primary origin and are accelerated in cosmic ray sources along with protons. It is believed that their loss rate due to synchrotron radiation becomes comparable to that due to escape from the galaxy at energies about 30 GeV, and then one deduces the confinement time in the galaxy as about 10^7 years. This model has been discussed by various authors. However, there were several well known problems[(62)] which are noted here. Firstly there were separate measurements of the positrons only up to 20 GeV; so it is of importance to measure e^- and e^+ fluxes and energy spectra separately up to the highest energies. Only recently this has been attempted. Secondly, it is necessary to derive independently the source spectrum of electrons, rather than the ad hoc assumption that it is same as that of nuclei. Thirdly, the confinement time of electrons derived as $\sim 10^7$ yrs is in agreement with the ^{10}Be data below 1 GeV/n. If an energy dependent path length is used, the agreement may not be applicable.

For these and other problems concerning cosmic ray electrons and their origin, the superconducting magnet experiments being conducted by collaborative efforts of a number of groups are expected to provide new information.

7.4.2.9 Ultra Heavy Cosmic Ray Nuclei (UH)

Ultra heavy (UH) nuclei (Z = 30-92) of cosmic rays have the special significance that the relative abundances of some groups of nuclei provide signatures for the nucleo synthesis processes in cosmic ray sources. It is well known that abundances of most of the nuclei of the elements $Z > 30$ could be explained by neutron capture on seed nuclei with either a slow time scale (of $\sim 10^3$ yrs) i.e. S-process, or a rapid time scale (of a few second) i.e. r-process. The S-process in general occurs in normal stars during advanced stages of evolution. The r-process is generally attributed to supernova explosions. The early studies of UH-nuclei in cosmic rays in balloons and in Skylab together with later results of the High Energy Astrophysics Observatory-3 (HEAO-3) experiment are given by the review by Israel[(63)]. HEAO-3 results showed that for $30 < Z < 60$ nuclei the cosmic ray source composition is similar to that

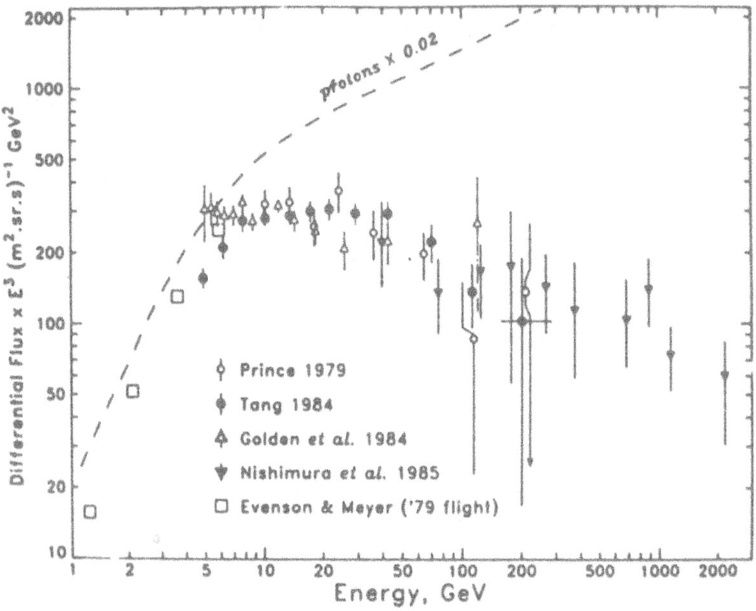

Figure 7.4.12: Cosmic Ray Electron fluxes multiplied by E^3, plotted against energy (compiled by Müller, ref. 62)

of the solar system which is a mixture of r and s process synthesis with the bias of the first ionization potential. In the Pt-Pb group $74 \leq Z \leq 87$, the Pt peak indicates the significant r-process contribution. However the definitive results are provided by the ratio, R, of Actinides to Pt-Pb groups i.e. $R = [(Z \geq 88)/(74 \leq Z \leq 87)]$, of which there was very little data in the HEAO-03 results.

The Long Duration Exposure Facility (LDEF) experiment results provided about 3000 nuclei of $Z > 65$. About 30% of the data was analyzed till 1996. The details of the experiment, calibrations and initial results of the experiment on more than 900 UH nuclei of $Z > 60$, are given by D O'Sullivan of Dublin[(64)]. Based on two types of calibrations, the ratios of actinides to Pt-Pb groups are obtained as $R = 0.016 \pm 0.004$, and $R = 0.023 \pm 0.005$. These can be compared with the ratios R of about 0.08 in freshly synthesized r-process material which after billions of years formed the solar system, and $R \sim 0.01$ of the solar composition at the time of formation of the solar system, and R $\sim$ 0.005 of the solar system at the present time (due to decay of uranium). The best results available at present indicating R $\simeq$ 0.016 to 0.023 suggests that UH cosmic ray nuclei represent a mixture of r-process material with solar system like composition. The nature of the mixing and its implications may be available with further studies.

The flux of UH-nuclei of $Z > 70$ of $E > 1$ GeV in LDEF experiment is about $\lesssim 6{\cdot}10^{-7}$ particles/m^2·sr·sec.

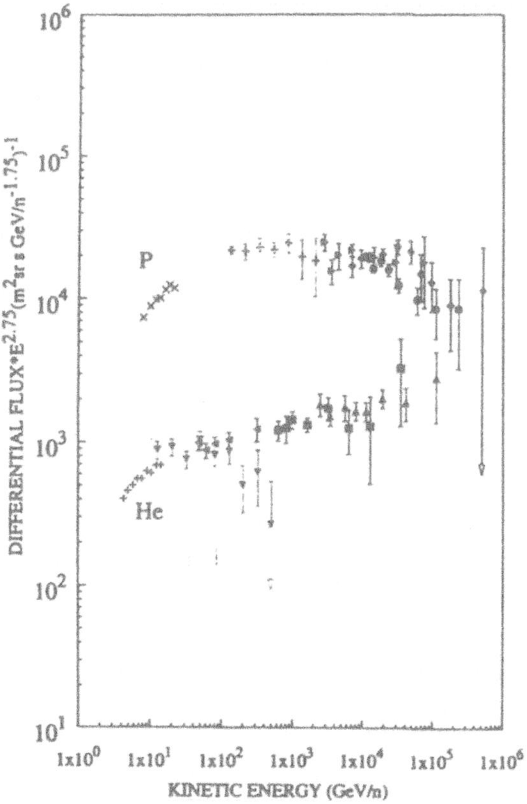

Figure 7.4.13(a): P and He spectra as compiled by Swordy (ref. 66)

7.4.3 High Energy Cosmic Rays (10^{12} - 10^{14} eV)

7.4.3.1 Protons and He: E $\sim 10^{12}$ - 10^{14} eV

The energy spectra and composition of hydrogen, helium and heavy nuclei in high energy cosmic rays of $> 10^{12}$ eV (i.e. $>$ 1 TeV) contain the signatures of their acceleration process and their escape mechanisms from the galaxy. We firstly note the main observational results of H and He. The first direct measurements of these particles were made with the 'PROTON' satellite of the USSR in early 1970 when Grigorov et al.(65), using a scintillation counter and an ionization calorimeter, measured all particle fluxes and energy spectrum in the $5{\cdot}10^{11}$ to 10^{15} eV energy range. Since then high energy proton and helium spectra have been measured by several groups, notably in the JACEE collaboration who observed that P and He differential spectra were almost the same in the 1 to 500 TeV/N energy range, the index being -2.72 $\pm$ 0.11 for Protons and -2.77 $\pm$ 0.09 for He nuclei(26). A number of new measurements were conducted in the 1990's including new studies with JACEE 7 and 8 in very long duration balloon flights in Antarctica, the results of which were compiled by Swordy(66) with some of the earlier results, as shown in Figure 7.4.13a. Here the differential fluxes have been multiplied by $E^{2.75}$ so as to reveal clearly the small changes in the spectral slopes. The results shown in the figure show two striking features: (1) Unlike previous results with almost the same spectral slopes of protons and helium nuclei, in the 50 GeV/N to $\sim 10^{14}$ eV/N energy range the He/P ratios seem to be increasing continuously. This is a new surprising result. (2) The proton spectrum has a spectral of index 2.75 in 10^2-10^3 GeV and above 10^3 GeV the slope appears to be increasing with some sign of a spectral cut-off at $\sim 10^{14}$ eV. If this observation is confirmed it would constitute a major discovery. This is due to the suggestion that supernova shock acceleration may become less effective for energies around 10^{14} eV/N.

7.4.3.2 Composition and Energy Spectra: CNO to Fe ($> 10^{12}$ eV)

Since 1985 or so a number of studies on the composition and energy spectra of cosmic ray heavy nuclei of carbon to iron of energies $> 10^{12}$ eV/N were carried out including Spacelab-2(62), JACEE 7 and 8 and SOKOL satellite experiments. Results from these compiled by Swordy(66) and presented at the 23rd ICRC, Calgary are shown in Figure 7.4.13(b). In this figure some data of E $>$ 10 GeV/N is also shown. The fluxes are multiplied by $E^{2.75}$ so as to discriminate small changes in the spectral slope. Nuclei are grouped as

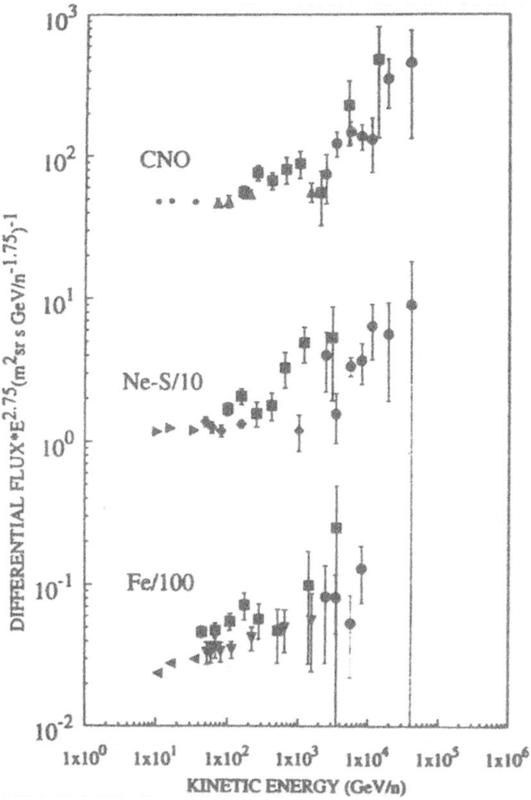

Figure 7.4.13(b): CNO, Ne-Si and Fe spectra as compiled by Swordy (ref. 66)

CNO, Ne - S and Fe. The results bring out the following main features:

1. Both the CNO and Ne-S energy spectra have a spectral slope of -2.75 up to about 100 GeV/N and in 100 to 10^4 GeV/N the spectra flatten significantly to a slope of about -2.5.

2. The Fe spectrum at high energies and its difference of slope with CNO spectrum which have been the subject of much controversy in earlier years, appears to have been resolved as follows. In the 10-100 GeV/N range the slope of the Fe spectrum is 2.5 which is flatter than that of 2.75 for CNO; and in 100-10^4 GeV/N the Fe spectral slope remains the

same i.e. 2.5, nearly the same as that of CNO.

3. The Silicon spectrum which has a slope of 2.5 in the 5-50 GeV/N energy range, which is about the same as that of CNO and Fe, becomes steeper with a slope of ~ 2.75 in the 100-1000 GeV/N energy range. The reason for this change of slope is not known.

The above new findings are to be understood in terms of the acceleration of cosmic ray nuclei and their propagation and loss from the galaxy. In particular it is yet to be understood to which extent the changes are related to propagation effects and loss from the galaxy with increasing energies or to source effects.

As the extensive air shower studies at $E > 10^{14}$ eV measure energy/particle instead of energy/nucleon, it is useful to derive the composition of incoming nuclei with energies of about 10^{14} eV/particle in order to compare these with EAS observations. Using figures 7.4.13(a) and (b), the relative abundances at a mean energy of $\sim 10^{14}$ eV/particle are estimated(66) as P:He:CNO:Ne-S:Fe as 20%:36%:19%:12%:13%. In this representation the CNO flux is almost equal to the proton flux and $Z \geq 6$ nuclei constitute 44% of the incoming particles.

7.4.4 Ultra High Energy Cosmic Rays ($E \simeq 10^{14}$-10^{21} eV)

Flux and Energy Spectrum:

The main features of the fluxes and energy spectra of ultra high energy cosmic rays are shown in the compilation by Hara et al.(67) in ICRC, 1983, and in Figure 7.4.14. Here the differential flux dN/dE has been multiplied by $E^{2.5}$ so as to reveal the changes in the slopes more clearly. The distinctive aspects of the flux and the energy spectrum are:

1. The all particle flux and energy spectrum measured by Grigorov et al(65) in the Russian 'Proton' satellite show that from $5 \cdot 10^{11}$ to 10^{15} eV, the spectrum is well represented by $E^{-2.7}$.

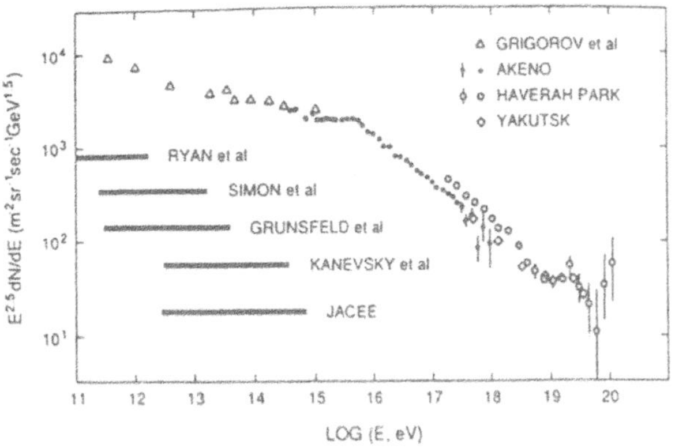

Figure 7.4.14: Fluxes and energy spectra of high energy cosmic rays, compiled by Hara et al. (ref. 67)

2. Just above $2{\cdot}10^{15}$ eV the all particle spectrum shows a sharp steepening from $E^{-2.7}$ to $E^{-3.3}$. This is known as the 'Knee' of the spectrum, which has been extensively, but inconclusively discussed in the literature.

3. The steepened spectrum continues from $\sim 10^{15}$ eV with slope 3.3 which slightly flattens to $E^{-3.0}$ at 10^{17}. Beyond 10^{17} there are more flattenings and bends which are more clearly established in recent studies as discussed below.

In the highest energy of 10^{17} to 10^{20}, data have been obtained over a decade or so with the four Giant Air shower Arrays (GASH), of Akeno of Japan, Fly Eye of the USA, Haverah Park of England and the giant array of Yakutsk of Russia. All the four spectra show general similarities including steepening, dip and then flattening. However there are some differences in the energy values of the locations of the spectral breaks which are ascribed by Teshima[(68)] as due to uncertainties in energy measurements. By normalizing the energy scales by small amounts and normalizing these to the Akeno energies, Teshima[(68)] compiled the unified energy spectra as shown in Figure 7.4.15 which indicates a good agreement between the four major groups of investigators. In this unified spectrum we note the following major features:

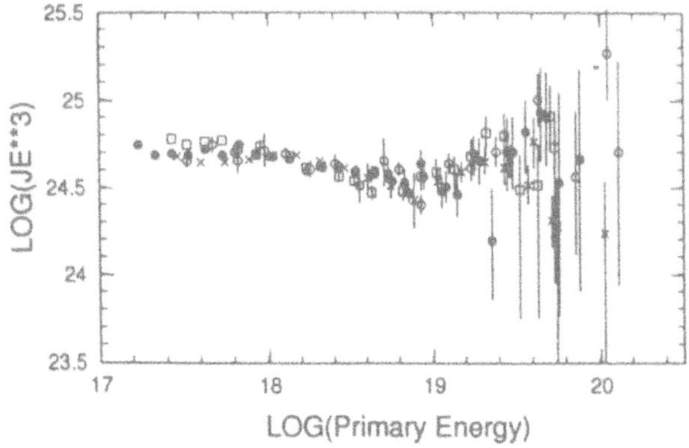

Figure 7.4.15: Unified energy spectra of ultra high energy cosmic rays, as compiled and normalized by Tereshima (ref. 68)

1. From 10^{17} to $10^{17.8}$ eV, the spectral slope is 3.02 ± 0.03, which slightly steepens to a slope of 3.20 ± 0.05 in $10^{17.8}$ to $10^{18.9}$ eV.

2. At $10^{18.9}$ eV, there is a sharp bend and the spectrum flattens significantly and the slope changes from 3.20 to 2.6 ± 0.1, for $E > 10^{18.9}$.

3. Above 10^{20} eV there are the three highest energy events till 1996. The $3{\cdot}10^{20}$ eV event of Fly's Eye, the $2{\cdot}10^{20}$ eV event of the Akenos array and the $3{\cdot}10^{20}$ eV event of the Yakutsk array. Whether these represent a cut-off or not is an open question.

Ultra High Energy Events: Combining the data of the four Giant Air Shower Arrays, till 1973, the total number of ultra high energy events of energies, $\geq 10^{19}$ eV, $\geq 10^{20}$ eV and $\geq 2{\cdot}10^{20}$ eV observed, were found to be 881, 7 and 1 respectively. By 1996 the three highest energy cosmic ray particle events with an energy $> 2{\cdot}10^{20}$ eV were observed. The event measured by Fly's Eye seems to indicate that if it is a proton it is coming from a distance further than that allowed by the interaction and loss of energy by the 2.7 K radiation by inverse Compton effects (Greisen-Zatsepin cut off). Several features of ultra high energy cosmic rays are discussed by Wolfendale[(70)], some of which are given here.

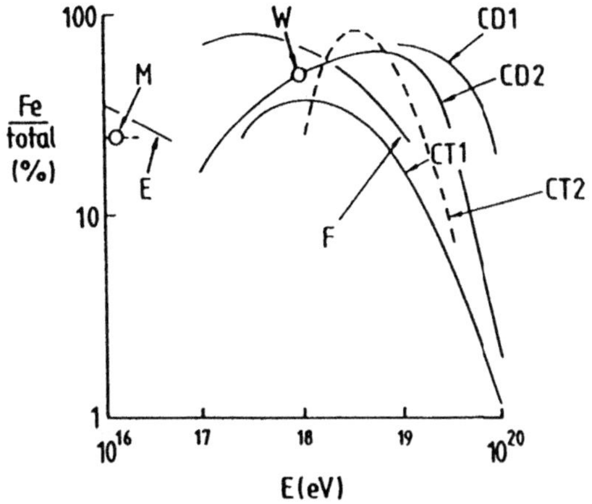

Figure 7.4.16: Estimates of the percentage of heavy nuclei ("Fe") in UHE cosmic rays from various methods, as given by Wolfendale (ref. 70)

Mass Composition:

The mass estimates of ultra high energy particles are very difficult and can be done only on a statistical basis. A compilation of results from recent studies made by Wolfendale(70) is shown in Figure 7.4.16. Here Fe denotes heavy nuclei mainly iron. The results from various techniques show that: (1) at 10^{15}-10^{16} eV there is a mixed composition of protons and heavy primaries; (2) the mixed composition changes in the 10^{17}-10^{19} eV range to enhanced heavy primaries mainly of Fe; this predominance of "Fe" primaries is supported by research groups, but with minor variations; (3) above 10^{19} eV and up to $3{\cdot}10^{20}$ eV, the mass composition strikingly changes to exclusively proton primaries. As discussed later (in Section 7.7.3) these 10^{19}-$3{\cdot}10^{20}$ eV particles are ascribed to an extra galactic origin.

The other features such the anisotropies of ultra high energy particles are examined in Section 7.7.3 in relation to their source distribution.

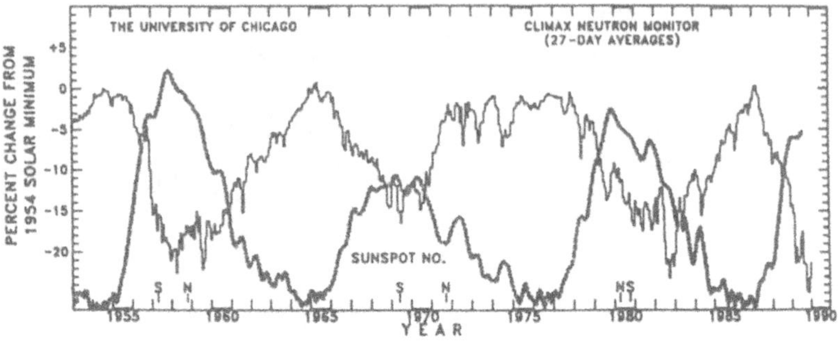

Figure 7.5.1: 27 day average of the Climax neutron monitor intensity and the smoothed sunspot numbers (adopted from Simpson, ref. 71)

7.5 Cosmic Ray Propagation in Interplanetary Space

7.5.1 General Features

The propagation of cosmic rays in interplanetary space is strongly influenced and controlled by the solar wind plasma and the interplanetary magnetic field, specially for cosmic ray particles of energies less than about 10 GeV/N. The several types of variation which occur fall in two main classes:

(A) The anisotropic or Spatial Variations: These include the periodic variations of diurnal and semi-diurnal intensity changes. The diurnal (24 hr) variations are due to the small anisotropy of the cosmic ray intensity in certain directions in space. A cosmic ray detector on the surface of the earth records particles coming from a particular direction and it samples slightly different intensities in some directions during the 24 hr. rotation of the earth. The amplitude of the diurnal variation is about 0.1-0.5% and the phase is at a maximum at about 14 hrs. local time and it is at a minimum at $\sim$ 4 hrs. The general features of the diurnal variations can be understood in terms of the spiral structure of the interplanetary field at the orbit of the earth and the flow pattern of cosmic rays. The details of these and other types of variations are discussed by several authors, e.g. by Dorman of Moscow State University.

(B) The second class of variations include periodic and isotropic temporal variations such as the 27 day and 11 year solar cycle variations of cosmic ray intensities. The 27 day variation of cosmic ray intensity is related to the 27 day rotation period of the Sun and to the passage of high speed solar wind streams in the vicinity of the earth. The 11 year solar cycle variation is of major interest and we shall briefly discuss its main features and its origin.

7.5.2 11 Year Solar Cycle Variations of Cosmic Ray Intensities

The 11 year variation of cosmic ray intensities has been studied by many investigators over several solar cycles using neutron monitors at different locations on the earth at sea level and at mountain altitudes. As an example we show in Figure 7.5.1, the data of the 27-day average of the cosmic ray intensity measured by the Neutron Monitor of the University of Chicago at Climax, Colorado, USA at an altitude of $\sim$ 4300 m, for the period of 1954 to 1990, covering nearly four solar cycles[(71)]. The smoothed sunspot number is also plotted showing the inverse correlation of cosmic ray intensities with the sunspot number. We also see the presence of a lag in the cosmic ray maximum relative to the time of solar minimum, which is known as the 'hysteresis' effect.

The studies of solar modulation of the flux and energy spectra of cosmic ray protons, helium and heavy nuclei have been carried out by many investigators using detectors in balloons, satellites and space vehicles. Proton and helium spectra at different levels of solar modulations are shown in Figure 7.4.5. We shall discuss some of the basic ideas of the theoretical models of the 11 year solar modulation of cosmic ray intensities. The understanding of the solar modulation process can lead us to the determination of the cosmic ray energy spectrum at the local interstellar space and also the relative abundances of elements and isotopes outside the heliosphere.

7.5.2.1 Solar Wind Modulation of Cosmic Rays

Diffusion-Convection Model of Parker: When cosmic particles of density ρ (particles/cm^3) penetrate into the solar system, the outward sweep of the interplanetary magnetic field carried in by the solar wind streaming with velocity $\vec{v}$, tends to remove the particles from the solar system. This can be considered as a convection process and the convected flux is $\vec{v}\rho$. If K is the diffusion coefficient, then the flux of particles diffusing inwards will be K $\vec{\nabla}\rho$. So the flux at any point in space is given by $\vec{F}$ = -K $\vec{\nabla}\rho + \vec{v}\cdot\rho$. From elementary diffusion theory the net rate of change of particle density is $\frac{\partial\rho}{\partial t}$ and this is given by - $\vec{\nabla}\cdot\vec{F}$. Therefore

$$\frac{\partial\rho}{\partial t} = \vec{\nabla}\cdot(K\cdot\vec{\nabla}\rho) - \vec{\nabla}\cdot(\vec{v}\rho) \tag{7.5.1}$$

Let us assume steady state conditions so that $\partial\rho/\partial t = 0$. We also assume that the particle density is spherically symmetric around the Sun. Let the diffusion coefficient K be assumed isotropic and the radial velocity v of the solar wind be uniform up to a distance R (i.e. K and v are independent of radial distance). Beyond R the diffusion coefficient is infinite and the solar wind vanishes, i.e. R is the boundary of the modulating region. With these conditions, equation 7.5.1 can be integrated to yield the density at any point r in space as

$$\rho(r) = \rho_o exp\Big\{-\frac{v(R-r)}{K}\Big\} \tag{7.5.2}$$

where ρ_o is the cosmic ray density beyond R. Thus the cosmic ray intensity N(E, r, t)dE is given by

$$N(E,R,t)dE = N_\infty(E)exp - \Big\{\frac{v(R-r)}{K}\Big\}dE \tag{7.5.3}$$

Knowing v, R-r and K, one can estimate the change of intensity. v is taken as $\sim$ 400 km$\cdot$s^{-1} and R is assumed to be $\sim$ 100 A.U., K is determined from the spectral index α of the power spectrum of irregularities in the interplanetary magnetic field; for cosmic ray particles of velocity v(= βc), and rigidity P, K is given by K = cβ $P^{2-\alpha}$, where $\alpha = 1$ for P> P_o and $\alpha \simeq 2$ for P < P_o. Thus the modulation factor in equation 7.5.2 is given by

$$exp\Big\{-\frac{v(R-r)}{c\beta P}\Big\} \quad for \quad P > P_o \tag{7.5.4}$$

or

$$exp\Big\{-\frac{v(R-r)}{c\beta P_o}\Big\} \quad for \quad P \leq P_o \tag{7.5.5}$$

So denoting $\frac{v(R-r)}{c}$ as η, the modulation effect is represented by

$$N(E,r,t)dE = N_\infty exp\Big\{-\frac{\eta(t)}{P\beta}\Big\}dE \tag{7.5.6}$$

7.5.2.2 Solar Modulation with Energy Loss

As the solar wind undergoes radial divergence along with the interplanetary magnetic field, cosmic ray particles which are convected outward, suffer energy loss due to adiabatic deceleration. So all the three effects - diffusion, convection and adiabatic deceleration are taken into account in the generalized Fokker-Planck equation as given by Parker[72] (1965):

$$\frac{v}{r^2} \cdot \frac{\partial}{\partial r}(r^2 \rho) - \frac{2v}{3r}\frac{\partial}{\partial E}(\alpha E \rho) - \frac{1}{r^2}\frac{\partial}{\partial r}\left(r^2 K \frac{\partial \rho}{\partial r}\right) = 0 \qquad (7.5.7)$$

where ρ(r,E) is the steady state number density of particles of energy E and $\alpha = (E + 2E_o)/(E + E_o)$. It is seen that when energy loss effects given by the second term are neglected the above equation reduces to the diffusion-convection equation 7.5.1. The analytical solution of equation 7.5.7 for the general case is very difficult and so some modifications are made to obtain the solution to the above equation.

The force-field solution of equation 7.5.7 was given by Gleeson and Axford[73] in 1967, of which one obtained an expression, corresponding to equation 7.5.7, for the current density of the streaming particles, S(r,E) as

$$S = v\rho - K \cdot \frac{\partial \rho}{\partial r} - \frac{v}{3} \cdot \frac{\partial}{\partial E}(\alpha E \rho) \qquad (7.5.8)$$

When the modulation is small, the streaming can be neglected and equation 7.5.8 can be solved analytically for S = 0, assuming that K is a separable function of position and rigidity. The solution is given by Gleeson and Axford[73] in 1968, as

$$N(r, W, E) = \frac{N(r \cdot W + \phi)(W^2 - \phi^2)}{(W + \phi)^2 - E_o^2} \qquad (7.5.9)$$

where W is the total energy of the particle whose rest energy is E_o. As a result of this approximation the expression for solar modulation given by equation 7.5.9 is similar to that obtained by assuming modulation by an electric field of potential ϕ. The examples of modulation P and He are given in figures 7.4.5 a,b.

Further details of solar modulation including the drift are given by Jokipii,[74] 1989.

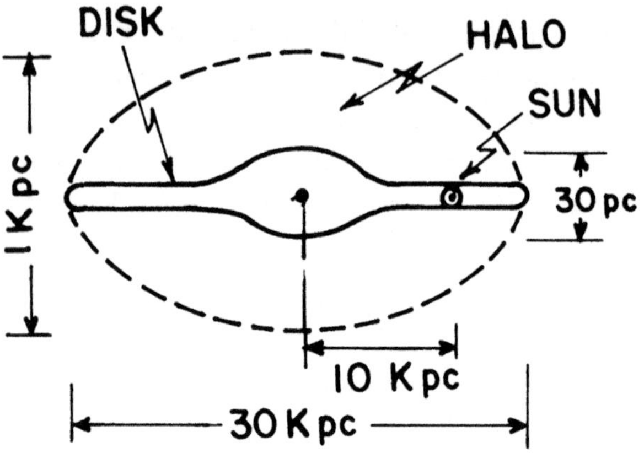

Figure 7.6.1: A schematic diagram of the disc and halo of the galaxy

7.6 Cosmic Ray Propagation in Interstellar Space

7.6.1 General Features and Models

Cosmic ray particles of energy $< 10^{15}$ eV accelerated in the sources in the galaxy are injected into the interstellar medium where they are confined for a long time until they are lost by nuclear interactions with interstellar gas or they escape from the galaxy. The confinement volume includes the galactic disc and the halo which surrounds the disc of the galaxy (Figure 7.6.1). The size of the halo is problematical and still uncertain, as some authors suggest a small halo of thickness < 800 pc, while others advocate a large halo of thickness ~ 3 kpc. This may be compared with the diameter of ~ 30 kpc of the disc of the galaxy and its thickness of ~ 30 pc. The average density of gas in the disc of the galaxy is ~ 1 atom/cm^3 and in the halo it is uncertain (it may be 10^{-2} or less). The average magnetic field in the disc is $\sim 3{\cdot}10^{-6}$ gauss and is about the same in the halo. In the interstellar magnetic field the gyroradius, ρ, of a proton of rigidity R in GV is given by $\rho = 10^{12}$ R_{GV} cm. So a 10 GV proton has $\rho \sim 1$ AU.

As the Sun is located near the periphery at ~ 10 kpc from the galactic center, the observed isotropy of cosmic rays is sought to be explained by the long confinement and propagation time of cosmic ray nuclei in the galaxy.

The models of cosmic ray propagation are mainly as follows:

1. **Leaky Box Model:** In this model (LB), cosmic rays are trapped within reflecting boundaries surrounding the galaxy, but they have a finite probability of leaking out of the galaxy into the intergalactic medium. A further modification of the model is called the Nested Leaky Box (NLB) model in which particles are first trapped in the source region and then leak out from the source to ISM.

2. **Closed Galaxy Model:** In this model cosmic rays can not leave the galaxy and they are confined within the galaxy until they lose their energy by nuclear interactions with the interstellar gas particles.

3. **Diffusion Model:** In this picture cosmic rays leave the galaxy by diffusing through magnetic field irregularities.

In models (1) and (2) the density of particles is uniform throughout the confinement volume. We shall mainly discuss model (1) here.

Leaky Box Model

In this simplified model, neglecting energy loss, the loss of cosmic ray particles by nuclear interactions and by escape are balanced by the production in the source and by fragmentation of heavier nuclei, so that an equilibrium state is present. This state is given by

$$\frac{dJ_i}{dX} = 0 = -\frac{J_i}{\lambda_e} - \frac{J_i\sigma_i^t \cdot N}{\bar{A}} + \sum \frac{N\sigma_{ij}J_i}{\bar{A}} + Q_i \qquad (7.6.1)$$

where J_i = differential flux of the nuclei of species, i; λ_e = the escape mean free path; σ_i^t = total interaction cross-section of species i; σ_{ij} = cross-section for the production of species i, from fragmentation of species j; Q_i = source flux, N = Avogadro number and $\bar{A}$ = mean atomic weight of interstellar gas (95% H and 5% He). The first two terms of the equation 7.6.1, represent the losses by escape and by nuclear interactions, and the third and fourth represent the production terms. This set of equations, one for each species, can be solved analytically so as to derive source abundances from the measured data of the composition for E > 1 GeV/N. The set of source abundances shown in Table 7.4.4 for E $\geq$ 1 GeV/N are derived from the above model.

For a cosmic ray nuclei energy < 1 GeV/N, the energy loss term of interstellar propagation is to be included. This effect together with radio active decay of unstable nuclei are taken into account in the computation process. Here we start with a long set of abundances J_i(O,E) at the source of species i, of energy E(MeV/n), and compute the change ΔJ_i for traversal of $\Delta x (= 0.1\ \mathrm{g \cdot cm^{-2}})$ of matter in ISM, arising from each of the following terms. Such iterations are made over a large number of increments in path length till the mean path length traversals are achieved:

$$\frac{\partial J_i}{\partial x} = \frac{\partial}{\partial E}\Big(\frac{dE}{dX}J_i\Big) - \frac{N\sigma_i^t J_i}{\bar{A}} - \frac{J_i}{\rho\beta C \cdot \gamma T_i} + \sum_{j \neq i} \frac{N\sigma_{ij} J_i}{\bar{A}} + \sum_{j \neq i} \frac{J_i}{\rho\beta C \cdot T_{ij}} \quad (7.6.2)$$

where matter traversed in time T is given by

$$x = \rho \cdot \beta C \cdot T \quad (7.6.3)$$

ρ being the interstellar gas density, βC the velocity of the particle and T is the rest life time and γ, the Lorenz factor.

The first term on R.H.S. represents the ionization energy loss factor; the second term the loss by nuclear interactions; the third term the loss by radio active decay; the fourth term represents gain by the fragmentation of species j to i, and the last term the production of species i by the radio active decay of species, j. For stable nuclei only the 1st, 2nd and 4th terms are relevant. Thus the local interstellar spectrum can be derived.

Path Length Distributions (PLD)

Cosmic ray nuclei arrive at the heliosphere from their source after traversing a large variety of path lengths in the galaxy. Thus the equilibrium spectrum or the local interstellar flux of species i, F_i is obtained by averaging the results of equation 7.6.1 at different values of x having path length distribution P(x), as

$$F_i = \int_o^\infty J_i(x) P(x) dx \quad (7.6.4)$$

Early models of confinement assumed a constant "slab" of material traversed by cosmic rays of about 6 g·cm^{-2}. As measurements of secondary to primary ratios, e.g. B/C ratio improved and extended over larger energies, it became clear that the cosmic ray path length distribution is a rather complex phenomenon. Firstly experimental studies led to the exponential distribution of path lengths given by (Figure 7.6.2a)

$$P(x) = \frac{1}{x_o} exp(-x/x_o) \tag{7.6.5}$$

x_o being the mean path length. Secondly the mean path length was found to be decreasing with increasing rigidity of the particle as in an inverse power law, i.e.

$$x_o \propto R^{-0.6} \tag{7.6.6}$$

above rigidity $\sim$ 5 GV. For lower rigidities the mean path length decreases with decreasing energies, as $x_o \sim E^2$. The set of parameters which fitted the B/C ratios in the entire energy range yielded mean path length $x_o = 8$ g·cm^{-2} and mean matter density $\sim$ 0.2 atoms·cm^{-3} and life time T $\approx 20 \cdot 10^6$ years using ^{10}Be/Be data.

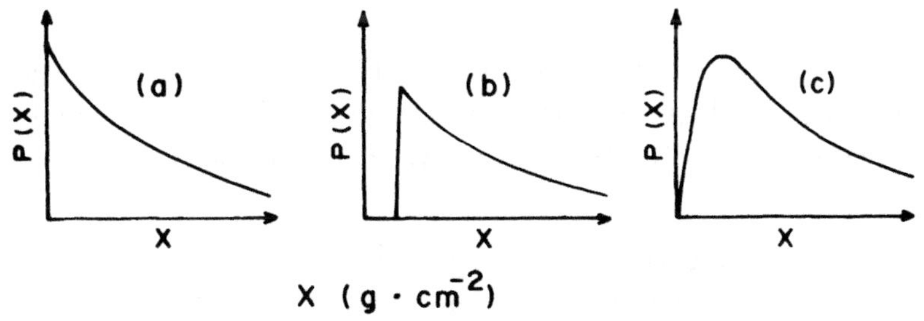

Figure 7.6.2: Sketches of three types of pathlength distributions (see text)

Truncated Path Length Distributions (PLD)

The results of Sub-Fe/Fe i.e (Sc-Cr)/Fe ratios show that while the data at high energies is in good agreement with the calculated curve using pure exponential PLD and the parameters used for fitting B/C ratios, the agreement becomes progressively worse towards lower energies, indicating a need for more fragmentations of iron at low energies. This can be achieved by reducing the number of short path lengths in PLD, like truncation with zero short path lengths (Figure 7.6.2b), or by a gradual decrease in the short path lengths as in Figure 7.6.2c. This would cause more iron to interact and produce fragments. This process of truncation produced no significant effect in the B/C ratios because of the long interaction mean free path of carbon so that it is not affected by the cut off at short path lengths in PLD.

Thus two forms of truncated exponential PLD have been quantitatively studied; (1) the zero short path length in PLD, and (2) the double exponential in PLD as in Figure 7.6.2c. Here we show the calculated results using truncation with zero short path length in exponential PLD, which is given by

$$0 \;\; for \;\; x < x_c, \quad p(x) = (1/x_o) \cdot exp(-x/x_o) \; for \; x > x_c \qquad (7.6.7)$$

where P(x) is the normalized probability distribution, and x_c and x_o are functions of energy which are determined from data.

In Figure 7.4.9 are shown the resulting B/C ratios and Sub-Fe (Sc-Cr)/Fe ratios both fitted with the same parameters of truncated zero short paths in PLD as in equation 7.6.7, and reasonable agreements are achieved. The physical meaning of equation 7.6.7 can be interpreted as follows. Cosmic ray particles may pass through a slab of matter before propagating in the galaxy with exponential path length distribution. This slab can be visualized as a shell of matter surrounding the cosmic ray source region.

The double exponential path length distribution corresponds to the "nested leaky box" model of Cowsik and Wilson[75] in 1973.

Details of the LB model of propagation of cosmic rays are given by several authors, e.g. Shapiro and Silberberg[76], 1974, Garcia-Munoz et al[77], 1987. The diffusion model is discussed by Ginzburg et al[78] 1980, and the closed galaxy model by Peters and Westergaard[79], 1975.

7.7 Origin of Cosmic Rays

7.7.1 Early Concepts and General Features

Ever since the discovery of very high energy cosmic rays, the problem of the origin and the acceleration of cosmic rays has been one of the major challenges in high energy astrophysics, and a large number of physicists proposed different models. In the past decades as our knowledge of astrophysical phenomena progressed it was realized that the presence of high energy particles is a common feature in many regions of space. For example, it was discovered that solar flares are associated with the generation of high energy particles; the envelopes of supernova and their remnants contain a large amount of relativistic electrons which produce synchrotron radiation and the non-thermal radio-emission from the galaxy originates from high energy electrons spiraling in the galactic magnetic field. On the larger scale of the universe large amounts of energies are generated in radio galaxies, in quasi-stellar objects and in active galactic nuclei. So some of these processes, it is thought, are associated with the origin of cosmic rays.

On looking back we note that Baade and Zwicky in 1934 first suggested that high energy cosmic rays may originate from discrete events of supernova explosions. In 1949 Fermi proposed a radically different, diffuse origin whereby cosmic rays are slowly accelerated by random collisions with magnetized clouds in the interstellar medium during millions of years. However it was found that the energies of magnetized clouds were not sufficient to produce the cosmic ray energy spectrum. So the origin came back to supernova remnants such as, the Crab nebula which was found to contain strong magnetic irregularities which accelerate particles to high energies by the Fermi-type mechanism. These ideas were developed extensively by Ginzburg and Syrovatsky(80) in 1953 onward. But still many problems remained. To account for some of these problems, it was proposed in the past decades, in 1975-78 by Axford et al., Bell and Ostriker and others that acceleration of cosmic rays take place firstly in the turbulent magnetic field inside the supernova remnant by a second order Fermi process and secondly, the injected particles, mixed with interstellar gas are further accelerated in the shock front of the remnants. Other variations have been proposed. In order to investigate the problem of the origin of cosmic rays in a systematic manner, we divide the problem in two parts. First, we discuss the basic acceleration processes which may be applicable, and secondly, the likely objects where conditions are favorable for such acceleration mechanisms.

7.7.2 Basic Acceleration Mechanisms

7.7.2.1 Betatron Process

This process is understood by considering the adiabatic invariant of motion of a charged particle in a slowly varying magnetic field, as given by

$$\frac{p_1^2}{H} = \frac{p^2 \cdot \sin^2 \alpha}{H} = Const. \tag{7.7.1}$$

where P_1 is the component of the particle momentum perpendicular to the magnetic field H. In a time varying magnetic field, the particle gains energy in the increasing phase of the field H, and it would slow down if H returns to the initial value. So there is no net acceleration. The energy gain would occur if by some mechanism, like a gradient in H, the particle is removed from the region with an increasing field before it begins to decrease. The energy gain, E, of the particle in time, Δt, is given by

$$E = \mu \frac{\partial H}{\partial t} \cdot \Delta t \tag{7.7.2}$$

where μ is the magnetic moment (Chapter 2).

7.7.2.2 Fermi Acceleration of the First Order

This mechanism of cosmic ray acceleration proposed by Fermi[(81)] in 1954 is based on a fairly regular magnetic field bounded by regions with stronger fields. This is likely to occur in the spiral arm of the galaxy. The charged particles, spiraling in the magnetic field with pitch angle, θ will be reflected back on approaching the stronger field and thus may be trapped between two regions of high field, traveling back and forth between the magnetic mirrors (Figure 7.7.1). If the magnetic mirrors i.e. the jaws in the field are approaching one another, the particles will gain energy whereas in the receding trap, deceleration of energy will occur. In the energy gain phase, the component of the particle momentum parallel to the field, p_{11}, will gradually increase so that its pitch angle will decrease resulting in the escape of the particle from the trap. Irregularities in the magnetic field may bring back the pitch angle of the particle to a large value and the particle may be trapped again in another trapping region and the process may be repeated. As the average number of collisions with approaching magnetic mirrors is larger than those from receding mirrors, there is a net gain of energy for the particle. In this process when the magnetic mirror is moving along the axis of the magnetic field with velocity V_m, the energy gain, ΔE, of the particle on reflection from an approaching mirror is given by

$$\Delta E = 2\gamma^2 V_m (v_{11} + V_m) \cdot E/C^2 \tag{7.7.3}$$

where E is the total energy of the particle, v_{11} the parallel component of velocity of the particle and γ, the Lorenz factor [$= 1/(1 - V_m/C^2)^{1/2}$]. The equation 7.7.3 is derived by the relativistic velocity addition theorem.

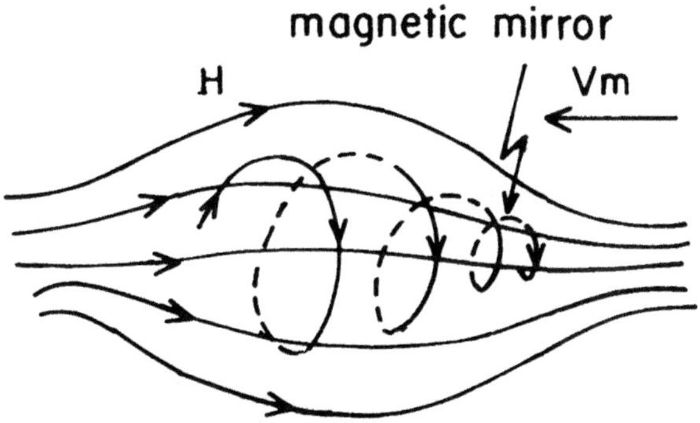

Figure 7.7.1: A sketch of the collision of a particle with moving magnetic mirrors

7.7.2.3 Statistical Process of Fermi Acceleration

The mechanism of acceleration of cosmic ray particles by collisions with magnetized clouds was proposed by Fermi[(82)] in 1949. In this process it is assumed that plasma clouds with frozen in magnetic field moving in a random direction with velocity, V_m, collide with cosmic ray particles and exchange energy. Although Fermi developed this theory in relation to the motion of cosmic ray particles in interstellar space, the process is a general one and is applicable in other regions in space, such as in supernova remnants.

Let us consider a cosmic ray particle that moving with velocity $V(= \beta C)$ collides with a magnetic cloud moving with velocity $V_m (= \beta_m C)$, at an angle

θ (Figure 7.7.2). Then if W is the total energy of the particle and p its momentum, its total energy in the rest frame of the cloud W*, is given by

$$W^* = \frac{W - \beta_m p \cos\theta}{\sqrt{1-\beta_m^2}}$$

or

$$W^* = W \cdot \frac{1 - \beta_m \beta \cos\theta}{\sqrt{1-\beta_m^2}} \tag{7.7.4}$$

as $p/W = (m_o \beta\gamma)/(m_o\gamma) = \beta$.

After the collision the energy in the frame of the cloud is the same as before because in this frame, the magnetic field of the cloud is a static field. The momentum is distributed at random and has an average value of zero. Transforming back in the frame of the galaxy, the average energy W' after collision is

$$W' = \frac{W^*}{\sqrt{1-\beta_m^2}} = W \cdot \frac{1 - \beta_m \beta \cos\theta}{(1-\beta_m^2)} \tag{7.7.5}$$

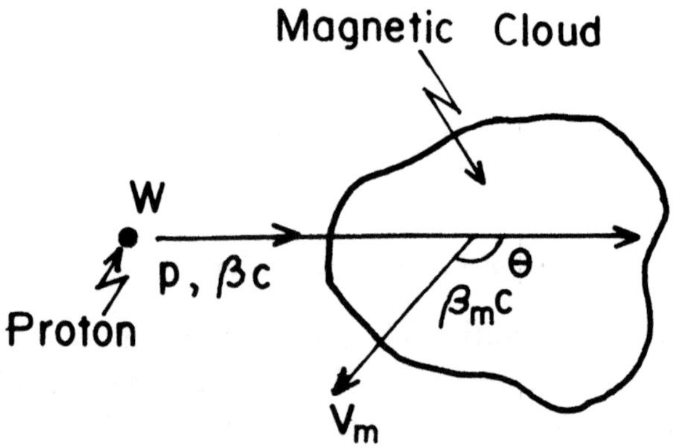

Figure 7.7.2: A sketch of the collision of a particle with a moving magnetized cloud

The probability of a particle to collide with a cloud whose velocity $\vec{\beta}_m$ forms an angle between θ and $\theta+d\theta$ with that of the particle is proportional to the number of clouds and the relative velocity. Thus the probability with appropriate normalization for $\beta_m \ll \beta$, is

$$\frac{1}{2}\sin\theta\Big(1-\frac{\beta_m}{\beta}\cos\theta\Big)d\theta$$

By computing the average over all possible angles, we get

$$< W' >_{av} = \frac{1}{2}\frac{W}{(1-\beta_m^2)}\cdot\int_o^\pi (1-\beta_m\beta\cos\theta)\Big(1-\frac{\beta_m}{\beta}\cos\theta\Big)\sin\theta d\theta$$

$$= \frac{1+\frac{1}{3}\beta_m^2}{1-\beta_m^2}\cdot W$$

$$\simeq \Big(1+\frac{4}{3}\beta_m^2\Big)\cdot W \qquad (7.7.6)$$

Therefore the average energy gain per collision, α, is

$$\alpha = \frac{\Delta W}{W} = \frac{< W' >_{av} - W}{W}, \quad and \;\; \Delta W = \alpha W,$$

Therefore

$$\alpha \simeq \frac{4}{3}\beta_m^2 \qquad (7.7.7)$$

From relation 7.7.7, it is seen that there is no limit to the energy gain, as long as the particle is present in the accelerating region, and its energy increases exponentially. A particle with initial total energy W_o, will have after a time t, an energy $W(t) = W_o\, e^{\alpha t/\tau}$, where τ is the mean collision time. If N is the number of collisions, we have $t = N\tau$.

Therefore

$$W = W_o e^{\alpha N}, \quad N = \frac{1}{\alpha}ln\frac{W}{W_o}t = \frac{\tau}{\alpha}ln\frac{W}{W_o} \qquad (7.7.8)$$

If T is the mean life time of the particle determined by nuclear collisions and leakage from the galaxy, as given by

$$\frac{1}{T} = \frac{1}{T_c} + \frac{1}{T_l} \qquad (7.7.9)$$

the number of accelerated particles after time t is given by

$$N(> t) = N_o e^{-t/T}$$

Therefore

$$N(>W) = N_o e^{-t/T} = N_o exp\Big(-\frac{\tau}{\alpha T} ln\frac{W}{W_o}\Big)$$

Therefore

$$ln\frac{N(>W)}{N_o} = -\frac{\tau}{\alpha T}\cdot ln\Big(\frac{W}{W_o}\Big)$$

Therefore

$$N(>W) = N_o\Big(\frac{W}{W_o}\Big)^{-\tau/\alpha T}$$

Therefore

$$N(>W) = Const.\cdot W^{-\tau/\alpha T} = K\cdot W^{-\gamma} \tag{7.7.10}$$

where

$$\gamma = \tau/\alpha T = 1/(\bar{a}\cdot T) \tag{7.7.11}$$

$\bar{a}$ being the energy gain per sec.

Thus the exponent of the integral energy spectrum of cosmic rays, γ, is related to the rate of energy gain by relations 7.7.10 and 7.7.11. Knowing the value of γ and T the rate of energy gain required can be calculated.

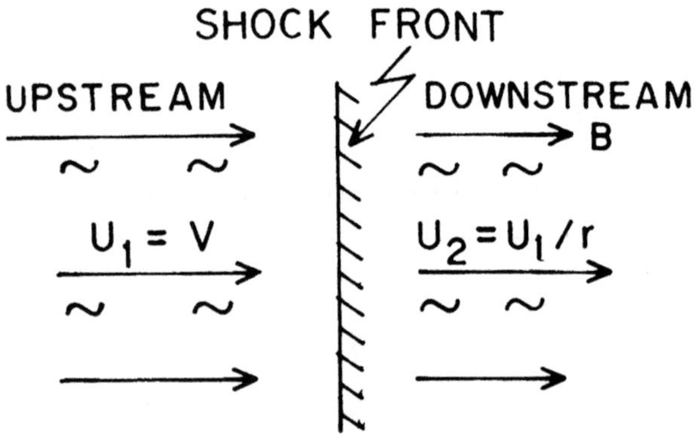

Figure 7.7.3: A sketch of shock front acceleration

7.7.2.4 Shockwave Acceleration

This mechanism formulated[83,84] in 1977 has found many applications. We consider a strong shock propagating with a velocity V in the direction of the magnetic field, $\vec{B}$, called parallel shock in a scattering medium. We assume that V $\gg$ V_A where V_A is the Alfven velocity [$V_A = B/(4\pi\rho)^{1/2}$], where ρ is the density of the plasma particles. In the frame of the shock, the gas flows with velocity u_1 = V. At the shock, the gas is compressed by a factor r, so that the velocity down stream, is U_2 = V/r, relative to the shock (Figure 7.7.3). We assume the presence of scattering centers in the magnetic field so that cosmic rays diffuse on both sides of the shock. The scattering centers cause the cosmic ray particles to reflect back and forth across the shock a large number of times. Every passage through the shock is equivalent to a collision with a magnetic wall moving with velocity V = u_1-u_2 = V(1-1/r). When we average over incident angles, the mean energy gain per passage is obtained as

$$\Delta E = (4/3)\frac{V}{c}(1 - 1/r)E \qquad (7.7.12)$$

This is similar to equation 7.7.7. We consider particles of energy E_o entering the shock and continuously gain energy during repeated passages. The escape of a particle from the region is determined as follows. As the diffusion mean free path of cosmic rays is much larger than the width of the shock the cosmic ray density is continuous across the shock. The cosmic ray distribution at the shock is isotropic in first order in V/c. So, the flux of cosmic rays crossing the shock at a given time is given by (1/4) N_o v, N_o being the cosmic ray density at the shock, and V the velocity of the cosmic rays. The number of particles escaping from the system is derived from the number swept back down stream and is therefore given by $N_o u_2$. Thus the escape probability per cycle is

$$p_e = \frac{n_o u_2}{n_o v/4} = \frac{4u_2}{v} \qquad (7.7.13)$$

After m passages through the shock, the cosmic ray energy is

$$E_q = E_o exp\left\{\frac{4}{3}m\frac{V}{c}(1 - 1/r)\right\} \qquad (7.7.14)$$

The probability of completing m or more passages is

$$P_m = (1 - p)^m \qquad (7.7.15)$$

From 7.7.13 and 7.7.14, we get

$$N(E)dE \alpha E^{-\mu} dE, \quad \mu = (2 + r)/(r - 1) \qquad (7.7.16)$$

For strong shocks, r = 4, so $\mu = 2$.
For weaker shocks, the spectral slope is steeper.

The remarkable property of this shock acceleration mechanism is that the slope of the power law spectrum of cosmic ray depends only on the shock and not on the diffusion coefficient, or the dimension of the scattering region.

If cosmic rays extract large amounts of energy from strong shocks of r = 4, generating an E^{-2} spectrum, the cosmic ray pressure may broaden the shock, making it less effective as the shock width becomes larger than the mean free path of the particle. It is also noted that the process is a slow one. So in case of actual shocks with a finite life time, the theory predicts a high energy cut off. In the case of supernova shocks, the upper limit of cosmic ray energy is about 10^{15} eV.

Other acceleration processes:

Other mechanisms of cosmic ray acceleration include (a) hydrodynamic shocks produced immediately after a supernova explosion, and (b) acceleration by continuum trains of hydro-magnetic waves moving in opposite directions as proposed by Parker.

7.7.3 Possible Sources of Cosmic Rays

It is generally believed by many investigators that cosmic rays observed in the vicinity of the earth arise from at least three different sites, where a number of processes generate the cosmic rays in high, very high and ultra high energy bands. We discuss the basic features of these possible sites and of the sources of cosmic rays.

7.7.3.1 Cosmic Ray Sources for E $\lesssim 10^{15}$ eV

A. Supernova and Supernova Remnants:

The energy range of 100 MeV/N to $\sim$ 1000 GeV/N containing the bulk of cosmic rays, the energy spectra of P, He and heavy nuclei are nearly the same as their life time, T, in the galaxy is determined mainly by the escape mean free path (equation 7.7.9) in the Leaky box model or by the diffusion coefficient in the diffusion model. These features probably extend to the energy range of 100 TeV/N i.e. 10^{14} eV. The supernova (SN) and supernova remnant (SNR) model of the origin of cosmic rays in this energy band have been extensively discussed by Ginzberg[85] and his collaborators and others and we shall discuss some of the basic ideas. The main features of cosmic rays of energy ≥ 0.1 GeV/N and of the galaxy with halo model[85] are given in Table 7.7.1. It is seen that the intensity of cosmic rays incident in the vicinity of the earth corresponds to an energy density of $\sim 10^{-12}$ erg/cm^3 i.e. ~ 1 eV/cm^3. This is about the same as the energy density of star light.

Next we estimate the source power required, as follows. We assume, following Ginzberg, that cosmic rays are confined to a halo of the galaxy at the height of $\sim$ 10-15 kpc (Figure 7.6.1). The presence of the halo of our galaxy has been problematical and opinions differ about the size of the halo. Some authors e.g. Ginzberg et al. suggest a fairly large size halo while some others prefer weak halo. On comparing with other galaxies, it is generally believed that our galaxy has a halo, but its size is uncertain. Using the volume of the halo as $\sim 10^{68}$ cm^3, the total energy of cosmic rays is obtained as 10^{54} erg. Using a cosmic ray life time of 10^8 sec as obtained from the diffusion model, the power of cosmic ray energy density in the galaxy is estimated as $3{\cdot}10^{40}$ erg/s. In a supernova explosion the kinetic energy released is about 10^{49}-10^{52} erg, and SN explosions occur in the galaxy at the rate of once in $\sim$ 30 yrs or so. Therefore the power of energy release of supernova explosions is 10^{40}-10^{43} erg/s. Thus from an energetic point of view supernova can provide the necessary power for cosmic ray production.

In the 'Leaky Box' model with the mean path length ~ 7 g$\cdot$cm^{-2} (from B/C ratios), the mean life time of cosmic rays in the galaxy is obtained as $\sim 2{\cdot}10^7$ yr (from ^{10}Be radio active isotope measurements); thus the mean density of the confinement volume of cosmic rays is found to be ~ 0.2 hydrogen atoms$\cdot$cm^{-3}. So the confinement volume of the halo is estimated to have a thickness of ~ 500 pc and the total storage volume as $\sim 10^{67}$ cm^3. With cosmic ray energy density, $W_{CR} \approx 10^{-12}$ erg$\cdot$cm^{-3} the total C.R. energy becomes $\sim 10^{55}$ erg. With a mean cosmic ray life time of $\sim 2{\cdot}10^7$ yr, the required power of cosmic ray sources is determined as $W_{CR}/T_{CR} \approx 2{\cdot}10^{40}$ erg/s. This is the same as derived from the diffusion model as given in Table

Table 7.7.1: Cosmic Rays and Galaxy with Halo Model

Components	Parameters
Cosmic Rays	
Total intensity, I_{CR}	$\simeq 0.25$ particles/(cm^2·sr·s)
	$\approx 0.25 x 4\pi$ p/(cm^2·s)
	$\simeq 3$ p/(cm^2·s)
Mean energy E_{aV}	$\simeq 6$ GeV/n
Total energy flux, F_{CR}	$\simeq 3x6x10^9x1.6x10^{-12}$ erg/(cm^2·s)
	$\simeq 2.9x10^{-2}$ erg/(cm^2·s)
Energy density, W_{CR}	$\simeq F_{CR}/C = 2.9x10^{-2}/3x10^{10}$ erg/cm^3
	$\simeq 10^{-12}$ erg/cm^3 = 1 eV/cm^3
Energy density of electrons E_e	$\simeq 10^{-14}$ erg/cm^3
Galaxy with Halo model	
Size of cosmic ray halo	10-15 kpc = (3-5)x10^{22} cm*
Size of the disk of the galaxy	~ 30 kpc $\simeq 10^{23}$ cm
Volume of disk and halo V_h	$\sim 10^{68}$ cm^{3*}
Total energy of cosmic rays, W_{CR}	$W_{CR} \cdot V_h \simeq 10^{56}$ erg
Total energy of electrons W_e	$\approx 10^{54}$ erg
Lifetime of cosmic rays, T_{CR}	$\simeq 10^8$ yrs $\approx 3x10^{15}$ s
Power of cosmic ray sources, U_{CR}	$\simeq W_{CR}/T_{Cr} \approx 3x10^{40}$ erg/s

*From the LB model, the halo thickness is taken as $\sim$ 500 pc and the confinement volume of disk and halo as $\sim 10^{67}$ cm^3 (see Text.)

7.7.1.

The supernova theory was strongly supported by the observations of strong radio emission in the Crab Nebula in 1951-53. This radio emission was found to originate from synchrotron emission from relativistic electrons spiraling in the magnetic field of the nebula. The Crab nebula is the remnant of the supernova which exploded in 1054 A.D., and it has been studied extensively in radio frequencies of $\sim 10^8$-10^{11} Hz and the properties of high energy electrons in the nebula were estimated. It was inferred that high energy cosmic ray particles and electrons are accelerated in the collisions with turbulent magnetic fields in the nebula by the second order Fermi mechanism. Thus SN and SNR's are considered as powerful sources of cosmic rays. Stars of various types can also generate cosmic rays but these have a very small power. As an example we note that the mean power of cosmic ray production by the Sun is $\sim 10^{25}$ erg/s; so even if 10^{11} stars like the Sun generated cosmic rays, it would provide only a power of $\sim 10^{36}$ erg/s. Some stars like type O, WR etc, however, are much more active, but they are also fewer in number. They may contribute in a smaller energy band, as discussed later. Besides

Crab, a number of SNR's in our galaxy, have been studied in detail in radio and X-ray bands (Chapter 8, e.g. Cas A (remnant of SN explosion in 1660 AD), Tycho (SN 1572), Cygnus loop (remnant of a very old, ~ 20,000 years SN)) and several others. The radio emissions from the remnants have the differential spectrum $J(\nu) \sim \nu^{-\alpha}$, where α is related to the spectral index Γ of the source spectrum of electrons producing synchrotron radiation, as $\Gamma = 2\alpha + 1$. In case of the Crab nebula, $\alpha = 0.26$ and $\Gamma = 1.52$. In the Crab nebula, optical observations of bright wisps traveling outward and merging with stationary wisps are suggestive of hydromagnetic waves and the turbulent field which produces particle acceleration. Knowledge of the generation of turbulent fields and particle acceleration in supernova remnants, like in the crab received a big boost with the discovery of the 33 milli-second period Pulsar in the Crab nebula in 1968. Pulsars as storehouses of large energies were interpreted as the supplier of energy to the remnant and to the cosmic ray acceleration. Before we turn to Pulsars, we consider the very important cosmic ray sources: the shock wave acceleration in SN remnants.

B. Shock wave Acceleration in SNR in the Interstellar Medium:

New astrophysical observations of the interstellar medium and studies of shock wave acceleration in the past decades have provided strong support to cosmic ray acceleration in a diffusive manner in the interstellar medium in shock fronts associated with supernova remnants[86–88]. Most of the interstellar medium is found to be filled with hot gas with very low density with a temperature of $\sim 7{\cdot}10^5$ K and a number density of $\sim 3{\cdot}10^{-3}$ cm^3 having a magnetic field of ~ 1 μGauss (for details, see Chapter 9). The hot interstellar medium (HISM) is the result of repeated supernova explosions. In this hot, thin interstellar medium, supernova remnant shock fronts can travel a large distance due to a very small energy loss and the interstellar medium is thus filled up with many shock fronts. The cosmic ray particles injected by the supernova undergo shock wave acceleration and ultimately produce the cosmic ray source spectrum. It is calculated that shock waves which are largely responsible for determining the cosmic ray source spectrum, $\mu \sim 2.3$, (equation 7.7.16) have Mach number ~ 3.3, shock radius, $R_s \simeq 75$ pc and shock speed $V_s \sim 650$ km/s, and the age of $\sim 3{\cdot}10^4$ yrs; and these cover less than 1% of the space. From several points of view it appears that shock front acceleration in the hot, tenuous interstellar medium is the appropriate source of galactic cosmic rays because, (1) supernova shocks can provide the adequate power, (2) shock acceleration mechanism is an efficient one, the overall efficiency may be $\sim$ 10%, (3) low energy loss in hot ($\sim 7{\cdot}10^5$ K), low density (n $\sim 3{\cdot}10^{-3}\text{cm}^{-3}$) interstellar medium, (4) power law source energy spectra ($\mu \sim 2.3$) can be achieved under suitable conditions, (5) acceleration is relatively prompt in the sense that acceleration from thermal to cosmic ray energies can take place on a time scale much shorter than the cosmic ray mean life time, (6) injection of particles from supernova mixed with interstellar matter may provide the seed material from which the cosmic ray source composition is produced (see e.g. Sakurai[(89)]), and (7) it has been shown by several authors that shock wave acceleration in SNR in hot interstellar medium can generate cosmic ray particles of energies up to $\sim 10^{14}$ eV. Therefore, although there are some problems on the source composition, propagation and acceleration of cosmic rays, which are yet to be clarified, the origin and acceleration of the bulk of cosmic rays of E $< 10^{15}$ eV can be reasonably understood in the frame work of SN shocks in the ISM.

7.7.3.2 Cosmic Ray Sources of E $\sim 10^{15}$-10^{18} eV

Cosmic ray sources in the energy range of E $\sim 10^{15}$-10^{18} eV are unknown and are most difficult to understand at present. In this energy range, cosmic ray particles show two puzzling features: Firstly, the exponent, γ, of the differential energy spectrum, J(E) dE $\sim E^{-\gamma}$ dE changes sharply at $\sim 10^{15}$ eV; this is known as the 'knee' of cosmic ray spectra. Secondly, the composition inferred from the extensive air shower studies, indicates that the composition is a mixed one with a preponderance of heavy nuclei. Because of these two striking features, it is concluded by most of the investigators that these cosmic rays originate from different kinds of sources as compared to those of lower energies ($< 10^{15}$ eV). However there are no definitive clues on the sources.

There are some suggestions in the literature on this matter. Firstly, it has been proposed by Biermann that a supernova explosion shock occuring in the former stellar wind of the star may generate cosmic rays in these energy range. Vahia et al.[(90)] have proposed that high energy cosmic rays may arise from magnetically active stellar systems, such as flare stars, RS CV_n binaries and cataclysmic variables. Various features of these models need further examinations. The other hypothesis relate to Pulsars (see Chapter 8) as the possible source of these high energy cosmic rays. Pulsars, which are rotating neutron stars are well known as a large store house of magnetic field energy as well as of rotational energies, as they emit a variety of pulsed radiation in radio, X-ray and gamma-ray bands. It was proposed that pulsars which emit large amounts of low frequency electromagnetic waves could accelerate particles to cosmic ray energies. Other mechanisms suggested energy generation up to $\sim 10^{18}$ eV. However, basic difficulties of the Pulsar hypothesis are that it can not explain the heavy nuclei composition and that a very strong photon field present close to the Pulsar will photo-disintegrate the heavy nuclei. So the acceleration and origin of these cosmic rays are still open questions.

7.7.3.3 Origin of Cosmic Rays of E $\sim 10^{18}$-10^{20} eV

The observation of ultra high energy cosmic rays and their interpretations in terms of their sources and acceleration mechanisms represent a complex and fascinating area of high energy astrophysics. New results and analysis are unfolding new frontiers of the phenomena; the basic features of these are discussed here.

Anisotropy of 10^{18}-10^{20} eV Primaries

The results on the anisotropy of cosmic ray arrival directions above 10^{17} eV presented by the groups of Haverah Park (England), Fly's Eye (USA), Akeno (Japan) and Yakutsk(Russia) provided a total of about 26,000 and ~ 1000 events of $\geq 10^{18}$ eV and $\geq 10^{19}$ eV respectively, till 1993. The analysis of these events yielded [(68)] upper limits to the anisotropy of arrival direction as 5% and 30% respectively. In another study Wolfendale et al.[(69)] reported the presence of asymmetry in the arrival direction favouring the galactic plane, for $> 10^{18}$ eV particles, from the analysis of the four large arrays. However, the galactic plane enhancement is problematical because of two factors as pointed out by the authors: a predominant flux of heavy nuclei implies considerable deflections by the galactic magnetic field and the likely presence of extra galactic particles at $> 10^{19}$ eV.

Deflections in the Magnetic Field:

For the propagation in the galactic magnetic field the r.m.s. deflection, θ, of a proton of energy, pc (in units of 10^{20} eV) was calculated by the Durham group[(91)] and it is estimated as

$$\theta = \frac{510(L\lambda)^{1/2}}{pc} \cdot B \ (degrees) \qquad (7.7.17)$$

where L = source distance in Mpc; λ, the mean length of the magnetic field before reversal (in Mpc), which is estimated as ~ 100 kpc and B, the magnetic field (in μG) in the intergalactic medium, which is assumed as 10^{-2} μG. The terms λ and B are most difficult to estimate, as they depend on the region of the EG medium and the distance of the source L; it is estimated that $B\lambda \approx 1$. From equation 7.7.17, we obtain for protons of energy pc = 10^{20} eV originating at L = 15, 10 and 5 Mpc, the mean deflection θ as 22°, 6° and 3.6° respectively. Therefore, we may probably look for sources from the arrival directions of 10^{20} eV protons as far as ~ 10 Mpc and for extended clusters as far as 15 Mpc, e.g. the Virgo Cluster.

Clusters and Sources of > 10^{19} eV Particles

It is pointed out by several authors that definite proof of specific astronomical objects as sources of cosmic rays will come only when 'clusters' are observed in the arrival direction of the highest energies and these coincide with likely sources. With these in mind maps of cosmic ray 'clusters' were made [(91)] using world data till 1992 for > 10^{19} eV, as shown in Figure 7.7.5(a). In this figure, 'clusters' are plotted in galactic coordinates . A cluster was defined such that four particles had > 10^{19} eV energies and were within 6° of a particle of energy > $3{\cdot}10^{19}$ eV. Nine such clusters are plotted as o. Also plotted are three new close pairs (as +) observed by the Akeno group having energies > $4{\cdot}10^{19}$ eV. Four more clusters from the new data of Yakutsk are also plotted as o. In addition the three highest energy events of > $2{\cdot}10^{20}$ eV (mentioned earlier) are plotted as Δ. From these results, three regions of best clusters are shown as ringed areas whose l, b values are (1) 165°, 0°, (2) 70°, +20° and (3) 135°, -35°. From the analysis made by Stanev[(92)] et al., (Figure 7.7.5(b) the above three "regions of excess" were also identified; they also observed regions of excess in high latitude areas of (A) L ≈ 180°-300°, b ≈ 60°-80° and (B) l ≈ 130°-160°, b ~ 45-65°. For the discussion of sources, the clusters are grouped as those in the galactic plane, b ≤ 20°, and those with b > 20°, and these are discussed as follows.

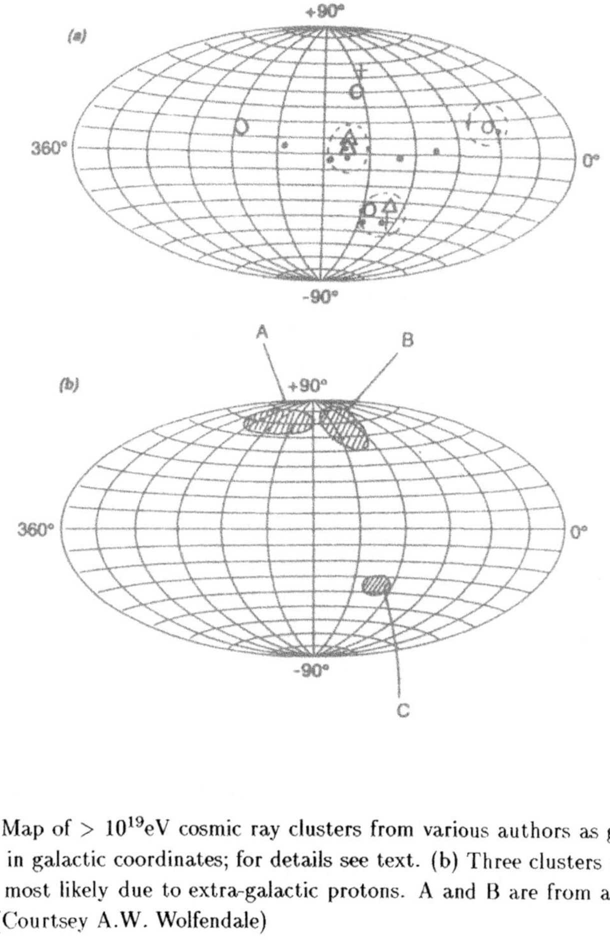

Figure 7.7.4: (a) Map of > 10^{19}eV cosmic ray clusters from various authors as given by Durham group(91), plotted in galactic coordinates; for details see text. (b) Three clusters are shown with b > 20°, which are most likely due to extra-galactic protons. A and B are from analysis of Stanev and C from (a). (Courtsey A.W. Wolfendale)

Clusters with b $\leq 20^o$

There are at least 6 clusters or regions of excess which are close to the galactic plane and it is likely, but by no means certain, that these are associated with galactic objects. However lack of obvious astronomical objects which could produce particles of ultra high energy remains the major problem. Pulsars have been suggested as a possible source, but the production of $> 10^{19}$ eV particles by them is unknown at present.

Clusters with B $> 20^o$

In Fig. 7.7.4(b) three clusters are shown with b $>$ 20; of these two clusters A and B are from the analysis by Stanev[(91)] and the cluster C is from figure 7.7.4a. The most reasonable interpretation of these clusters A, B and C is that they are due to extra-galactic protons of ultra high energy. Keeping in mind the discussions of the mean deflections of these particles in the magnetic field during propagation from the extra-galactic sources, the potential sources of extra galactic particles within 10 and 15 Mpc are searched by examining the maps of galaxies of various types and comparing them with cluster regions A, B, C as in Figure 7.7.5b. It is found that within 10 kpc colliding galaxies show correlations for A, B, C clusters. In the 10-15 kpc range, clusters A and B contain the Virgo cluster and the Ursa major cluster respectively. In summary: nearby ($\leq$ 10 Mpc) colliding galaxies are found to be strong candidates for cluster C, and the distant normal galaxy clusters Virgo and Ursa major may be associated with the cosmic ray clusters of A and B.

In case of extra-galactic protons details of their interactions with cosmic microwave background (CMB) were calculated by several authors. For $3{\cdot}10^{20}$ eV proton the mean free path for inverse Compton scattering by CMB is found to be $\sim$ 15 Mpc i.e. the distance to Virgo. Therefore this process would reduce the proton energy of $3{\cdot}10^{20}$ eV from Virgo Cluster to $\sim 1{\cdot}10^{20}$ eV.

7.7.3.4 Concluding Remarks

In conclusion we note that there is general support among the investigators (although there are some with different view points) that there are three main components in the origin of cosmic rays of high and ultra high energies

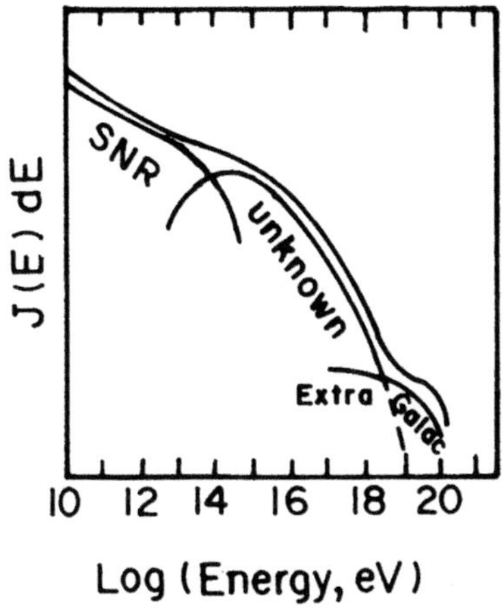

Figure 7.7.5: A sketch showing the three main components of cosmic rays

(Figure 7.7.5). The first component of E $\sim 10^{10}$-10^{14} eV is ascribed to acceleration in the supernova shock fronts in the interstellar medium. The second component of E $\sim 10^{14} \sim 10^{18}$ eV, composed of enhanced heavy nuclei is difficult to understand and its origin is unknown. Pulsars may explain the energy source but they can not explain the heavy elements, which will be lost by photo disintegration in the strong radiation field near Pulsars. The third component of E $> 10^{18}$ eV is ascribed to extra galactic protons whose origin is again uncertain. Colliding galaxies, active galactic nuclei or hot spots in radio galaxies are some of the possible candidates of these ultra heavy cosmic rays. Understanding of these complex and fascinating processes associated with the acceleration and propagation of ultra high energy cosmic rays remains one of the challenging areas of high energy astrophysics.

7.8 Problems on Chapter 7

1. Consider a proton and a ^{56}Fe nucleus, both of kinetic energy 1500 MeV/N. Calculate (a) their momentum/N in MeV/C, (b) rigidity in MV, and (c) their radius of curvature in cm, in an interstellar magnetic field of 10^{-6} gauss. [Ans. (a) 2250 MeV/C, (b) 2250, 4838 MV., (c) $7.5 \cdot 10^{12}$ and $1.6 \cdot 10^{13}$ cm.]

2. Explain briefly the main features of the ionization loss formula for cosmic ray nuclei and the significance of the important terms.

3. Derive the Compton scattering formula for the energy of scattered photons in terms of the incident energy and the angle of scattering.

4. Explain the basic processes in the energy loss of γ-rays in a medium in terms of photo-electric, Compton and pair production effects.

5. Describe the main features of the electron-photon cascade in the atmosphere by a high energy cosmic ray particle.

6. Briefly explain the two methods used in determining the ionization states of anomalous cosmic rays (ACR).

7. Explain the different steps in the model of the origin of ACR in terms of interstellar neutral atoms.

8. Describe how the mean lifetime of cosmic rays is determined from ^{10}Be and B/C measurements.

9. Explain the striking features of the composition and energy spectra of P, He, CNO, Ne-S and Fe in the 10^{12}-10^{14} eV region.

10. What are the distinctive aspects of the flux and energy spectrum of cosmic rays in the 10^{14}-10^{18} eV range?

11. State the equation for the equilibrium spectrum of relativistic cosmic rays in the simplified "Leaky Box" model and explain the meaning of the various terms.

12. What are the basic features of the Second Order Fermi acceleration of cosmic rays by statistical processes and show that the mean energy gain per collision, α, is given by, $\alpha = (4/3)\ \beta_m^2$, where β_m is the mean velocity of magnetic clouds.

13. What are the arguments supporting the model of shockwave acceleration in supernova remnants in ISM for cosmic rays of 10^9-10^{14} eV?

14. Describe the characteristic features of the highest energy cosmic rays of $> 10^{19}$ eV at high galactic latitudes in terms of "Clusters" and their possible identifications with extra galactic cosmic ray sources.

Chapter 8
SPACE ASTROPHYSICS

8.1 New Astrophysics from Space Observatories

8.1.1 Infrared Astronomy from Space

8.1.1.1 General Features

The infrared spectrum begins at a end of the visible spectrum at the wavelength of about 1 μm (i.e. 10,000 Å) and stretches over more than ten octaves to about 1 mm where it overlaps the shortest wavelengths of the microwave radio region of the electromagnetic spectrum (Figure 8.1.1). This large region is divided into roughly three regions, namely the near infrared (1-3 μm), the intermediate IR (3-5 μ) and the far IR (5 to $\sim$ 200 μm). In this broad region of wavelengths lie the characteristic black body radiation of the moon and the planets, cool stars, proto star clouds as well as the cosmic background radiation of the universe. Any object having a temperature in the region of 3K to 300K is suitable for observations at infrared wavelengths. In the field of molecular spectroscopy, the vibrational-rotational bands and lines of many chemically active molecules lie in the infrared region.

As a few selected bands of IR wavelengths are transmitted by the earth's atmosphere, these wavelengths can be studied with optical telescopes at mountain altitude using specific filters, for the terrestrial windows at 2.2 μm, 5 μm, 10 μm, 20 μm etc. These studies were made over several decades. The IR detectors which belong to the class of thermo-electric detectors are to be cooled to very low temperatures with liquid nitrogen temperature or liquid Helium temperature to reduce the atmospheric background radiation. In addition to mountain altitude observations, IR studies were carried out in the 1960's and 1970's with high flying aircrafts, stratospheric balloons and sounding rockets. These studies discovered a number of cool stars and pre-stellar objects; and also some highly luminous 'infrared' stars that are barely detectable with optical telescopes.

With the advent of *in situ* measurements in IR of planetary atmospheres by space missions like Mariner, Pioneer, Venera etc, basic studies of the atmospheric composition of the terrestrial planets and later of the giant

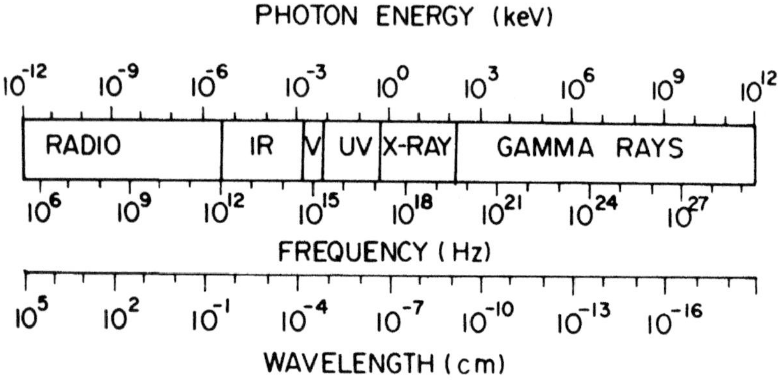

Figure 8.1.1: The electromagnetic spectrum

planets and their satellites were made. These studies made some dramatic discoveries, some of which were already discussed in Chapter 4. Here we shall discuss briefly the new results and discoveries made in stellar and galactic astrophysics made with recent satellite missions for infrared studies.

8.1.1.2 The IRAS and ISO Missions

The Infrared Astronomy Satellite (IRAS) was the first space mission dedicated to infrared studies particularly at the longer wavelengths 12 to 100 μm, which are not accessible from the ground. The Infrared Astronomy Satellite was developed, funded, built and operated by three national scientific communities of the USA, the Netherlands and the UK. The IRAS was launched into orbit in 1984, and provided high quality data for several years. The IRAS survey results have been used by scientists and astronomers all over the world. The IRAS mission made a complete survey of the sky at four wave bands centered on 12, 25, 60 and 100 μm. These provided a wealth of information on many areas of astrophysics such as the solar system, stellar and galactic astrophysics, interstellar medium and extra galactic studies. We shall briefly focus our attention to stellar and galactic phenomena with IR studies. The details of the results in different areas from IRAS are given by several authors in the book on 'Comets to Cosmology', edited by A. Lawrence.

The ISO Mission

A new landmark in space astrophysics was made with the successful launching and operation by the European Space Agency (ESA) of the Infrared Space Observatory (ISO) in November 1995. The ISO is a state of art space observatory operating in the wavelengths of 2.5-240 μm. ISO consists of a large cryostat containing about 2300 liters of liquid helium which maintained the IR-telescope and the scientific instruments at a temperature of 2-8 K. The telescope's primary mirror has a diameter of 60 cm. The pointing accuracy of the telescope is maintained to a few arc seconds by a three axis stabilization system. There are four main instruments in the ISO mission; (1) an Imaging Photopolarimeter (ISO PHOT) operating in 2.5-10 and 40-200 μm, (ii) a Camera (ISO CAM) in $\sim$ 2.5 to 20 μm, (iii) a Short wavelength Spectrometer (SWS) in $\sim$ 2.5-40 μm, and (iv) a Long wavelength Spectrometer (LWS) in $\sim$ 40-200 μm. The ISO mission has, compared to IRAS, a wider wavelength coverage, n improved spatial resolution, a higher sensitivity and a longer life time. The ISO mission was developed by the ESA member states France, Germany, the Netherlands and the U.K., and it was open to Guest observers. The ISO mission has an expected life time of about 2 years. The first set of results and their interpretations were published by a number of authors in Astronomy and Astrophysics of November, 1996.

8.1.1.3 Star Formation and IRAS

Star forming regions and star forming processes are the areas of major interest of stellar astrophysics and intensive studies were made in the past decades particularly with the IRAS mission. Here we present some basic ideas of this problem. There are three major problems which are basic to star formation processes. This involves a cloud with a gas density of $\sim 10^3$atoms$\cdot$cm^{-3} which collapses to form stars with densities of $\sim 10^{20}$ times greater. First one has to consider the problem of energetics. In order to form a stable star, the dense gas or proto-star must get rid of its gravitational binding energy. Secondly, a proto-star mass possesses some initial angular momentum. As the proto star collapses, its rotational energy will increase, because of the conservation of angular momentum. The increase in the rotational energy of the star will inhibit the infall of matter and will halt the collapse, unless some ways are found to get rid of the angular momentum. Thirdly, if there is some initial magnetic field in the collapsing cloud, the increase in the magnetic field in the star forming body may be sufficiently strong to halt the collapse particularly in the equatorial plane.

Let us now consider the first phase. A massive gas cloud when perturbed will tend to collapse under self gravitation which will be resisted by the internal pressure gradient. The criterion for collapse is therefore that the gravitational force should be greater than the internal gas pressure forces. Simple calculations show that the condition for collapse of an uniform gas cloud of mass M, density ρ and radius R, is $GM\rho/R^2 > p/R$, where p is the gas pressure. Using $(p/\rho) \simeq C_s$, the speed of sound in the gas, and $M \sim \rho R^3$, we get $R > R_j = C_s/(G\rho)^{1/2}$. This characteristic length R_j is known as the *Jeans length*. Thus we define the Jeans mass, M_j, as

$$M_j \sim \rho R_j^3 \simeq 10^5 \frac{T^{3/2}}{M^2 \cdot N^{1/2}} \tag{8.1.1}$$

where M is the average mass of the gas relative to hydrogen and N, the Avogadro number. The time scale of the collapse of the gas cloud, τ, is approximately given by

$$\tau \simeq R_j/C_s \approx (G\rho)^{-1/2} \tag{8.1.2}$$

More accurate derivations are given by Spitzer, 1978.

The values of the Jeans mass for different regions of interstellar gas (as given in Table 9.3.1 in Chapter 9) can be calculated from equation 8.1.1. These masses range from M_j = 10,000 to 20 $M_\odot$ for a diffuse HI cloud at T = 80K and the coldest molecular cloud at T = 10 K, respectively. M_j decreases as the density of the cloud increases and hence the instability would lead to continued fragmentation as the cloud collapses. The fragmentation stops when the cloud becomes optically thick to radiation, which is expected to occur at $M \sim 0.01\ M_\odot$. It is found that all stars observed by us have masses greater than about 0.1 $M_\odot$. This is due to the fact that stars with a mass less than $\sim 0.08\ M_\odot$ are not hot enough in their central region for nuclear reactions to take place.

The material of the collapsing cloud consists of molecular gas and dust. These materials of the cloud are heated during the collapse and the energy can be lost efficiently by radiation so long as the cloud remains optically thin. When the central region becomes optically thick the radiation is trapped and the core begins to heat up. The core forms a pressure supported structure, while accretion from the matter in the envelope to the core continues. The core of the proto star is, thus, expected to evolve independently of the accreting envelope. A sketch of the model for a typical accreting proto star according to Shu et al.[(47)] is shown in Figure 8.1.2. The outer envelope of the dust shell ha outer and inner radii of $r_c \sim 10^{14}$ and $r_d \sim 10^{13}$ cm respectively. Beyond the outer radius of the dust shell, $r \sim 10^{14}$ cm which is also called the dust photosphere, lies the gas envelope whose outer boundary is at $r \sim 10^{17}$ cm. In the outer envelope the gas and dust are optically thin and so they radiate away thermal energy very efficiently.

The collapsing gas and dust eventually reach increased densities so that the dust becomes optically thick and the region with optical depth unity, called the dust photosphere (Figure 8.1.2) is at $r \sim 10^{14}$ cm. Inside this radius the temperature of the dust envelope increases with decreasing radius until it becomes so hot, $T \approx 2300$ K that, the dust grains, such as graphite evaporates. The gas is then accreted onto the core, which acts as a 'solid body', and an accretion shock is produced; this dissipates the kinetic energy of the infalling gas and the radiation has to pass through the infalling matter. In passing through the dust envelope, this radiation is trapped, degraded, and then radiated away at the temperature of the dust photosphere. Model calculations indicate a typical temperature of the dust photosphere at about 100 K.

This picture clearly shows why proto-stars are expected to be intense infrared sources. This picture is supported by observations of star forming regions by IRAS. Furthermore the IRAS mission made striking discoveries of objects with spectral characteristics for each stage of stellar evolution. For example, objects at the earliest stage of evolution are deeply embedded in the dust clouds out of which they are formed, and are pure infrared sources. The emissions corresponding with later stages of development of the accretion discs are identified with objects in the IRAS mission.

The detailed evolution of collapsed objects from proto-star phase to hydrogen burning stars on the main sequence is a complex process depending on the mass of the star and its characteristic time scales. These time scales are noted briefly here. For a 1 $M_\odot$ star the collapsing time scale is about 10^6 years. For such stars the time scale for radiating away its gravitational binding energy, which is called the Kelvin-Helmholtz time scale, is about $2{\cdot}10^7$ years. Such a star undergoes a prolonged accretion phase. On the other hand, a star with a mass greater than 5 $M_\odot$ has a Kelvin-Helmholtz time shorter than the collapsing time so that the evolution and formation of this star are more rapid, about 10^5-10^6 yr. Thus for massive stars the binding energy is radiated away rapidly so that further accretion takes place directly on the star.

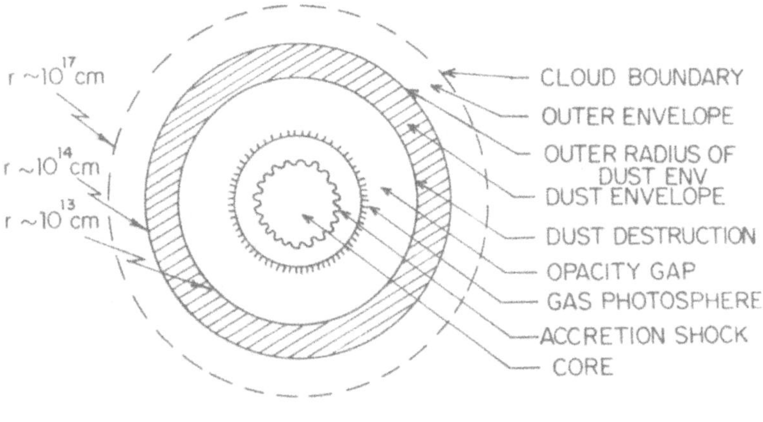

Figure 8.1.2: A sketch of the model of an accreting proto-star. (Based on Shu et al., 1980)[47]

It is found that the mean life time of massive stars on the main sequence is about the same as the time scale of the formation of stars in a large molecular cloud. As a result it is possible that some massive stars may complete their evolution and explode as type II supernova while other stars are still forming in the same large molecular cloud. Such events have been observed. Type II supernovae were found in OB associations and in the giant HI regions in the spiral arm. It is estimated that about half of the SNR's of type II in the outer parts of our galaxy is associated with large molecular clouds. Thus it is suggested by some authors that there is a distinct possibility that supernova remnants which produce compression of the gas in the molecular cloud may lead to a new burst of star formation. This is called sequential star formation, the evidence of which have been found in some star forming regions.

8.1.1.4 Extremely Luminous Far-Infrared Sources (ELFS)

Now we turn our attention to a new class of extra galactic sources found by the IRAS mission. These extra galactic sources were found to have very high luminosities in the far infrared band, for example, at 60 μm the luminosities exceeds 10^{11} L$\odot$. Some of the striking properties of these ELFS are as follow:

(a) ELFS mostly appear as colliding galaxies or some type of the peculiar galaxies.

(b) Their optical luminosities are generally orders of magnitude smaller than the FIR luminosities.

(c) Their radio fluxes are proportional to FIR fluxes.

(d) Their molecular hydrogen contents are high, $\sim 10^{10}$ $M_\odot$.

(e) The infrared emissions, seen at 10 and 25 μm with better resolution in ground level observations seem to come from the central region of the sources; these regions have typically sizes of ≤ 1 kpc, as estimated for the brighter ones.

(f) The luminosity function of these ELFS drops steeply toward 10^{13} $L_\odot$.

These new classes of extra galactic objects are far from understood. Several views have been proposed to explain some of the observed properties and some of these proposals are as follows:

1. It is proposed that they are star forming processes on a massive scale. In this view ELFS are called 'Star burst' galaxies.

2. The other view is of a central active source which may be a black hole which is fed by infalling matter.

3. The third view attributes the observed luminosity as due to the collision of two gas rich galaxies. Particularly in the model of two colliding disks.

These options need further theoretical and observational studies. More details of these phenomena are given in the review by Herwit and others in the book edited by A. Lawrence.

8.1.2 Astrophysics with the Hubble Space Telescope (HST)

8.1.2.1 HST in Orbit

After more than twenty years of research and development, the Hubble Space Telescope (HST) with it advanced instrument packages was launched into space in March 1990. This epoch making event in astrophysics, however, had initial set backs as was seen in the first series of images received in May 1990. It was found that the primary mirror had serious spherical aberration which was due to accidental errors in the grinding of the primary mirror. The solution of this serious problem was worked out by scientific and technical teams and as a result corrective optics and some other components were installed in the Hubble Space Telescope during an on-orbit operation in December 1993. After this refurbishment HST provided spectacular photographs of every object seen by it with exceptional details for the first time. It is well known that even very large telescopes at ground level can not achieve the theoretically expected resolution due to turbulence and density fluctuations in the atmosphere. This can be reduced to some extent by installing the optical telescopes in remote areas and at mountain altitude. The HST fully overcomes this problem. In the original design of the HST it was aimed that about 70% of the total light energy from a star would be concentrated into a circle with a diameter of 0.1 arc sec. After the installation of the corrective optics it was found that the actual performance value was about 84%. This was exceptionally good and is close to the theoretical limit of 85-87%. This made the HST teams and the astronomy and astrophysics community very happy.

8.1.2.2 HST Instruments

HST was designed with a 2.4 m (94.5") diameter mirror which, it was claimed, would enable scientists to see astronomical objects as far as $15 \cdot 10^9$ light years away in space. Objects about 50 times fainter than anything seen before by the most powerful earth based telescopes would be seen. This was largely achieved with observations by HST since 1995, the then most distant objects, a cluster of galaxies at a distance of 7 to $10 \cdot 10^9$ light years. The telescopes on Earth can typically observe about 2000 hrs per year and they lose appreciable observing time due to cloud covers etc; in contrast, HST would be able to reach the impressive observation time of about 4500 hrs per year.

The scientific instruments in HST are the following:

(1) Wide Field and Planetary Camera (WF/PC-2)

All HST's instruments overlap each other to some extent in function and wavelength range so that maximum information may be collected on an object. WF/PC-1 was replaced by WF/PC-2 during the first servicing mission. Both WF/PC-1 and its clone WF/PC-2 are sensitive to a wide spectral range of about 1200 Å in the UV to 11000 Å in the infrared. The resolution of WF/PC-1 and PC-2 is about 0.05 to 0.1 arc sec. The instrument can be operated in WF or PC modes. WF/PC-2 uses four CCD's as its main light receptors giving a full picture with an array of 1600x1600 pixels across four CCD's. WF/PC instruments have been widely used since the beginning and they provided excellent data, allowing its users to see stars of the 28th magnitude during long exposures. These instruments were designed and built by JPL/Cal Tech. supported by NASA.

(2) Faint Object Camera (FOC)

The FOC was contributed by the European Space Agency and was designed in European centers with Leiden Observatory as the team leader. FOC is sensitive in the wavelength range from 1150 to 6500 Å and has two complete detector systems with image intensifiers to gather as much light as possible from weak and distant objects. Its spatial resolution is high, about 0.05 arc sec or better.

(3) Faint Object Spectrograph (FOS)

FOS is sensitive in the wavelength range of 1100 to 8000 Å, and more extensive than GHRS (item 4). FOS operates in two resolution modes, low and high. In low resolution mode it can image 26th mag objects with 1 hr exposure, and in high resolution mode 22nd mag objects with 1 hr exposure. FOS was built by the MMA group of Denver, Colarado.

(4) Goddard High Resolution Spectrograph (GHRS)

HST has two spectrographs of which GHRS looks for finer details, while FOS covers a wide range. GHRS is sensitive to 1150 to 3200 Å, and operates in three modes, low, medium and high. At low resolution it measures spectral features 1 Å wide, while at high resolution features of 0.02 Å wide are measured. During the first service mission some electronic packages were installed which increased the efficiency of the GHRS to its original value.

(5) The High Speed Photometer (HPS)

HPS was built at the University of Wisconsin and was designed to measure the high speed fluctuations of light from energetic objects. HPS was able to measure light intensity changes every 1/50,000 second. HPS functioned flawlessly during the first three years. However in order to accommodate the corrective optics in the main mirror, the HPS instrument was removed from the HST, during the first servicing mission in December 1993.

HST, Protoplanetary Disks and Beta Pectoris

The Hubble Space Telescope has been providing spectacular new details of the objects in the solar system, nearby stars in the galaxy, interstellar medium and the intriguing features of extra galactic objects and galactic clusters. Here we present the highlights of a few new discoveries by HST, as an introduction. Firstly we consider the observations of "proplyds" and Beta Pectoris.

In the Orion nebula, HST discovered several objects which showed accretion disks around new born stars. These proto-planetary discs were called 'proplyds' by O'Dell. When a star is born a great deal of material of gas and dust is still present in the form of a circumstellar disk, out of which eventually proto-planets are formed in regions with high densities. These proto-planetary disks are very difficult to see because they are composed of cold gas and dust which are out shone by the nearby bright star. HST with its exceptional power was able to see some of them as they were illuminated with UV light by the neighbouring hot new stars.

The best known example of a proto planetary disk is the star Beta Pectoris which was studied earlier by ground based telescopes and by IRAS. This star is a nearby one located at a distance of 18 pc in the constellation Pector. GHRS in HST discovered three components of the gas and dust disk of the star. The first component is highly variable, changing over hours or less and is probably produced by evaporating bodies (FEB's). The second one is slowly variable over several days; this may be due to varying showers of the FEB's; and the third component is the stable one formed by evaporation of many comet-like bodies at several A.U. Thus it is at present presumed that the gas present in the circumstellar disk is secondary and it originates from the evaporation of already formed compact bodies.

8.1.2.3 HST, Extra-galactic Astrophysics and the Hubble Constant

The well known Hubble Law was postulated by E. Hubble in 1929 and states that the radial velocity V and the distance D of extra-galactic nebulae can be written as $V = H_o D$, where H_o is a constant, now known as the Hubble constant. From the redshifts of galaxies, their radial velocities are determined; and from the absolute and apparent magnitudes, their distances are estimated. There are several problems and uncertainties in the determination of H_o. Applying the corrections as known at that time, A. Sandage, a student of Hubble, determined in 1956-58 H_o as 100 km/s/Mpc. As new astronomical measurements were made and various improvements were made during 1975-95, the improved value of H_o determined by Sandage et al came down to 50 ± 5 km/s/Mpc. Other groups of observers, however, after careful analysis determine a higher value of H_o as $\sim$ 100 km/s/Mpc. Thus the enigma of the value of H_o continues.

In 1995 two groups of observers made independent measurements of the distances using the Virgo cluster of galaxies. One group used a Hawaii based observatory and obtained the value of H_o as 87 km/s/Mpc. The group of Freedman et al used HST to observe Cepheids directly in M 100 located near the center of the Virgo cluster. These are not observable with ground based telescopes. In addition, calibrations were made with the more distant Como cluster. Thus the peculiar proper motions of nearly galaxies, which were the sources of some uncertainties in earlier studies, were avoided. Thus, HST measurements yielded the value of H_o as 80 ± 17 km/s/Mpc, supporting the high value of the Hubble constant.

Thus the uncertainty of the value of H_o being either $\sim$ 50 or $\sim$ 100 km/s/Mpc remains an enigma. The anomaly in the H_o value is also reflected in the estimated value, T_o, the age of the universe in the standard model of cosmology. Using a H_o of 100 km/s/Mpc, we get an age, T_o, of the universe of $10 \cdot 10^9$ years and for 50 km/s/Mpc, the corresponding value is $20 \cdot 10^9$ years, for the density parameter $\Omega_o = 1$. However we find the anomaly that the HST estimate of H_o and hence the age of the universe, $T_o \simeq 10{\cdot}10^9$ yrs from the standard model of cosmology is lower than the estimated value of the age of our galaxy from cosmochronology as 15-20$\cdot10^9$ yrs. Also T_o from HST results is lower than the estimated age of the globular clusters in our galaxy which lie in the range of 12-18$\cdot10^9$ yrs. These and other discrepancies in the standard model of cosmology are discussed by several authors.

In the standard model, the universe began with the big bang at T_o and the fireball at 10^{-43} sec consisted of energetic elementary particles and photons. At about 10^{-34}-10^{-30} sec the universe passed through a rapid expansion and at about 10^{-6} sec, the quarks combined to form protons and neutrons. At about 1000 sec after the big bang, the universe cooled and the formation of major elements, such as hydrogen, deuterium, helium and lithium, took place. Their proportions were determined by the density parameter, Ω_o which is taken as unity. At some point the matter and radiation decoupled and thereafter they evolved separately. The matter cooled down; and the radiation cooled from $\sim 10^4$ K to the present day temperature of 3 K, as measured by Cosmic Background Explorer, which also measured its small fluctuations.

The enigma of the value of H_o and the discrepancies of other observations with the standard model are yet to be understood. Further details of HST discoveries are given in several books, e.g. Hubble Vision, and others.

8.2 New Windows to the Universe

With the advent of satellites and spacecraft dedicated to the study of energetic electromagnetic radiation from space, streams of new knowledge came through many windows and these expanded into new branches of space astronomy. Ultra violet (UV), extreme ultra violet (EUV) and Infrared (IR) astronomies from space opened up new vistas in space astrophysics, and X-ray and Gamma-ray astronomies unfolded new and spectacular scenarios in high energy astrophysics. In this chapter we shall briefly discuss some of the highlights of UV, EUV and IR observations in space; then, with a brief introduction to observational methods, we focus our attention to some of the new findings and discoveries in high energy astrophysics as revealed by X-ray and Gamma-ray studies in space.

8.2.1 The Ultra Violet (UV) Astrophysics

The IUE Space Mission

The International Ultra Violet Explorer (IUE) satellite was launched in January 1978 after years of joint studies by NASA of the USA and the European Space Agency (ESA). A remarkable feature of this mission was that this provided uninterrupted, high quality UV results to the international scientific community for more than ten years. The IUE observatory consisted of a 45 cm diameter telescope equipped with two spectrographs with spectral ranges of 1900-3200 Å and 1150-1950 Å. The spectrographs were coupled to secondary electron conduction (SEC) vidicon cameras as detectors. The spacecraft was placed in an elliptical geosynchronous orbit and was operated from NASA-GSFC, Maryland, USA and from the ESA ground station near Madrid, Spain. Some of the principle findings are summarized as follows:

Stellar Chromospheres:

Chromosphere emissions were observed from many types of stars from early A type to late M types, from which correlations of various stellar parameters such as stellar age and rotation were determined. Chromospheric properties of single stars were found to be of three kinds: solar-type, non-solar-type and mixed-type. A solar type star has a hot transition region ($\sim$ a few 10^5 K) and a corona of $\sim 10^6$ K. A non-solar type star does not have a hot transition region, but a massive and cool stellar wind. A mixed type star has a hot transition region as well as a massive, cool stellar wind. Chromospheres in close binary stars differ from those of single stars, due to loss of angular momentum through stellar winds causing a slower rotation. In a binary system a late type star rotates faster than a single star, because the loss of angular momentum is compensated by the large angular momentum transfer. Faster rotation in binaries, increase stellar dynamo action producing a more active chromosphere.

The structures of the outer atmospheres of cool stars were explored by the observations of binary systems in which such stars have hot companions.

Binary Systems, Accretion and Mass Loss

It is estimated that at least two-third of all stars are members of binary or multiple systems. Furthermore a large fraction of special astrophysical objects such as novae, X-ray sources etc. involve binary systems. So the studies of binaries are of special importance in the understanding of stellar evolution. The temperature and density of the flowing gas are such that their signatures are readily observed in UV - in the resonance absorption lines of singly or multiply ionized atoms, e.g. Mg II, Fe II, Al III, C IV, Si IV and N V. From the radial velocities of the absorption lines, it is observed that a fraction of the gas is being lost from the binary. At the same time a fraction of the gas flowing from the mass losing star is accreted onto the companion providing the signature of high temperature spectral lines such as Si IV, C IV and N V. Otherwise these lines would not appear in single stars of the same spectral type.

The observation of the absorption lines which are Doppler shifted in short wavelengths indicate that in most of the close binaries matter is lost by the system. In a number of binaries, hot companions were found. Algol type binaries i.e. eclipsing binaries were found to exhibit extremely active mass flows in periodic intervals. In this type of binaries such as, U cep, optically thick plasma engulfs the less evolved B component. Studies on U cep's active gas flow period in 1986 with a high resolution UV spectrometer showed an optically thick plasma originating from the G-giant covering much of the B-star companion suppressing much of the UV flux from the hot star. The observations indicated that most probably multiple gas streams produced numerous absorption lines at different velocities seen in the Mg II resonance doublet. The dynamic mass of the flow event measured again 10 days later showed a substantial change in the features; numerous gas streams were found to coalesce and so to produce the appearance of an extremely broad absorption band. It appeared that the gas streams covered $\sim$ 70% of the B-star hemisphere.

Stellar Wind

Stellar winds from an early type of stars were extensively studied with IUE. The studies included O, B stars in the galaxy, in central stars of planetary nebulae and even in some highly luminous hot stars in a few galaxies of the local group, such as LMC, SMC. Hot central stars of many planetary nebulae were found to have fast stellar winds. Terminal velocities of stellar winds from these central stars were found to be in the range of 1000 to 5000 km·s^{-1}. The analysis of the results indicated that central stars with temperatures less than 63,000 K almost always have a wind and those with radii larger than 0.32 $R_\odot$ have invariably stellar winds. These observations supported the theory of the origin of planetary nebulae from stellar winds.

Stellar winds and their relationship with stellar evolution were studied by several authors. It was found that the rate of mass loss in young, massive stars seems to be about two order of magnitude higher than from the main sequence to the hydrogen-shell burning phase. Wolf-Rayet stars, which are in a more advanced evolutionary stage showed mass loss rates which may be another order of magnitude higher. It was found that in O-type stars, the ratio of terminal wind velocity to the stellar escape velocity depended on the evolutionary history; for stars in the main sequence this ratio is about ≤ 3, whereas for evolved stars the ratio is > 3.

Details of these and other results from UV observations are given in reviews by several authors (e.g. Kondo et al. 1987).

8.2.2 Extreme Ultra Violet (EUV) Astrophysics

Introduction

The extreme ultra violet (EUV) band of wavelength is commonly referred to as the region from 100 Å to 900 Å, which lies between the well studied UV band of 912-3000 Å and the soft X-rays of 1-100 Å. This region of EUV was not systematically studied until recently. This was due to the common belief that the EUV region is opaque because of interstellar absorption. Figure 8.2.1 shows the mean free path of photons, assuming the interstellar density, n, as 1 atom per cm^3, vs. photon energy in keV. In Figure 8.1.1 we have shown the scales of the electromagnetic spectrum in wavelength λ, frequency ν and photon energy in keV. It is seen that for EUV radiation of 100 Å, i.e. of energy 0.124 keV, the mean free path for absorption is about 10 pc for a hydrogen density in the ISM of $n = 1$ cm^{-3}. For $n = 0.01$, the m.f.p is 1000 pc, and the column density of hydrogen is $3 \cdot 10^{19}$ cm^{-2}. Therefore it is argued that EUV of 100 Å is not observable in any direction having a column density

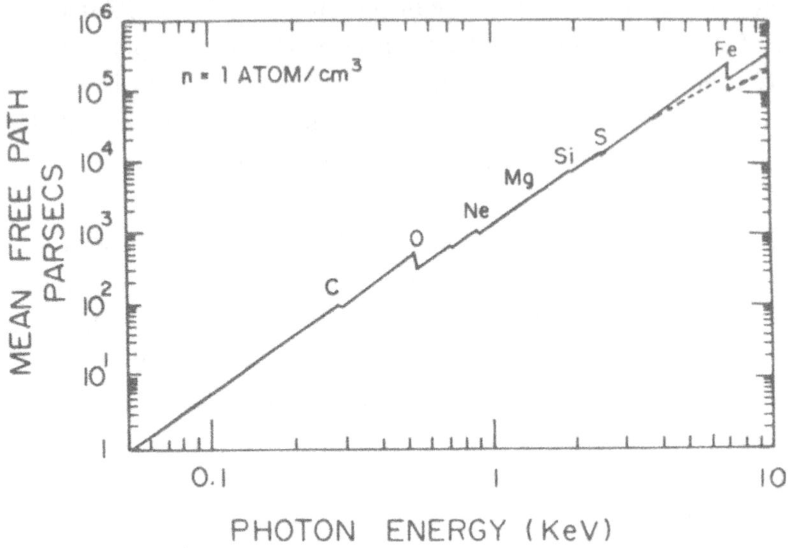

Figure 8.2.1: Mean free path of photons in parsec in the ISM (for n = 1 atom·cm^{-3}) vs photon energy

of $\geq 10^{20}$ cm^{-2}. As most of the regions of the ISM have a column density of $\geq 0.6{\cdot}10^{20}$ cm^{-2}, very few EUV sources were expected to be observable. However several initial experiments, e.g. with a small EUV telescope in the Apollo-Soyuz mission several EUV emitting stars were discovered and the feasibility of EUV astronomy was demonstrated. Based on these observations the scientific mission of the EUV astronomy Explorer satellite was proposed by the Berkeley group of the USA and it was selected for development in 1976 by NASA. After a long waiting EUV Explorer was flown in space in June 1992. Several experiments in this EUVE satellite revealed spectacular results of EUV sources. In the meantime, as the EUVE mission was delayed, an experiment on EUV with the wide field camera (WFC) was carried out in the European Röntgent satellite (ROSAT) which was launched in mid-1990. The WFC experiment of the University of Leicester, UK, in ROSAT discovered and studied a number of new sources in EUV wavelengths.

We briefly present some of the observational aspects of the WFC and EUVE experiments and then discuss the highlights of the new discoveries made in EUV with these two space missions. The details of these experiments and the results are given in the papers by several authors in the book on "Astrophysics in the extreme ultra violet" edited by S. Bowyer and R.F. Malina.

Observational Features

In the EUVE mission, an all sky survey was made in the entire EUV band covering 90-820 Å, in four sub-bands or colors with specially designed band pass filters. These band pass filters were (1) 100 Å(90-150 Å), (2) 200 Å(170-250 Å), (3) 450 Å(400-600 Å) and (4) 600 Å(550-750 Å). Following the sky survey, deeper surveys over a limited portion of the sky were made which gave more insights into the types of sources. Spectroscopic measurements of targeted sources were made.

The Wide Field Camera (WFC) experiment in ROSAT performed the first all sky survey during 1990-91 at EUV wavelengths. The survey was conducted in two 'colors' using broad band filters covering ranges of 60-140 Å and 112-200 Å. The WFC experiment was fully imaging with a resolution of $\sim$ 3 arc min.

Identification of EUV Sources

The first bright source catalogue of EUV sources in the WFC experiment in ROSAT listed 383 sources in 1993. This was supplemented by a complete survey of the data base and the new catalogue of EUV sources called '2 RE' was produced which contained 479 EUV sources. Most of the sources, 444 out of 479 have proposed identifications. There were 387 2 RE sources which were detected in both survey wave bands. The identification of 444 EUV sources in the 2 RE catalogue is given in Table 8.2.1 (Pyle et al.[(48)]).

The EUV Explorer data is summarized in the second EUVE sources catalogue (Bowyer et al.(49)). This included data from the all-sky survey of the first six month period, special programmed observations, deep survey sources and the targeted spectroscopic observations. A total of 734 EUV sources were recorded, of which 489 sources were identified. The classification of these sources are given in Table 8.2.1.

Sources in the ROSAT 2RE catalogue were studied for variability on time scales of $\sim$ 1.5 hr and 31 sources were found to be variable. Cataclysmic variables accounted for most of them, $\sim$ 50%, while 10% were late-type stars. About 40% of G-type RS CV_n systems were found to be variable and other variable sources were X-ray binary Her x-1, eclipsing binaries, Algol; and active galactic nuclei e.g. Markarian 478. The nomenclature of some variable binary stars are as follows.

Table 8.2.1: Identification of EUV Sources

Source Type	Number of ROSAT(2RE)	Sources EUVE(2nd C)
White dwarfs	140	104
Early-type stars (A,B)	10	24
Late-type stars (F-M)	251	273
Cataclysmic variables	18	14
Low-mass X-ray binaries	3	2
Extra galactic objects	8	37
Others	15	35
No identification	34	245
Total	279	734

Variability of EUV Sources

Cataclysmic variables are binary systems in which the compact companion is a white dwarf. The primary star is usually a late-type star near or on the main sequence in which Roche lobe overflow takes place onto the compact white dwarf star. The periods of these binaries range from 80 min. to several hours.

RSCV$_n$ binaries: These are cataclysmic variables with primary stars of F or G type of the main sequence and a G or K type sub-giant secondary.

Algols: These are eclipsing binaries with strong mass flows occuring periodically.

Contact Binaries: In these binaries the common envelope of the binary lies outside the Roche lobe. The stars are identified by their very short periods of $\sim$ 10 hrs or so and by their distinctive light curves, e.g. W UMa-type binaries.

EUV and Active Galactic Nuclei

About three dozen sources of active galactic nuclei identified by the WFC and EUVE instruments provided a number of surprises. These were observed in the shortest wavelength band pass of both the experiments in the directions of galactic neutral hydrogen 'windows' along which the column density does not exceed $\sim 1{\cdot}10^{20}$ cm^{-2}, and for which the HII/HI ratio is favourable.

EUVE observations identified 14 AGN's of which 8 are Seyferts and 5 BL Lac objects, and 1 quasar (3C 273). The WFC bright source catalogue identified 7 AGN of which 4 are Seyfert, and 3 are BLO's.

Radio Quiet QSO's and Seyfert 1 Galaxies

EUV observations of radio-quiet QSO's and spectroscopically Seyfert 1 galaxies can provide valuable clues to the nature of these sources. The peak power of radio quiet QSO's is known to be emitted in the EUV region i.e. in the several hundred of Å range. Studies of these QSO's in EUV and soft X-rays (0.2-2 KeV) revealed a steep spectrum with spectral index $\alpha_x = 1.5$, that flattens by $\Delta\alpha x = 0.5$ above $\sim$ 2 KeV. Similar observations are indicated in Seyfert 1 galaxies. The origin of the 'soft excess' in the EUV-soft X-rays (0.2-2 KeV) spectrum in these on average radio-quiet AGN's are not fully understood. Results analyzed so far seem to point to optically thick emission by a hot ($\sim 10^6$ K) gas, possibly associated with the inner most region of a nuclear accretion disk.

Blazars and BL Lac Objects

Unlike radio-quiet QSO's and similar Seyfert 1 galaxies whose emissions are dominated by a thermal component, BL Lac Objects (BLO's) and other members of the 'Blazar' class (i.e. high polarization and optically variable QSO's) are distinguished by the dominance of a non-thermal emission component. The 'Blazars' are believed to have a jet of relativistic gas which is the source of non-thermal emission in featureless continuum extending from radio to hard X-ray wavelengths. This emission is interpretated as synchrotron, and also as inverse Compton radiation in the relativistic jet. The axis of the jet, in this picture, is oriented towards the observer so that the intensity is boosted by Doppler beaming.

EUV measurement made on QSO 3C 273, BLO Mrk 421 and BLO PKS 2155-304 show variabilities supporting the jet model. In the latter source it was observed that EUV flux variations occurred with a lag of a few hours after the soft X-ray flux change. These observations are interpreted in terms of a shockwave propagating in the jet.

BLO's are characterized by a featureless continuum without emission lines, indicating the absence of dense gas clouds. They are thought to be bare core matter of AGN. Measurements of PKS 2155-304 which is one of the brightest X-ray selected BLO's, showed this object as the brightest AGN in EUV, and so it was selected for spectroscopic EUV studies in the range of 70-110 Å with EUV-Explorer. Measurements made one month apart showed an intensity change of 50%. When the energy spectrum is fitted with the exponent $\alpha = 1.65$ corresponding to that of the soft X-ray spectrum, it is found that it fits in 85-110 Å, but it indicates a strong absorption in the source in 70-85 Å. This absorption is not due to continuum opacity, but is attributed to localized ionized hydrogen clouds of high radial velocity formed by the beamed radiation of the relativistic jet. The EUV and X-ray studies together provide support to this BLO identification as a galaxy with redshift, Z = 0.116.

EUV and the Virgo Cluster

The EUV studies of the central region of the Virgo cluster of galaxies, led to the discovery of warm gas in the galaxy M 87 which is centrally positioned in the cluster. From soft X-ray ($\sim$ 3 keV) measurements by the Einstein and ROSAT missions, it is well known that M 87, the central galaxy of the cluster has a halo extending to $\sim$ 100'. Results of EUVE and ROSAT showed that in the EUV band of 65-248 eV, there was an excess flux within the central 30' radius which can not be associated with the cluster gas at X-ray temperature (kT $\sim$ a few keV), and one needs a second gas component of T $\sim 5{\cdot}10^6$ K (kT $\sim$ 0.1 keV). Thus a two component, warm and hot, gas model is required. These features are being investigated.

In this brief review we have focussed our attention to some of the highlights of the new discoveries in these areas, the details of which are reviewed by Konig and other authors (see gen. text ref. 2).

8.2.3 X-ray Astronomy

8.2.3.1 Introduction and General Features

The beginning of X-ray astronomy was a dramatic event with the discovery of the first X-ray star by B. Rossi and his co-workers(1) of M.I.T. in the USA. In 1962 they made a rocket flight with Geiger counters to measure the reflected solar X-rays from the moon; in this experiment they discovered an intense X-ray source in the constellation Scorpius, called Sco X-1. This strong X-ray source, however, was not associated with any bright stellar object; and later studies showed that it was associated with a faint blue star of the 18th magnitude. It was noted that its X-ray output relative to its optical emission was fantastically high and was about 10^6 times higher than that of the Sun! Soon a number of X-ray sources, such as supernovae remnants, like the Crab Nebula, CasA etc. and other objects like Cyg X-1, Cyg X-2 etc were discovered. In the following decade this new area of astronomy and astrophysics expanded with phenomenal speed so that by the early 1980's a large number of X-ray sources of many varieties were discovered.

By cosmic X-rays we generally include photons of energy of about 0.1 to 300 keV. The lower limit arises from the absorption of X-rays in the interstellar medium. In Figure 8.2.1 the absorption mean free path of X-rays in the interstellar medium is shown as a function of energy, where the ordinate is in parsec assuming the average density of the interstellar gas as 1 atom per cm^3. It is well known that the interstellar gas is completely transparent in the radio wave region. At 912 Å i.e. at photon energy of 13.5 eV (which is the ionization energy of hydrogen) the interstellar gas becomes suddenly opaque. At higher photon energies of $\geq$ 100 eV, interstellar gas becomes progressively transparent. For example at 0.1 keV the absorption mean free path is $\sim$ 6 pc for a gas density of $\sim$ 1 hydrogen atom per cm^3, which is a typical average value for the galactic plane. Thus we can observe soft X-rays of $\sim$ 100 eV from sources and sky background from distances $\leq$ 10 pc from the Sun. For observing X-ray sources in the galactic center region at d = 10,000 pc, the minimum X-ray energy is $\sim$ 3 keV. When we look in the direction of a galactic pole, where the mean gas density is much less, $\sim$ 0.02 atoms/cm^3, extra-galactic X-ray sources of energy > 0.3 KeV are detectable, e.g. M87. The Crab nebula is located at a distance of $\sim$ 1600 pc from the Sun in the anti-center direction where the mean gas density is $\sim$ 0.3 atoms per cm^3. Thus the Crab nebula X-ray source is detectable in X-rays with an energy > 0.8 keV which can be estimated from the figure. The absorption edge for X-rays by the K-shell of oxygen occurs at 0.52 keV which is shown.

The X-ray emissions from sources are generally grouped as soft X-rays in 0.1-1 keV, medium, 1-10 keV and hard X-rays > 10 keV. In addition to discrete sources a diffuse X-ray background was observed. The high energy X-ray background (> 10 keV) will be discussed in 8.3.2. In general X-ray emission takes place in hot plasma at a temperature of 10^6 to 10^8 by thermal bremsstrahlung i.e. the free-free transition process. In this process, the flux of photons with energies between E and E+dE keV as a function of energy (keV) is given by the exponential form

$$j(E)dE = K exp\Big(-E/E_o\Big)dE \quad photons/(cm^2 \cdot sr \cdot s \cdot keV)$$

where E_o is the characteristic energy, $E_o = KT_o$, T_o being the plasma temperature. Thus from the measurements of E_o the effective plasma temperature T_o can be determined. Numerically

$$T_o = 1.2 \times 10^7 \times E_o \quad (in \;\; keV)K$$

8.2.3.2 Observational Methods

X-ray astronomy detectors consist of a diversity of instruments, for the soft X-rays 1-20 keV and for hard X-rays of > 20 keV. The main elements of the instruments are: (1) a collimater, which may be baffles, wires, and a focusing device, which defines the field of view of the instrument; (2) a beam conditioner, which may be filters, gratings etc. and (3) a detector, which may be a Be-window proportional counter or a photoelectric detector. These units are rigidly mounted in the payload of a spacecraft together with an orienting device so that the X-ray telescope may be pointed to a given direction in the sky. An attitude sensor provides the information on the celestial direction of the telescope, i.e. its right ascension and declination. Then there are electronic systems to collect, store and telemeter the data to the ground.

One early successful X-ray astronomy instrument was flown in NASA's Small Astronomy Satellite (SAS-1) in early 1970. This instrument surveyed the sky with a sensitivity of $\sim 10^{-3}$ times Crab in the 1-10 keV energy range using proportional counters. The positional accuracy was $\sim$ 1' for strong sources. The Einstein X-ray Observatory launched in 1978 greatly expanded the knowledge of the X-ray sources using grazing incidence focusing devices. The mostly common used configuration is two surface focusing, first by a hyperboloid surface and then by a paraboloid reflecting surface to a focusing point. These grazing incidence focusing devices have a resolution of $\sim$ 5 arc sec or less. Japan's first X-ray astronomy satellite HAKUCHO launched in 1979 further expanded the field. The second X-ray satellite of Japan, TENMA, was flown in 1983 with an improved larger detector system. EXOSAT of ESA was also launched in 1983. In the 1990's highly sophisticated X-ray telescopes and detectors were built. These included the Granat mission of Russia, and the Broad Band X-ray Telescope (BBXRT) with the new X-ray optics of the USA. These types of high resolution optics were employed on the Advanced Satellite for Cosmology and Astrophysics (ASCA) jointly by Japan and the USA. The joint German-UK-USA mission of the Röntgen Satellite (ROSAT) is another landmark in this field. The missions in the 1980's and 1990's made numerous discoveries and advanced the understanding of the X-ray sources in the galaxy and in the extra-galactic region.

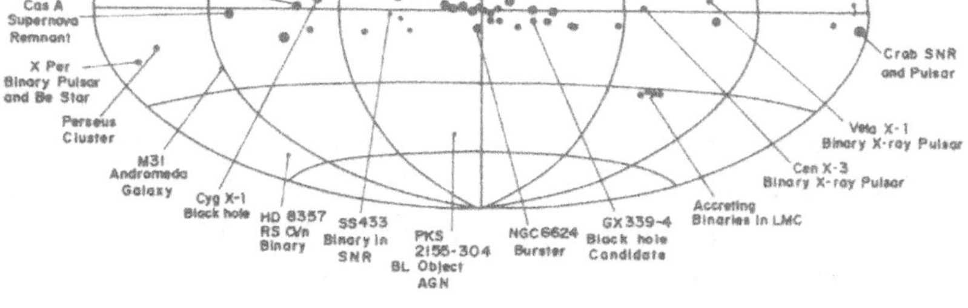

Figure 8.2.2: A sample of X-ray sources in galactic coordinates from different types of objects. (Adopted from the compilation of K.S. Wood of NRL)

8.2.3.3 Properties of X-ray Sources and Implications

Types and Distribution of X-ray Sources

In the decades of the 1980's and 1990's, spectacular achievements were made in the studies of wide varieties of X-ray sources and their diverse properties. The X-ray objects known till about the end of the 1980's are shown in Figure 8.2.2 in galactic coordinates. Bright sources are shown in larger circles. It is noted that there is a large concentration of sources in the plane of the galaxy, as well as in the galactic center region. A number of sources are indicated by their catalogue members and by the source type. Galactic X-ray sources include, (1) Binary X-ray sources such as Sco X-1, the strongest X-ray sources, (2) Supernova remnants, such as Crab, CasA, Tycho, etc, (3) Pulsars and X-ray binaries, such as Vela X-1, Cyg X-3, Her X-1, X-Per, etc., (4) Cataclysmic variables which consist of a white dwarf and a normal star, e.g. AM Her, EX Hya, etc. (5) RS CVn binary, e.g. HD 8357, (6) Contact binary e.g. VW Cep, (7) Black hole candidates, Cyg X-1, GX-339-4, (8) Flare stars, e.g. YZ CMi, and (9) Unusual systems like SS 433, which was identified as consisting of an accreting binary in a SN remnant. Extra galactic X-ray sources includes: (1) Nearby galaxy, M31 (Andromeda galaxy), Large Magelanic Cloud (LMC), etc, (2) Galactic Clusters, such as the Virgo Cluster, Coma cluster, Perseus Cluster, (3) Seyfert galaxies, e.g. NGC 4151, (4) Active Galactic Nuclei AGN, e.g. PKS 2155-304 which is a BL Lacertae object, (5) Bursters e.g. NGC 6624, MXB 1659-29, etc, (6) Quasars, 3C 273.

Spectral Features

We look back to the early years of the subject and note that one of the early determinations of spectral shape was made on the Crab nebula. This was the first X-ray source identified with the optical object of the supernova remnant by a lunar occultation technique by the NRL group(2) of the USA in 1964. The hard X-rays from this source were first detected from balloons(3) in 1965. Since then it was extensively studied in the 1960's and 1970's and the spectral shape measured by rockets and balloons is shown in Figure 8.2.3(a). The striking feature is the constancy of the power law spectrum of $dN/dE \sim K{\cdot}E^{-2.3}$ protons/cm^2·s·keV in the entire range of 1 to 500 keV(4); the fluxes in the 1-10 keV range were measured from rockets and in the 10-500 keV range

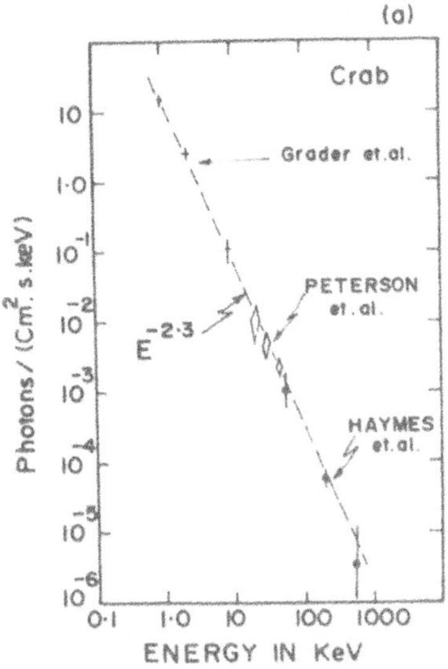

Figure 8.2.3(a): Spectrum of the Crab nebula

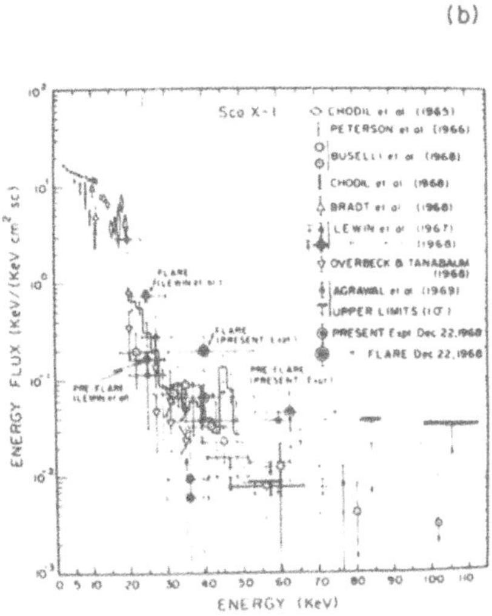

Figure 8.2.3(b): Spectrum of SCO X-1

from balloons. The flux and the spectral shape were found to be constant in time as measured for several years. The power law spectrum suggests that X-rays are produced by synchrotron radiation in a 10^{-4} gauss field of the nebula by energetic electrons. However the energy loss of these fast electrons producing 50-100 keV X-rays is so large that their lifetime would be only of the order of a few hours. The steadiness of the source indicates that there is some unseen source of energy which replenishes the loss. It was a major puzzle in the early and mid 1960's. The discovery of the radio pulsar in the Crab nebula and the identification of its optical pulsar[(5)] NP 0532 and the X-ray pulsar[(6)] added new dimensions to this field of high energy astrophysics. The pulsar is thus identified as the source of high energy electrons. The pulsars are discussed later.

The earliest and strongest X-ray source in the Sky at 3 keV, Sco X-1, showed major emission in the exponential form of $J(E) \sim C \cdot \exp(-E/E_o)$ in the energy range 15-50 keV, where $E_o \sim 5$ keV, suggesting thermal emission from hot thin gas at a temperature of $\sim 60 \cdot 10^6$ K. The exponent E_o was found to be variable by a factor of ~ 2, as observed by many investigators. Sco X-1 was identified with a 13th magnitude blue star which undergoes random fluctuations in intensity, some of the flickerings appearing as flare like events. A striking feature of Sco X-1 shown in Figure 8.2.3(b), was the sudden change in the intensities in the 30-50 keV band as observed by Lewin

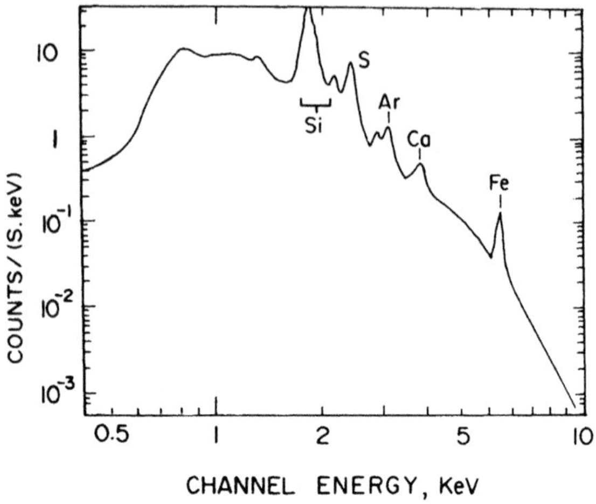

Figure 8.2.4: Spectrum of them Tycho SN-remnant dominated by lines of Si, S, etc to Fe, from ASCA

et al.[7] and Agrawal et al.[8] when the intensities changed by a factor of 3 to 4 during periods lasting for 15-30 min; these were denoted as X-ray flares. The model of Sco- X-1 suggests that the source is composed of hot gas at $\sim 50{\cdot}10^6$ K and is a compact object of diameter $\sim 10^8$ cm and has a density of $\sim 10^{16}/\text{cm}^3$.

With the above historical examples of the Crab nebula and Sco X-1 X-ray sources, we turn our attention to the recent decades of the 1980's and 1990's when the spectroscopy of X-ray sources progressed very rapidly both in resolution and sensitivity. With the new technologies of the past two decades, investigators can now identify the emission lines of several heavy elements such as Mg, Si, Ca, Fe etc. in X-ray sources. For example, the Broad Band X-ray Telescope (BBXT) in 1993 employed large area thin foil mirrors, which enabled studies of a large number of sources with high resolution. This technique was further used in the Advanced Satellite for Cosmology and Astrophysics (ASCA) launched in 1993. A typical example of the high resolution spectroscopy of the Tycho Supernova Remnant is shown in Figure 8.2.4 as obtained by ASCA[9]. Tycho is the remnant of SN 1572 of type Ia whose X-ray spectrum has been studied by many investigators. The X-ray spectrum shows the continuum emission of hot plasma and the emission lines from heavy elements of Mg, Si, S, Ar, Ca and Fe. The abundances of these elements in the SNR, are now estimated from this data and on comparing them with the solar system abundances, it is found that they are ~ 0.5

times the solar values. The X-ray spectrum can be understood in terms of a shock front propagating into the ISM and a reverse shock propagating into the uniform density ejecta and heating up of the gas. Narrow band X-ray maps of Tycho by ASCA suggest well mixed ejecta and uniform heating and expansion.

X-ray Variations and X-ray Pulsars

Most of the X-ray binaries show various type of intensity variations or pulsations, including the X-ray pulsars. An interesting example of X-ray pulsation is provided by the X-ray source, Hercules X-1, which was studied in the 15-125 keV range by many investigators. The source Her X-1 discovered in 1972 is a low mass X-ray binary (LMXRB) whose optical counter part is the main sequence, late type star HZ Her. The X-ray source shows a variety of temporal variations: (a) 1.24 sec pulsation which represents the X-ray pulsar period of the rotating neutron star; (b) 1.7 day period which is the rotation period of the binary system, and (c) a 35 day period on-off cycle which is interpreted as due to the accretion disk corona surrounding the rotating neutron star. In addition the Her X-1/Hz Her binary system shows a long term variations on a time scale of 1000-3000 days which is interpreted as due to a variable accretion rate. The X-ray pulsar is interpreted as due to a highly magnetized neutron star of which the magnetic dipole is misaligned with the axis of rotation, so that the accretion is funneled by the magnetic lines of force into the magnetic pole. Thus two hot spots produce X-ray pulses according to the neutron star rotation period.

The X-ray spectrum is best fitted with a thermal continuum of $KT > 5$ keV with an imposed iron fluorescence line at 6.5 keV and a prominent spectral feature at 60 keV which was first observed and interpreted by Trümper[(10)] as due to the cyclotron line. A variability of the cyclotron line on a time scale of 30 min was found by Manchanda et al[(11)]. The measurement of the cyclotron line energy in Her X-1 provided the first direct evidence of the magnetic field of the neutron star. In case of Her X-1, the magnetic field of the neutron star is estimated as $\sim 10^{12}$ gauss. However the observations were not suitable for the discrimination between an absorption feature at $\sim$ 35 KeV and an emission one at $\sim$ 50 KeV, although many prefer it as an absorption feature.

As X-ray pulsars are perfect astronomical clocks, an accurate Doppler

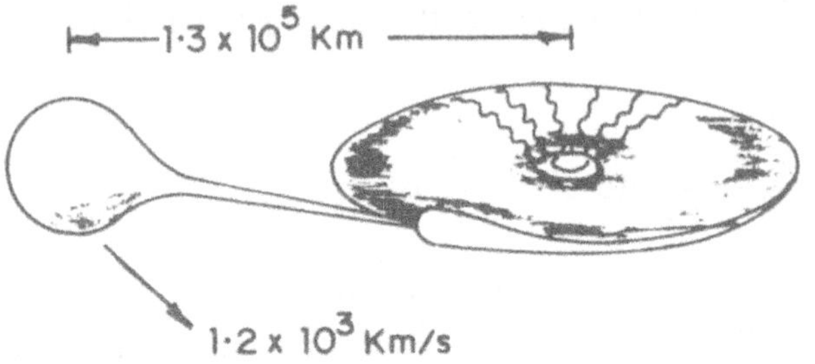

Figure 8.2.5: A sketch of the binary system, neutron star and accretion disk of the typical X-ray binary 4U 1820-30

measurement enables us to determine the orbital parameters and the mass of the companion stars. The binary systems were determined for a number of X-ray pulsars, all of which yield the neutron star mass as $1.4 \pm 0.2\ M_\odot$. Figure 8.1.7 shows the sketch of 4U 1820-30, neutron star, binary system and the accretion disk. The large gravitational attraction of the central object, a neutron star or a black hole, pulls matter out of the companion star forming an accretion disk surrounding the central object. Matter spiraling onto the neutron star heats up the gas by friction to several 10^6 K generating strong X-ray emission. Eclipsing of the accretion disk by the companion star causes the orbital periodicity of the X-ray emission. The details of X-ray binaries are discussed by Joos and Rappaport(12), (1984) and Tanaka(13) (1983) of earlier studies, and of recent works in the book edited by Agrawal and Vishwanath (1988).

X-ray Bursters

Many low mass X-ray binaries show X-ray burst phenomenon. In these the X-ray intensity suddenly increases to a luminosity level exceeding 10^{38} erg/sec, followed by a gradual decrease with a time constant of the order of 10 sec. Energy spectra of bursts are well represented by black body radiation and the black body temperature decreases with declining intensity. Thus X-ray bursts are interpreted as due to eruptive heating of the neutron star atmosphere which gradually cools down by black body radiation. It was found that the black body radius remained essentially constant at $\sim$ 10 km as expected for a neutron star. This implies that the entire surface of the neutron star is involved in the burst.

X-ray Emission and Black Holes

Cyg X-1 has been the earliest known black hole candidate noted by Oda[13]. The most convincing observation is the lower limit of the compact object as 8 $M_\odot$. The X-ray source in the Large Magelanic Cloud LMC X-3, is another good candidate as the lower limit of the mass of the compact object is 7 $M_\odot$. The source LMC X-1 indicates the mass of the compact body as > 3 $M_\odot$. Thus all these compact X-ray sources have masses well above the limit of the neutron star mass of 1.6 $M_\odot$ and are good black hole candidates.

In addition to the direct mass measurements these sources show other special features. The Cyg X-1 source exhibits: (1) very rapid intensity variations over a wide time scale from sub milliseconds to seconds, such rapid fluctuations are rarely seen in other sources; (2) a peculiarity of its bimodel spectral behaviour; Cyg X-1 shows two distinct states of its spectrum, a "low" state of the hard spectrum and a "high" state of the soft spectrum. The low state hard spectrum is represented by a power law, whereas the high state spectrum is distinctly softer and has a high energy tail above 10 keV, but with lower intensity, (3) the rapid intensity change seems to be suppressed in the "high" state and characteristic of the hard component.

The source GX 339-4, located near the galactic center exhibits similar behaviour as Cyg X-1. It shows striking bimodal spectral forms, rapid intensity changes in times of 100 m sec or less, and hard X-ray spectrum extending to 100 keV or more. This source has been extensively studied, including with hard X-ray imaging techniques.

Although several features of bimodal spectra of X-rays from the accreting black holes are not yet understood, the possible interpretations are given as follows. It is found that the observed spectrum of low mass binaries, such as, Sco X-1, GX 349+2, etc. are softer than X-ray pulsars and this is interpreted as due to the bremsstrahlung spectrum of hot gases at temperatures of (5-10)$\cdot 10^7$ K. In these sources it is found that if one plots the difference of the two spectra for high and low intensity levels, the difference spectra fits well with a black body spectrum of KT = 2 keV (Tanaka, 1983). Based on the accretion disc model, the 2 keV black body component can be identified with the emission from the neutron star surface and the soft component is considered to originate from the inner part of the optically thick disc. In this model X-ray emission from an accreting black hole, in the high state is interpreted as due to the absence of 2 keV black body spectrum (from the neutron star) resulting in the ultra soft spectrum with the tail of the hard component from an inverse Compton process. In the low state the spectrum lacks the ultra soft component, and the hard spectrum extending to 100 keV or more is presumed to originate from inverse Compton scattering of soft photons within some hot, optically thin region near the inner edge of the accretion disk. Several questions remain unanswered such as what causes the transition from one mode to the other, and whether the rapid flickering is a characteristic feature of a black hole X-ray source. Some of these problems are expected to be explored in the recent X-ray Timing Explorer (XTE) spacecraft.

8.2.4 Gamma Ray Astronomy

8.2.4.1 Early Years and Observational Methods

Soon after the discovery that the primary cosmic rays are composed of high energy protons, helium and heavier nuclei, it was suggested in the 1950's that high energy cosmic ray primaries, mainly protons would interact with interstellar hydrogen and produce π^+, π^-, π^o-mesons, and π^o mesons would decay to produce high energy γ-rays. The attempt to measure these γ-rays was made by W. Kraushaar and G.W. Clark of M.I.T. (USA) with the Explorer 10 satellite of NASA flown in 1961. They detected a small number of primary γ-ray events with energies $>$ 50 MeV coming from different directions in space. This was followed by Orbiting Solar Observatory-3 (OSO-3) launched by NASA in 1968, and the improved Small Astronomy Satellite-2 (SAS-2) by NASA in 1972 which detected about 8000 γ-ray events during its lifetime. The European Cos B mission in 1975 mapped the galaxy with γ-rays of $\geq$ 100 MeV with an angular resolution of $\sim 2^o$. The French-Russian SIGMA γ-ray telescope in the GRANAT satellite launched in December 1989 was the first γ-ray imaging telescope in orbit with a high angular resolution of $\sim$ min of arc. The Compton Gamma Ray Observatory (CGRO) has been the major great observatory of NASA, USA, following the Hubble Space Telescope. The CGRO was launched by the Space Shuttle Atlantis in April 1991. We discuss some of the major features and observations of this large

γ-ray mission.

CGRO is a 16,000 kg spacecraft containing four γ-ray experiments covering a wide spectrum of γ-ray energies from about 30 keV to 30 GeV. The Z-axis of the spacecraft could be pointed to any direction in the sky with an accuracy of 0.5^o. The four gamma-ray experiments were as follows:

1. Burst and Transient Source Experiment (BATSE) has been designed to identify and measure the γ-ray bursts in the energy range 30 keV to 2 MeV and to measure their intensity variations up to 1 m sec.

2. Oriented Scintillation Spectrometer Experiment (OSSE) measures spectra of γ-ray sources in the energy range of 0.1 to 10 MeV.

3. The Compton Telescope (COMPTEL) experiment has been designed to perform a sensitive survey in the range of 1 to 30 MeV γ-rays. The γ-rays are detected by two successive interactions, the first one is a collision in a low Z material and the second one in a high Z detector material. This is a wide field of view instrument with good angular resolution.

4. The Energetic Gamma Ray Experiment Telescope (EGRET) covers the highest and broadest energy range of about 30 MeV to 30 GeV. This instrument has a wide field of view as well as a high angular resolution and at the same time a very low background. The γ-rays are detected by e^+-e^- pair productions using a number of spark chambers and plastic scintillators, and their total energies are measured by a NaI crystal.

8.2.4.2 Emission Processes

Gamma-rays in the galaxy are produced by three important processes. These are, (a) nucleon-nucleon collisions, (b) bremsstrahlung of fast electrons and (c) Compton interactions. Diffuse radiation in the galaxy dominates over the contributions of discrete sources. Synchrotron emission by cosmic ray electrons spiraling around the galactic magnetic field is negligible for γ-rays.

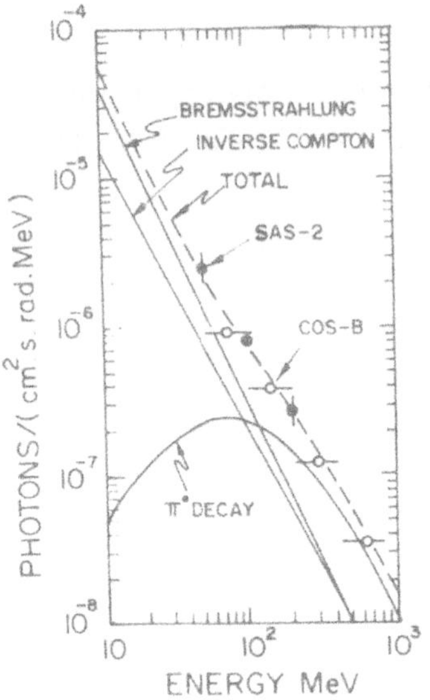

← 8·2·6

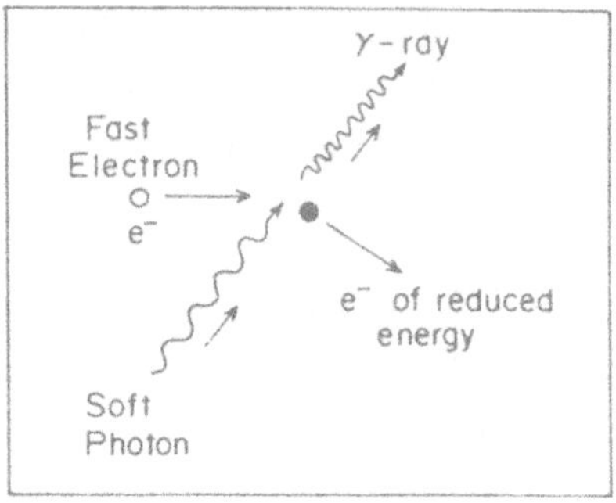

8·2·7

Figure 8.2.6: Calculated γ-ray fluxes from three processes, and measured values for the galactic center region
Figure 8.2.7: A sketch of the Inverse Compton process

(a) The nucleon-nucleon collisions:

High energy cosmic ray protons interacting with nuclei produce $\pi^{\pm}$ and π^{o} mesons. The π^{o}'s decay in 10^{-16}s to two γ-rays with energies of 68 MeV in the rest system of the π^{o} mesons. Several other types of K-mesons and hyperons also produce π^{o} mesons of lesser number. Detailed calculations of γ-ray intensities have been made by many authors. The γ-ray spectra from π^{o}-decay and the total are shown in Figure 8.2.6 as calculated for the galactic center region.

(b) Bremsstrahlung:

High energy electrons interacting with protons (and nuclei of heavier ions) of the interstellar medium produce photons with a wide range of energy (Chap. 7.). The largest γ-ray intensity originates from this process as shown in Figure 8.2.6. The calculated flux of γ-rays of less than 100 MeV from this process is not well known because the interstellar electron spectrum below 100 MeV is rather uncertain due to the strong solar modulation process.

(c) Inverse Compton Process:

Cosmic ray electrons also interact with star light photons of both the optical and the infrared region (or with the 2.7 K microwave black body radiation) to produce Compton γ-rays (Figure 8.2.7). In this process a high energy electron in its rest system sees an approaching optical photon as a γ-ray due to Lorentz contraction. The Compton collision between the γ-ray and the electron at rest takes place in the rest system of the electron. On transforming back to the laboratory system we have an electron with reduced energy and a γ-ray, called Compton γ-rays (Figure 8.2.7). It is shown by Ginzburg and Syrovatskii (1964) that a differential power law of the electron spectrum with index α will produce a γ-ray spectrum with index $(\alpha + 1)/2$. In case of Compton γ-rays, the energy of the parent electron E_e is related to the energy of γ-rays E_γ by the relation

$$E_\gamma \simeq \left(E_e/mc^2\right)^2 E_{ph} \qquad (8.2.1)$$

where E_{ph} is the photon energy before collision and mc^2, the rest energy of electron. Thus if energetic electrons interact with star light photons of a few eV, in order to produce γ-rays of about 100 MeV, the electron energy E_e should be several GeV. For 3 K radiation, the typical photon energy is about $8{\cdot}10^{-4}$ eV. Therefore the parent electron energy should be about 200 GeV for producing 100 MeV γ-rays.

The composite γ-ray production spectrum in the galactic region is shown in Figure 8.2.6. The primary electron fluxes were assumed to be the double of the measured values (ref. gen. text: 6).

8.2.4.3 High Energy Diffuse γ-ray Background in the Galaxy

Results from SAS-2[15], CosB[16] and EGRET in Compton GRO[17] showed the intense celestial radiation of high energy γ-rays from the galactic plane. It is found that the intensities of γ-rays from the galactic plane are strongly correlated with the physical features of the galaxy. The high resolution studies of EGRET in CGRO of the high energy, $>$ 100 MeV diffuse γ-ray background of our galaxy is shown in Figure 8.2.8. In this plot the known strong γ-ray sources have been subtracted from the total radiation. The measured flux in the galactic plane as shown in the figure matches very well with the calculated fluxes using the column densities of neutral hydrogen.

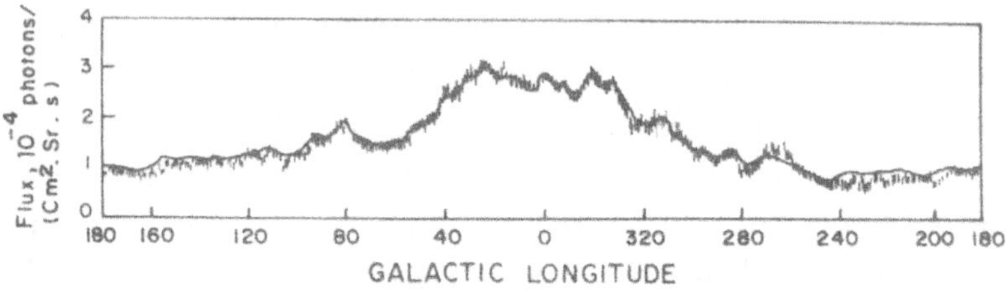

Figure 8.2.8: Measured and calculated fluxes of high energy (> 100 MeV) diffuse γ-rays within 10° of the galactic plane, vs galactic longitude. (Based on Fichtel and Trombka)

For example the radiation is particularly intense in the quadrant centered around the galactic center, and least intense in the general direction of the anti-center. The distribution in latitude for the longitude band of 210° is given by Fichtel and Trombka which shows that the bump at b = -10° to -20° matches the occurance of a large molecular cloud. The strikingly good agreements between the calculated and observed fluxes conclusively show that cosmic rays are of galactic origin, and the diffuse γ-ray flux can be well accounted for by the concept of cosmic ray interactions with matter and with photons. Further details are given in book by Fichtel and Trombka, 1997.

Cosmic Ray Distribution and Dynamic Balance

Studies of the high energy diffuse γ-ray background thus indicate that cosmic rays are primarily of galactic origin and are confined to the galaxy by the galactic magnetic field. The local cosmic ray energy density is about 1 eV/cm^3 which is about the same as the energy density of the galactic magnetic field and of the randomly moving clouds. Thus the total pressure of expansion of the cosmic ray gas, magnetic field and the random motion of gas cloud, is approximately balanced by the gravitational attraction, so that a rough equilibrium exists. Thus cosmic rays exert nearly the maximum pressure for the galaxy to be near the equilibrium state. Furthermore the magnetic field in the galaxy must have only closed lines of force, otherwise cosmic rays would leak out of the galaxy. The cosmic rays thus, have a larger scale height relative to matter in the disk and they spend only a small time

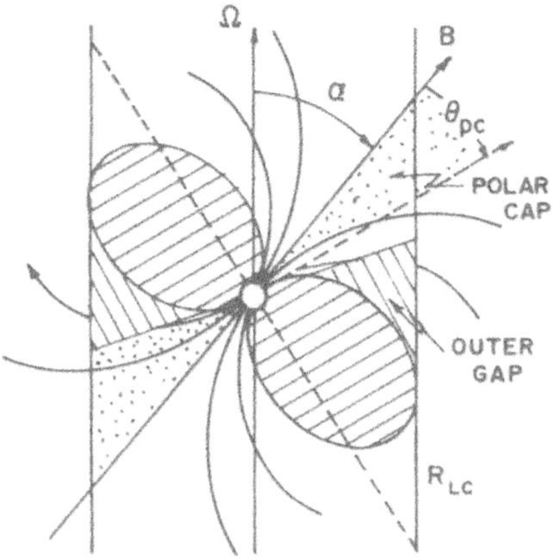

Figure 8.2.9: Schematic diagram of a Pulsar. α is the angle between the spin axis Ω, and magnetic axis $\vec{B}$. See text for details.

in the dense gas clouds and more time in the halo, which is important in determining the Compton contribution to diffuse emission. The large energy density of cosmic rays in the galaxy suggests copious production from cosmic ray sources.

8.2.4.4 High Energy Gamma Ray Sources

We shall briefly discuss the basic features of high energy γ-ray sources such as, (A) Pulsars, (B) γ-ray Burst Sources, (C) Nearby Galaxies, e.g. LMC, (D) Quasars and (E) Active Galactic Nuclei. Details are given by several authors, see e.g. the book by Fichtel and Tromka (1996).

(A) GAMMA-RAY PULSARS

Since their discovery in 1967, more than 500 radio pulsars were detected by 1996. Among these seven γ-ray pulsars have been discovered till now, and there are ten pulsars with pulsed X-ray emission. Pulsars belong to a special class of astrophysical objects which are believed to be highly magnetized, rapidly rotating neutron stars with magnetic axis aligned at an angle to the spin axis, as shown in Figure 8.2.9. In this sketch the magnetic axis is along

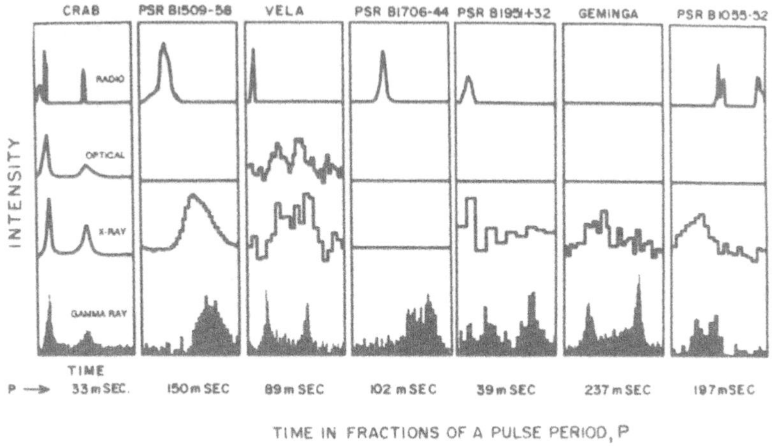

Figure 8.2.10: Diagrams showing pulsed emission of 7 gamma-ray pulsars. (Based on Ormes, 1996).

$\vec{B}$ and the spin axis is along Ω. The spin period varies from 33 msec for the Crab pulsar, and 89 msec for the Vela pulsar which are the two youngest ones to 237 msec for the old pulsar Geminiga. Both Crab and Vela show pulsed emissions up to several GeV γ-rays and the Crab has unpulsed emission up to several TeV or more. The pulsations look different when the pulsars are viewed in different wavelengths of radio, optical, X-ray and γ-rays as shown in Figure 8.2.10, for seven γ-ray pulsars. The time on the X-axis is plotted as a fraction of the pulse period i.e. for Crab the time scale is 0 to 1.0 times 33 msec, and the ordinate shows intensity on an arbitrary scale. The Crab shows the main pulse followed by interpulses in radio, optical X-ray and γ-ray bands; whereas such correspondance in different energy bands are not present in others. In Geminiga, there is no pulsation in radio and optical wave bands. This pulsar was first discovered in γ-rays and was named 'Geminiga' (meaning 'it is not there' in Italian) as there was no optical or radio counterpart.

The characteristic properties of seven γ-ray pulsars are given in Table 8.2.2. The age of the pulsar is estimated from the rate of slowing down of the spin period, as given by the formula, $\tau \approx P/2\dot{p}$. The luminosity shown in the Table refers to the γ-ray energy emission of > 100 MeV, assuming an emission cone angle of one steradian. The gamma-ray efficiency shown in the table is determined from the fraction of dipole spin down power which is radiated in pulsed γ-rays of energy > 100 MeV, i.e. η (> 100 MeV) $\equiv L_\gamma$ (> 100 MeV)/$\dot{E}$. In these calculations a large uncertainty arises from our lack of knowledge of the solid angle of emission. Here one steradian is used as emission angle. Inspite of this there is a noticeable trend that the efficiency η increases with increasing apparent age τ. This trend is consistent with the polar cap emission model of γ-ray pulsars.

Table 8.2.2: Characteristics of Seven γ-ray Pulsars

Pulsar	Period(s)	Age(yr)	Luminosity (erg·s^{-1})	Efficiency
Crab	0.033	1300	4.5x10^{35}	0.001
B1509-59	0.150	1550	1.9x10^{35}	0.010
Vela	0.089	11,000	2.3x10^{34}	0.003
B1706-44	0.102	17,000	3.7x10^{35}	0.011
B951+32	0.040	110,000	1.4x10^{34}	0.004
Geminga	0.237	340,000	2.6x10^{33}	0.080
B1055-52	0.197	530,000	5.0x10^{33}	0.150

(A) Models of Pulsar γ-rays

The models for pulsar γ-ray emission fall into two classes: The Polar Cap and the Outer Gap models. The polar cap model assumes that acceleration takes place near the surface of the neutron star, at the magnetic pole in the region denoted by θ_{pc} in Figure 8.2.9. Electromagnetic cascades develop whereby a large number of electron-positron pairs are produced which radiate most of the energy of the observed γ-rays. The e^+-e^- pairs produced in the cascades may be responsible for emission at radio, optical and X-ray bands. Radio emission properties seem to support the polar cap model in which polar cap cascades are produced by the primary accelerated electrons of $\gamma \geq 10^6$.

In the outer gap model it is assumed that the particle acceleration of most electrons takes place in the outer magnetosphere in the vacuum gaps that develops near the speed of light cylinder, RLC as shown in Figure 8.2.9. Electromagnetic cascades then develop as noted in the polar cap model. It may be noted that in the strong magnetic field in the polar cap model fast electrons radiate γ-rays by curvature radiation and by inverse Compton scattering or by both and these photons then produce pairs in the strong magnetic field which is called magnetic pair production.

This brief description is given as an introduction to the pulsar mechanism, the details of which involve various complexities. These are given in books on pulsars and their magnetospheres. The models for high energy acceleration and emission processes are still being developed and as yet no clear picture is available of these acceleration and emission processes.

In summary we note that

1. Pulsars must accelerate electrons to high energies at least to several GeV, and in principal can accelerate to 10^{16} eV.

2. The beaming geometry of pulsars is complicated and is not yet clear.

3. Models of γ-ray pulsars do not explain all observed features.

(B) Gamma Ray Bursts

One of the very surprising and unexpected results from the BATSE instrument in the Compton Gamma Ray Observatory is the observations of γ-ray bursts. Since the time of its launch on April 21, 1991 till May 27, 1994, thus experiment observed 1000 γ-ray bursts of energies 50-300 keV, in its all sky survey. The locations of the γ-ray bursts determined from the comparison of the response of the 8 detectors. The distribution of the first 1000 bursts in galactic co-ordinates is shown in Figure 8.2.11, which indicates isotopic distribution. Burst times range from milli sec. to several seconds, and some bursts have a complicated profile while others are simple. If the bursts are uniformly distributed in space their number, N vs intensity, I, distribution should follow a power law with an index of -1.5, e.g. $N(>I) = KI^{-1.5}$, where I is photon/cm^2·sec integrated over 50 to 300 keV. This hold for large bursts of $I \geq 10$, but not for smaller ones. How can one reconcile these two properties of γ-ray bursts? Both the solar system and the galaxy have boundaries which would lead to a curve like observed number as a function of intensity, but neither of these would be isotropic. The smaller bursts would be very anisotropic, which is not seen. In the galactic model there is no obvious galactic boundary to invoke. Thus we face a perplexing problem. It means that γ-ray bursts are either very local or cosmological. If the sources are galactic, the solar system should be situated near the center of the distribution, i.e. sources should be distributed through a larger region compared to 10 kpc, the distance of the Sun to the galactic center. On this scale size one may see the effects of merging of sources of our galaxy with that of Andromeda, our nearest galaxy. Such an effect is not seen. The distribution of sources could be very local, associated with our solar system, but there are no known objects as candidates.

If the burst sources are located at cosmological distances, the isotropy as well as log N vs log I relations are satisfied. Since there is no observed correlations with Virgo clusters, burst sources should be located beyond, at redshift Z=1 or more. In this case energy requirements of the burst events become enormously large of the order of 10^{51} erg.

In some recent observations it is found that a subset of γ-ray bursts, called soft repeaters, are associated with neutron stars and supernova remnants. One has a pulsar that has been apparently ejected from a nearby neutron star. It is hypothesized that such high velocity ($\sim$ 1000 km/s) neutron stars may be formed by a slight asymmetry in the explosion of the star which becomes a supernova, and perhaps such high velocity neutron stars may be common and may form a 100 kpc halo around our galaxy. This may interact with the halo of the nearby Andromeda galaxy. This hypothesis of high velocity neutron stars is yet to be verified. There were some reported observations in the Ginga satellite of Japan of the presence of 20 and 40 keV spectral features suggesting possible cyclotron lines from high magnetic fields. These features were not confirmed by the BATSE results.

The unusual observation of a delayed high energy γ-ray burst was made by EGRET in CGRO on February 17, 1994. It was a large burst seen also by ULYSSES. The striking feature was the continuous emission of bursts even after one hour when the satellite could see the source again as it emerged from the earth's eclipse, and there were photons with energies as high as 25 GeV which is the highest in γ-ray bursts. This example shows the high complexity of γ-ray bursts. Therefore the origin of γ-ray bursts remains a fascinating problem in high energy astrophysics.

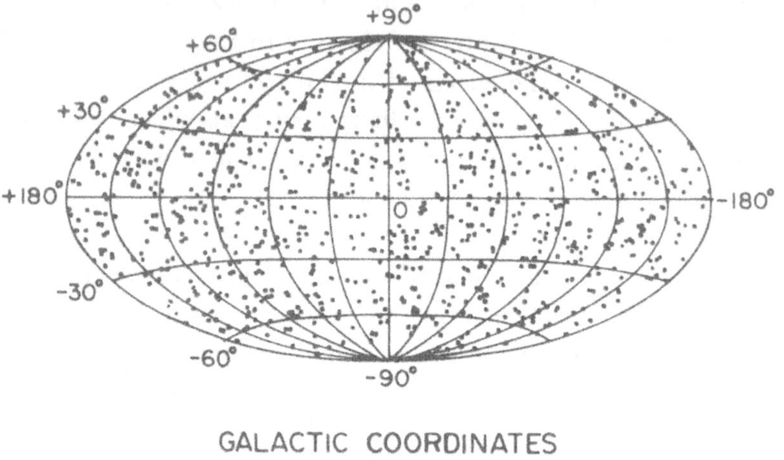

Figure 8.2.11: The distribution of 1000 γ-ray bursts observed by BATSE in the Compton Gamma Ray Observatory.

(C) Nearby Galaxies

The Magellanic clouds are the nearest galaxies to our Milky way and they provide suitable opportunity to study several astrophysical problems of high energy γ-ray astronomy. The Large Magellanic Cloud (LMC) has a stellar mass of about $1{\cdot}10^{10}$ $M_\odot$, (compared to the $1.4{\cdot}10^{11}$ $M_\odot$ of our galaxy) and is located at a distance of 52 kpc. The Small Magellanic Cloud (SMC) has an estimated mass of $1.7{\cdot}10^{9}$ $M_\odot$ at a distance of about 63 kpc. The EGRET instruments detected high energy γ-rays from LMC for the first time and the integrated flux of γ-rays above 100 MeV was found(18) to be $(1.9 \pm 0.4){\cdot}10^{-7}$ photons/(cm^2·s). This value agrees well with the calculated flux for the quasi-equilibrium intensity of cosmic rays. Using LMC γ-ray results and calculations of γ-ray emission by cosmic rays, it was shown by Fichtel et al. (1996) and Chi and Wolfendale(19) that cosmic rays seen near the earth are of galactic origin.

One interesting feature of LMC is the observation of a Supernova explosion in 1987 in LMC. The 122 keV line from Co^{57} radio-active decay was still visible after 5 years.

In case of SMC, only on upper limit of γ-ray flux of $0.5{\cdot}10^{-7}$ photons/($cm^2{\cdot}s$) of $>$ 100 MeV could be determined. This indicated that the quasi-equilibrium between cosmic ray intensity and other parameters did not exist in SMC.

(D) Active Galactic Nuclei

A rich field of study opened with the first detection and measurements by the EGRET instrument in the Compton Gamma Ray Observatory of high energy γ-rays of $>$ 100 MeV from a number of Active Galactic Nuclei (AGN). The term AGN includes a variety of objects such as, Quasars, BL Lacertae objects, Seyfert galaxies, and radio galaxies. The characteristic feature of all these objects is the unusually bright central region compared to the Milky Way Galaxy. The most familiar members of AGN are the Quasi-stellar objects, in short, the Quasars which are the brightest members of AGN. Since their discovery Quasars presented an enigma. What makes Quasars so exceptional is that an enormous power exceeding a thousand times that of the entire galaxy is emitted by a region that is rather small compared to galactic standards and whose size is about that of the solar system. Because the Quasars are very far away and much brighter than the host galaxy, it is almost impossible to detect the associated galaxy which is much fainter; it appears that quasars represent a naked source of energy. Seyfert galaxies are much nearer objects whose luminosities are about $\sim 10^{-2}$-10^{-3} times that of Quasars. The absolute luminosity of Quasars is about 10^{14}-10^{15} $L_{\odot} \simeq 10^{45}$-10^{49} erg/s; and that of Seyfert galaxies $\sim 10^{41}$-10^{44} erg/s.

The requirements of such tremendous power output signifies only one choice of source of power for AGN which is gravity. Thus the standard model of AGN, although not accepted by all astrophysicists, is that it consists of a super-massive black hole of 10^6-10^9 solar mass surrounded by an accretion disc formed by materials spiraling into the center. Often a "jet" of matter and radiation emanuate in the direction perpendicular to the accretion disc, along the rotation axis of the black hole. It is generally agreed that the awesome power of AGN originates from matter accreting into a black hole, so that an efficiency of about mc^2 could be achieved, in contrast to nuclear reactions where efficiencies of 10^{-3} mc^2 are achieved. However despite about two decades of studies, the details of energy release and accretion mechanisms remain obscure and elusive.

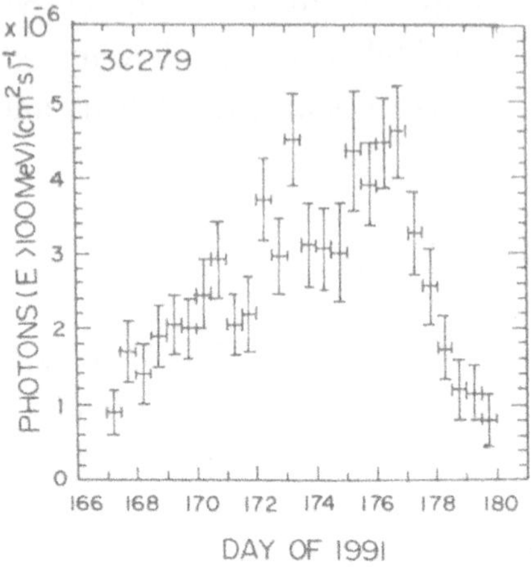

Figure 8.2.12: Variability of the γ-ray flux of > 100 MeV of Quasar 3C 279.

Gamma-ray results opened new windows for studies of AGN which fall in two groups as follows:

(A) The OSSE instrument in CGRO identified several Seyfert galaxies which emit soft γ-rays of 0.1-10 MeV. The well studied Seyfert NGC 4151 showed variable fluxes in the MeV range and the spectrum steepened beyond 0.3 MeV in the OSSE results, so that these Seyfert sources are not seen in higher energy γ-rays in the COMPTEL and EGRET instruments in CGRO. The energy spectra of these objects seem to suggest that γ-ray emission takes place in the hot gas in the accretion disc. Beaming effects are unlikely from these objects. The γ-ray luminosities of these are in the range $\sim 10^{41}$-10^{44} erg/s.

(B) The strikingly new discovery by EGRET in CGRO is that a large number of active galactic nuclei show strong emissions of high energy γ-rays of > 100 MeV. In the survey of the sky in > 100 MeV γ-rays more than 50 AGN were found to be strong emitters of high energy γ-rays. These Active Galactic Nuclei include Quasars which are Optically Violent Variables (OVV), BL Lacertae objects and several radio galaxies with super luminal components (i.e. which emit very high velocity gas jets with apparent speeds greater than the velocity of light). A list of these AGN objects and their γ-ray emission properties are given by Fichtel et al.[(17)]. Here we present striking features of some of them.

A remarkable feature of the observed AGN is that nearly all of them are not close objects, but have broad distribution in the redshift index, Z, and many of them have values greater than 1.0. In Figure 8.2.12 the results on the variability of γ-ray fluxes of energy $>$ 100 MeV as a function of days are shown for the Quasar 3C 279, as measured in June 1991. The surprising feature of this Quasar is the observation of significant changes in the γ-ray flux of $>$ 100 MeV on a time scale of 1-2 days. The observed variability of the source in a time scale of ΔT indicate that the size of the emitting region(R), cannot be larger than that which light can travel in a time ΔT, i.e.

$$R \leq C(\Delta T)/(1+z)$$

where z is the redshift of the object. The term (1 + z) denotes the Doppler shift due to the motion of the object away from us. Thus for ΔT $\sim$ 1 day as seen in the figure, and z = 0.158, the value of R is about 10^{15} cm. Thus the size of the emitting region is estimated to be about 10^{10} km i.e. about the size of the orbit of Pluto. This relatively small size suggests that γ-rays are produced near the central black hole of the AGN. Such a region has a high density of X-ray and other photons. If the γ-rays are produced isotropically their absolute intensity would be very high and this also leads to other factors such as absorption in the source. Therefore the solution to these problems is given by the hypothesis that γ-rays probably, originate in jets. This class of AGN are called 'BLAZARS'.

Thus BLAZARS are the AGN objects in which the γ-ray emissions take place in the "jet" which is directed towards the earth. Based on this model the beaming effect reduces the absolute luminosity requirements by a factor of 10^2-10^3, to the range of 10^{45}-10^{46} erg/s. Even then this energy is extra ordinary. The theoretical models of Blazars require that the acceleration of particles to very high energies take place in the jet, and these high energy particles produce high energy γ-rays by several probable mechanisms. These are yet to be worked out in detail.

A striking AGN object is Quasar 1633+382 whose γ-ray spectrum was measured by EGRET from 50 MeV to 5 GeV as shown in Figure 8.2.13. When such a spectrum, e.g. of Blazar PKS 0208-512 is plotted as energy flow spectrum vs wavelength such as log [$h\nu \cdot F(\nu) d\nu$] vs Log ν and compared with similar plots for X-ray, optical and radio regions of frequency we see the surprising aspect as shown in Figure 8.2.14. It is seen that the power per

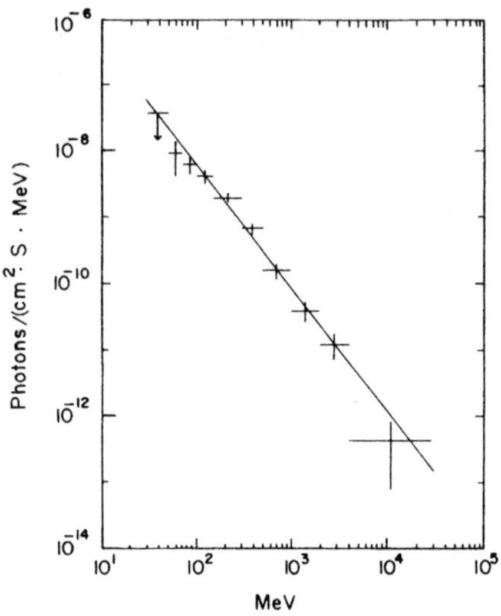

Figure 8.2.13: γ-ray photon spectrum of AGN 1633+382 as measured by EGRET.

decade of frequency in γ-rays exceed those of X-ray, optical and radio by an order of magnitude or more. If the γ-radiation is isotropic the gamma-ray luminosity $L_\gamma = 4\pi\ F_\gamma\ D^2 \approx 10^{48}$ erg/s which is extra-ordinary large.

The AGN Markarian 421 (1101 + 384) which is a BLac object with super luminal features has a photon spectrum with power law index -1.9, as seen by EGRET. The Whipple observatory ground level telescope measured with atmospheric Cerenkov radiation method, a positive intensity signal at a γ-ray energy of $5{\cdot}10^{11}$ eV, which is consistent with the extrapolation of the energy spectrum seen by EGRET. It is also found that this AGN showed large day to day variability in the TeV range and γ-rays during the burst of May 7, 1996 extended from 0.3 to 8 TeV.

Thus it is found that AGN and their γ-ray emissions represent several new fascinating problems in high energy astrophysics which are yet to be understood.

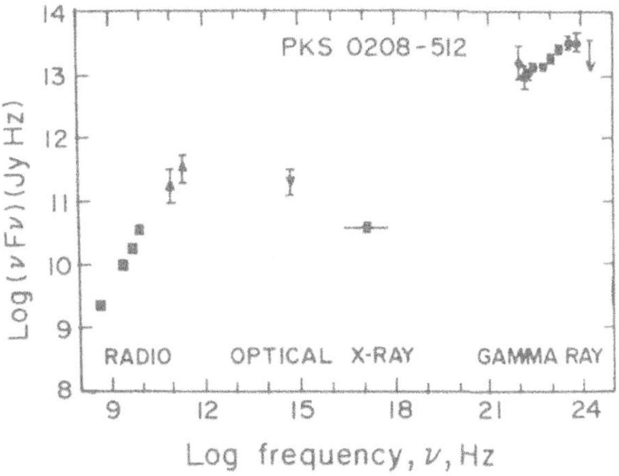

Figure 8.2.14: Log observed power $\nu F(\nu)$ vs frequency ν for AGN PKS 0208-512 showing the multi-wavelength power spectrum per decade of frequency.

8.3 Cosmology from Space

8.3.1 Cosmic X-ray Background and Cosmology

Early Observations

The existence of the isotropic cosmic X-ray background (CXB) in the 2-20 keV energy range was established since the time of the discovery of discrete X-ray sources. The early studies with the UHURU satellite in 1974 confirmed the extra galactic origin of the CXB and its cosmological significance from the observation that fluctuations in CXB were less than about 3% on an angular scale of 10° (Giacconi and Gursky, 1974). Studies of a number of groups using balloons rockets and satellites, yielded the energy flux spectrum of cosmic X-ray and Gamma-ray background, as compiled by Gruber[(20)] in 1992, which is shown in Figure 8.3.1. Here the photon flux J(E) dE is multiplied by photon energy E, to derive the photon energy flux I(E)dE, and this is best represented by two power law fits in the form

$$I(E)dE = 8.5E^{-0.40} \quad for \quad E = 1 - 20 \ keV$$
$$167E^{-1.38} \quad for \quad E \approx 20 - 100 \ keV \tag{8.2.1}$$

In 1-2 keV, the separation between the galactic and extra galactic components are difficult. But the galaxy is transparent to X-rays of ≥ 2 keV. Therefore it is concluded that the isotropic X-ray background in the 2-200 keV

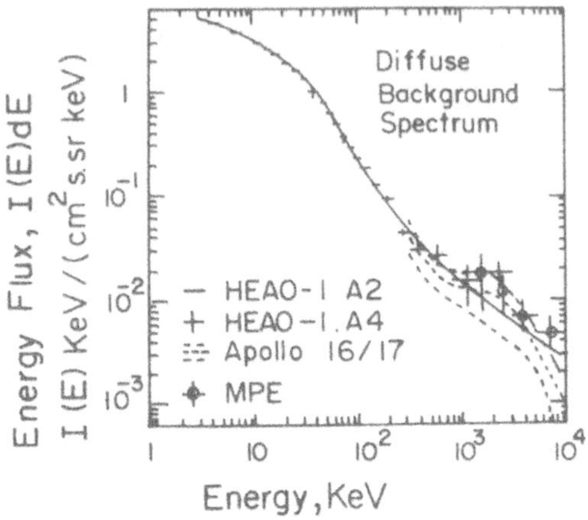

Figure 8.3.1: Diffuse extra-galactic background spectrum of 1 keV to 10 MeV.

range is of extra-galactic origin. Then one asks the question whether the CXB originate from the superposition of emissions from many unresolved X-ray sources such as AGN, Quasars etc, or due to diffuse X-rays originating from the hot intergalactic medium. These bring up several cosmological questions relating to the origin of CXB.

On the Origin of the Cosmic X-ray Background

If one asks the question from what ranges of distances cosmic rays from discrete sources may be coming to us, we note the early estimation by Swartz and Gursky in 1976 as follows. For the uniform emissivity function (in co-moving co-ordinates) throughout the universe, one finds that about 80% of the CXB would come from distances with redshift $Z \leq 1$. If we assume some type of evolution such as, in the simplest form, with an additional factor of $(1+Z)^3$, as would be expected for emissivity depending on ρ^2, then it is found that only $\sim$ 10% of the CXB would come from a distance $Z < 1$ and most of the background must originate at $Z > 1$. Thus cosmological evolution is intimately related to the origin of CXB.

The early interpretation of CXB as of diffuse origin was made by Cowsik and Kobetich[(21)] in 1972 who proposed that CXB originates from thermal bremsstrahlung of hot intergalactic gas. By fitting the model to the observed spectral data, they computed the temperature of $3 \cdot 10^8$ K and density of hydrogen atoms as $\rho \simeq 3 \cdot 10^{-6}$ atoms/cm^3 with $\rho = \rho_c$ for a closed galaxy and the Hubble constant as 55 km/s·Mpc. Several authors made critical studies of the model and concluded[(22)] that the required temperature is too high as this would make the energy content too large. Furthermore this high temperature would evaporate the HI clouds found to exist in the intergalactic medium (IGM). Also the question arises how such a large heating of the gas in IGM could occur.

Since 1970 new observational results became available on the number counts of extra galactic sources as a function of detected fluxes and X-ray emissions for extra galactic sources. These findings shifted the question of the nature of CXB to whether they could truly be due to unresolved contributions of individual sources or are partly due to a diffuse process. UHURU sources of luminosity up to 10^{-11} erg/cm^2·s in the 0.3-3.5 keV range indicated the number density distribution of X-ray sources at high galactic latitudes ($b > 20^o$), showed a power law in the Log N(>S) vs Log S plot with an index of -1.5. Therefore speculations were made on the types of sources below the UHURU limit, which could contribute to diffuse CXB. At that time only one Quasar 3C 273 was known as an extra-galactic X-ray source; and it was suggested that perhaps the integrated flux of all quasars could explain the CXB.

Direct measurements of discrete sources and their intensities by the Einstein Observatory (HEAO-2) largely advanced the knowledge of fainter sources and their contribution to the isotropic X-ray background. The Einstein observatory extended the lower limit of intensity of X-ray sources observed to $1.3 \cdot 10^{-14}$ erg/cm^2·s in the 1-3 keV range, which was 3-orders of magnitude lower than the UHURU limit. Thus a Log N vs Log S distribution could be made for much fainter X-ray sources up to the Einstein limit and the distribution of AGN and clusters could be made in Log N - Log S plots to fainter sources of $S \sim 10^{-13}$ erg/cm^2·s. From these and other studies the following conclusions are derived on the origin of CXB:

1. It is established that a substantial[(23)] fraction, ∼ 50%, of cosmic X-ray background originates from the contributions by individual unresolved sources such as ∼ 1% from normal galaxies, 6% from cluster of galaxies, 18% from Seyfert galaxies, and a significant fraction from Quasars.

2. There is no direct evidence for the existence of a truly diffuse component of the cosmic X-ray background. Present studies place the upper limit as 10-20%. These limits are derived from the present estimates of baryonic matter from the existence of diffuse HI clouds in IGM, and from the survival of the clusters. Also it may be noted that as about 50% of CXB is known to originate from individual sources, it seems unlikely that the remainder of CXB can be due to thermal bremsstrahlung of a very hot and thin gas.

3. It is found by several authors[(23)] that the spectrum of the remainder of CXB, after subtraction of the contributions of known sources, (which is softer than CXB), becomes very hard. A power law index of $\leq$ 0.2 is required. It appears that the remainder of CXB originates in compact extra galactic sources of some kind capable of producing the required hard spectrum. This requires a different emission spectrum from members of populations such as QSO's at an early epoch or evolutionary effects, or a new class of objects (e.g. proto galaxies). This subject of study thus offers challenging ideas for new findings and discoveries.

Isotropy of the Cosmic X-ray Background

At high galactic latitudes the cosmic X-ray background is essentially isotropic on large angular scales. The small scale fluctuations of the CXB surface brightness may be ascribed to statistical variations in the population of unresolved discrete sources. The population of sources represented by the HEAO-2 sample seems to account for about 20% of CXB in 3-10 keV; hence it is concluded that CXB itself must be dominated by sources other than those responsible for most of the fluctuations. This means that the number of sources fainter than in the HEAO-2 deep space survey should be high.

The large scale brightness variations have been studied in the HEAO-1 (A2) Sky survey in the X-ray band of 2-40 keV. For this purpose the region of Sky analyzed was the ecliptic latitude band, β = -24° to +24° and ecliptic longitudes λ = 0-360°. Of particular interest are some specific directions which cluster in the longitude band, $\lambda \simeq 160^o$-190°, in the equatorial region of the ecliptic, i.e. $\beta \simeq \pm 24^o$. These directions are: (A) at $\lambda = 164^o$, the direction of the dipole anisotropy of the microwave background measured by Smoot et al.[24]; (B) the same at $\lambda = 185^o$ as measured by Cheng et al.[25]; (C) the direction of the intersection of the plane of super galaxy at $\lambda = 195^o$, and (4) the direction $\lambda = 170^o$ which represents the direction of motion of the Sun with respect to distant galaxies. Figure 8.3.2. shows the results of measurements in which the circle represents isotropy and percent deviations up to 1% are plotted as a function of the ecliptic longitude λ. Each interval correspond to a region of 800 sq.deg. for which the average surface brightness is determined with a statistical accuracy of 0.1 to 0.2% from photon counting noise. The galactic plane deviations are shown by a dashed line and are not relevant here. Considering only the solid line portions of the plot where galactic effects are negligible, a residual large scale asymmetry of CXB becomes evident. The specific preferred directions are: (i) the dipole anisotropy direction of the microwave background "A" at $\lambda = 164^o$, as given by Smoot et al.; and "B" at 185° as given by Cheng et al.; (ii) the solar velocity direction relative to distant galaxies, "C" at $\lambda = 170^o$, and (iii) the longitude at which the super galaxy plane crosses the ecliptic plane at "D" at $\lambda = 195^o$. From the results shown in Figure 8.3.2. the following conclusions are derived:

I Based on the HEAO-1-A2 results, the large scale anisotropy of CXB is consistent with the direction of motion of the Sun relative to distant galaxies. This direction is also compatible with the dipole anisotropy of the microwave background radiation.

II Within experimental uncertainties, the anisotropy of CXB is consistent with the interpretation of a possible component of extra-galactic background correlated with the direction defined by the physical center of super galaxy at $\lambda = 184^o$ in the direction of the Virgo cluster.

III Using the anisotropy direction of the cosmic microwave background which provides the precise direction of the observers velocity with respect to the proper frame of CXB, the investigators have determined the forward/backward asymmetry of CXB vs energy, i.e. $\Delta I/I$ vs E as given by the Compton-Getting effect. The results of $\Delta I/I$ for CXB are obtained as $\sim 10^{-2}$ for 1-10 and 10-20 keV which are consistent with the Compton-Getting effect calculations, as given by

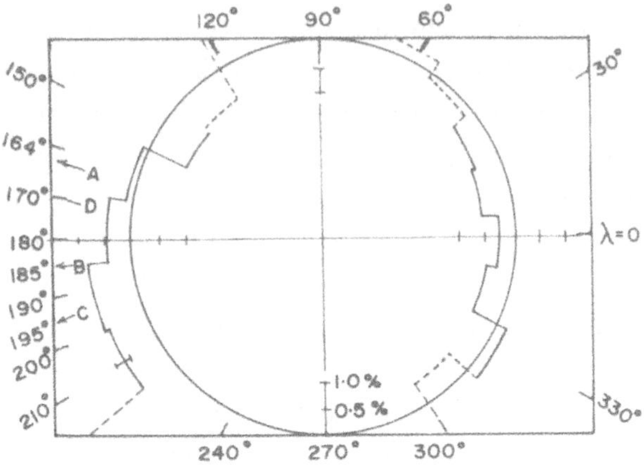

Figure 8.3.2: Anisotropy of CXB. Percent deviations from average vs ecliptic longitude λ, for $\beta = 0 \pm 24^o$ are shown. See text for details.

$$\Delta I/I = (v/c)(1 + \cos\delta)(3 + \alpha(E/B))$$

where δ is the half cone angle defining the opposite region of the sky relative to the axis of motion, $\alpha = 0.29$, the spectral index of CXB of energy E, and B = 40 keV.

IV However an anisotropy in the foreground distribution of faint unresolved extra galactic sources, e.g. correlated with super galaxy, exhibiting spectral index $\alpha \leq 0.7$ could not be ruled out as a major cause for the asymmetry observed for the CXB [Holt(26), 1987]. Possible enhancements of CXB associated with the Super Galactic Plane have been investigated by Jahoda(27) (1993).

8.3.2 Extra Galactic Gamma Ray Background and Cosmology

8.3.2.1 General Features

Diffuse and isotropic celestial radiation was found to exist from the X-ray region of ~ 2-20 keV to the low energy γ-rays of about 200 keV from several spacecraft observations (see e.g. Figure 8.3.1). The early evidence of a celestial or Cosmic Gamma Ray Background in the 1 MeV range of the spectrum was obtained by Arnold et al.(28) in 1962. Above 10 MeV isotropic and extra-galactic gamma radiation was first detected by Kraushaar and

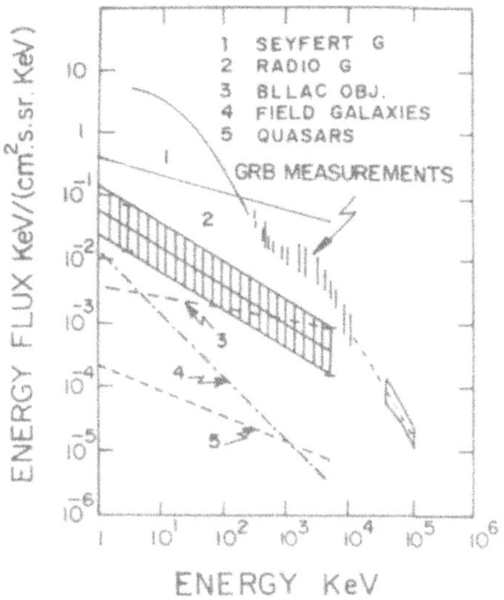

Figure 8.3.3: Diffuse extra-galactic γ-ray background, measured by several groups and theoretical estimates from several sources.

Clark[29] in 1962 with the Explorer II spacecraft. High energy (> 30 MeV) diffuse γ-rays were first measured with Orbiting Solar Observatory (OSO-3) by Kraushaar et al.[30] in 1972. The SAS-II γ-ray experiment in 1977 clearly established[31] a high energy extension of the diffuse extra galactic γ-rays of 30-200 MeV with a steep spectrum. The Compton GRO experiment[32] yielded high quality extra-galactic γ-ray fluxes and the energy spectrum from 30 MeV to 8 GeV.

Spectral features of the diffuse γ-ray background as measured by a number of investigators are shown in Figure 8.3.3. as compiled by Bassani et al. in 1985. The SAS-II results analyzed by Fichtel et al. in 1978 yielded a steep spectrum of extra-galactic γ-rays of 30-150 MeV.

The extra-galactic GRB is of special interest in cosmology because the universe is transparent for high energy γ-rays back to redshifts of 100 or more. A determination of the isotropy of GRB can provide sensitive tests of its cosmological origin and of different cosmological models.

8.3.2.2 Composite Spectra of GRB

The composite spectra of GRB including the recent results from the COMPTEL and EGRET experiments in the Compton Gamma Ray Observatory are shown in Figure 8.3.4. Here the ordinate is plotted as E·I(E) which represents the energy intensity I(E) multiplied by energy E, i.e. intensity per logarithmic energy interval. This plot allows us to distinguish small changes in the slope of the spectrum more easily. The energy intensity I(E) is derived from the photon flux of energies between E and dE, multiplied by the photon energy E.

The low energy part ($\sim$ 10-60 keV) of the energy flux spectrum is characterized by a bremsstrahlung spectral shape which could also be approximated by a power law with a spectral index of -0.4. Above 60 KeV it becomes a power law with index 1.6 and then flatters out with an index of $\sim$ 0.7 around 1 MeV. It steepens again above several MeV to an index $\sim$ 1.7 that extends to well above 100 MeV (Figure 8.3.3). When plotted in intensity per logarithmic energy interval EI(E), the GRB spectrum (Figure 8.3.4) shows two peaks, one at $\sim$ 30 KeV due to thermal bremsstrahlung and the other at a few MeV, called "the MeV bump". Between 10 to 30 MeV GRB data have large uncertainties because of the high atmospheric background. The COMPTEL data provides information in this energy band and the MeV bump is revised downward. Above 100 MeV, improved data is available from EGRET up to $\sim$ 5 GeV.

8.3.2.3 Models of the Origin of Extra Galactic GRB

Several theoretical models have been proposed to interpret the extra-galactic gamma ray background (GRB). A brief summary is given here for some of these models:

[A] Low Energy GRB's from Thermal Bremsstrahlung

The empirical fitting of the bremsstrahlung spectrum in the low energy region led to the early models of its origin in a hot, diffuse intergalactic gas at a temperature of $\sim 4{\cdot}10^8$ K. These models were however ruled out by the recent observations of the microwave background of 3K black body radiation. The COBE results rule out the possibility that the intergalactic medium is filled with hot gas.

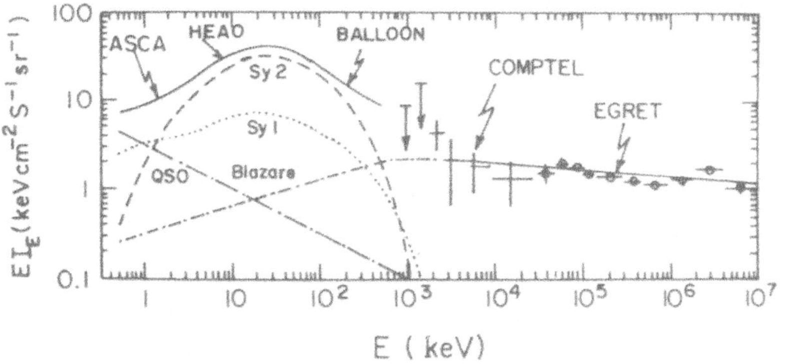

Figure 8.3.4: Extra-galactic Cosmic Gamma Ray Background intensity in energy flux per logarithmic Energy interval ($E.I_E$) vs Energy. Measured results are shown by ASCA, HEAO, Balloons, COMPTEL and EGRET. Estimates from the sources are shown. (See text for details).

[B] The other models attempt to explain the low energy GRB spectrum by the integrated fluxes of extra-galactic sources. The difficulties of these models are that Active Galactic Nuclei (AGN) such as Seyferts and Quasars, which are likely contributers, produce continuum spectra that are significantly different from that of GRB. The average AGN spectrum has a power law index of ~ 0.7 in contrast to that of the GRB spectrum, 0.4 below 60 KeV. Taking into account possible evolutionary effects of AGN sources deduced from X-ray studies, it is found that Seyferts are likely the sources which make the most important contribution to GRB at low energies. Radio galaxies, quasars, field galaxies and BL Lac objects contribute to a small extent. Seyfert contribute $\sim$ 20% at 40 keV which increases to $\sim$ 160 keV where they can account for almost 100% as seen in Figure 8.3.4. After subtracting the Seyfert contribution, a significant residual flux of GRB remains. CGRO studies propose 'reprocessed' AGN spectra to explain the low energy GRB. These models can produce a spectral break in the Seyfert spectrum at $\sim$ 80 keV. As the MeV bump is greatly reduced in the CGRO results, it seems that the AGN model can explain a good part of the GRB spectrum in the MeV range.

[C] High Energy GRB from $p\bar{p}$ annihilation

Stecker[33] proposed in 1971, 1985 a diffuse emission model in which GRB arises in a baryon symmetric Big-bang cosmology from matter-antimatter annihilations. The annihilations occur at the boundaries between regions at all redshifts producing π^o which decays to high energy γ-rays. This model produces a maximum contribution to GRB in the 1-10 MeV range.

[D] GRB from Primordial Black Holes

Page and Hawkings[(34)] postulated in 1976 that a population of primordial black holes (PHB) created in the early universe could explain GRB. PBH's created with an initial mass less than the critical mass of $5{\cdot}10^4$ g would have completely evaporated by now. PBH's of slightly larger mass would be radiating energy at the rate of $2.5{\cdot}10^{17}$ erg$\cdot$s^{-1}. Assuming these PBH's to be of mass $\geq 5{\cdot}10^4$ g, their contributions are calculated on cosmological time scales. Thus the authors derive a GRB photon spectrum with power law index $\sim$ 3 for $>$ 100 MeV, which is consistent with the slope of high energy GRB in the 100 MeV range.

[E] GRB from SN Type I

Clayton and Silk[(35)] proposed in 1969 that the cumulative effect of γ-rays from Type I SN during the history of the universe can contribute to the GRB. SN of Type I emits γ-ray lines with energies of 0.8 to 3.2 MeV from the radio active decay of ^{56}Ni $\rightarrow ^{56}$Co $\rightarrow ^{56}$Fe. This is confirmed by observations of SN 1987a. The contribution of this process is calculated by integrating over cosmological time scales. It is found that this process may provide an appreciable flux in the GRB in the MeV range, which is not supplied by AGN.

[F] High Energy GRB from AGN and Discrete Sources

Compton GRO conclusively showed AGN as strong sources of γ-rays and AGN have been proposed as sources of low energy GRB. In the high energy region, contributions of Blazars i.e. AGN with beaming properties have been calculated in the energy region of $\geq$ 30 MeV by several authors. Gamma ray luminosity functions for Blazars are estimated from the radio luminosity distributions and the GRB fluxes in the 30-100 MeV range and fall significantly short of the SAS-2 results. To rectify this situation, it has been proposed that Blazars with hard spectra as seen in the EGRET-2 catalogue should be used and their spectra is to be extended to high energies. Thus considering variable Blazars with hard spectra it is found[36] that the entire GRB in the 100 MeV to a few GeV range can be explained as due to contributions of unresolved Blazars on a cosmological time scale. The details of these studies are given in the book by Fichtel and Trombka (1996).

Thus the extra-galactic gamma ray background provides a fascinating field where many problems are yet to be resolved.

8.3.3 Cosmic Microwave Background, COBE and Cosmology

8.3.3.1 Historical Background

It has been well known that the strong background radio emissions at meter and decimeter wavelengths are produced by our galaxy. The extrapolation of this radiation to the wavelengths of a few centimeter leads to almost insignificant intensities in the centimeter band. When Penzias and Wilson[37] of Bell Laboratories of the USA made their first measurements in 1965 at $\sim$ 7 cm they were surprised to find a significant intensity which corresponded to $\sim$ 3 K black body emission. During this time Dicke and his collaborators at Princeton University, had unknown to Penzias and Wilson, inferred from their cosmological model the existence of a black body radiation of a few Kelvin and they were planning experimental testing. By combining the P-W results at $\sim$ 7 cm with their measurements at 3 cm, they supported the cosmological hypothesis. It was then remembered that George Gamow of Washington University, USA had predicted almost 20 years ago, in 1948, the existence of an universal black body radiation as a relic radiation from the Big Bang model of the origin of the universe, although it was estimated to be about 10 K instead of $\sim$ 3 K.

Since 1965 several measurements were made by many groups of investigators of this microwave background radiation and they concluded that it was consistent with black body radiation at $\sim$ 3 K. The peak of the 3 K Planck spectrum occurs at about 2 mm wavelength; this and shorter wavelengths are completely obscured by water vapour in the earth's atmosphere so that no ground level measurements were possible. Although measurements in the late 1960's and 1970's were found to be consistent with the 3 K black body radiation, it could not be claimed that the cosmic microwave background was established. Therefore it was essential to measure this radiation with space borne detectors and to establish its cosmological origin.

8.3.3.2 COBE and the Cosmic Microwave Background (CMB)

The Cosmic Background Explorer (COBE) satellite was launched by NASA on November 18, 1989 to an altitude of 900 km in a Sun-synchronous orbit with an inclination of 99° to the equator. The purpose of the COBE mission was to make definitive measurements on the cosmic microwave background (CMB) and the cosmic infrared background (CIB) radiations. This was NASA's first satellite designed primarily for observational cosmology and it took many years of work to develop the scientific instruments with the desired precision. There were three scientific instruments in COBE, which, are as follows: (1) The Far Infrared Absolute Spectro-Photometer (FIRAS), which was designed to make precise measurements of the CMB radiation from 1 cm to 100 μm; (2) The Differential Microwave Radiometer (DMR) which searched and measured the variations in the temperature of CMB on an angular scale of $\sim 7^o$ at 31.5, 53 and 90 GHz; and (3) The Diffuse Infrared Background Experiment (DIRBE) which searched for a CIB radiation by making absolute brightness measurements of the diffuse infrared radiation in 10 photometric bands from 1 to 240 μm. The FIRAS and DIRBE instruments provided excellent data for ten months until the liquid helium was exhausted. The other 50% of the instruments did not require liquid helium and continued to provide high quality science results. Details of the instruments, data analysis and results are given by E.L. Wright[38].

8.3.3.3 COBE Results and Implications

[A] Cosmic Microwave Background Radiation (CMBR)

The measurements made by the COBE instruments of the CMBR at the north galactic pole are shown in Figure 8.3.5 in the wavelength range of 0.5 to 10 mm[39]. The measured data is shown as square symbols and the full line black body spectrum. The black body temperature is determined, with high accuracy as 2.728 $\pm$ 0.002 K. The agreement is striking and this is the finest example of naturally occuring black body radiation.

These results on CMB radiation is in agreement with the prediction of the standard, hot, Big-Bang model suggesting that the universe was once

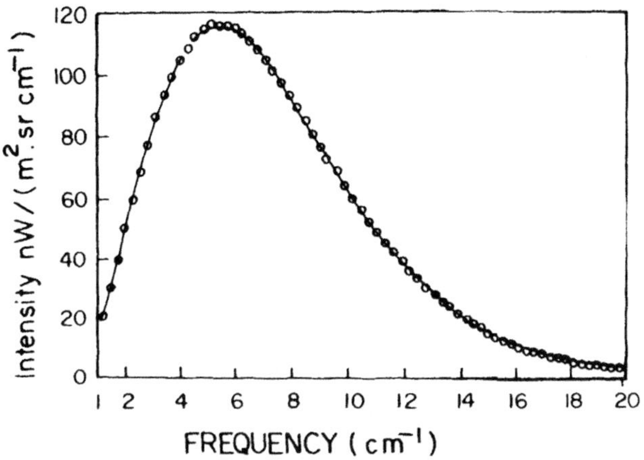

Figure 8.3.5: Cosmic Microwave Background radiation measured by COBE.

hot, dense and isothermal, giving a background radiation spectrum that is close to a perfect black body. The need to preserve this black body spectrum to match the measured spectrum by COBE, places strong limits on events occuring later than one year after the Big-Bang. In the standard model, the universe was initially a Primordial Fireball, i.e. $\sim 10^{-43}$ s after the Big-Bang, when it consisted of energetic quarks and photons. In this domain cosmology meets elementary particle physics. At $\sim 10^{-34}$ to 10^{-30} s the universe is thought to have passed through a stage of rapid inflation. Around 10^{-6} s the elementary particles (quarks) combined to form neutrons and protons. By about 1000 sec. the universe has cooled and the formation of major elements in the primordial fireball was complete. These elements were hydrogen, ^{2}H, ^{4}He, ^{6}Li. At some point of time, $\sim$ sec. the matter and photons no longer interacted, and they evolved separately. The matter cooled separately and as the universe expanded the electromagnetic radiation cooled from $\sim 10^4$ K to the current 3 K of the almost isotropic background radiation at about 3 K. The FIRAS instrument measured this black body temperature as 2.73 ± 0.01 K.

One of the most significant features of the CMB is that it uniformly pervades the whole universe. The energy density of CMB radiation is ($= \sigma$ T^4) equal to about $6{\cdot}10^{-13}$ erg·cm^{-3}, which is comparable to the galactic cosmic ray energy density of 10^{-12} erg·cm^{-3} and with the starlight energy density in the solar neighbourhood of $\sim 10^{-12}$ erg·cm^{-3}. It is remarkable that these quantities are of same order of magnitude.

Secondly the CMB radiation is undiluted black body radiation. When interpreting the FIRAS spectrum on the effects that could have distorted the CMB, we note the following basic features. Theoretical analysis of light element abundances gives limits to the density of baryons in the Big-bang nucleo-synthesis. The ratio of baryons to photons is

$$\eta = (3.5 \pm 0.7) \times 10^{-10}$$

As a result the effect of Lyman α photons from recombination is negligible simply because there are so few hydrogen atoms per photon. The number density of photons, N, is determined as $N_{ph} = 410 \pm 5\ cm^{-3}$; thus the baryon density is determined as $N_b = 1.6 \cdot 10^{-7}\ cm^{-3}$. Using a primordial helium abundance of $Y = 23.5\%$ by mass one has 13 protons and 1 He in 17 baryons. This gives an electron density $n_e = 1.43 \cdot 10^{-7}\ cm^{-3}$. For a fully ionized universe the optical depth for electrons over a path equal to the Hubble radius can be shown to be very small. Therefore distortion of the black body spectrum is likely to be negligible. Details of the calculations are given by Wright[38].

[B] Fluctuations of the Cosmic Microwave Background

One of the major discovery of COBE with the DMR instrument is fluctuations in the temperature of the cosmic microwave background which are intrinsic of anisotropic CMB radiation. The microwave map of the whole sky based one on a year of COBE data[40] was made. In galactic coordinates the radiation from the galaxy is seen as a bright band along the galactic equatorial region. The fluctuations seen at high galactic latitude include instrument noise. When computer analysis was applied to separate the instrument noise, the cosmic fluctuations were determined as $\sim 30\ \mu K$. The pattern of cosmic fluctuation is consistent with some models of galaxy formation.

The observation of intrinsic fluctuations of CMB by COBE indicates the magnitudes of gravitational potential fluctuations that existed $\sim 3 \cdot 10^5$ years after T_o. It is inferred that these potential fluctuations were produced in the expansion phase sometime during the first 10^{-12} sec. after the Big-Bang. The gravitational forces suggested by the observed potential can have induced the observed clustering of galaxies. This can take place in the case where the density of the universe is dominated by some forms of dark matter which are 'transparent' to photons, i.e. interact weakly with photons.

[C] Dipole Anisotropy

As CMB radiation is isotropic on an universal scale it may be considered as a standard frame with respect to which the motion of another frame of reference such as the motion of the solar system can be taken. This is expected to produce a Doppler shift of the black body signal. In the early studies of CMB, Penzias and Wilson estimated that CMB is isotropic within about 10%. Using instruments in a high flying U2 plane, Berkeley group scientists Muller and Smoot determined in 1976 the dipole temperature variation of 3 mK arising from the motion of the solar system with respect to the rest frame defined by the CMB. The COBE results greatly improved this measurements and yielded a velocity of the solar system in the CMB frame of 350 km/s in the direction of $\alpha = 11^h 3^m$, and $\delta = 7^o.5$.

Since its discovery CMB played a central role as one of the corner stones of standard hot, big-bang cosmology. The details of this are given by several authors, see e.g. the book by R.B. Patridge (1995) and H. Kragh 1996 on Cosmology and controversy.

8.3.4 Non-standard Model of Cosmology and CMB Radiation

Although CMB is extensitively interpreted in terms of the standard, hot Big-Bang cosmology (HBBC), not all astronomers agree with the standard model of cosmology. The quasi-steady state cosmology (QSSC) has been developed recently in a series of papers by Hoyle, Burbidge and Narlikar[41] in 1993-95 with the objective of offering a theory with viable alternatives to standard HBBC.

The motivation for the QSSC, as the authors remark, "was to replace the singular event of big-bang cosmology, which has no formal mathematical description within conventional physics, by a rigorous formalism that describes the creation of matter". This formalism, known as the C-field formalism in the steady state cosmology in the sixties, was revived in the context of QSSC by Hoyle et al. in 1993. A more elegant representation of the idea was made by them[42] in 1995 in a conformally invariant framework and this incorporates the Mach principle. This leads, in the many particle approximation to Einstein-like equations of general relativity including additional terms for the creation of matter and a negative cosmological constant. The dynamics of the QSSC was expected to follow from these field equations. Approximate solutions of these general equations were found by using a scale factor S(t) in the usual Robertson-Walker line element

$$ds^2 = c^2dt^2 - S^2(t)\Big[\frac{dr^2}{1 - Kr^2} + r^2\Big(d\theta^2 + \sin^2\theta d\phi\Big)\Big]$$

where r,θ,ϕ are the co-moving coordinates, t is the cosmic time and $K = 0, \pm 1$ for the three possible curvatures of space at t = constant. The scale factor for QSSC was given as

$$S(t) = exp\Big(\frac{t}{P}\Big)\Big(1 + \eta\frac{2\pi t}{Q}\Big)$$

Here P is the longer time scale for steady state expansion and Q is the period of a typical oscillatory fluctuation from the steady state. η is a parameter whose magnitude lies between 0 and 1. The authors[43] explain the physical picture of the universe as follows. The physical picture of the universe is imagined as a universe which has a smoothed out background of the creation field whose intensity falls short of the threshold required for the creation of a typical primary particle. The typical created particle is the so-called Planck particle which is short lived and decays to more stable particles like baryons. The intensity of the field may rise above the background and above the threshold near a collapsed massive object which therefore becomes a typical site for a creation event. Since the creation field has negative stresses it blows outward not only the matter it created, but also space itself. The long term steady state with time scale P thus represents an overall expansion of space produced by several such mini creation events acting together. As the expansion slows down, the creation matter also slows down. At this stage the negative cosmological constant brings in the contraction of the universe. As the universe contracts, the creation field rises and eventually converts the expansion into contraction.

The authors remark that the QSSC developed so far[42] can explain the temperature, spectrum and anisotropies of the CMB and offers a different theory for the origin of the light elements H and He and is consistent with large scale observations of discrete sources. The QSSC is also able to account for the very old as well as for the very young galaxies and is consistent with the baryonic option for dark matter. This interprets the observed emissions of active galactic nuclei (AGN) to mini-creation events where matter and energy are produced in an explosive manner. Detailed accounts of these features and of CMB radiation are discussed by the authors.

The CMBR which pervades the entire universe can provide new insights into the origin of the universe in the standard and Nonstandard models.

8.3.5 Search for Exotic Dark Matter

8.3.5.1 General Features

The first clue on the existence of dark matter was obtained by F. Zwicky(44) in 1933 who noted that the total gravitational mass of the coma cluster of galaxies, estimated from the measured dispersion of velocities of the galaxies in the cluster, was about 8 times larger than the total luminous i.e. visible mass of the galaxies in the cluster. This method was progressively improved and it was convincingly demonstrated that the dynamical mass of galactic clusters was 8 to 10 times larger than their visible mass. Thus it is well established that 90% of the matter out of which the universe is made is unknown to us and the presence of the dark, unseen matter has a profound influence on the dynamics of the universe.

For example, with only 10% of the mass accounted for as visible mass, the universe will indefinitely expand. The cosmologists calculated that when the density of luminous matter is $\rho_l \simeq 10^{-30}$ g·cm^{-3}, which is much smaller than the critical density $\rho_c = 10^{-29}$ g·cm^{-3}, the cosmological constant Ω $(= \rho/\rho_c)$ is $\ll (= 0.1)$. This will cause an infinitely expanding universe. Theoreticians favour from several consideration, a cosmological constant $\Omega = 1$ i.e. the actual mean density of the universe ρ is equal to ρ_c. Therefore it follows that the dark matter should be about 10 times more than the visible mass of the universe. The other evidence for dark matter is as follows. By confining oneself to the bright regions of the galaxy the measured value of the rotational velocity v_c, yields, by balancing the gravitational force on the stars with the centrifugal force, the luminous mass of the galaxy. From galaxy counts one gets the number density of galaxies and the product yields. $\Omega_{lum} \leq 0.01$. Thus the questions: what is the nature of dark matter and how can we look for them are asked? Here two branches of research i.e. particle physics and cosmology are independently looking for new and as yet undetected types of matter. Details are given in the review by Cowsik[45].

Possible Types of Dark Matter

It is believed that dark matter particles were created at the very early phases of the evolution of the universe and are relics of early energetic events. The dark matter should be such that it emits or absorbs very little radiation. Therefore the dark matter is very difficult to detect. Some of the possible types of dark matter (DM) candidates which have been hypothesized are as follows:

[A] Baryonic DM:

From the primordial nucleo-synthesis theory which explains correct the abundances of He, D and Li, it is predicted that the ratio of baryons to photons in the universe is $\eta = (3 \text{ to } 5)\ 10^{-10}$. From these calculations it is found that we are yet to find baryons which would give rise to the correct ratios of elements synthesized in the big-bang. An interesting possibility is that these baryons may be present in the form of massive compact objects distributed in the halos of galaxies, including our own. These objects are referred to as brown dwarfs and their mass is estimated to be between $\sim 10^{-7}$ to $10^{-3}\ M_{\odot}$. These objects are also called MACHOS i.e. massive compact halo objects.

[B] Hot, Dark Matter:

It was hypothesized that weakly interacting particles, such as neutrinos, left over from the big-bang, could form a cloud of gravitating matter around the cluster of galaxies, if the neutrinos had a rest mass of a few eV. Computations by Cowsik and Mc Clelland[46] in 1972 showed that one expects about 110 neutrinos per cm^3 in the universe today as compared to a baryonic density of about $10^{-7}\ cm^{-3}$. Thus even if the neutrino mass is only a few eV, they could contribute significantly to the gravitational mass of the galaxies. These particles are called hot DM, signifying that at the time of thermodynamic decoupling, they were relativistic particles. Details of various aspects of this model are given by Cowsik[45].

[C] Cold Dark Matter:

The cold dark matter candidates include the hypothetical weakly interacting massive particles (WIMP's), which have an estimated mass, M_x > 20 GeV. Direct searches for WIMP's have been made in cosmic rays with semi-conductors and only an upper limit can be placed. The annihilations of WIMP's $x \cdot \bar{x}$ in various sites producing $p\bar{p}$, γ's etc have been speculated.

[D] Axions:

Axions are very light, neutral particles which were invented to resolve a mathematical problem in the theory of quarks. The axion mass is expected to be in the range of 10^{-3} to 10^{-6} eV, and it is speculated that they could substantially contribute to the dark matter.

[E] Other Exotic Particles:

(a) **GUT Monopole:** The grand unification theories predicts the existence of magnetic monopoles with a super heavy mass of $\sim 10^{16}$ GeV produced in the early universe. Their magnetic charge is $g = n(\hbar e/2e)$, where $n = 1, 2, 3$ etc and their velocity is in the range of $\sim 10^{-2}$ to 10^{-4} c. Many experimental searches made in the last decade place rather stringent limits on their fluxes.

(b) **Strange Quark Matter (SQM):** SQM particles are believed to be made of roughly equal numbers of up, down and strange quarks, and are expected to be stable. Searches for these have been made in cosmic rays by looking for anomalous particles e.g. a nucleus with $Z = 14$ and mass $A = 370$.

(c) **Charged Massive Particles (CHAMP's):** These charged massive particles are expected to have masses in the range of (10-1000) x 10^{12} eV. These have been looked for using different techniques.

8.4 Problems on Chapter 8

1. Describe and explain the processes involved in the formation of a proto-star and its infrared

emitting accretion envelope.

2. What are the striking properties of extremely luminous infrared sources found in IRAS mission 1 and their possible interpretations?

3. Explain the enigma in the determination of the Hubble constant with HST and the standard model of cosmology.

4. Show that extreme UV sources of 100 Å are not observable in the directions in the galaxy which have a column density > 10^{20} atoms/cm^2. [Use Figure 8.2.1]

5. Explain why (1) we can observe soft X-rays of 100 eV from sources and the sky background only from distances < 10 pc from the Sun; and (2) the minimum X-ray energy is about 3 keV for observing X-ray sources from the galactic center region.

6. Describe and explain the main types of periodicities in a typical X-ray binary such as Her X-1.

7. Explain with a sketch the formation of a binary system with a neutron star and an accretion disk of a typical X-ray binary. In which regions do the X-ray emissions take place?

8. Describe the characteristic features of X-ray sources which are considered as black hole candidates.

9. Explain briefly the three basic processes of the production of γ-rays in the galaxy and their relative contributions.

10. Describe the main properties of γ-ray pulsars which are also studied well in X-ray and radio emissions. What are the unusual properties of the γ-ray pulsar "Geminiga"?

11. Describe the main features of γ-ray bursts and the possible sources.

12. What are the objects which are usually included in the term Active Galactic Nuclei (AGN)? What are the striking features of the γ-ray emissions and their time variations in case of the AGN's 3C 279 and PKS 0208-512?

13. What types of discrete sources are likely to make significant contributions to the extra galactic γ-ray background and why?

14. What are the features of the dipole anisotropy direction of the cosmic microwave background radiation and what is its relation to the motion of the solar system in the galaxy?

Chapter 9
THE INTERSTELLAR MEDIUM (ISM)

9.1 General Features

Our ideas on the interstellar medium (ISM) have changed radically during the past two decades with the rapid advancement of our knowledge of space astrophysics. Originally the interstellar medium was thought to be a simple, quiescent medium; but now it is well known that the ISM is a complex and dynamical medium and plays a key role in the cycle of birth to death of stars. Out of this medium new stars are born and it is being continuously replenished by the mass loss from stars. The ISM is a dynamic medium which is continuously being blown by stellar winds, supernova explosions and the expanding envelopes of supernova remnants. The interstellar medium has four main constituents: (a) Gas in all phases, - molecular, atomic and ionized gases, (b) Dust particles, which obscures several regions from visible and UV radiations, (c) High energy cosmic ray particles (which were discussed in detail in Chapter 7), and (d) the Galactic magnetic field. Here we shall discuss the basic features of the first two aspects.

The interstellar medium contains about 5% of the visible mass of our galaxy. Most of the gas which is mainly hydrogen, is confined close to the disk of our galaxy and the overall density of gas in the disk is about 1 atom of hydrogen per cm^3. It must be remembered that very large variations in gas density and temperature occur in regions throughout the galaxy. We consider now briefly some of the large scale dynamical processes which contribute to the main physical properties of the interstellar medium.

9.2 Large Scale Dynamical Processes

9.2.1 Large Scale Rotation

The interstellar gas and dust rotate in circular orbits around the galactic center. The gravitational forces on the gas are determined by the mass distribution of stars and galactic dark matter. Thus the interstellar gas and the velocities of interstellar neutral hydrogen and molecules provide important information on the distribution of matter in the galaxy. These observations established that there exists a differential rotation of the galactic disk i.e. its outer regions do not rotate like a rigid body. This data provided evidence for the presence of dark matter in the outer regions of the galaxy.

The evidences for spiral arm structures in the galaxy have been provided by the traces of O and B stars and by hot ionized hydrogen in the interstellar gas by their radio emission as shown in Figure 9.2.1. Spiral arm structures are identified also by the 21 cm radiation from atomic hydrogen in the ISM. The spiral structures of the galaxy contain giant molecular clouds. How such large molecular clouds with gas densities of 100-1000 atoms/cm^3 are formed in the spiral arms, and how spiral arms of the galaxy form a dynamically stable structure have been discussed by astrophysicists in terms of density wave theories.

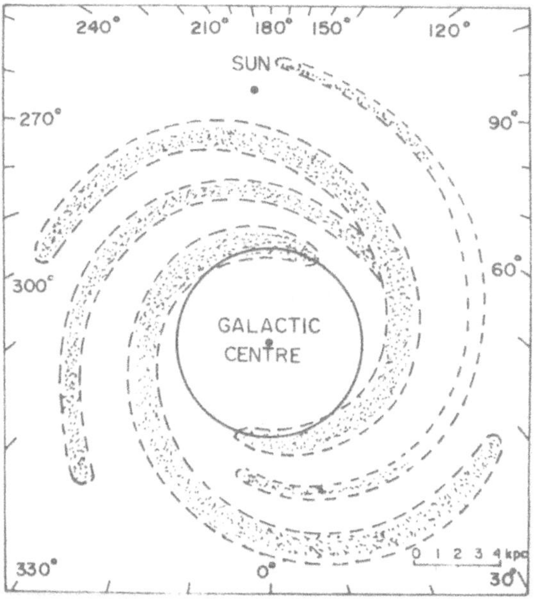

Figure 9.2.1: A sketch of the spiral structure of the galaxy as determined from optical and radio observations of ionized hydrogen

9.2.2 Heating of the ISM

The interstellar medium is heated by starlight, stellar winds and by supernovae. Some of the energy is injected as cosmic rays. Stars inject matter into the ISM in the form of stellar winds, planetary nebulae and supernovae.

The ISM recycles this mass in the process of star formation in regions where enough gas accumulates to become gravitationally unstable.

i) Heating of the ISM by SN

The dominant sources of the energy of the ISM are starlight and supernovae. The luminosity of the galaxy is about $3{\cdot}10^{43}$ erg·s^{-1}, almost all of which is radiated directly or indirectly after absorption by interstellar dust. The injection of energy by supernovae takes place at the rate of about 10^{42} erg·s^{-1}, corresponding to an average rate of one supernova of 10^{51} erg occuring every 30 years. This energy is not injected uniformly, but at random points in the disk. After an initial stage in which the SN ejecta generates a shockwave in the surrounding ISM, the supernova remnant (SNR) evolves into an adiabatic blast wave. In a uniform medium the mean pressure and temperature inside SNR falls of as R^{-3} where R is the radius of the SNR. Young SNR's fill up a very small fraction of the volume with hot gas at T $\sim$ 10^8 K, whereas older SNR's fill up a large fraction of the volume with gas at a lower temperature. When the temperature falls below $\sim 10^6$ K, radiative cooling takes place and a dense, radiative shell forms at the outer boundary of the SNR. The shell encloses a hot bubble of low density gas at T $\sim 10^6$ K. This radiative stage ends when the SNR has expanded to the point that it is in pressure equilibrium with the ambient gas. It is calculated that this hot gas inside the old SNR's occupies a substantial fraction of the volume of the ISM.

Observational results support the model of hot gas from SNR's pervading a large fraction of interstellar space. Analysis of the soft X-ray background established that the region within $\sim$ 500 pc or so around the Sun is filled with hot gas of $\sim 10^6$ K. The wide-spread nature of the hot gas comes from the observation of UV absorption lines of O^{+5} ions seen in in many stars, most of which are beyond 100 pc. The ionization potential of O^{+4} is 114 eV; so the temperature indicated by the presence of O^{+5} is several times 10^5 K.

ii) Heating of the ISM by UV

Ultra violet radiation of young stars constitutes the second important agency of heating the ISM. Very young stars, still embedded in the gas clouds from which they are formed contribute to this process. The heating of the gas by UV is identified by strong emission lines of hydrogen and oxygen. The gas temperature is determined by the balance between photo ionization of the neutral gas and recombination of the ionized component. The typical temperature is of the order of 10^4 K. Older blue stars which do not have gas envelopes, heat and ionize the surrounding medium which is observed in the UV spectra of O and B type stars.

iii) Heating by Cosmic Rays

Energetic cosmic ray protons and heavier nuclei traversing the interstellar medium undergo ionization loss, which heats up the interstellar gas. In order to calculate the heating of gas in the ISM one needs the fluxes and energy spectra of cosmic ray protons and heavier nuclei in the ISM after correcting for the solar modulation effects in the 100 MeV to $\sim$ GeV energy range (Chapter 7.4). For energies > 5 GeV, modulation effects are negligible. Several authors have derived the local interstellar spectrum of cosmic ray protons and He nuclei and calculated their energy loss and heating of the interstellar gas. The discovery of anomalous cosmic rays in the energy range 1-100 MeV/n (Chapter 7.3.1) has added another factor in the energy loss process which is applicable in the vicinity of the Sun. For relativistic cosmic rays the ionization rate in the interstellar gas is estimated as $\sim 4{\cdot}10^{-23}$ n eV/s when n is the number of hydrogen atoms per cm^3. Using a demodulated non-relativistic spectrum, the energy loss is estimated a little higher. Not all energy loss goes to heating as significant part goes to atomic excitation. So the effective heating of ISM gas is not larger than the above estimates and it is much less than by SNR's. Heating by cosmic rays becomes an important factor in large molecular clouds as will be discussed later.

9.3 Physical Features of the ISM

9.3.1 Neutral and Molecular Hydrogen and their Distributions

The distribution of neutral hydrogen in the interstellar medium was determined in 1950 and that of molecular hydrogen more recently. As it is known now from many studies both neutral and molecular hydrogen are closely confined to the plane of the galaxy with half widths of about 120 and 60 pc respectively. Their distributions as function of distance from the galactic center, are very different. The atomic hydrogen extends mainly from about 3 kpc to $\sim$ 18 kpc from the center (Figure 9.3.1(A)). The molecular hydrogen which occurs in molecular clouds, is mainly confined in the region

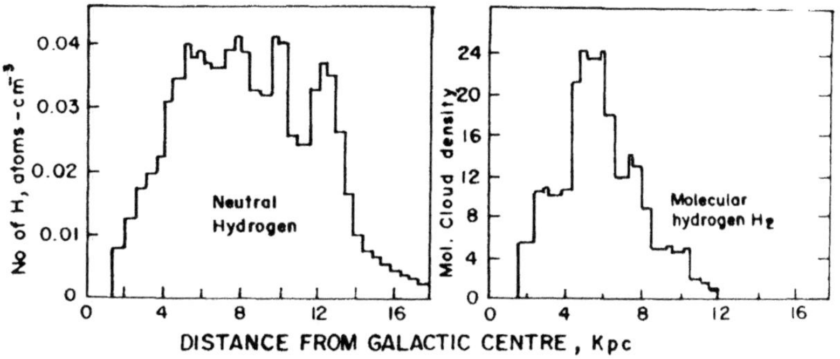

Figure 9.2.2: (A) Radial distribution of atomic hydrogen (HI) as obtained from radio observations of the 21 cm line of HI. (B) Radial distribution of molecular hydrogen from the molecular emission lines of CO in the mm region. The ordinate scale is radial density in 10^{-3} km·s^{-1}(kpc)$^{-1}$

at about 3 to 8 kpc from the galactic center (Figure 9.3.1(B)). The radial distribution of atomic hydrogen was determined from radio surveys of the galaxy using the 21 cm line of atomic hydrogen, as discussed below. The molecular H_2 distribution is obtained from CO emission lines as given in 9.4.

Neutral hydrogen emits line radiation at wavelength $\lambda_o = 21.1$ cm (frequency $\nu_o = 1420.4058$ MHz) in an almost totally forbidden hyperfine transition. In this process the spins of the proton and the electron change from being parallel to anti-parallel. The spontaneous transition probability, $A = 2.85 \cdot 10^{-15}$ s^{-1}, for the ground state of hydrogen is very small, so that collisions and other processes have time to establish an equilibrium distribution and the populations of upper and lower levels, denoted by, n_2 and n_1, are given by the Boltzmann distribution, $n_2/n_1 = (g_2/g_1) \exp(-h\nu_o/KT)$ where T is the equilibrium excitation temperature and g_2 and g_1 are weight factors. In the case of 21 cm HI radiation, the excitation temperature is called the spin temperature, T_s, and in cosmic conditions $h\nu_o/K \ll T_s$; so that $n_2/n_1 = g_2/g_1 = 3$. Thus for an optically thin region the emissivity of the 21.1 cm radiation by spontaneous emission is given by ϵ.

$$\epsilon = \frac{g_2}{g_2 + g_1} \cdot N_H \cdot Ah\nu_o = \frac{4}{3} N_H Ah\nu_o \tag{9.3.1}$$

where N_H is the number density of neutral atoms of hydrogen.

As HI atoms are distributed along the line of sight of the observer, the flux density S, received within a solid angle, ω, of the beam of the radio telescope, is given by

$$S = \int \frac{\epsilon(l)}{4\pi l^2} \cdot w \cdot l^2 dl$$

or

$$J = \frac{S}{w} = \frac{3}{16\pi} A \cdot h\nu_o \cdot \int N_H dl \tag{9.3.2}$$

where J is the intensity in the particular direction in which the column density of neutral hydrogen is $\int$ Ndl. J is measured in watts per cm^2.

Because of the very small transition probability the line width of the 21 cm line is very narrow. When the neutral hydrogen is moving, the Doppler shifts of the 21 cm line provide important information on the velocities of neutral hydrogen gas in our galaxy and in other galaxies.

Supernova remnants and other extra-galactic sources which have smooth continuum radio spectra due to synchrotron radiation, show absorption features at 21 cm when neutral hydrogen clouds lie along the line of sight to the radio source. These features provide information on the structure and velocity dispersion of neutral hydrogen in the gas cloud.

The distribution of neutral hydrogen in the galaxy as a function of distance from the galactic center is shown in Figure 9.3.1(A).

9.3.2 The Main Phases of the Interstellar Medium

Based on estimates of the heat inputs in the ISM and the cooling processes, it is found that the interstellar medium is composed of several phases:

i) The Hot 'Coronal' Gas

It is believed that hot gas from supernova explosions expands so as to form a hot halo in our galaxy with a scale height much larger than the stars in the disc. This is inferred from observations of lines of highly ionized carbon (C^{+4}) and (O^{+5}) by the International Ultra violet Explorer. It is also derived from the observations with the far UV wide field camera on the ROSAT satellite and from the studies of soft X-ray emissions from the galaxy that the solar system is embedded within a temperature of $\sim 10^6$ K. It is believed that hot gas from old supernova remnant shells is reheated by new SNR's and the heated gas tunnels through the ISM. It is estimated that about 50% of the interstellar medium including the halo may be filled with the hot gas T $\sim$ 10^6 K.

ii) The Warm Inter-Cloud Medium

The warm inter-cloud medium heated by SNR's expanding envelopes, is estimated to occupy about 40% of the volume of the ISM. The average temperature of this region is about 8000 K; and it is composed of atomic and ionized hydrogen and electrons, the ionization fraction is 10-20%. So this hot region is studied from the emissions of HI and HII and their absorptions as noted earlier. The estimated number density of hydrogen in this WIM is 10^{-1}-1 cm^{-3}. The warm inter-cloud regions contain about 20% of the mass of the ISM.

Table 9.3.1: Main Phases of Interstellar Gas

Component	Composition	Temp. K	Density $n \cdot cm^{-3}$	Volume fraction	Mass fraction
Hot "Coronal" Gas	H^+,e^-, O^{5+},C^{4+} etc.	$\sim 10^6$	10^{-3}	$\sim$ 50%	0.1%
'Warm' Inter cloud Medium	H,H^+,e^-	8000	10^{-1}-1	$\sim$ 40%	20%
Diffuse HI Clouds (Cold neutral Medium)	H,C,O	80	1-10^2	$\sim$ 5%	40%
Molecular Clouds	H_2,CO, CS etc.	10-30	$\geq 10^3$	$\sim$ 0.5%	40%

iii) Diffuse HI Clouds

The diffuse HI clouds are also called cold neutral media and are cold shells of clouds at a temperature of about 80 K, having a density of $\sim$ 40 cm^{-3}. These cold clouds are surrounded by the warm inter-cloud gas at $\sim$ 8000 K noted above. The cold HI clouds are composed mainly of H, C, O etc and occupy a volume of $\sim$ 5% of the ISM and have a mass fraction of $\sim$ 40%. These clouds are detected due to 21 cm emissions and absorptions as described earlier.

iv) Molecular Clouds

Molecular clouds are composed mainly of H_2, CO and many other complex molecules discussed later. The large molecular clouds have high densities of $n_H \geq 10^3$ atoms$\cdot cm^{-3}$, at a temperature of $\sim$ 10-30 K. In giant molecular clouds H_2 and other molecules can survive as these are shielded by the high density gas from optical and UV radiation. These giant molecular clouds although they occupy a small fraction of $\sim$ 0.5% of the volume of the ISM, they constitute about 40% mass of the ISM. The molecules in these clouds are identified from molecular lines.

9.4 Interstellar Dust

Interstellar dust can survive in the ISM at temperatures less than 10^3 K and dust grains play an important role at low temperatures. Although it is well established that there are large amounts of dust in the ISM as seen from the obscuration of visible and UV images of the galaxy, we have very limited knowledge, as yet, on the composition and other properties of the dust grains. It is presumed that dust grains must contain a large fraction of heavy elements which are seen to be depleted in interstellar gas.

Electromagnetic waves are strongly absorbed by dust grains when their wavelengths are shorter than the grain diameter. The longer waves are transmitted by the grains. The absorption curve in the ISM is shown in Figure 9.4.1, where the optical depth τ, is plotted against wave number λ^{-1}. The smooth absorption of interstellar dust grains from UV to IR indicate that there must be wide range of grain sizes present in the ISM. It is seen

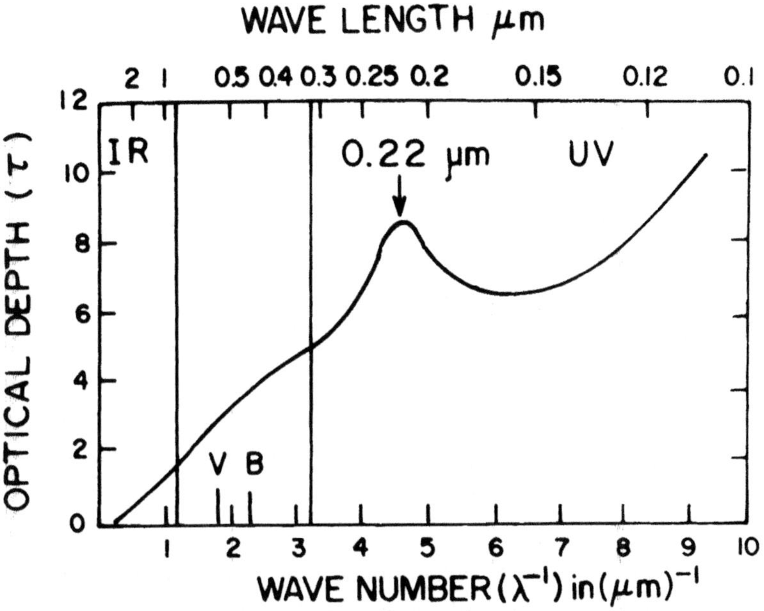

Figure 9.4.1: Absorption coefficient of interstellar dust grains. A broad absorption feature is seen at 0.22 μm wavelength. (Based on J.M. Greenberg (1982))

that the optical depth $\tau \propto \lambda^{-1}$, indicating that infrared radiation can easily penetrate the dust grains whereas optical and UV radiations are strongly absorbed.

It is seen in the figure that super-imposed on the smooth curve there is a strong and broad absorption feature at $\lambda \approx 0.22$ μm. This feature corresponds to resonance band of graphite and it is therefore commonly presumed that graphite is present in interstellar grains. Several other absorption features have been discussed by many authors, e.g. infrared pronounced absorption and emissions observed at 9.7 and 18 μm are ascribed to silicates. Other features indicate ice and ammonia molecules. Hence a favored model of interstellar grains is that of a graphite or silicate core surrounded by ice mantle. One of the most important role of dust grains is the formation of molecules. Atoms and molecules are adsorbed on the surface of the grains where they can migrate and combine with other species and then return to the interstellar medium. Thus grains act as intermediaries in the formation of complex and organic molecules.

9.5 Interstellar Molecules

9.5.1 Types and Abundances

In 1963 the first interstellar molecule was detected at radio frequencies when the hydroxyl radical OH was observed in absorption against the bright radio source Cassiopeia A. Soon OH lines were observed in emission with strong and variable intensities of compact sources, which indicated, to much surprise of the astronomers, some kind of maser action in the ISM. In 1968 Charles Townes discovered the presence of NH_3 in the ISM. In the following years major discoveries were made when water vapour, H_2O molecules and formaldehyde (H_2CO) were detected. The carbon monoxide CO molecule was detected with high intensities in 1970. Since then the number of interstellar molecules detected rapidly grew in number.

The readers are referred to excellent accounts of these areas of research and discoveries in the books and papers by Greenberg[(1)] on astrochemistry, Interstellar dust and related matters. By 1993, some 100 entries have been made in the list of molecules detected in the ISM. This includes from the simplest molecule H_2 to the highly complex $HC_{11}N$, consisting of 13 atoms. The molecules are composed of hydrogen (and deuterium) nitrogen, carbon, sulphur, silicon and oxygen and their isotopes. Table 9.5.1 shows the list of 102 interstellar molecules as given in the review paper by Winnewisser and Herbst, arranged in the number of atoms from 2 to 13. These are gas phase molecules.

The most abundant interstellar molecule is H_2 and the second most abundant one is CO which has a fractional abundance of 10^{-4} relative to H_2. Remaining molecules are organic or carbon containing ones. These have much smaller abundances of $< 10^{-8}$ (relative to H_2) and are considered as trace constituents. These interstellar molecules are in general found in large and dense molecular clouds which have a tremendous size of about 1-100 light years and gas densities of 10^2-10^3 atoms/cm^3. Because of the vast sizes of the giant clouds, the absolute amounts of interstellar organic molecules exceed their terrestrial counter parts.

9.5.2 Molecular Emission Processes

Interstellar molecules occur in the dark and cold regions (T = 10-60 K) of large molecular clouds where they are shielded from optical and ultra violet

Table 9.5.1: Interstellar Molecules*

1	2	3	4	5	6
H_2	NS	H_2O	NH_3	SiH_4	
OH	HCl	H_2S	H_3O^+		
SO	NaCl	N_2H^+			
NO	KCl	SO_2			
SiO	AlCl	HNO			
SiS	AlF	SiH_2?			
SiN	PN	H_2D^+?			
	NH				
	CH^+	HCN	H_2CO	HC_3N	CH_3OH
	CH	HNC	HNCO	C_4H	CH_3CN
	CC	C_2H	H_2CS	H_2CNH	CH_3NC
	CN	C_2S	HNCS	H_2C_2O	CH_3SH
	CO	SiC_2	C_3N	NH_2CN	NH_2CHO
	CSi	HCO	C_3H^{lin}	NCOOH	H_2CCH_2
	CS	HCO^+	C_3H^{ring}	CH_4	C_5H
	CP	HOC^+?	C_3O	$H_2C_3^{ring}$	HC_2COH
		OCS	C_3S	$H_2C_3^{lin}$	$H_2C_4^{lin}$
		HCS^+	$HOCO^+$	CH_2CN	
		CO_2	HCCH	C_4Si	
		C_2O	$HCNH^+$	HCCNC	
		MgNC	HCCN	HNCCC	
		CaNC	H_2CN	CCCCC	
		CCC			

7	8	9	10	11	12
HC_5N	$HCOOCH_3$	HC_7N	CH_3C_5N?	HC_9N	$HC_{11}N$
CH_3CCH	CH_3C_3N	$(CH_3)_2O$	$(CH_3)_2CO$?		
CH_3NH_2		CH_3CH_2OH			
CH_3CHO		CH_3CH_2CN			
H_2CCHCN		CH_3C_4H			
C_6H					

*As compiled by Winnewiser and Herbst (1993)

radiation which would otherwise dissociate them.

The interstellar molecules emit line radiation due to transitions between (a) electronic, (b) vibrational and (c) rotational levels. The highest energy line emissions are due to transitions of electronic levels and these lines generally occur in the optical region of the spectrum. The vibrational transitions are associated with the molecular binding between atoms of the molecule which can be represented by a simple harmonic oscillator. The transitions between these vibrational levels are in general of the order of $h\nu \approx 0.2$ eV and lie in the infrared region. The lowest energy lines are due to transitions between rotational energy levels. The frequencies of these rotationals are worked out from the quantization of angular momentum. For example in case of CO, the lowest frequency rotational transition, $j = 1 \rightarrow 0$ has a frequency of 115 GHz, or $\lambda = 2.6$ mm. The next transition in the rotational levels $j = 2 \rightarrow 1$ produces line emission at $f = 230$ GHz, and so on. Thus diatomic molecules composed of atoms of masses m_1 and m_2 produce rotational transition lines which are equally spaced. The transition probability depends on the net electric dipole moment; so symmetrical molecules like H_2 do not emit electric dipole radiation. Asymmetrical molecules e.g., CO, NO, etc are the sources of line emissions in the millimeter wavelength band. Other molecules such as the hydroxyl radical OH and formaldehyde emit radiation in the radio wave band.

9.5.3 Molecular Synthesis in GAS Phase Reactions

As interstellar molecules occur in deep depths in high density, large molecular clouds where optical and UV radiation can not penetrate, the energy source is supplied by cosmic ray particles. Cosmic rays are nuclei, mainly protons moving at relativistic speeds and producing positive ions either directly or via secondary electrons. The most important ionization process is

$$H_2 + Cosmic\ \ ray \longrightarrow H_2^+ + e^- + Cosmic\ \ ray \qquad (9.5.1)$$

There is no direct measurement of the fractional ionization in dense molecular clouds, it is estimated that typically the fractional abundance of positive ions is of the order of 10^{-7}-10^{-8}. The negative charge abundance, residing mainly on electrons is assumed to be nearly equal to the positive charge abundance, as the negative ion formation probability is in general small. In more diffuse clouds, penetration of electromagnetic radiation occurs and photo ionization produces a larger fractional abundance of positive ions. For example the photo ionization of neutral C_2 leads to a C_2^+ fraction of $\sim 10^{-4}$.

Positive ion-molecule reactions synthesize molecular ions via several routes. Interstellar ion chemistry deals with these phenomena. Here we mention a few basic molecular synthesis processes in the gas phase. Following the cosmic ray ionization of H_2 to produce H_2^+, the hydrogen ion reacts with H_2 by a well studied reaction to produce the simplest polyatomic ion H_3^+,

$$H_2^+ + H_2 \longrightarrow H_3^+ + H \tag{9.5.2}$$

This ion can react with a number of neutral atoms such as O, C, S via proton transfer

$$\begin{aligned} O + H_3^+ &\longrightarrow OH^+ + H_2 \\ C + H_3^+ &\longrightarrow CH^+ + H_2 \\ S + H_3^+ &\longrightarrow SH^+ + H_2 \end{aligned} \tag{9.5.3}$$

Afterwards OH^+ (and CH^+ etc) can consecutively add additional hydrogen atoms via two types of reactions with H_2,

$$\begin{aligned} OH^+ + H_2 &\longrightarrow OH_2^+ + H \\ CH^+ + H_2 &\longrightarrow CH_2^+ + H \end{aligned} \tag{9.5.4}$$

and by radiative association

$$\begin{aligned} OH^+ + H_2 &\longrightarrow OH_3^+ + h\nu \\ CH^+ + H_2 &\longrightarrow CH_3^+ + h\nu \end{aligned} \tag{9.5.5}$$

Subsequent recombination reactions with e^- lead to neutral species such as H_2O, CH_4, OH, CH_3, CH_2 and CH. These molecules can react with ions to produce other molecules. In addition some neutral-neutral reactions take places,

$$CH + O \longrightarrow CO + H \tag{9.5.6}$$

The formation of CO can lead to the production of C^+ ions

$$CO + He^+ \longrightarrow C^+ + O + He \tag{9.5.7}$$

in which He^+ is formed by the ionization of He by cosmic rays. Other atomic species e.g. N^+, O^+ are formed in similar processes. These atoms act as precursors of other poly atomic molecules. Unlike H_2O and CH_4, the synthesis of ammonia (NH_3) can only be initiated by N^+, because the proton transfer reaction (9.5.4) is endothermic and so cannot take place. The production of organic molecules, starting with CH_4 is discussed later.

9.5.4 Molecular Synthesis on Dust Surfaces

Compared to gas phase chemistry, molecular synthesis on the surfaces of dust grains is less supported by laboratory evidences. The standard view is that "classical" dust grains have radii of about 0.1 μm and are composed of refractory cores of silicates and carbonaceous material, upon which mantles are assumed to grow as the gas cloud evolves in time. To produce a molecule on a grain surface, the precursor molecules are needed to be adsorbed and to migrate along the surface until they are close enough to react. Adsorption of neutral gaseous species at low temperature by solid dust particles is thought to be an efficient process, while it is generally assumed that positively charged ions will be repelled by the dust grains. However a gas phase molecule will need some time to find a dust particle.

Synthesis of polyatomic molecules on dust particles occur in several ways. The most common process seems to be the conversion of heavy atoms such as C, N, O and S into their saturated forms such as CH_4, NH_3, H_2O and H_2S via successive association reactions between intermediate radicals and the most abundant atomic hydrogen, e.g. $C \rightarrow CH \rightarrow CH_2 \rightarrow CH_3 \rightarrow CH_4$. The saturated molecules thus formed are not reactive and do not lead to more complex species. Photo emission on dust grains followed by photo dissociation of methane and ammonia probably lead to a diatomic radical like NH and CH^+ which are not produced efficiently in the gas phase.

Another important synthesis process on grain surfaces is the formation of water when the abundance of atomic hydrogen is too low for standard hydrogenation of oxygen. This involves several reactions with activation energy

$$O + O \longrightarrow O_2$$

$$
\begin{aligned}
O_2 + O &\longrightarrow O_3 \\
O_3 + H &\longrightarrow OH + O_2 \\
OH + H &\longrightarrow H_2O \\
OH + H_2 &\longrightarrow H_2O + H
\end{aligned}
\tag{9.5.8}
$$

Above reactions allows water to be formed rapidly even when the surface H_2 abundance is much higher than the surface abundance of atomic hydrogen.

9.6 Organic Molecules in Large Molecular Clouds

9.6.1 Formation of Large Hydrocarbon Molecules

Once a saturated hydrocarbon molecule such as CH_4 is formed, more complex hydrocarbons can be produced in three ways. The first process is called carbon insertion in which C^+ or C and a neutral or ionic hydrocarbon reacts and inserts itself into the hydrocarbon to produce a larger carbon structure. The reaction studied in the laboratory is as follows:

$$
\begin{aligned}
C^+ + CH_4 &\longrightarrow C_2H_2^+ + H_2 \\
&\longrightarrow C_2H_3^+ + H
\end{aligned}
\tag{9.6.1}
$$

The reaction with neutral C is assumed to be similar, but has not actually been studied in the laboratory. The second mechanism is called condensation which has been well studied. In this a hydrocarbon ion reacts with a neutral one such as

$$CH_3^+ + CH_4 \longrightarrow C_2H_5^+ + H_2 \tag{9.6.2}$$

The third process is radiative association between hydrocarbon ions and neutral ones. Of this process there are only a few laboratory studies. For faster reactions, this process needs larger reaction partners, e.g.

$$C_4H_2^+ + C_2H_2 \longrightarrow C_6H_4^+ + h\nu \tag{9.6.3}$$

The carbon insertion reaction seems to be the most efficient synthetic mechanism as long as the hydrogen lost in the formation of a more complex structure is replenished by the hydro-generating reactions (9.5.4) and (9.5.5).

As the hydrocarbon neutrals are synthesized, it may be noted that these are depleted by ion molecule reactions by the predominant interstellar ions such as H_3^+, HCO^+, H^+, He^+, C^+ etc. In addition reactions of neutral hydrocarbon radicals with oxygen atoms contribute to the loss process.

Hydrocarbon precursors contribute to the production of organic molecules involving atoms other than carbon and hydrogen. An example of the production of an organo-sulphur compound in which S^+ inserts itself into the hydrocarbon C_2H_2 is as follows:

$$S^+ + C_2H_2 \longrightarrow HC_2S^+ + H \tag{9.6.4a}$$

This is followed by dissociative recombination

$$HC_2S^+ + e^- \longrightarrow CCS + H \tag{9.6.4b}$$

and CCS is produced. This is a well known interstellar radical. Laboratory studies show that organonitrogen molecules are probably formed by the reactions of N and hydrocarbon ions

$$N + C_3H_3^+ \longrightarrow H_2C_3N^+ + H$$

and

$$N + C_3H_5^+ \longrightarrow H_3C_3N^+ + H_2 \tag{9.6.5}$$

followed by dissociative recombination or neutralization to form HC_3N, the simplest cyanoacetylene or cyanopolyne.

The production of oxygen containing organic molecules probably takes place by radiative association reactions with H_2O or CO. The formation of methanol occurs from its ionic precursor $CH_3OH_2^+$

$$CH_3^+ + H_2O \longrightarrow CH_3OH_2^+ + h\nu$$

$$CH_3OH_2^+ + e^- \longrightarrow CH_3OH + H \tag{9.6.6}$$

The less saturated species such as C_3O and HC_3O are formed by the reaction

$$C_2H_2^+ + CO \longrightarrow H_2C_3O^+ + h\nu$$

$$H_2C_3O^+ + e^- \longrightarrow HC_3O + H \tag{9.6.7}$$

Interstellar grains, as noted earlier, act as 'seedlings' in the formation of complex organic molecules. M. Greenberg[1] of the University of Leiden, Netherlands conducted a series of experiments for over a decade and showed the phenomena in the laboratory. He cooled a mixture of water, ammonia, methane and other simple molecules in a vacuum chamber to about -260 ^oC and thus the gases were frozen on the surface of a metal finger. The frozen mixture were irradiated with UV light and it was found that a complex residue of organic matter was formed. Greenberg thus showed that similar processes take place on the surface of interstellar grains, where UV rays from stars break up some molecules and the fragments recombine to form new and more complex compounds.

In the gas phase synthesis of larger organic molecules there are several uncertainties at present in the understanding of the balance between the production rates and the destruction rates. Continuing laboratory and theoretical studies are expected to provide answers to these problems.

Table 9.5.1 shows the list of interstellar molecules including large hydrocarbon molecules as compiled by Winneswisser and Herbst in 1993.

9.6.2 Polycyclic Aromatic Hydrocarbons (PAH)

The list of molecules in gas phase in large molecular clouds and circumstellar sources in interstellar medium, given in Table 9.5.1, was obtained with high resolution spectral techniques and this list stops at molecules of thirteen atoms, $HC_{11}N$. However, the question of even larger molecules is rather intriguing. The evidences of these have come in 1984-85 from two groups, French (Leger and Puget[2]) and American (Allamandola et al.[3]) from the studies of previously unidentified interstellar emission lines in infrared. These lines, found only near UV sources such as new born stars were identified as due to vibrational emission frequencies in molecules known as polycyclic aromatic hydrocarbons or in short PAH. These large planer molecules are composed of a dozen or more fused benzene rings forming hexagonal structures. These are, however, not fullcrenes, which are polyhedral dehydrogenate structures of which there is as yet no evidence in space.

The PAH hypothesis is based on the observation that infrared features can be assigned to localized vibrational bond motions in aromatic molecules which are studied well in the laboratory. However these lines are observed in absorption in the laboratory. The most prominent interstellar emission feature occurs at 3.29 μm which is ascribed to the stretching of C-H bonds. Other features at 6.2, 7.6-8.0, 8.7 and 11.2 μm have been ascribed to the stretching of aromatic C-C and aromatic C-H bending. It may be noted that no PAH molecule has an infrared absorption spectrum equivalent to the interstellar emission spectrum. The interstellar emission features are not narrow since rotational substructures can not be resolved; most probably many different molecules are present. In figure 9.6.1 we show in high resolution the infrared spectra of Orion bar emission in 3.2-3.6 μm and 5-13 μm and these are compared with the laboratory infrared spectra of PAH molecules, as given by Allamandola et al.[(4)].

The mechanism of the excitation of PAH molecules is believed to be UV absorption leading to an excited electronic state followed by conversion of at least a part of the energy into the vibrational energy of the ground electronic state. Laboratory experiments on PAH material showed that visible and UV excitation do lead to infrared emission; however quantitative determinations are not yet available.

The fractional abundance of the interstellar PAH component is estimated as 10^{-7} relative to H_2. As there may be 1000 different molecules and isomers each type may have a fractional abundance of 10^{-10}; this is comparable to fractional abundances of the largest molecules observed by high resolution spectral techniques.

9.7 The Origin of Life: Terrestrial vs Extra Terrestrial

9.7.1 General Concepts

The search for the origin of life in terrestrial and extra-terrestrial space has been a subject of fascinating interest among scientists of several disciplines such as astronomers, physicists, chemists and biologists. This problem has many dimensions and one asks the basic questions: Did the early form of life originate from the chemical evolution on the earth, such as from the "primeval soup", hypothesized many years ago; or did the primeval form of life travel from space to the earth in the early epochs of Earth's evolution? We do not know the full answers for these questions yet and these are of

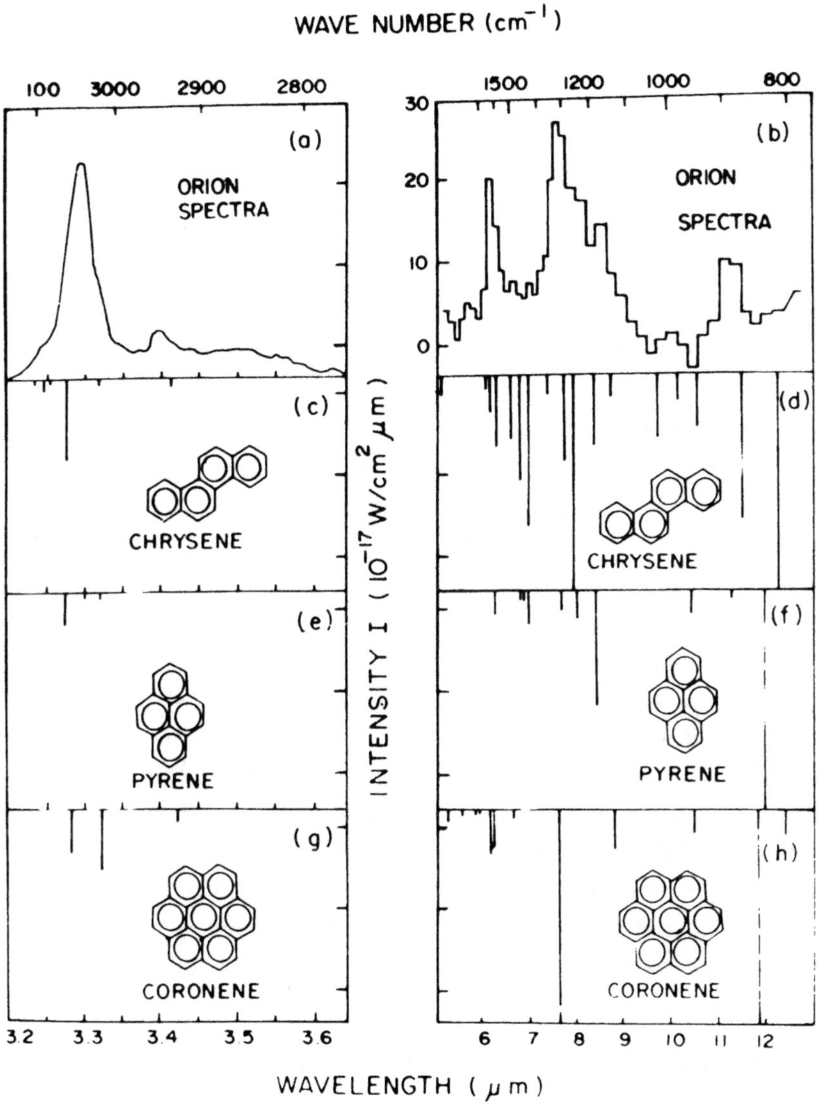

Figure 9.6.1: Infrared spectra of the Orion bar in (a) and (b), and laboratory spectra in infrared of some PAH compounds in (c) to (h). (Based on the review by L.J. Allamandola et al.[4])

profound interest to scientists and the mankind as a whole.

In the 1950's it was proposed by Miller and Urey that the earth's early atmosphere was rich in methane, water vapour and ammonia and sparking in such a gas mixture, simulating the lightning discharge in the atmosphere led to the 'primeval soup' which was found to contain about 20 amino acids found in protein. This was the starting point of the chemical evolution, as believed a few decades ago. The difficulty of this model is as follows. Now the atmospheric chemists are of opinion that the primitive atmosphere of the earth was not rich in CH_4, and most of the carbon in the atmosphere was in the form of CO_2. In such an atmosphere it is not possible to synthesize the variety of amino acids needed for living organisms. Therefore scientists looked for other sources.

Some astronomers hypothesized that primitive forms of life in the form of interstellar grains may have traveled a long distance in space to reach the earth in the early evolution-phase. This view is not favoured by scientists in general. The earlier picture of the origin of the ingredients of life in the primeval soup had undergone many changes. Astronomical and astrophysical studies during the past 2-3 decades have revealed new observations which provided new directions in the theories of the origin of life by planetary biologists. There is some support for the ideas of some astronomers and exo-biologists that it may be possible that organic ingredients being the precursors of life i.e the building blocks, may have been brought to the earth by comets, meteorites etc in the distant past and these might have led to the chemical and organic evolution under favorable conditions. We shall discuss highlights of some of the recent observations and the interpretation of these aspects of the problem. It may be noted that there is no unanimity of scientist on this hypothesis; but there is no agreement on any other theory either.

9.7.2 Meteorites and Organic Molecules

It has been noted earlier (Chapter 5) that several meteorites containing chondrites are rich in organic molecules. A well studied case is the Murcheson meteorite, which is a 82 kg sample found in Murcheson, Australia in 1969. It was established that interstellar material in the form of grains is hidden in chondrules in the carbonaceous chondrites. Several scientists studied the organic nature of chemicals in these samples. In 1987 it was established by

Epstein (Caltech) and Cronin (Arizona) in the vacuum distillation process that there were about 55 different amino acids in the Murchison fragments. Further surprises were revealed on these striking results that the deuteron abundances in these compounds were more than twice the level found on Earth. These results confirmed that the organic compounds did not come from the earth and these amino acids were of extra-terrestrial origin. Recently it was found that these amino acids were leavo-rotatory in contrast to the dexo-rotatory properties of terrestrial amino acids. Thus it was confirmed that these amino acids were made in the presolar nebula.

9.7.3 Interplanetary Dust and Organics

Interplanetary dusts were studied in the past decades by collecting them by high flying planes at altitudes of 70,000 ft by a greased collecting surface. These particles were then individually plucked one by one and examined by the group of Walker at Washington University in St Louis. They revealed the surprising fact that some dust particles had embedded grains which contained carbonaceous materials. These materials had an extremely enriched deuterium abundance indicating that the grains were of interstellar origin. Further studies of these interstellar grains by excitation and emission of radiation by Sandford and Allamondola revealed the presence of polycyclic aromatic hydrocarbons (PAH) - the same compounds which the astronomers found to be present in dense molecular clouds. These striking findings lead to the argument that if the presolar cloud had contained complex molecules such as PAH of the ISM, it may also had precursor molecules of amino acids. This indicates that when the solar system was formed the presolar nebula was already seeded with complex molecules of biogenic interest. In addition to the above more copious sources of interstellar matter are provided by the comets.

9.7.4 Comets as Couriers

Comets are known to be a primary source of interplanetary dust particles. The observations made in the past decade with the European satellite Giotto and the Russian satellite Vega, revealed the presence of organic materials in the dust streaming out of Halley's nucleus. The surprising fact that the quantity of these organic materials was exceedingly large - may be as much as 20% of Halley's mass. Some of the organic compounds were not simple. The signals from the mass spectrometer in Giotto seem to indicate the presence of broken chains of polymers - from a long chains of formaldehyde molecules, each one having formula H_2CO. These polymers were discovered by W. Huebner of Los Alamos National Laboratory. A further striking feature was that formaldehyde polymers were distributed throughout the nucleus of Halley, indicating that the organic material was already present there when the comet was formed. Some of these new findings are in general agreement with the hypothesis, proposed by Greenberg of Leiden observatory, that comets are formed near the Oort cloud at a very low temperature, out of the interstellar grains, which contain organic compounds. It is generally agreed now that comets contain very large amounts of organic materials and some of these were produced in interstellar space.

Furthermore astronomers believe that the earth and the moon system was heavily bombarded by meteorites and comets until about $4{\cdot}10^9$ years ago. The earliest single cell organisms appeared on the earth during $3.5\text{-}3.8{\cdot}10^9$ years ago. The time interval of 200-500 million years is believed to be too short for the earth to evolve complex organic molecules.

This problem can be partly resolved if it is assumed that comets brought the complex organic materials with them to the earth. Scientists calculated that in the first 10^9 years of Earth's history, about 10^5 comets had impacted on the earth. Thus it is argued by J. Oro of Houston University that comets have imported about 10^{16} tons of organic matter to the earth in the distant past. This is more that 10,000 times the present total biomass on Earth. This scenario envisages the manufacture of complex organic molecules in interstellar space and their compilation and transportation to the earth by the comets; subsequent chemical evolution took place in terrestrial sites. However the problems and their complexities persist.

9.7.5 The Martian Enigma

In August, 1996 E.K. Gibson and D.S. McKay of NASA's Johnson Space Center, Houston announced the starting discovery that they found possible evidence of life in a meteorite from Mars. The rock which was found in an ice field on Antartica was chemically identified as originating from Mars, and this Martian meteorite is called ALH 84001. These authors reported[(5)] that several types of amino acids and carbon compounds found inside the sample suggest that Martian microbes once lived inside the rock and left behind fossil-like blobs and organic chemicals. These results created great excitements because, if confirmed this would become the discovery of the century as the first evidence of extra-terrestrial life. However some other groups of scientists who afterward analyzed the samples from the meteorite ALH 84001, reported their disagreement with the interpretation of the above results and suggested the above observations as due to earthly contaminations. The NASA scientists, however, do not accept these interpretations and they stick to their original hypothesis. So the Martian enigma continues, and further studies would be needed to resolve the questions.

9.7.6 The Riddle of Life

The complex organic molecules which we referred to as the building blocks of life are, in fact, the building blocks of the building blocks of life. How these building blocks of life were formed is a highly complex and difficult question for which there is no satisfactory answer. The problems arise as follows. Starting with the amino acids as precursors, one has to string them together to form proteins, the purine and pyrimidine bases which carry genetic information in DNA. One also needs the sugars deoxyribose and ribose to be formed as the framework of DNA and RNA. These are the hardest part of the chemical evolution where the right kinds of monomers are to be agglomerated in the correct sequence to make the right kind of polymer. One basic ingredient that is required is water in which the monomers are to be dissolved. These processes could not possibly take place in the ISM or in comets where water is in the form of ice. It is believed by many scientists that comets brought to the earth a sizeable fraction of the water in the earth's oceans, during the intense bombardment in the early epoch. Secondly the monomers should not be too far apart in order to form biopolymers. Thus the biopolymers could not be formed in open oceans. Therefore some scientists think that these biopolymers could have been formed in some kind of "little warm ponds" on the early earth where the primeval soup was created by comet-borne chemical molecules and water. Thus arguments came back to a different from of primeval soup. Thus there is strong consensus that biogenic polymers were formed on the earth, rather than in space. How these polymers could evolve into living organisms constitutes the profound mystery of the riddle of life.

9.8 Problems on Chapter 9

1. Describe the main features of the spiral arm structure of the galaxy as obtained from interstellar hydrogen and hot stars.

2. What are the main processes which contribute to the heating of the interstellar medium? Describe briefly the heating of the ISM by supernovae and their remnants.

3. Describe the principal features of the distribution of (a) neutral or atomic hydrogen, and (b) molecular hydrogen, in the galaxy, and the main methods used for mapping these.

4. Discuss the properties of the hot 'coronal' component of the ISM, and how this component was detected.

5. What are the four major components of interstellar gas? Describe their main features such as composition, temperature, density and volume.

6. Briefly describe the principal features of absorption of electromagnetic radiation, by interstellar dust grains, and how this gives clues to the composition of these dust grains.

7. In which region of space are interstellar molecules mostly found and why? What are their emission processes which are used for identifying them?

8. Explain briefly how cosmic rays assist in the formation of large interstellar molecules.

9. Describe how the hydrogen ion reactions are initiated by cosmic rays and how the interactions of these with O, C, S etc. lead to the formation of compounds such as H_2O, CH_4, OH etc. How CO and hence C^+ are formed by ionization of He by cosmic rays?

10. Describe the basic steps of the reaction of C^+ and C with hydrogen, CH_4, etc so as to form large carbon structures.

11. Explain the hypothesis of the presence of complex molecules such as polycyclic aromatic hydrocarbon (PAH) in the Orion bar nebula to interpret the emissions observed.

Chapter 10
CONCLUDING REMARKS

Summary and Future Scenario

As we come near the end of the twentieth century at the time of writing, we may look back to get a perspective view of the achievements in the exploration and research in space physics during this century - in particular in the past fifty years. Although research in space physics was initiated in a modest way in the 1930's and 1940's by use of balloons, the exploration of space expanded in leaps and bounds with the advent of earth satellites in the 1950's. These earth satellites which constituted a major milestone in human achievements of man-made instruments - and later astronauts - could orbit the earth for prolonged periods and thus could make *in situ* measurements of fields and particles in space for the first time. Also the earth satellites provided new platforms in space from which we could make observations unhindered by the earth's atmosphere and thus obtain spectacular views of the Sun, the planets, the stars, the galaxy and beyond - for the first time. The universe could be studied with new windows of electromagnetic radiation leading to the opening of new branches of Infrared astronomy (in the wavelength region longer than that of visible light) and of ultra-violet, Extreme UV, X-ray and gamma ray astronomies (at the shorter wavelengths). The capabilities of rockets rapidly increased so that they could take heavy payloads and complex instruments into space. Space stations were established and laboratories in space were operated. Historical events took place as spacecraft landed astronauts on the moon. In planetary missions spacecraft landed instruments on the nearby planets - Mars and Venus, and others explored in fly-by mode the distant planets, - Jupiter, Saturn, Uranus and Neptune. Thus the "grand tour" of the planets was accomplished after a long period of interplanetary travel by spacecraft for over 20 years.

These spectacular technological and scientific achievements, in which scientists of many countries contributed, (particularly in cooperative endeavors of international collaborative missions), led to new and striking discoveries in space physics and space astrophysics. For example, the Space Shuttle, constructed by the USA, accommodated the Spacelab jointly built by several countries of Europe combined in the European Space Agency, and this Space Shuttle-Spacelab assembly provided numerous national and international scientific space missions participated by scientists of many countries for over 25 years. The long duration space studies were conducted by groups of scientists of many countries in the Space Stations - Soyuz and Mir, built and operated by the USSR (Russia) for over twenty years. Thus the effectiveness of international collaborations is best demonstrated in space exploration and research.

These new space research investigations opened up new horizons of knowledge in many directions. In the near earth space new discoveries were made on the structure of the earth's magnetosphere. The startling discoveries of the intense radiation belts around the earth were one of the major early achievements of space exploration. New vistas were revealed in the studies of the Sun and the interplanetary medium with the discoveries of the solar wind and the interplanetary magnetic field originating from and controlled by the Sun. The discoveries of the coronal holes, coronal mass ejection and solar wind in 3-dimensions provided the understanding of several new facets of solar and heliospheric phenomena. The studies of the moon and moon-rocks gave new clues to the origin of the solar system, and the spacecraft studies of the planets opened up new dimensions of the planetary atmospheres and magnetospheres. The multi-spacecraft exploration of the comet Halley and the studies of meteorites and the asteroid belt led to new insights into the formation of the solar system. Studies of energetic particles from the Sun at the orbit of the earth and in the interplanetary medium revealed the direct knowledge of the elemental and isotopic composition of solar energetic particles and these gave clues to the nucleosynthetic processes and acceleration mechanisms in the Sun.

A new type of low energy cosmic rays was discovered which originate from interstellar neutral atoms drawn into the solar system, and are probably accelerated at the heliospheric boundary region. These have an unusual composition and energy spectra, and hence, are called Anomalous Cosmic Rays (ACR).

Investigations of high energy particles from the galaxy, the galactic cosmic rays have yielded series of new knowledge of their composition and energy spectra, which led to a new understanding of the propagation and acceleration processes and on possible sources. These also revealed many new features on the origin and acceleration of very high energy cosmic rays in the galaxy and beyond. Studies made from platforms in space have established surprising features of the new branches of space astronomies and space astrophysics. Infrared astronomy revealed striking observations of star forming processes, whereas ultraviolet and extreme UV studies showed several new aspects of stellar evolution. Cosmic X-rays and gamma rays have established new phenomena in high energy astrophysics in objects like binary systems, pulsars, supernova remnants and others. Gamma ray astronomical studies revealed striking new discoveries of emissions of very large amounts of high energy gamma rays in the form of γ-ray bursts, from unknown sources. Very distant objects like Quasars, and others called Active Galactic Nuclei, were found to emit immense amounts of high energy γ-rays originating from as yet unknown processes. These include objects called 'Blazars' in which it is presumed that enormous amounts of high energy gamma-rays are emitted in the form of oppositely directed jets from the central object believed to be a supermassive black hole. The observations of X-ray and γ-ray background radiation as well as the cosmic microwave background are found to be related to interesting new aspects of observational cosmology. In the interstellar space, surprising evidences were found on the presence of a large number of simple and complex molecules - including large organic molecules which are synthesized in large molecular clouds. These discoveries led to renewed studies on the riddle of the origin of life; some physical aspects of these are briefly outlined in relation to terrestrial and extra-terrestrial models of the origin of life. These are some of the highlights in the studies of physics and astrophysics in space, during the past fifty years or so, as discussed in this book.

On looking to the future scenarios we visualize that the new frontiers of knowledge in space physics and space astrophysics which were opened up during the past decades will expand in many dimensions in the coming decades using new developed technologies, space instruments and space vehicles. In these ventures scientists of many countries are expected to participate individually as well as in collaborative manner. The Sun and the planets will be studied in much greater detail leading to the understanding of the complex processes of the planetary atmospheres which in turn would throw light on the evolution of Earth's atmosphere. Preparations which are in progress may culminate in the dramatic space event of landing astronauts on Mars in an International collaboration mission 'Man on Mars' in the first decade of the 21st century. This will be an epoch making event in human history. A novel space mission may take place in which astronauts may successfully land on an asteroid and make a direct exploration of these primeval objects. Space astrophysics is expected to get a further boost with the new generation of space observatories for IR, UV and optical wavelengths that will explore the presence of planetary systems by many stars. These are likely to open new questions on the conditions suitable for the presence of life elsewhere on other planetary objects. Future high power space observatories are expected to extend our knowledge of the distant objects in deep space such as Quasars, Blazars and Active Galactic Nuclei, which are emitting vast amounts of energies in X-rays and high energy gamma-rays. The understanding of the puzzling nature of these sources of enormous power may provide us with new and vital clues to the major questions on the origin and evolution of the universe.

In addition new discoveries may be made in the coming decades that are beyond our present day imagination and these may revolutionize our concepts and understanding of the universe around us. As the universe has evolved over billions of years and we have explored only a small part of it, it seems that it has innumerable secrets and mysteries which are yet to be unlocked and deciphered. These will present fascinating new adventures for future generations of space scientists. It is appropriate to conclude with a quotation from an ancient text:

"Jnanam Anatam"
"The knowledge is infinite"

- *Rig Veda*

REFERENCES

Chapter 1: Introduction to Space Research

General Texts

1. LeGally, D.P., 'Space Explorations', in Space Science, Ed. D.P. LeGally, John Wiley, New York, 1963.

References

1. Hess, V.F., 1912, Physik Zentschr., 13, 1084.
2. Millikan, R.A., Neher, H.V. and Pickering, W.H., 1942, Phys. Rev., 61, 397.
3. Freier, P.S., Lofgren, E.J., Ney, E.P., Oppenheimer, F., Bradt, H.L. and Peters, B., 1948, 74, 213.
4. Bradt, H.L. and Peters, B., 1948, Phys. Rev., 74, 1828; 1950, Phys. Rev., 77, 54.
5. Powell, C.F., Fowler, P.H., Perkins, D.H., The Study of Elementary Particles by the Photographic Method, Pergamon Press, London 1959.
6. Biswas, S., George, E.C. and Peters, B., 1953, Proc. Ind. Acad. Sci. A., 38, 418.
7. Lal, D., Pal, Y. and Peters, B., 1953, Proc. Ind. Acad. Sci. A., 38, 398.
8. Damle, S.V. 1990, in "Scientific Ballooning", Ind. J. Radio and Space Phys., Vol. 20, Special Issue, CSIR, New Delhi.

Chapter 2: The Magnetosphere of the Earth

General Texts

1. Ness, N.F., 1965, Proc. 9th ICRC.[1], London, 1, 14.
2. Van Allen, J.A., 1963, Dynamics of the Radiation Belt, in Space Science, Ed. D.P. LeGally, John Wiley, New York, p.226.
3. Vernov, S.N., 1969, The Earth's Radiation Belts, Invited Paper in 11th ICRC., Budapest, p.85.

References

1. Ness, N.F., 1965, J. Geophys. Res., 70, 517.
2. Ness, N.F., 1965, J. Geophys. Res., 70, 2989.
3. Parker, E.N., 1985, Phys. Fluids, 1, 171.
4. Axford, W.A., Petschek, H.E. and Sisoco, G.L., 1965, J. Geophys. Res., 70, 1231.
5. Dessler, A.J., 1964, J. Geophys. Res., 69, 3913; with Judy, R.D., Plan. Space Sci. 13, 65.

[1]ICRC denotes: International Cosmic Ray Conference

6. Störmer, C., 1955, The Polar Aurora, Oxford Univ. Press, London.

7. Katz, L., Meyer, P. and Simpson, J.A., 1958, Nuovo Cim., 8, Suppl.II. 277.

8. Queenby, J.J. and Webber, W.R., 1959, Phil Mag., 4, 90.

9. Kellogg, P.J. and Schwartz, M., 1959, Nuovo. Cim., 13, 761.

10. Chapman, S. and Bartels, J., Geomagnetism I and II, 1940, Oxford Univ. Press.

11. McCraken, K.G., Rao, U.R. and Shea, M.A., Trajectories of Cosmic Rays in a high degree simulation of the geomagnetic field, Tech. Rep. 77, MIT, 1962.

12. Shea, M.A. and Smart, D.F., McCraken, K.G. and Rao, U.R., Asymtotic directions, variational coefficients and cut-off rigidities, Tech. Rep. AFCRL-68-0030, 1968.

13. Shea, M.A., Smart, D.F., McCall, J.R. and Guma, B.S., 1975, Tables of Asymp. directions etc., Tech. Rep. AFCRL-TR-75-0042.

14. Peddie, N.W., 1982, J. Geomag. Geoelectr., 34, 309.

15. Singh, R.K., 1990, Studies of Ionization States of Anomalous and Galactic Cosmic Rays, Ph.D. Thesis, (Bombay University).

16. Dutta, A., 1991, Studies of Low Energy Anomalous and Galactic Cosmic Rays in Near Earth Space, Ph.D. Thesis, (Gujarat University).

17. Störmer, C., 1907, Archives Sci. Phys. 4e Period Vol.24, Geneva; 1911-12, ibid, 32, 33.

18. Chapman, S. and Ferraro, Y.C.A., 1932, Terr. Mag., 37, 142.

19. Meredith, L.H., Gottlieb, M.B. and Van Allen, J.A., 1955, Phys. Rev., 97, 201.

20. Vernov, S.N., Grigorov, N.L., Logachev, Y.I. and Chudakov, a.E., 1958, Dok. Akad. Nauk., SSSR., 120, 1231.

21. Van Allen, J.A., Ludwig, G.H., Ray, E.C. and McIlwain, C.E., 1958, Jet Propulsion, 28, 588.

22. Vernov, S.N. and Chudakov, A.E., 1958, Trans. 2nd Intl. Atomic Energy Conf., United Nations.

23. Vernov, S.N., Chudakov, A.E., Vakulov, P.V., Logachev, Y.I. 1959, Dok. Acad. Nauk-SSSR, 2, 304.

24. Van Allen, J.A., McIlwain, C.E. and Ludwig, G.H., 1959, J. Geophys. Res. 64, 271; Van Allen, J.A. and Frank, L.A., 1959, Nature, 183, 430; ibid. 184, 219.

25. Van Allen, J.A., 1959, J. Geophys. Res. 64, 1683.

26. Vernov, S.N. and Chudakov, A.E., 1960, Space Res., North-Holland, Amsterdam, p. 751.

27. Vernov, S.N., Vakulov, P.V., Logachev, Y.I., Nikolaev, A.G., Sosnovets, E.N. and Stolpovsky, V.G., 1967, Space Res. 7, 577.

28. Vernov, S.N., 1967, Review in Proc. 11th ICRC., Budapest, p.85.

29. Van Allen, J.A., Randall, B.A. and Krimigis, S.M., 1970, J. Geophys. Res., 75, 6085.

30. McIlwain, C.E., 1961, J. Geophys. Res., 66, 3681.

31. Singer, S.F., 1958, Phys. Rev. Letters, 1, 181.

32. Vernov, S.N. and Lebedinski, A.I., 1958, IGY 1st Gen. Assembly Report Moscow; Also Sov. Phys. Dokl., 4, 154, 1959.

33. Hess, W.N., 1959, Phys. Rev. Lett., 3, 11.

34. Freden, S.C. and White, R.S., 1960, J. Geophys. Res., 65, 1377.

35. Shoklovsky, I.S., Krassovsky, V.I. and Galperin, G.I., 1959, Akad. Nauk SSSR ser Geophys., 12, 1799.

36. Holly, F.E., Allen, L.A. and Johnson, R.G., 1961, J. Geophys. Res., 66, 1627.

37. Armstrong, A.H., Harrison, F.B., Heckman, H.H. and Rosen, L., 1961, J. Geophys. Res., 66, 361.

38. Naugle, J.E. and Kniffen, D.A., 1961, Phys. Rev. Letters, 7, 3.

39. Cladis, J.B., Chase, L.F., Imhof, W.L. and Knecht, D.J., 1961, J. Geophys. Res., 66, 2297.

40. Arnoldy, R.L., Hoffman, R.A. and Winckler, J.R., 1960, J. Geophys. Res., 65, 1361.

41. Axford, W.I., 1967, Space Sci. Rev., 7, 149.

42. Tversky, B.A., 1969, Rev. Geophys., 7, 219.

43. Biswas, S., Durgaprasad, N., Nevatia, J., Venkatavaradan, V.S., Goswami, J.N., Jayanthi, U.B., Lal, D. and Matto, S.K., 1975, Astrophys. Space Sci., 35, 337.

44. Biswas, S. and Durgaprasad, N., 1980, Space Sci. Rev., 25, 285.

45. Blake, J.B. and Freisen, L.M., 1977, Proc. 15th ICRC., Plovdiv, 2, 341.

46. Grigorov, N.L., Bobrovskaya, V.V., Zhuravlev, D.A., Kondratyeva, M.A., Papina, L.P., Podgurskaya, Tatko, M.G., Tretyakova, Chi, A. and Tretyakova, S.P., 1990, Proc. 21st ICRC, Adelaide, 6, 176.

47. Cummings, J.R., Cummings, A.C., Mewaldt, R.A., Selesnick, R.S., Stone, E.C. and Von Rosenvinge, T.T., 1993, Proc. 23rd ICRC, Calgary, 3, 428.

48. Biswas, S. Durgaprasad, N., Mitra, B., Singh, R.K., Vahia, M.N., Yadav, J.S., Dutta, A. and Goswami, J.N., 1988, Astrophys. Sp. Sci., 149, 357.

49. Singh, R.K., Mitra B., Durgaprasad, N., Biswas, S., Vahia, M.N., Yadav, J.S. Dutta, A. and Goswami, J.N., 1991, Astrophys. J., 374, 753.

50. Selesnick, R.S., Cummings, A.C., Cummings, J.R., Mewaldt, R.A., Stone, E.C. and von Rosenvinge, T.T., 1965, J. Geophys. Res., 100, 9503; see also, Proc. 24th ICRC, Rome, 1995, paper SH.6.

51. Biswas, S., 1996, Space Sci. Rev., 75, 423.

52. Fan, C.Y., Gloeckler, G. and Hovestadt, D., 1975, Phys. Rev. Letters, 34, 495.

Chapter 3: The Sun and the Heliosphere

General Texts

1. Parker, E.N., 1963, Interplanetary Dynamical Processes, Interscience Publishers, New York.

2. Alfven, H. and Fälthamer, C., 1963, Cosmical Electrodynamics (2nd Ed.) Clarendon Press, Oxford.

3. DeJager, C., (Ed.), 1969, Solar Flares and Space Research, North Holland, Amsterdam.

4. Svestka, Z., 1976, Solar Flares, D. Reidel, Dordrect, NL.

5. Kundu, M.R. and Gergely, T.E. (Ed.), 1980, Radio Physics of the Sun, Proc. IAU Symp. No. 86, D. Reidel, Dordrect.

6. McLean, D.J. and Labrum, N.R. (Ed.), 1985, Solar Radio Physics, Cambridge University Press, London.

7. Sakurai, K., 1974, Physics of Solar Cosmic Rays, Univ. of Tokyo Press.

8. Chupp, E.L., 1984, Ann. Rev. Astron. Astrophys. 22, 359.

References

1. Chapman, S. and Ferraro, V.C.A., 1931, Teres, Magnetism and Atmospheric Elc. 36 77, 71.

2. Alfven, H., 1950, Cosmical Electrodynamics, Claredon Press, Oxford.

3. Biermann, L., 1951, Z. Astrophys., 29, 274; and 1957, Observatory, 107, 109.

4. Parker, E.N., 1958, Astrophys. J., 128, 664.

5. Parker, E.N., 1963, (see ref. to Books No.1).

6. Shklovskii, I.S., Moroz, V.I. and Kurt, V.G., 1960, Astron. Zh., 37, 931.

7. Bridge, H., Dilworth, C., Lazarus, A.J., Lyon, E.F., Rossi, B. and Scherb, F., 1961, Proc. 16th ICRC, Kyoto, J. Phys. Soc. Japan, 17, Suppl. A-II.

8. Mustel, E.R., 1961, 16th IAU Symp. Cloudcraft, New Mexico.

9. Synder et.al., 1963, J. Geophys. Res., 63, 6861.

10. Hundausen, A.J., 1975, Astrophys. Space Sci., 6, 1.

11. Hundausen, A.J., 1972, Coronal Expansion and Solar Wind, Springer Publishers.

12. Hirsburg, J., 1973, Rev. Geophys. Space Phys., 11, 115.

13. Bame, S.J., Hundausen, A.J., Asbridge, J.R. and Strong, I.B., 1968, Phys. Rev. Let., 20, 393.

14. Bame, S.J., Asbridge, J.R., Hundausen, A.J. and Montgomery, M.D., 1970, J. Geophys. Res., 75, 6360.

15. Geiss, J., 1972, Solar Wind, Ed. C.P. Sonett, P.J. Coleman and J.M. Wilcox, Proc. of Asilomar Conf. NASA Publ. p. 559, 1972.

16. Boltenkov, B.S., Gartmanov, V.N., Kocharov, G.E., Mamyrin, B.A. and Naidenov, V.O., Space Research XII, Proc. 14th COSPAR Conf. Seatle, 1971, Ed. Bowhil, S.A., Joffe, L.D., Roycraft, M.J., Academic Verlag, Berlin, p. 1487, 1972.

17. Ness, N.F., Scearce, C.S., Seek, J.B. and Wilcox, J.M., 1965, Space Research VI.

18. Ness, N.F., 1968, Ann. Revs. Astron. Astrophys., 6.

19. Fairfield, D.H., 1969, J. Geophys. Res., 74, 3451.

20. Burlaga, L.F. and Ness, N.F., 1968, Can J. Phys., 46, 962.

21. Ness, N.F. and Wilcox, J.M., 1965, Science, 148, 1592.

22. Fairfield, D.H. and Ness, N.F., 1967, J. Geophys. Res., 72, 2379.

23. Coleman, Jr., P.J., Davis, Jr., L., Smith, E.J. and Jones, D.E., 1967, J. Geophys. Res., 72, 1637.

24. Ness, N.F. and Wilcox, J.M., 1967, Solar Phys., 2, 351.

25. Ness, N.F. and Wilcox, J.M., 1964, Phys. Rev. Lett., 13, 461.

26. Schatten, K.H., Ness, N.F. and Wilcox, J.M., 1969, Solar Phys., 6, 442.

27. Pizzo, V.J., 1978, J. Geophys. Res., 83, 5563.

28. Iucci, N., Parisi, M., Storini, M. and Villoresi, G., 1979, Lett. Nuovo Cimento, 24, 225.

29. Burlaga, L.F., Schwenn, R. and Rosenbauer, J.H., 1983, Geophys. Res. Lett., 10, 413.

30. AS & E Rocket results, See ref. 31, p 90-91.

31. NASA Publication, The New Sun, Ed. Ise, R., NASA-SP-402, 1979, p. 88-105.

32. Smith, E.J. and Wolf, J.H., 1979, Space Sci. Rev., 23, 217.

33. Alfven, H. and Fälthamar, C., 1963, Cosmical Electrodynamics (2nd Ed.), Clarendon Press, Oxford.

34. Tang, F. et.al., 1984, Solar Phys. 91, 75

35. Carrington, R.C. in Geomagnetism, p. 334, Cited in Störmer, p. 205.

36. Appleton, E.V. and Hey, J.S., 1946, Solar Radio Noise, Phil. Mag. Ser. 7, 37, 73; Nature, 158, 339.

37. Kundu, M.R., 1965, Solar Radio Astronomy, Interscience, New York.

38. Wild, J.P. and McCready, L.L., (1950), Aust. J. Sci. Res. Ser A., 3, 387.

39. Melrose, D.B., 1980, Space Sci. Rev., 26, 3.

40. Kundu, M.R., 1961, Astrophys. J., 134, 96.

41. Pick, M., 1961, Ann. Astrophys., 24, 183.

42. Takakura, T. and Kai, K., 1961, Publ. Astrono. Soc. Japan, 13, 94.

43. Kundu, M.R. and Haddock, F.T., 1960, Nature, 186, 610.

44. Kundu, M.R., 1983, Adv. Space Res., 12, 159.

45. Peterson, L.E. and Winckler, J.R., 1959, J. Geophys. Res., 64, 694.

46. Winckler, J.R., May, T.C. and Masley, A.J., 1961, J. Geophys. Res., 66, 316.

47. Van Beek, H.F., Hoyng, P., Lafleur, B. and Scimnett, G.M., 1980, Solar Phys., 65, 39.

48. Orwig, L.E., Frost, K.J. and Dennis, B.R., 1980, Solar Phys., 50, 153.

49. Dijveman, A., Hoyng, P., Machado, M.E. et al., 1982, Solar Phys., 81, 137.

50. Kaufmann, P., Correia, E., Costa, J.E.R., Zadi Vz, A.M. and Dennis, B.R., 1985, Nature, 305, 291.

51. DeJager, C., 1987, Proc. 20th ICRC, Moscow, 7, 66.

52. Lingenfelter, R.E. and Ramaty, R., 1967, in High Energy Nuclear Reaction in Astrophysics, W.A. Benjamin, New York.

53. Chupp, E.L., Forrest, D.J., Ryan, J.M., Heslin, J., Reppin, C., Pinkau, G., Kanbach, G., Reiger, E. and Share, G.H., 1982, Astrophys. J. (Letters), 263, L95, and Chupp, et al., 1983, Proc. 18th ICRC, Bangalore, 4, 74.

54. Biermann, L., Haxel, O. and Schulter, A., 1951, Z. Natur forsch., 6a, 47.

55. Chupp, E.L., Debrunner, H., Flückiger, E., Forrest, D.J., Golliez, F., Kanbach, G., Vestrand, J.T., Cooper, J. and Share, G., 1987, Astrophys. J., 318, 913.

56. Gosling, J.T., 1975, Rev. Geophys. Space Phys., 13, 1053.

57. Burlaga, L.F., Klein, L., Sheeley, Jr., N.R., Michels, D.J., Harward, M., Koomen, J., Schwenn, R. and Rosenbauer, H., 1982, Geophys. Res. Lett., 9, 1317.

58. Phillips, J.L., Bame, S.J., Feldman, W.C., Gosling, J.T., Hammond, C.M., McComas, D.J., Goldstein, B.E. and Neugebauer, M., 1995, Adv. Space Res., 16, (9) 85.

59. Lazarus, A.J., Belcher, J.W., Paularena, K.I., Richardson, J.D. and Steinberg, J.T., 1995, Adv. Space Res., 16, (9) 77.

60. Smith, E.J., Balogh, A., Lepping, R.P., Neugebauer, M., Phillips, J. and Tsurutani, B.T., 1995, Adv. Space Res., 16, (9) 165.

Chapter 4: The Moon and the Planets

General Texts

1. Taylor, S.R. (Ed.), 1975, Lunar Science: A Post Apollo View, Pergamon Press, London.

2. Hartman, W.K., Phillips, R.J. and Taylor, G.J. (Eds.), 1986, Origin of the Moon, Lunar and Planetary Inst., Houston.

3. Kieffer, H.H., Jakosky, B.M., Synder, C.W. and Mathews, M.S. (Eds.), 1992, Univ. of Arizona.

4. Venus, Spl. issues Icarus, Vol. 51, No. 2, August 1982, and Vol. 52, No. 2, November 1982.

References

1. Runcorn, S.K., Collinson, D.W., O'Reilly, W., Stephenson, A., Greenwood, N.N. and Buttey, M.H., 1970, Science, 167, 695.

2. Strangway, D.W., Larson, E.E. and Pierce, G.W., 1970, Proc. Apollo 11 Lunar Sci. Conf., 3, 2213.

3. Helsley, C.W., 1970, Proc. Apollo 11 Lunar Sci. Conf., 3, 2213.

4. Runcorn, S.K., 1994, Phil. Trans : Phys. Sci., 394, 181.

5. Rabinowitz, D.L., 1993, Astrophys. J., 407, 412.

Chapter 5: Comets, Meteorites and the Origin of the Solar System

General Texts

1. Kerridge, J.K. and Mathews, M.S. (Eds.), Meteorites and the early solar system, Univ. of Arizona Press, Tuscon, 1988.

2. Black, D.C. and Mathews, M.S. (Eds.), Protostars and Planets II, Univ. of Arizona, Tuscon, 1985.

3. Levy, E.H. and Lunine, J.I. (Eds.), Protostars and Planets III, Univ. of Arizona, Tuscon, 1993.

4. Lissauer, J., Ann. Rev. Astron. Astrophys. 31, 1993.

5. Ferlet, R. and Vidol-Madijar, A. (Eds.), Circumstellar dust disks and planet formation, Inst. d'Astrophys., Paris, Editions Frontier, 1994.

6. Lewis, J.S., The Sun and the Solar Nebula and Meteorites and Asteroids, in Physics and Chemistry of the Solar System, Academic Press, New York, 1985.

7. Krishnaswamy, K.S., 1996, Physics of Comets (2nd Ed.), World Scientific, Singapore.

8. Schramm, D.N., 1979, Proc. 16th ICRC, Kyoto, 14, 51.

References

1. Sagdeev, R.Z., Eliasberg, P.E. and Moroz, V.I., 1987, Astron. Zyh. Lett., 13, 621.

2. Oort, J., 1950, Bull. Astron. Inst. Neth., 11, 91.

3. Hills, J.G., Astron. J., 86, 1730.

4. Weissmann, P., 1985, Space Sci. Rev., 41, 299.

5. Marochnik, L.S., Mukhin, L.M. and Sagdeev, R.Z., 1988, Science, 242, 547.

6. Weissmann, P.R., 1990, Nature, 344, 285.

7. Hughes, D.W., 1994, On Comets and Asteroids in Contemporary Phys., 35, 75.

8. Ramadurai, S., 1994, in Circumstellar dust disks and Planet Formation p. 271 (ref. no. 5 of General Text)

9. Clayton, R.N., Grossman, L. and Mayeda, T.K., 1973, Science, 182, 485.

Chapter 6: Solar Energetic Particles

General Texts

1. Sakurai, K., 1974, Physics of Solar Cosmic Rays, Univ. of Tokyo Press.

2. Ancient Sun, 1980, Eds. Repin, R.O. et al., Pergamon Press.

References

1. Forbush, S.E., 1946, Phys. Rev., 70, 771.

2. Meyer, P., Parker, E.N. and Simpson, J.A., 1965, Phys. Rev., 104, 768.

3. Pomerantz, M.A. and Duggal, S.P., 1974, Rev. Geophys. and Space Phys., 12, 343.

4. Smart, D.F. and Shea, M.A., 1985, in Handbook of Geophys. and Space Environment, Ed. Jursa, A.S., Air Force Geophysics Laboratory, Bedford, Mas, USA.

5. Freier, P.S., Ney, E.P. and Winckler, J.R., 1959, J. Geophys. Res., 64, 684; Ney, E.P., Winckler, J.R. and Freier, P.S., 1959, Phys. Rev. Lett., 3, 183.

6. Charakhchian, 1960, Space Research, (COSPAR) Ed., Bijl, H.K., 1960, North Holland.

7. Biswas, S., Freier, P.S. and Stein, W.A., 1962, J. Geophys. Res., 66, 1023.

8. Fichtel, C.E. and Guss, D.E., 1961, Phys. Rev. Lett., 6, 495.

9. Biswas, S., Fichtel, C.E. and Guss, D.E., 1962, Phys. Rev., 128, 2756.

10. Biswas, S., Fichtel, C.E., Guss, D.E. and Waddington, C.J., 1963, J. Geophys. Res., 68, 3109.

11. Biswas, S. and Fichtel, C.E., 1964, Astrophys. J., 139, 941.

12. Biswas, S., 1963, Proc. 8th ICRC, Jaipur, 1, 43.

13. Biswas, S. and Fichtel, C.E., 1965, Space Sci. Rev., 4, 709.

14. Bryant, D.A., Cline, T.L., Desai, U.D. and McDonald, F.B., 1962, J. Geophys. Res., 67, 4983; 1965, Astrophys J., 141, 478.

15. Fichtel, C.E. and McDonald, F.B., 1967, Ann. Rev. Astron. Astrophys., 5, 351.

16. Meyer, P. and Vogt, R., 1962, Phys. Rev. Lett., 8, 387.

17. McGuire, R.E., Von Rosenvinge, T.T. and McDonald, F.B., 1979, 16th ICRC, Kyoto, 5, 61.

18. Anglin, J.D., Dietrich, W.F. and Simpson, J.A., 1975, Astrophys. J., 198, 733; Proc. 15th ICRC, Plovdiv, 5, 43, 1977.

19. Mogro-Campero, A. and Simpson, J.A., 1972, Astrophys. J., 171, L5.

20. Dietrich, W.F. and Simpson, J.A., 1979, Proc. 16th ICRC, Kyoto, 5, 85.

21. Cook, W.R., Stone, E.C., Vogt, R.E., Trainor, J.R. and Webber, W.R., 1979, Proc. 16th ICRC, Kyoto, 12, 265.

22. Scholar, M., et al, 1978, J. Geophys. Res., 83, 3349.

23. Von Rosenvinge, T.T. and Reames, D.V., 1979, Proc. 16th ICRC, Kyoto, 5, 68.

24. Lutsenko, V.A. and Ficher, S., 1979, Proc. 16th ICRC, Kyoto, 5, 102.

25. Lingenfelter, R.E. and Hudson, H.S., 1979, in Ancient Sun, Ed. R.O. Repin et al., Pergamon Press, N.Y., p. 69.

26. Van Hollebeke, M.A.I., Ma Sung, L. and McDonald, F.B., 1975, Solar Phys., 41, 189.

27. Biswas, S., Durgaprasad, N. and Vahia, M.N., 1983, Solar Phys., 89, 163.

28. Freier, P.S. and Webber, W.R., 1963, J. Geophys. Res., 68, 1605.

29. Lin, R.P. and Hudson, H.S., 1976, Solar Phys., 50, 153.

30. Bertsch, D.L., Biswas, S. and Reames, D.V., 1974, Solar Phys., 39, 479.

31. Nevatia, J. and Biswas, S., 1975, Proc. 14th ICRC, Munich, 5, 1586.

32. Crawford, H.J., Price, P.B., Cartwright, B.G. and Sullivan, J.D., 1975, Astrophys. J., 195, 231.

33. Vahia, M.N., 1987, Astron. Astrophys., 173, 361.

34. Waddington, C.J. and Freier, P.S., 1964, Phys. Rev., 136, B1535.

35. Fireman, E.L., DeFelice, J. and Tiles, D., 1961, Phys. Rev., 123, 1935.

36. Tiles, D., DeFelice, J. and Fireman, E.L., 1963, Icarus, 2, 258.

37. McDonald, F.B., Von Rosenvinge, T.T., Serlemitsos, A.J. and Balasubramanian, V.K., 1975, Proc. 14th ICRC, Munich, 5, 1619.

38. Hempe, H., Muller-Mellin, R., Kunow, H. and Wibbernez, G., 1979, Proc. 16th ICRC, Kyoto, 5, 95.

39. Hurford, G.J., Stone, E.C. and Vogt, R.E., 1975, Proc. 14th ICRC, Munich, 5, 1624.

40. Ramadurai, S., Vahia, M.N., Biswas, S., Sakurai, K., 1984, Pramana, 24, 305.

41. Ibragimov, J.A. and Kocharov, G.E., 1977, Proc. 15th ICRC, Plovdiv, 11, 340.

42. Fisk, L.A., 1978, Ap. J., 224, 1048.

43. Ramaty, R. and Kozlovsky, B., 174, Ap. J., 193, 729.

44. Mewaldt, R.A., Spadling, J.D., Stone, E.C. and Vogt, R.E., 1981, Proc. 17th ICRC, Paris, 3, 131, 1981.

45. Dietrich, W.F. and Simpson, J.A., 1979, Proc. 16th ICRC, Kyoto, 5, 85.

46. Gloeckler, G., Weiss, H., Hovestadt, D., Ipavich, F.M., Klecker, B., Fisk, L.A., Scholar, M., Fan, C.Y. and O'Gallagher, J.J., 1981, Proc. 17th ICRC, Paris, 3, 136.

47. Jordon, C., 1969, Mon. Not. Roy Astr. Soc., 142, 501-521.

48. Van Hollebeke, M.A.I., Ma Sung, L.S. and McDonald, F.B., 1975, Solar Phys., 41, 189.

49. Devicheva, E.A. et al. and Gombosi, T. et al., 1979, Proc. 16th ICRC, Kyoto, 5, 170 and 163.

50. Shea, M.A. and 25 others, 1979, 16th ICRC, Kyoto, 5, 229.

51. Webberenz, G., 1979, Proc. 16th ICRC, Kyoto, 14, 236.

52. Sanderson, T.R., Marsden, R.G., Heras, A.M. and Wenzel, K.P., 1991, Proc. 22nd ICRC, Dublin, 3, 173.

53. Mandzhavidze, N., 1983, Proc. 23rd ICRC, Calgary, 1993, Ed. Leahy et al., World Scientific, p. 157.

54. Reedy, R.C., 1979, in The Ancient Sun, Ed. Pepin, R.O. et al., Pergamon Press, New York, p. 365.

55. Goswami, J.N., Lal, D., Macdougall, J.D., 1979, in The Ancient Sun, Ed. Pepin, R.O. et al., Pergamon Press, New York, see also Crozaz, G., 1979, ibid, p. 331.

56. De Jager, C., 1969, in Solar Flares and Space Research, Ed. De Jager, C., North Holland, Amsterdam, p. 1.

57. Syrovatskii, S.I., 1969, ibid; 1981, Ann. Rev. Astron. Astrophys., 19, 163.

58. Sturrock, P.A., 1968, IAU Symp., 35, 471.

59. Lin, R.P. and Hudson, H.S., 1976, Solar Phys., 50, 153.

60. De Jager, C., 1987, Proc. 20th ICRC, Moscow, 7, 66.

61. Shibata, et al., 1994, Astrophys. J. Lett., 431, L51.

62. Ramaty, R. and Foreman, M.A., 1985, in Essays in Space Science, NASA Publication, 2464, p. 47.

Chapter 7: Galactic Cosmic Rays

General Texts

1. Janossy, L., 1950, Cosmic Rays, Clarendon Press, Oxford.

2. Heitler, W., 1947, Quantum Theory of Radiation, Oxford Univ. Press (2nd Ed.).

3. Powell, C.F., Fowler, P.H. and Perkins, D.H., 1959, Studies of Elementary Particles by Photographic Method, Pergamon Press, London.

4. Wolfendale, A.W., 1993, Cosmic Rays and the Universe, Proc. 13th ICRC, Calgary, World Scientific, p. 143.

5. Shapiro, M.M. (Ed.), 1983, Composition and Origin of Cosmic Rays, Reidel, D., Dordrecht.

6. Ginzburg, V.L. and Syrovatskii, S.I., 1965, The Origin of Cosmic Rays, Pergamon Press, London.

7. Fleisher, R.L., Price, P.B. and Walker, R.M., 1975, Nuclear Tracks in Solids, Univ. of California Press, Berkeley.

References

1. Silberberg, R. and Tsao, C.H., 1973, Astrophys. J. Suppl., 25, 315 and 335.

2. Silberberg, R., Tsao, C.H. and Letaw, J.R., 1983, in Composition and Origin of Cosmic Rays, Reidel, D., p. 321.

3. Webber, W.R. and Brautigam, D.A., 1982, Astrophys. J., 260, 894.

4. Perron, C., 1976, Phys. Rev., 14C, 1108.

5. Bhabha, H.J. and Heitler, W., 1936, Proc. Roy. Soc., 159, 432.

6. Carlson, J.F. and Oppenheimer, 1936, Phys. Rev., 51, 220.

7. Shapiro, M.M. et al., 1973, Proc. 13th ICRC, Denver, 1, 578.

8. Waddington, C.J., 1977, in Fundamentals of Cosmic Physics, Vol. 3, p. 1-85, Gordon and Breach Publ., UK.

9. Biswas, S., Peters, B. and Rama, 1955, Proc. Ind. Acad. Sci. A, 41, 154.

10. Shirk, E.K. and Price, P.B., 1978, Astrophys. J., 220, p. 719.

11. Biswas, S. and Durgaprasad, N., 1980, Space Sci. Rev., 25, 285.

12. Biswas, S., Durgaprasad, N., Mitra, B. and Dutta, A., 1993, Space Sci. Rev., 62, 3.

13. Singh, R.K., 1990, Ph.D. Thesis, Bombay University.

14. Dutta, A., 1991, Ph.D. Thesis, Gujarat University.

15. Grigorov, N.L. et al., 1990, Proc. 21st ICRC, Adelaide, 6, 176.

16. O'Sullivan, D., et al., 1996, Rad. Measurements, 26, 889.

17. Webber, W.R. et al., 1973, Astrophys. J., 183, L81; 1983, Proc. 18th ICRC, Bangalore, 8, 40.

18. Müller, D. et al., 1995, Proc. 24th ICRC, Rome, 3, 440.

19. Engelmann, J.J. et al. and Lund, N. et al., 1985, Astron. Astrophys., 148, 12; 1981, Proc. 17th ICRC, Paris, 12, 21.

20. Binns, W.R. et al., Stone, E.C. et al. and Waddington, C.J. et al., 1988, Astrophys. J., 324, 1106; 1981, 17th ICRC, Paris, 12, 53.

21. Bogomolov, E.A. et al., 1979, Proc. 16th ICRC, Kyoto, 1, 330; 1987, 20th ICRC, Moscow, 2, 72.

22. Golden, R.L. et al., 1979, Phys. Rev. Lett., 43, 1196; 1984, Astrophys. J. Lett., 24, 75.

23. Ormes, J.F., Yodh, G.B. et al., 1981, Proc. 17th ICRC, Paris, 9, 303.

24. Nishimura, J. et al., 1980, Astrophys. J., 238, 394; 1981, Proc. 17th ICRC, Paris, 2, 94.

25. Müller, D. and Tang, J., 1981, Proc. 17th ICRC, Paris, 9, 142; 1983, Proc. ICRC, Bangalore, 2, 60.

26. Parnell, T.A. et al., 1989, in JACEE collaboration, Adv. Space Res., 9, (12) 45.

27. McDonald, F.B., Teegarden, B.J., Trainor, J.H. and Webber, W.R., 1974, Astrophys. J., 187, L105.

28. Hovestadt, D., Vollmer, O., Gloeckler, G. and Fan, C.Y., 1973, Phys. Rev. Letts., 31, 650.

29. Garcia-Munoz, M., Mason, G.M. and Simpson, J.A., 1973, Astrophys. J., 182, L81.

30. Biswas, S., Durgaprasad, N., Nevatia, J., Venkatavaradan, V.S., Goswami, J.N., Jayanti, U.B., Lal, D. and Matto, S.K., 1975, Astrophys. Space Sci., 35, 337.

31. Klecker, B. et al., 1977, Astrophys. J., 212, 270.

32. Mewaldt, R.A. et al., 1994, Trans. Am. Geophys. Un., 85, 185, and 193.

33. Klecker, B. et al., 1995, Astrophys. J. Lett., 442, L69.

34. Mewaldt, R.A. et al., 1996, Geophys. Res. Lett., 23, 617.

35. Selesnick, R.S. et al., 1995, J. Geophys. Res., 100, 9503.

36. Christian, E.R. et al., 1995, 24th ICRC, Rome, Paper SH5.

37. Cummings, A.C. et al., 1991, Proc. 22nd ICRC, Dublin, 3, 362.

38. Leske, R.A. et al., 1996, Caltech, preprint.

39. Hoyle, F. and Clayton, D.D., 1974, Astrophys. J., 191, 705.

40. Biswas, S., Durgaprasad, N. and Trivedi, S., 1981, Proc. Ind. Acad. Sci. (Earth Planet Sci.), 90, 337.

41. Fisk, L.A., Kozolovsky, B. and Ramaty, R., 1974, Astrophys. J. Lett., 190, L35.

42. Singh, R.K., Mitra, B., Durgaprasad, N., Biswas, S., Vahia, M.N., Yadav, J.S., Datta, A. and Goswami, J.N., 1991, Astrophys. J., 374, 753.

43. Dutta, A., Goswami, J.N., Biswas, S., Durgaprasad, N., Mitra, B. and Singh, R.K., 1993, Astrophys. J., 411, 418.

44. Biswas, S., Durgaprasad, N., Mitra, B. and Dutta, A., 1993, Space Sci. Rev., 62, 3.

45. Biswas, S., Durgaprasad, N., Mitra, B., Singh, R.K., Vahia, M.N., Yadav, J.S., Dutta, A. and Goswami, J.N., 1988, Astrophys. Space Sci., 149, 357.

46. Adams, J.H., Garcia-Munoz, M., Grigorov, N.L., Klecker, B., Kondratyeva, M.A., Mason, G.M., McGuire, R.E., Mewaldt, R.A., Panasky, M.I., Tretyakova, Ch. A., Tylka, A.J. and Zhuravlev, 1991, Proc. 22nd ICRC, Dublin, 3, 358; also, 1991, Astrophys. J., 375, L45.

47. Klecker, B., McNab, M.C., Blake, J.B., Hamilton, D.C., Hovestadt, D., Kästle, H., Looper, M.D., Mason, G.M., Mazur, J.E. and Scholar, M., 1995, Astrophys. J., 442, L69.

48. Klecker, B., 1995, Space Sci. Rev., 72, 419.

49. Mewaldt, R.A., Selesnick, R.S., Cummings, J.R., Stone, E.C., McDonald, F.B., Webber, W.R. and Lal, N., 1987, Proc. 20th ICRC, Moscow, 3, 413, 417, 421.

50. Cummings, A.C. et al., 1995, J. Geophys. Res. Lett., 22, 341.

51. Cummings, A.C. and Stone, E.C., 1990, Proc. 21st ICRC, Adelaide, 6, 202.

52. Seo, E.S., Ormes, J.F. et al., 1991, Proc. 22nd ICRC, Dublin, 1, 627.

53. Ferrando, P., Lukasiak A., McDonald, F.B. and Webber, W.R., 1991, Proc. 22nd ICRC, Dublin, 1, 586.

54. Lukasiak, A., Ferrando, P., McDonald, F.B. and Webber, W.R., 1991, Proc. 22nd ICRC, Dublin, 1, 584.

55. Lukasiak, A. et al., 1993, Proc. 24th ICRC, Rome, 2, 576.

56. Connell, J.J. and Simpson, J.A., 1995, Proc. 24th ICRC, Rome, 2, 602.

57. Garcia Munoz, M. et al., 1987, Astrophys. J. Suppl., 64, 269.

58. Evenson, P. et al., 1995, Proc. 24th ICRC, Rome, 2, 585; Du Vernois et al., 1995; ibid, 2, 589; Thayer, M.R., 1995, ibid, 2.

59. Streitmatter, R.E. et al., 1989, Adv. Space Res., 9 (12), 65.

60. Stephens, S.A., 1996, in High Energy Astronomy and Astrophysics, Ed. Agrawal, P.C. and Vishwanath, P.R., Universities Press, Hyderabad, p. 315.

61. Tarle, G., Barwick, S.W. et al., 1995; and Müller D., Barwick, S.W. et al., 1995, Proc. 24th ICRC, Rome, 3, 17; and 3, 13.

62. Müller, D., 1989, Adv. Space Res., 9 (12), 31.

63. Israel, M., 1983, in Composition and Origin of Cosmic Rays, Ed. Shapiro, M.M., Reidel, D., p.47.

64. O'Sullivan, D. et al., 1995, Adv. Sp. Res., 15, 25; 1996, Rad. Measurements, 26, 889.

65. Grigorov, N.L. et al., 1971, Proc. 12th ICRC, 5, 1746.

66. Swordy, S., 1993, Proc. 24th ICRC, Rome, Rapporteur paper.

67. Hara, T. et al., 1983, Proc. 18th ICRC, Bangalore, 9, 198.

68. Teshima, M., 1993, Proc. 23rd ICRC, Calgary, Rapp. paper.

69. Wolfendale, A.W., 1993, Nucl. Phys. B. (Proc. Suppl.), 33, A,B, 34.

70. Wolfendale, A.W., 1995, Nucl. Phys. B. (Proc. Suppl.), 39A, 251.

71. Simpson, J.A. and Lopate, C., 1971, Proc. 22nd ICRC, Dublin, 2, 593.

72. Parker, E.N., 1965, Planet Space Sci., 13, 9.

73. Gleeson, L.J. and Axford, W.I., 1968, Astrophys. J., 154, 1011.

74. Jokipii, J.R., 1989, Adv. Space Res., 9 (12), 105.

75. Cowsik, R. and Wilson, L.W., 1973, Proc. 13th ICRC, Denver, 1, 500.

76. Shapiro, M.M. and Silberberg, R., 1974, Phil. Trans. Roy. Soc., London, A277, 319.

77. Garcia-Munoz, M. et al., 1987, Astrophys. J. Suppl., 64, 269.

78. Ginzburg, V.L. et al., 1980, Astrophys. Space Sci., 68, 295.

79. Peters, B. and Westergaard, N.J., 1977, Astrophys. Space Sci., 48, 21.

80. Ginzburg, V.L. and Syrovatskii, S.I., 1965, Origin of Cosmic Rays, Pergamon Press, London.

81. Fermi, E., 1954, Astrophys. J., 119, 1.

82. Fermi, E., 1949, Phys. Rev., 75, 1169.

83. Axford, W.I. et al., 1977, Proc. 15th ICRC, Plovdiv, 11, 132.

84. Krymsky, G.F., 1977, Dok. Akad. Nauk, SSSR, 234, 1306.

85. Ginzburg, V.L., 1987, Proc. 20th ICRC, Moscow, 7, 7.

86. Axford, W.I., 1981, Proc. 17th ICRC, Paris, 12, 155.

87. Blandford, R.D. and Ostriker, J.P., 1978, Astrophys. J., 211, 148.

88. Bell, A.R., 1978, MNRAS, 182, 147 and 43.

89. Sakurai, K., 1989, Adv. Space Res., 9 (12), 149.

90. Vahia, M.N., Rao, A.S. and Singh, R.K., 1991, Astron. Astrophys., 250, 424.

91. Al-Dargazelli, S.S., Wolfendale, A.W. et al., 1996, J. Phys. G. Nucl. Part. Phys., 22, 1825.

92. Stanev, T. et al., 1995, Phys. Rev. Lett., 75, 3056.

Chapter 8: Space Astrophysics

General Texts

1. UV Astronomy : Kondo, Y., Wamsteker, W., Boggess, A., Grewing, M., deJager, C. et al., (Ed.), 1987, Exploring the Universe with the IUE, Satellite, D. Reidel, Dordrecht, Kondo, Y. et al., 1989, Ann. Rev. Astron. Astrophys., 27, 397-420.

2. EUV Astronomy : Bowyer, S. and Malina, R.F. (Ed.), 1996, Astrophysics in the Extreme Ultraviolet, Kluwer Acad. Dordrecht.

3. IR Astronomy : Lawrence, A. (Ed.), 1988, Proc. 3rd IARS Conf., 'Comets to Cosmology' Springer-Verlag, Berlin, Proc. 1st ISO Conf., Astron. Astrophys., Vol. 315, Nov. 1996.

4. X-ray and γ-ray Astronomy : Agrawal, P.C. and Vishwanath, P.R. (Ed.), 1998, High Energy Astronomy and Astrophysics, Universities Press (India) Ltd., Hyderabad.

5. Giacconi, R. and Gursky, H., (Ed.), 1974, X-ray Astronomy, D. Reidel Publishing Co., Dordrecht.

6. Fichtel, C.E. and Trombka, J.I., Gamma-ray Astrophysics: New Insight into the Universe, 1997, NASA Publication, 1386.

7. Bassani, L. et al., 1985, Active Galactic Nuclei, Ed. J.E. Dyson, Manchester Univ. Press.

8. Partidge, R.B., 1995, 3K : Cosmic Microwave Background Radiation, Cambridge.

9. Kragh, H., 1996, Cosmology and Controversy, Princeton Univ. (N.J.).

References

1. Rossi, B., Giacconi, R. et al., 1962, Phys. Rev. Lett., 9, 439.

2. Bowyer, S., Byram, E.T., Chubb, T.A. and Freidman, H., 1964, Science, 146, 912.

3. Clark, G.W., 1965, Phys. Rev. Lett., 14, 91.

4. Peterson, L., 1969, in Non-Solar X- and γ-ray Astronomy, (Ed.) Gratton, L., D. Reidel, Dordrect.

5. Cocke, W.J. et al., 1969, Nature, 221, 525.

6. Fritz, G. et al., 1969, Science, 164, 3880 and 709.

7. Lewin, W.H.G., Clark, G.W. and Smith, W.B., 1968, Astrophys. J. Letters, 152, L49.

8. Agarwal, P.C., Biswas, S., Gokhale, G.S., Iyenger, V.S., Kunte, P.K., Manchanda, R.K. and Sreekantan, B.V., 1969, IAU Symp. No. 37, Non-Solar X- and γ-ray Astronomy, Gratton, L. (Ed.), D. Reidel, p.94 and 104.

9. Hwang, U. and Gotthelf, E.V., 1997, Astrophys. J., 475, 665

10. Trümper, J. et al., 1978, Astrophys. J. Lett., 219, L105.

11. Manchanda, R.K. et al., 1984, Astrophys. J. Lett., 24, 27.

12. Joos, P.C. and Rappaport, S.A., 1984, Ann. Astron. Astrophys., 22.

13. Tanaka, Y., 1983, Proc. 18th ICRC, Bangalore, 12, 91.

14. Oda, M., 1977, Space Sc. Rev., 20, 757.

15. Fichtel, C.E. et al., 1975, Astrophys. J., 198, 163.

16. Mayer-Hasselwander, H.A. et al., 1980, Annals. New York Acad. Sci. Proc. 19th Texas Symp., 336, 211.

17. Fichtel, C.E., Ožel, M., Stone, R. and Sreekumar, P., 1991, Ap. J., 374, 134.

18. Sreekumar, P. and Fichtel, C.E., 1991, Astron. Astrophys., 251, 447.

19. Chi, X. and Wolfendale, A.W., 1993, J. Phys. G: Nucl. Part. Phys., 19, 795.

20. Gruber, D.E., 1992, in X-ray background, Cambridge Univ. Press, p. 44.

21. Cowsik, R. and Kobetich, E., 1972, Astrophys. J., 177, 585.

22. Field, G.B. and Perrenod, S.C., 1977, Astrophy. J., 215, 717.

23. Schmidt, M. and Green, R.F., 1986, "Counts, Evolution and Background Contribution of X-ray Quasars and other extra galactic objects".

24. Smoot, G., Gorestein, M. and Müller, R., 1977, Phys. Rev. Lett., 39, 898.

25. Cheng, E., et al., 1979, Astrophys. J., 232, L139.

26. Boldt, E.A., 1987, in Essays in Space Science, NASA Conf. Publ., 2464, p. 339.

27. Jahoda, K., 1993, Adv. Space Res., 13, (12) 231.

28. Arnold, J.R. et al., 1962, J. Geophys. Res., 67, 4876.

29. Kraushaar, W.L. and Clark, G.W., 1962, Phys. Rev. Lett., 8, 106.

30. Kraushaar, W.L. et al., 1972, Astrophys. J., 177, 341.

31. Fichtel, C.E. et al., 1977, Astrophys. J., 217, L9.

32. Kniffen, D.A. et al., 1993, Astrophys. J., 411, 133

33. Stecker, F.W., 1985, Nuclear Phys., B252, 25.

34. Page, D.N. and Hawkins, S.W., 1976, Astrophys. J., 206, 1.

35. Clayton, D.D. and Silk, J., 1969, Astrophys. J., 158, L43.

36. Stecker, F.W., Salamon, M.H. and Malkan, M.A., 1993, Astrophys. J., 410, L71.

37. Penzias, A.A. and Wilson, R.W., 1965, Astrophys. J., 142, 419.

38. Wright, E.L., 1993, Proc. 23rd ICRC, Calgary, Inv. Paper.

39. Mather, J.C. et al., 1990, Astrophys. J., 354, L37.

40. Smoot, G.F. et al., 1992, Astrophys. J. Lett., 396, L1.

41. Hoyle, F., Burbidge, G. and Narlikar, J.V., 1993, Astrophys. J., 410, 439.

42. Hoyle, F., Burbidge, G. and Narlikar, J.V., 1995, Proc. Roy Soc. A, 448, 191.

43. Sachs, R., Narlikar, J.V. and Hoyle, F., 1996, Astron. Astrophys., IUCAA, Preprint 19/96.

44. Zwicky, F., 1933, Helv. Phys. Acta, 6, 110.

45. Cowsik, R., 1991, Proc. 22nd ICRC, Dublin, 5, 82.

46. Cowsik, R. and McClelland, 1972, Phys. Rev. Lett., 29, 669.

47. Shu, F., et al., 1980, Ap. J., 241, 641.

48. Pyle et al., 1975, M.N.R.A.S., 274, 1665

49. Bowyer, S., et al., 1995, Ap. J. Suppl. 96, 461

Chapter 9: The Interstellar Medium (ISM)

General Texts

1. Greenberg, J.M., 1987, Astrochemistry, (Ed.) Vardya, M.S. and Tarafdar, S.P., D. Reidel, Dordrecht.

2. Greenberg, J.M. and Van de Hulst (Ed.), 1973, Interstellar dust and Related topics, D. Reidel, Dordrecht.

3. Winnewiser, G. and Herbst, E., 1993, Interstellar Molecules, Rep. Prog. Phys., 56, 1209.

4. McKee, C.F., 1981, The Nature of Interstellar Medium, Proc. 17th ICRC, Paris, 12, 69.

5. 'Role of Solar UV in origin, evolution and distribution of Life' in COSPAR Symposium on Life Sciences : Space and Mars Recent Results, Adv. Space Res., Vol. 18, No. 12, 1996, p. 5-74.

References

1. Greenberg, M. (see general text refs. 1, 2).

2. Lager, A. and Puget, J.L., 1984, Astron. Astrophys., 137, L5.

3. Allamondola, L.J., Tielens, A.G.G.M. and Barker, J.R., 1985, Astrophys. J., 290, L25.

4. Allamondola, L.J. et al., 1989, Astrophys. J. Suppl., 71, 773.

5. McKay, D.S., Gibson, Jr., E.K., Thomas-Keprta, K.L., Vali, H., Romanek, C.S., Clementt, S.J., Chilliar, X.D.F., Maechling, C.R. and Zare, R.N., 1996, Science, 273, 924.

Index

GPSR Compliance

The European Union's (EU) General Product Safety Regulation (GPSR) is a set of rules that requires consumer products to be safe and our obligations to ensure this.

If you have any concerns about our products, you can contact us on ProductSafety@springernature.com

In case Publisher is established outside the EU, the EU authorized representative is:

Springer Nature Customer Service Center GmbH
Europaplatz 3
69115 Heidelberg, Germany

Batch number: 10755248

Printed by Printforce, the Netherlands